ANTIOXIDANTS:
in science, technology,
medicine and nutrition

GERALD SCOTT, DSc (Oxon)

Gerald Scott read Chemistry at Balliol College, Oxford, and after taking a first class honours degree in 1952 carried out research with Dr W.A. Waters on free radical reactions of biologically important polyhydroxyphenols. He joined ICI (Dyestuffs Division) where he was appointed Manager of Polymer Auxiliaries Research in 1961. In 1965 he published his first book *Atmospheric Oxidation and Antioxidants* covering both technological and biological materials.

In 1967, Scott left industry to take the Chair of Chemistry at the recently chartered University of Aston in Birmingham, where he developed an internationally renowned research group concerned with the oxidative deterioration and stabilisation of organic materials. His publications number some 300 scientific papers, mainly in the field of antioxidant mechanisms in recognition of which Oxford University awarded him a DSc in 1983. In 1985 he published (with Professor Grassie of Glasgow University) *Polymer Degradation and Stabilisation* for post-graduate students and researchers in the polymer industries. Professor Scott also edited and contributed to a series of eight books titled *Developments in Polymer Stabilisation* from 1979-1987, and in 1990 a book entitled *Mechanisms of Polymer Degradation and Stabilisation*. In 1993 Scott edited a new edition of his earlier book *Atmospheric Oxidation and Antioxidants* in three volumes, covering both chemical technology and biology, and in 1995 he edited (with the late Dan Gilead) *Degradable Polymers: Principles and Applications*.

Some 40 international patents resulted from Professor Scott's research, the first group concerned with antioxidants in the time-controlled degradation of plastics which are being used widely in many countries, particularly in the control of plastics litter, in agricultural applications. His second group of patents are concerned with the chemical attachment of antioxidants to polymers and are finding application in polymers subjected to aggressive conditions (high temperatures, solvent extraction, etc.). His third group of patents is concerned with a new class of antioxidants derived from spin-traps (including nitric oxide) which have application in both the deterioration of organic substrates and in biological peroxidation.

Professor Scott's research has increasingly turned in recent years to his early interest in mechanisms of biological antioxidants. He has shown that α-tocopherol and its oxidative transformation products are among the most powerful antioxidants yet examined in polymers at sub-ambient oxygen concentrations and he believes that the antioxidant mechanism of the quinonoid products is relevant to the behaviour of α-tocopheroquinone *in vivo*.

Professor Scott is a Fellow of the Royal Society of Chemistry, Fellow of the Institute of Materials, and a member of the New York Academy of Sciences. In 1988 he was elected honorary life member of the Japanese Materials Life Society and in 1992 he was elected to membership of the International Academy of Creative Endeavours.

ANTIOXIDANTS
in science, technology, medicine and nutrition

Gerald Scott, DSc (Oxon)
Professor Emeritus in Chemistry
Aston University
Birmingham

Albion Publishing
Chichester

First published in 1997 by
ALBION PUBLISHING LIMITED
International Publishers
Coll House, Westergate, Chichester, West Sussex, PO20 6QL England

COPYRIGHT NOTICE
All Rights Reserved. No part of this publication may be reproduced, stored in a retrieval system, or transmitted, in any form or by any means, electronic, mechanical, photocopying, recording, or otherwise, without the permission of Albion Publishing, International Publishers, Coll House, Westergate, Chichester, West Sussex, England

© Gerald Scott, 1997

British Library Cataloguing in Publication Data
A catalogue record of this book is available from the British Library

ISBN 1-898563-31-4

Printed in Great Britain by Hartnolls, Bodmin, Cornwall

Dedicated to the memory of

WILLIAM ALEXANDER WATERS

who fired me with his own enthusiasm
for free radical research

PREFACE

The history of antioxidants goes back to the 19th century when it was realised that the deterioration of natural rubber was not caused by biological processes, as had previously been assumed, but by peroxidation. However, the biological terms "ageing", "perishing", "poisoning", "fatigue", etc., continue to be used by polymer technologists to the present day to describe specific aspects of rubber deterioration. It was discovered empirically that certain chemicals used in vulcanisation were able, in very low concentration, to improve the durability of rubber products. Subsequent research into "antioxidants" or "antioxygens" falls into two distinct although overlapping phases. The first, concerned with the protection of technological materials from oxidative deterioration, began in the latter part of the 19th century and reached a peak in the 1960s. The second phase began in the 1950s with the recognition of the importance of biological antioxidants in some diseases. Food chemists had earlier shown that peroxidation was primarily responsible for the rancidification of polyunsaturated oils and fats and that this was inhibited by naturally occurring vitamin antioxidants. In parallel studies, oil chemists showed that the "drying" of paint films, catalysed by transition metal ions was due to the oxidative cross-linking of polyunsaturated oils to macromolecules. To the chemist it is evident that the same free radical chemistry must be potentially capable of peroxidising lipids *in vivo* to both carbonyl breakdown products and to oligomers. However, due to the effectiveness of the endogenous biological antioxidants, the analogies between the deterioration of edible oils and the pathological effects of lipid peroxidation (e.g. in atherosclerosis) were only slowly recognised and it took the emergence of large scale epidemiological studies into antioxidant vitamin deficiency to provide the impetus for the present major upsurge of interest in free radical peroxidation *in vivo*.

Antioxidant research in the life sciences has not yet reached its zenith since some of the more important conclusions arising from "classical" chemical studies of antioxidant mechanisms have not yet crossed the disciplinary barrier. To many biochemists and materials technologists "antioxidant" is still synonymous with "radical trap" and although this is part of the truth, it is not the whole truth. In 1965, in a review of peroxidation and antioxidants, I drew attention to the importance of hydroperoxides as ubiquitous initiators of peroxidation and reviewed the available evidence on synthetic antioxidants with "catalase-like" activity (*Atmospheric Oxidation and Antioxidants*, 1965). I described the peroxidolytic antioxidants as "preventive" because, by preventing the formation of free radicals from hydroperoxides, they provide a complementary mechanism to the kinetic chain-breaking antioxidants. Other preventive antioxidant mechanisms identified were metal ion deactivation, UV absorption and deactivation of reactive oxygen species. Preventive mechanisms have nothing to do with "radical trapping" and yet they are frequently as important and sometimes more important in peroxidising systems than radical traps. The use of synergistic combinations of antioxidants acting by different mechanisms became the basis of polymer stabilisation technology in the 1960s but the concepts developed at that time have still to impact fully on antioxidant research in the life sciences.

The emphasis in this book is therefore on mechanisms. A primary intention is to demonstrate the relevance of antioxidant mechanisms to current studies of antioxidants in both technology and biology. In principle it is also possible to develop from the present body of knowledge predictive theories to point the way more clearly to potential therapeutic applications of antioxidants *in vivo*. I make no excuse then for returning in this book to the chemical foundation of antioxidant theories which began with the very thorough science-based studies of peroxidation by Bolland, Bateman, Gee and others in the mid-decades of this century. There is no doubt that the antioxidant classification that resulted from this early work is germane to the understanding of how antioxidants function in biology and medicine. My hope is that the principles outlined and the more speculative extensions of them proposed will serve to stimulate a vigorous and fruitful debate between "classical" oxidation chemists and researchers in the life sciences.

I am grateful to my many collaborators past and present who have contributed to the understand of how antioxidants act and I am particularly indebted to the following who have provided me with detailed information about their own recent work: Professor Paul Addis, Dr Bruce Ames, Dr Fiorelle Biasi, Dr Norman Billingham, Dr Rod Bilton, Dr Walter Bottje, Dr Joan Braganza, Dr Richard Cottrell, Professor Evgueni Denisov, Dr Gary Duthie, Professor Fred Gey, Dr Michael Golden, Dr Edward Hall, Professor Barry Halliwell, Professor Philip James, Professor Robert Hider, Dr Frank Kelly, Dr Tim Key, Dr C.J.N. Lacey, Dr Jan Malik, Dr Michael Marmot, Dr M.J. Mitchinson, Dr Simon Maxwell, Mr J.B. Park, Dr Hilary Powers, Dr Jan Pospísil, Professor Russel Reiter, Dr Roland Stocker, Professor Alexander Tkác, Professor Walter Willett, Dr Paul Winyard, Dr R.J. Woodward and Dr Hans Zweifel. I am also grateful to the following organisations for details of their products: Dairy Crest Ltd (Colleen Amos), Kraft Jacobs Suchard Ltd (Simon Kane), Mattews Foods plc (G. Burrows), MD Foods Ltd (Søren Madsen), Safeway Stores plc (Anna Sinclair), St. Ivel Ltd (Sarah Waterfield), Van den Bergh Foods (Sarah Nolan).

I am grateful to Ellis Horwood, MBE and Rosmary Harris for the excellent collaboration in the production of this work.

Finally, I am deeply indebted to my wife, Gwen, for her help and patience.

Gerald Scott, 1997

Table of Contents

1 Peroxidation in Chemistry and Chemical Technology 1
  1.1 Peroxidation 1
  1.2 Effect of substrate structure on peroxidation rate 4
  1.3 Initiation of peroxidation 8
    1.3.1 Autoinitiation by ground-state dioxygen 8
    1.3.2 Peroxidation induced by ionising radiation 8
    1.3.3 Initiation by reactive oxygen species 9
    1.3.4 Initiation by physical stress 11
  1.4 Termination 11
  1.5 Products of peroxidation 13
    1.5.1 Hydroperoxides and their decomposition products 13
  1.6 The technological effects of peroxidation 17
    1.6.1 Surface coatings by peroxidation of polyunsaturated fatty esters 17
    1.6.2 Thermal oxidation of polymers 18
    1.6.3 Photooxidation of polymers 26
    1.6.4 The effect of polymer morphology in polymer degradation 27
  1.7 Measurement of polymer peroxidation 28
    1.7.1 Characterisation of polymer deterioration 28
    1.7.2 Accelerated testing 29

2 The Biological Effects of Peroxidation 36
  2.1 Causes of peroxidation in biological substrates 36
  2.2 Products of lipid peroxidation 38
  2.3 Rancidification of fats and oils 41
  2.4 Pathological effects of peroxidation 44
    2.4.1 Recognition and measurement of biological peroxidation 45
  2.5 Atherosclerosis 47
    2.5.1 Diet and heart disease 49
    2.5.2 Smoking and heart disease 50
  2.6 Cancer 51
    2.6.1 Diet and cancer 51
    2.6.2 Chemical carcinogens 54
    2.6.3 Smoking and cancer 55
    2.6.4 Alcohol abuse 56
  2.7 Inflammation 56
    2.7.1 Rheumatoid arthritis 56
    2.7.2 Hypoxia-reperfusion injury 57
    2.7.3 Pancreatitis 58
    2.7.4 Cystic fibrosis 58
    2.7.5 Disorders of prematurity 59
    2.7.6 Adult respiratory distress syndrome 59
    2.7.7 Inflammatory bowel disease 59
    2.7.8 Disorders of severe malnutrition 60

2.8 Iron overload 60
  2.8.1 Idiopathic haemochromatosis 60
  2.8.2 Thalassaemia 60
2.9 Environmental damage 61
  2.9.1 Age-related cataract 61
  2.9.2 UV skin damage 61
  2.9.3 Effects of ionising radiation 62
  2.9.4 Lung damage 63
  2.9.5 Physical exercise 64
2.10 Ageing 65
  2.10.1 Chemical evidence for peroxidation during ageing 66
  2.10.2 Metabolic rate, peroxidation and ageing 66

3 Chain-breaking Antioxidants 80
3.1 What are antioxidants? 80
3.2 The chain-breaking donor mechanism 83
  3.2.1 Structure-activity relationships in CB-D antioxidants 87
  3.2.2 Physical aspects of antioxidant effectiveness in polymers 93
3.3 The chain-breaking acceptor mechanism 93
3.4 The catalytic chain-breaking mechanism 96
3.5 Applications of the catalytic CB process in polymers 100
  3.5.1 Mechanoantioxidants 100
  3.5.2 Photoantioxidants 115

4 Preventive Antioxidants, Synergism and Technological Performance ... 126
4.1 Peroxidolytic mechanisms 129
  4.1.1 Aliphatic and aromatic sulphides 129
  4.1.2 Heterocyclic thiols, aliphatic dithioic acids and their derivatives ... 134
4.2 Metal deactivators 149
4.3 UV absorbers and screens 155
  4.3.1 Pigments 155
  4.3.2 Organo-soluble nickel complexes 156
  4.3.3 Phenols 158
4.4 Synergism and antagonism 159
  4.4.1 Homosynergism 160
  4.4.2 Heterosynergism 161
  4.4.3 Autosynergism 167
4.5 Physical aspects of antioxidant performance 172
  4.5.1 Effect of molecular size on antioxidant activity 172
  4.5.2 Evaluation of oligomeric antioxidants and stabilisers 175
  4.5.3 Polymer-bound antioxidants 177

5 Antioxidants in Biology 191
5.1 Antioxidant mechanisms *in vivo* 191
5.2 Naturally occurring chain-breaking donor (CB-D) antioxidants ... 193
  5.2.1 Vitamin E 193
  5.2.2 Vitamin C (ascorbic acid) 201
  5.2.3 Tetrahydropterins and dihydropterins 202

	5.2.4	Uric acid 203
	5.2.5	Ubiquinones (co-enzyme Q) and ubiquinols.......................... 204
	5.2.6	Bilirubin 206
	5.2.7	Melatonin and serotonin.. 206
	5.2.8	Oestradiol 207
	5.2.9	Oxides of nitrogen... 208
	5.2.10	Polyhydroxyphenols 208
	5.2.11	Herbiforous antioxidants 213
	5.2.12	Carotenoids and retinoids 215
5.3	Naturally occurring preventive antioxidants and synergists 219
	5.3.1	Superoxide dismutase (SOD).. 220
	5.3.2	Catalase 221
	5.3.3	Glutathione peroxidase... 222
	5.3.4	α-Lipoic acid... 225
	5.3.5	Metal chelating agents 226
	5.3.6	Photoantioxidants.. 227
5.4	Nutritional aspects of antioxidants 229
5.5	The antioxidant potential of drugs... 238
	5.5.1	Drugs with chain-breaking antioxidant activity 239
	5.5.2	Drugs with preventive antioxidant activity 247

6 Antioxidants in Disease and Oxidative Stress 262
- 6.1 Epidemiological studies 262
- 6.2 Epidemiological studies of cardiovascular disease 265
  - 6.2.1 Diet-based descriptive studies...................................... 265
  - 6.2.2 Plasma-based descriptive studies 265
  - 6.2.3 Case-control studies 267
  - 6.2.4 Prospective (cohort) studies... 268
- 6.3 Antioxidant intervention and supplementation.......................... 269
  - 6.3.1 Naturally occurring antioxidants 269
  - 6.3.2 Synthetic antioxidants 274
- 6.4 The effects of antioxidants on CVD in animals......................... 274
- 6.5 Epidemiological studies of cancer... 275
  - 6.5.1 Cancer and diet 275
  - 6.5.2 Melatonin and cancer... 279
- 6.6 Effects of antioxidants on cancer *in vivo* 279
- 6.7 Antioxidants and ageing 280
- 6.8 Parkinson's disease 283
- 6.9 Alzheimer's disease 284
- 6.10 Antioxidants in inflammation 284
  - 6.10.1 Rheumatoid arthritis.. 284
  - 6.10.2 Hypoxia-reperfusion.. 286
  - 6.10.3 Pancreatitis 287
  - 6.10.4 Cystic fibrosis.. 288
  - 6.10.5 Disorders of prematurity.. 288
  - 6.10.6 Inflammatory bowel disease.. 289
  - 6.10.7 Kwashiorkor.. 289
  - 6.10.8 Iron overload... 289

6.11 Antioxidants in environmental damage 290
    6.11.1 Cataract 290
    6.11.2 Sunburn 291
    6.11.3 Respiratory inflammation and atmospheric pollution 292
    6.11.4 Exercise and hyperoxygenation 293
    6.11.5 Protection against ionising radiation 294
6.12 Therapeutic potential of antioxidants 294
    6.12.1 Antioxidants in surgery 294
    6.12.2 Antioxidants in chemical toxicity and drug overdose 295

1

Peroxidation in Chemistry and Chemical Technology

1.1 Peroxidation
The reactions of dioxygen with organic materials are arguably among the most important of all chemical reactions. The oxidation of carbon-based nutrients is the basis of life energy and the combustion of hydrocarbon minerals is the primary source of domestic and industrial energy. Directed oxidation of primary oil-based hydrocarbons is currently the method of choice for the manufacture of intermediates in the chemical industry; a large number of the primary intermediates for the polymer and fine organic chemicals industries are based upon products derived from hydrocarbons by oxidation. Very often, as in the manufacture of phenol from isopropyl benzene, oxidation occurs (see Scheme 1.1) by the classical radical chain reaction first proposed by Bolland, Bateman and co-workers [1,2] and which is common to all peroxidation reactions [3,4]. In the case of cumene and other highly peroxidisable hydrocarbons, the mild oxidation conditions allow the intermediate hydroperoxide to be isolated.

In general, the more reactive is the substrate to autoxidation, the higher is the yield of hydroperoxide. However, when the desired product is not the hydroperoxide itself but one of its decomposition products, the oxidation is carried out under conditions where the intermediate hydroperoxide undergoes thermolysis to give a stable end-product. Thus, in the oxidation of cyclohexane to adipic acid or of ρ-xylene to ρ-toluic acid, the intermediate hydroperoxides are not isolated (see Scheme 1.2).

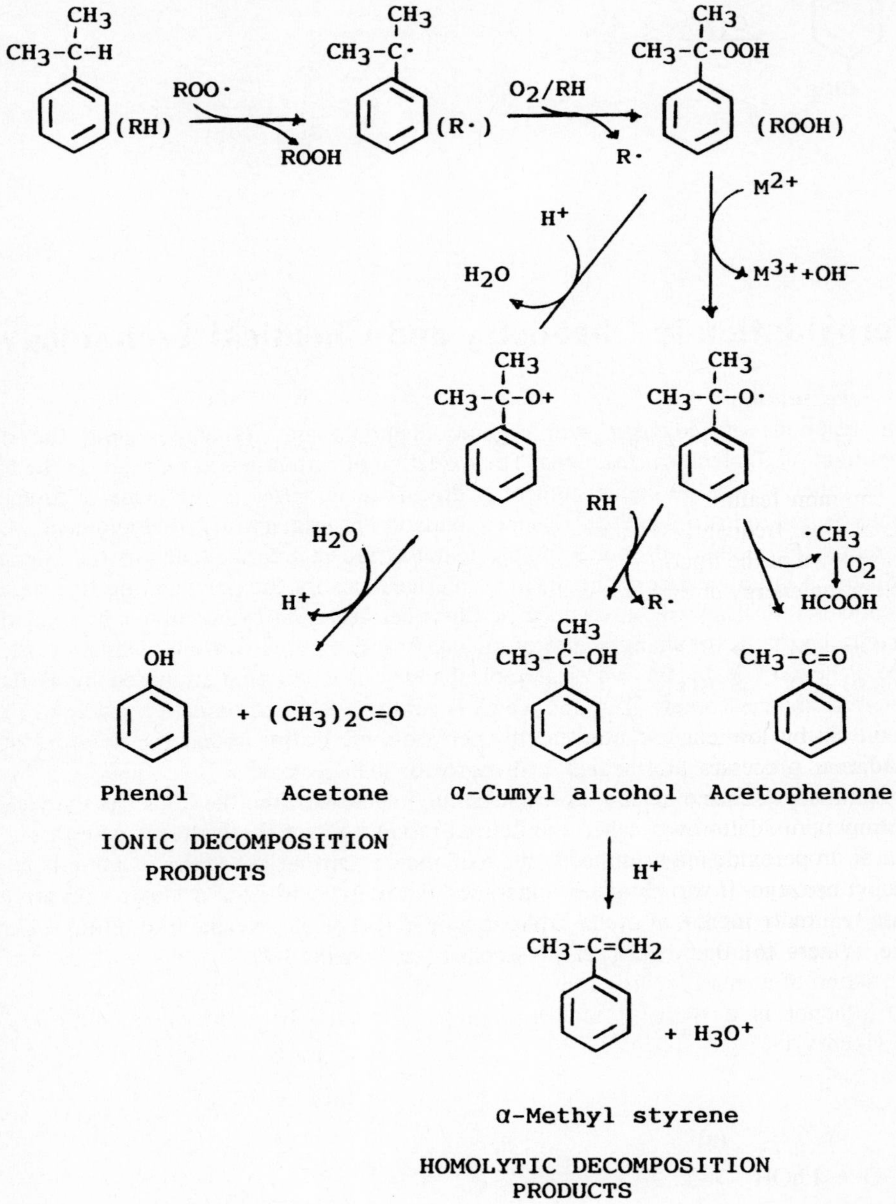

Scheme 1.1 Peroxidation of *iso*-propyl benzene (Cumene)

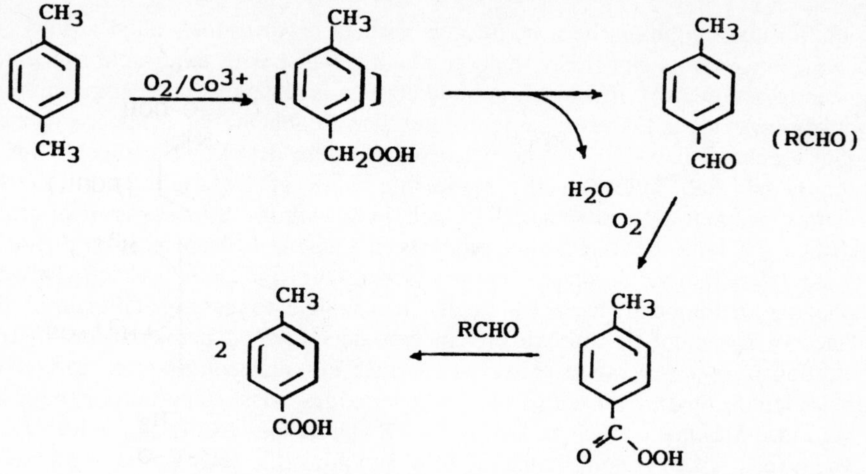

Scheme 1.2 Oxidation of ρ-xylene to ρ-toluic acid

A common feature of many industrial autoxidation processes is the use of transition metal ions, frequently cobalt carboxylates, which in small amount catalyse the radical breakdown of the intermediate hydroperoxide by reactions (1) and (2), thus reducing the activation energy of this process and increasing the rate of the overall reaction:

$$ROOH + M^{2+} \rightarrow RO \cdot + {}^-OH + M^{3+} \tag{1}$$

$$ROOH + M^{3+} \rightarrow ROO \cdot + H^+ + M^{2+} \tag{2}$$

Catalysis by low concentrations of transition metal ions is a common feature of all oxidation processes and is particularly important in biological systems due to the ubiquitous presence of iron in both complexed and ionic form.

Many peroxidation processes are autoinhibiting; that is the rates of oxygen absorption and hydroperoxide formation become slower as the reaction proceeds. In the case of *iso*-propyl benzene, it was discovered empirically that the addition of a small amount of a base markedly increased the yield of hydroperoxide [5]. It was suggested by Roberston and Waters [6] that the autoinhibition of alkyl aromatic oxidation was due to the formation of a small amount of phenol by the acid catalysed process shown in Scheme 1.1. Phenol is a weak antioxidant which functions by reducing the intermediate alkylperoxyl:

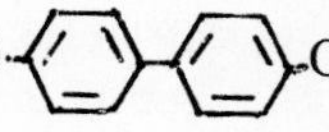

$$
\begin{array}{c}
\text{(a)} \\
ROO \cdot + PhOH \rightarrow ROOH + PhO \cdot \\
\text{(b)} \quad HO-\!\!\bigcirc\!\!-\!\!\bigcirc\!\!-OH, \text{ etc.} \\
\text{(c)} \quad RH \rightarrow PhOH + R \cdot
\end{array}
\tag{3}
$$

and is thus a retarder of the radical chain reaction which leads to the accumulation of hydroperoxide. The phenoxyl radical is relatively stable and readily undergoes dimerisation to dimers and trimers (reaction 3(b)) which are themselves weak chain-breaking antioxidants [7]. However, it is unlikely that reaction 3(a) is primarily responsible for autoinhibition, because phenols without *ortho tertiary* alkyl groups are not efficient antioxidants since they undergo chain transfer with oxidisable substrates like cumene (reaction 3(c)). It is much more likely that acid catalysed decomposition of cumene hydroperoxide is the main cause of inhibition in Scheme 1.1. This is a general antioxidant mechanism which will be discussed in more detail in Chapter 4, but it should be noted that, following the pioneering work of Oberright et al on the mechanisms of sulphur antioxidants in petroleum hydrocarbons [8], measurement of the ratio between the ionic decomposition products of cumene hydroperoxide (phenol + acetone) and the homolytic decomposition products (a-cumyl alcohol + a-methylstyrene + acetophenone) is frequently used as a diagnostic technique to identify and quantify the significance of peroxidolytic antioxidants in peroxide initiated peroxidations [9-16]. Autoinhibition is very general in peroxidation since low molecular weight carboxylic acids are widely formed by breakdown of hydroperoxides. Most important of these are formic acid (see Scheme 1.1) which has $pK_a = 3.8$ and malonic acid ($pK_a = 1.9$). Even benzoic acid ($pK_a = 4.2$) and acetic acid ($pK_a = 4.8$) are sufficiently acidic to provide a mild antioxidant effect, although benzoic acid and to a lesser extent other carboxylic acids can also induce radical decomposition of hydroperoxides at higher concentrations [17]:

$$ROOH + PhC(O)OH \rightarrow RO\cdot + H_2O + PhC(O)O\cdot \qquad (4)$$

In industrial practice, alkali is added continuously to autooxidising cumene in order to maintain the pH at 7.0 [18] and thus inhibit the acid catalysed decomposition of the hydroperoxide.

Many industrial chemicals undergo peroxidation under ambient conditions, particularly in the presence of light. The explosion danger of peroxidised diethyl ether during distillation is well known and warnings about their removal before distillation is an essential part of the safety advice given to chemistry students. Similar problems are encountered with other dialkyl ethers. For example, tetrahydrofuran is widely used in polymer characterisation (e.g. in size exclusion chromatography) and unless hydroperoxides are removed before using this solvent, their thermal degradation products may confuse the outcome of the analysis.

1.2 Effect of Substrate Structure on Peroxidation Rate

For most carbon-based substrates at ambient oxygen pressures, the reaction of an alkyl radical with dioxygen, reaction (5), occurs with zero activation energy. Consequently, the rate-controlling step in the radical chain oxidation process is the rate of reaction of alkylperoxyl with the substrate. Two alternative reactions have to be considered.

$$R\cdot + O_2 \rightarrow ROO\cdot \qquad (5)$$
$$ROO\cdot + RH \rightarrow ROOH + R\cdot \qquad (6a)$$
$$ROO\cdot + C=C \rightarrow ROOC-C\cdot (R'\cdot) \qquad (6b)$$

The first, reaction 6(a), is the rate of hydrogen abstraction of the most labile hydrogen atom in the substrate and the second, reaction 6(b) is the addition of alkylperoxyl to a reactive double bond. Both reactions may occur together if the activation energies are similar. For example, allyl benzene and indene copolymerise with ground state oxygen to give a low molar mass copolymer of oxygen and olefin [19]. This is illustrated for indene in Scheme 1.3 [20]. where it can be seen that hydroperoxide formation constitutes a chain-transfer process. 1,2-substituted double bonds do not normally participate in reaction 6(b) but olefins conjugated, either with carbonyl groups, with other olefinic double bonds or with aromatic rings may copolymerise with oxygen [19].

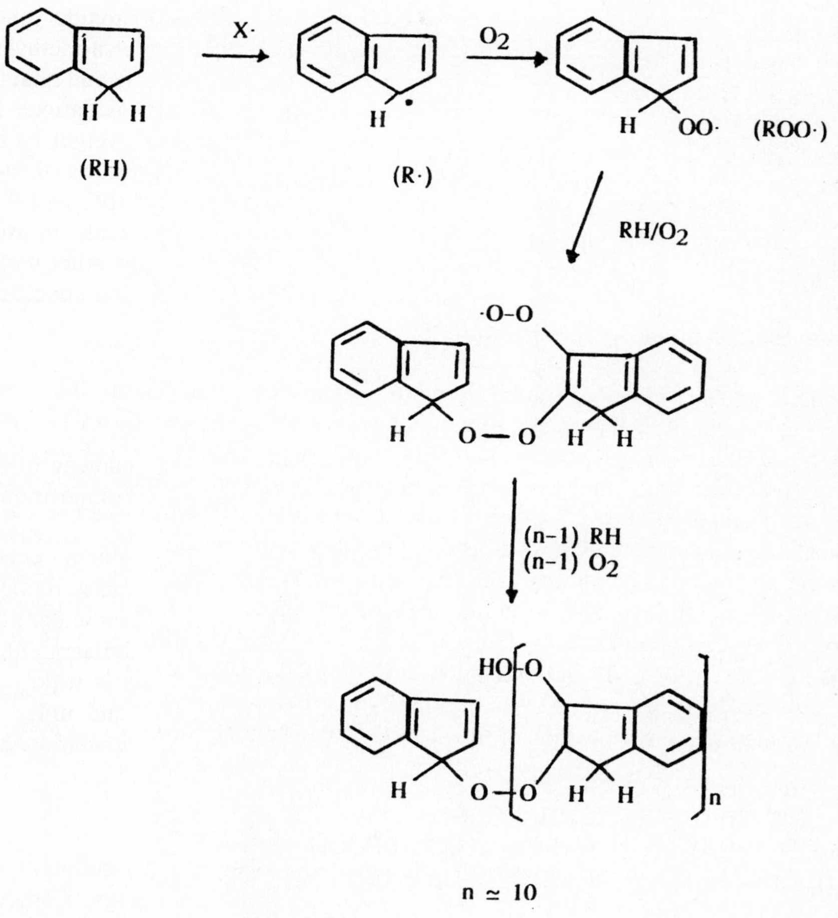

Scheme 1.3 Copolymerisation of indene with oxygen [20]

The driving force for both 6(a) and 6(b) is the stability of the radical produced. Many polyunsaturated allylic and benzylic compounds form delocalised radicals and autoxidise slowly at ambient temperature even in the absence of light. However, polar factors are also involved in the transition state since alkylperoxyl is electrophilic and it was elegantly shown by Russell many years ago that the relative oxidisabilities of the *para* substituted alkyl benzenes obey the Hammett relationship [21]. This is shown for substituted toluenes in Table 1.1. In general, substrates containing alkyl are activated to oxidation, whereas those containing halogens, nitryl, nitro and carboxyl groups are deactivated [21].

Table 1.1 Relative reactivities of substituted toluenes toward ROO· at 90°C (per hydrogen atom) [21]

Toluene substitution	Relative rate*
3,5-dimethyl	1.73
4-*tert*-butyl	1.73
4-methyl	1.53
4-*iso*-propyl	1.40
3-methyl	1.0
toluene	1.0
4-chloro	0.8
4-cyano	0.33
4-nitro	0.33

*Assumes constant rate of termination

The effect of electron delocalisation in the transition state is much more important than the effect of polarity. Alkanes and saturated fatty acids (e.g. stearic, C18:0 and palmitic, C16:0), unlike their polyunsaturated analogues (e.g. C18:2) do not oxidise at a significant rate at ambient temperatures and pressures. In general it is necessary to go to high temperatures (>150°C) with the use of transition metal ion catalysts to obtain appreciable oxidation rates in the case of saturated hydrocarbons. The introduction of just one double bond or an aromatic ring a- to methylene changes the position and it then become possible to isolate the intermediate hydroperoxides (see *iso*-propyl benzene above). The relative rates of oxidation of oleic, C18:1,(n-9), I, linoleic, C18:2(n-6), II, linolenic C18:3,(n-6), III and arachidonic, C20:4,(n-6), IV acids is 1:10:20:40. (Note that fatty acids are described by the number of carbon atoms:number of double bonds and, in parenthesis, the position of the nearest double bond to the terminal methyl.)

$C_8H_{17}CH=CH(CH_2)_7COOH$ (I)
$C_5H_{11}CH=CHCH_2CH=CH(CH_2)_7 COOH$ (II)
$C_3H_7CH=CHCH_2 CH=CHCH_2CH=CH(CH_2)_7COOH$ (III)
$C_5 H_{11}CH=CHCH_2=CHCH_2CH=CHCH_2CH=CH(CH_2)_3COOH$ (IV)

It is evident that the effect of two activating double bonds adjacent to methylene in II is significantly greater than the addition of a further 1,4-diene unit in III. The electron is delocalised over 5 carbon atoms in both cases and the activating effect of additional 1,4-diene groups is essentially statistical.

Some branched-chain substrates peroxidise intramolecularly by hydrogen abstraction by peroxyl from a neighbouring reactive methylene group. This is particularly true of the a-olefin polymers and copolymers. Polypropylene has been shown to form a sequence of hydroperoxide groups at vicinal *tertiary*-alkyl carbon atoms [22-24]. These are formed by the mechanism shown in Scheme 1.4. Propagation by hydrogen abstraction from a *tertiary* carbon atom is about six times faster than from a *secondary* carbon [25] and this is particularly facilitated in the transition state in Scheme 1.4 by intramolecular hydrogen bonding [23]. Furthermore, the vicinal hydroperoxides are less stable than isolated hydroperoxides due to internal hydrogen bonding which induces homolysis of the O-O bond, thus increasing radical formation. Finally, the rate of termination of *tertiary* peroxyl radicals is up to three orders of magnitude slower than the rate of termination of *secondary* peroxyl radicals (see Section 1.4). This combination of factors results in a much more rapid peroxidation of polypropylene than of polyethylene and the peroxidation kinetic chain length is approximately 100 for PP but only 10 for PE [26,27]. An important consequence of the greater oxidative sensitivity of PP than PE is that equilibrium hydroperoxide concentrations in the former are much higher than in the latter so that it is possible to follow the concentrations of hydroperoxide in PP during photooxidation whereas in PE concentrations are too low to be measured by standard spectroscopic methods [22].

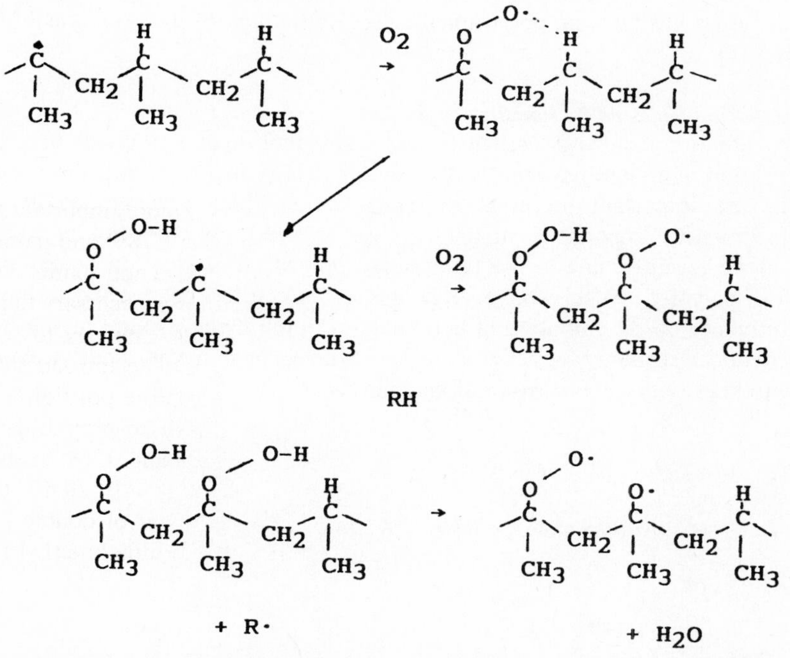

Scheme 1.4 Formation and decomposition of vicinal hydroperoxides in polypropylene

In the light of the above, it is not unexpected that blends of PE and PP decrease in photostability as the proportion of PP in the blend increases [28] and PP can be looked upon as a prooxidant for PE [22]. However, this is partly an effect of phase separation between the two components since a rubbery terpolymer of ethene, propene and a diene monomer improves both the compatibility of PE/PP blends and their resistance to photodegradation in spite of the fact that the terpolymer actually oxidises more rapidly [29]. Similar sensitising effects by polyunsaturated polymers have been observed in the "impact-modified" polymers which are generally polyunsaturated rubbery polymer in polystyrene [30-32] and PVC [33]. It is clear, then, that morphological effects can be as important as chemical peroxidisability in determining polymer durability and this will be discussed further in Section 1.6.4.

1.3 Initiation of Peroxidation
1.3.1 *Autoinitiation by ground-state dioxygen*
Substrates which contain reactive methylene groups are very readily hydroperoxidised by the chemistry discussed in the last section and, particularly in the presence of transition metal ions, it is experimentally difficult to exclude reactions (1) and (2) from the initiation process. Moreover, ground-state triplet oxygen is a relatively unreactive molecule, and direct involvement in initiation reactions is normally difficult to prove due to the widespread occurrence of hydroperoxides which are notoriously unstable to light and heat. However, styrene has no reactive methylene groups to form hydroperoxides and the normal hydroperoxide initiated rate equation: $r \propto (O_2 abs)^{\frac{1}{2}}$ which reflects that the rate of peroxidation is not followed in styrene peroxidation. Instead it is constant ($r = (1.4 \cdot 10^5)^{\frac{1}{2}}$). The most probable explanation is that this represents the rate of direct attack of ground state oxygen at the vinyl group [34] and a similar explanation has been proposed for the "self-initiation" of dihydrophenanthrene peroxidation [35].

1.3.2 *Peroxidation induced by ionising radiation*
High energy radiation can break carbon-carbon and carbon-hydrogen bonds in organic substates [36] and hydrogen-oxygen bonds in water [37] (Chapter 2) and the resulting free radicals are important initiators for peroxidation. This process has assumed considerable practical importance in polymers as a means of cross-linking through carbon but more recently, due to the increasing application of electron beams in the sterilisation of polymer artifacts for medical applications, it has been found to have an important influence on the initiation of peroxidation in the hydrocarbon polymers. The following chemical sequence is believed to be involved over a very short time-scale (given in parenthesis) in the formation of macroradicals;

$$\text{Polymer (P)} \xrightarrow{\Upsilon} e + P^+ \quad (\sim 10^{-18} \text{ s}) \tag{7}$$

$$P + e \rightarrow P^* \text{ (excited state)} \quad (10^{-17}\text{-}10^{-15}\text{s}) \tag{8}$$

$$P^* \begin{array}{c} \nearrow P\cdot \\ \searrow P\cdot + H\cdot \end{array} \quad (10^{-12}\text{-}10^{-3}\text{s}) \tag{9}$$

The subsequent reactions of macroradicals with oxygen depend on the availability of oxygen to the carbon radical centres, and the rate of peroxidation is markedly affected by oxygen diffusion rates and the presence or absence of antioxidants in the polymer. A practical consequence of this is that there is normally a pronounced lag time between irradiation and the onset of mechanical deterioration. Thus, in the case of polypropylene syringes sterilised by electron beam irradiation, there may be no discernable change in mechanical properties for many weeks before catastrophic disintegration occurs.

Competition between chain scission and cross-linking is very much dependent upon the structure of the polymer and the availability of oxygen [38], but the chemistry of peroxidation products follows very much the same pattern as that described in Section 1.2 for thermal oxidation of polymers. Consequently, initiation by high energy irradiation is a convenient technique for the study of the peroxidation chain reaction and termination processes at ambient temperatures [38]. However, in semi-crystalline polymers, some peroxyl radicals are long-lived due to incarceration in the crystalline phase and these play very little part in the subsequent propagation process [39,40].

1.3.3 *Initiation by reactive oxygen species*
In contrast to ionising radiation, light reaching the earth's surface (i.e. of longer wavelength than 295 nm), does not normally break C-C or C-H bonds in the absence of a suitable chromophore to form chemically reactive photoexcited states. However, singlet oxygen, ($^1\Delta gO_2$ or $^1O_2$) is formed readily under photoxidative conditions by energy tranfer to ground state oxygen from photoexcited states of sensitisers (Sens) such as rose bengal or methylene blue and some inorganic pigments. This topic has been comprehensively reviewed by Rabek [41] and the reader is directed to this book for an encyclopaedic overview of the voluminous literature on the effects of light in chemistry. The generation of singlet oxygen is described by reactions (10) and (11):

$$\text{Sens} \xrightarrow{h\nu} {}^1\text{Sens}^* \rightarrow {}^3\text{Sens}^* \quad (10)$$

$$^3\text{Sens}^* + {}^3O_2 \rightarrow {}^1O_2^*(\Sigma g) + {}^1\text{Sens} \quad (11)$$

Singlet oxygen is very reactive toward most olefins. Unlike ground state oxygen, it is not itself a diradical but it reacts instead like a dienophile. Thus in the case of a 1,2-dialkyl olefin, it adds to the double bond with simultaneous hydrogen abstraction in an "ene" reaction [42-45]:

$$\underset{-CH_2}{\overset{H_3C}{\diagdown}}CH=CH\underset{CH_2-}{\diagdown} \quad + \quad {}^1O_2 \quad \rightarrow \quad \underset{-CH_2}{\overset{H_2C}{\diagdown}}\overset{H\cdots O}{\underset{CH-CH}{\diagup}}\overset{\diagdown O}{\underset{CH_2-}{\diagup}} \quad \rightarrow \quad \underset{-CH_2}{\overset{H_2C}{\diagdown}}\overset{}{\underset{CH-CH}{\diagup}}\overset{OOH}{\underset{CH_2-}{\diagup}} \quad (12)$$

The hydroperoxides formed in reaction 12 are effective intititors for peroxidation.

1,3-Dienes readily form cyclic peroxides (endoperoxides) [46] with almost zero activation energy and the very fast reaction of $^1O_2$ with diphenyl*iso*-benzifuran (DBPF) without physical quenching [46,47] provides a useful quantitative method of measuring $^1O_2$ concentrations in solution;

$$\text{DPBF} \xrightarrow{{}^1O_2} \text{endoperoxide} \xrightarrow{\Delta} \text{dibenzoyl product} \quad (13)$$

High yields of hydroperoxides can be produced when singlet oxygen, made by microwave discharge or by reaction of hydrogen peroxide with hypochlorous acid [48], is reacted with olefins; but yields are much lower when produced by photosensitisation of ground state oxygen due to the photolysis of the hydroperoxide itself [49].

Singlet oxygen is relatively unimportant as an intitiator of photooxidation in saturated hydrocarbons [50], since it is physically quenched by more abundant molecules in the environment (e.g. water). However it may play a significant role in the peroxidation of unsaturated compounds, particularly in biological systems [46,51].

Ozone (O_3) also reacts rapidly with olefins [45,52] but very slowly with tetrahedral carbon. However, it has been shown [53] that aryl activated methylene groups do react slowly with ozone to give hydroperoxides. Ozonolysis is of considerable importance in tyre technology, where ozone cracking occurs at very low ozone concentrations. The chemistry of the ozonolysis of olefins is ionic rather than radical chain and has been very thoroughly investigated by Criegee [54] and by Bailey [55]. Its main significance to free radical peroxidation is the formation of dialkyl peroxides and particularly hydroperoxides which may under appropriate conditions act as initiators for free radical peroxidation. This chemistry is summarised in Scheme 1.5.

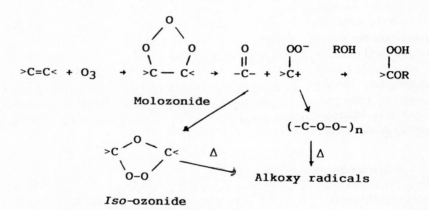

Scheme 1.5 Reactions of ozone with olefins

The technological effects of ozone have been extensively reviewed [3,52,56,57] and it should be noted that peroxides are primary products of ozonolysis of many unsaturated compounds *in vivo* as well as *in vitro* and it is likely that these species intensify asthmatic diseases in heavily industrialised atmospheres where damaging concentrations of ozone are frequently reported. This will be discussed in Chapter 2.

Superoxide ($O_2\cdot^-$) is readily formed by a variety of reducing agents/sensitisers in aqueous media, reaction 14, but it is not a powerful oxidising agent due to resonance stabilisation [52,58], and it frequently acts as a reducing agent for iron in biological systems, for example in the Fenton reaction, reaction 15(b):

$$O_2 \xrightarrow{+e} \cdot O\text{-}O^- \longleftrightarrow {}^-O\text{-}O\cdot \xrightarrow{+H^+} \cdot OOH \xrightarrow{+e} H_2O_2 \qquad (14)$$

$O_2\cdot^-$ has very low free radical reactivity [58] but, when protonated, it gives the much more reactive hydroperoxyl radical, $HOO\cdot$. This radical is more organo-soluble than superoxide, and like the latter it is readily reduced to hydrogen peroxide. In the presence of the reduced transition metal ions (notably iron), hydrogen peroxide is reduced to the highly reactive hydroxyl radical in the Fenton reaction:

$$\begin{align} \text{(a)} \quad & H_2O_2 + Fe^{2+} \rightarrow OH^- + \cdot OH + Fe^{3+} \\ \text{(b)} \quad & O_2\cdot^- + Fe^{3+} \rightarrow O_2 + Fe^{2+} \end{align} \quad (15)$$

Superoxide and hydrogen peroxide have not been found to play a very important role as initiators of peroxidation in technological media, almost certainly because the necessary hydrophilic conditions do not normally exist for the reduction of ground state oxygen to occur. However $O_2\cdot^-$ is extremely important as a source of hydrogen peroxide and hydroxyl radicals in biological systems and the consequence of hydrogen peroxide reduction *in vivo* will be discussed in in Chapter 2. Alkyl hydroperoxides on the other hand are the most common source of radicals formed in an analogous reduction process in hydrophobic systems by reactions (1) and (2) and this is of great significance in technological media, notably hydrocarbon oils [59], rubber [60] and plastics [60], where organo-soluble transition metal salts are frequently contaminants from their working environment. However some transition metal ions (notably Cu^{2+}) may undergo "inversion" from prooxidants to effective antioxidants as concentrations are increased (Chapter 3). Macroalkyl hydroperoxides are also very sensitive to UV light and are the primary source of radicals during the photooxidation of commercial polymers (Section 1.6.3).

1.3.4 *Initiation by physical stress*
Mechanical shear can break carbon-carbon bonds in solids [61] or in highly viscous liquids [62] leading to macroalkyl formation by carbon-carbon bond scission. Mechanooxidation is particularly important in polymer technology during processing (Section 1.6.2(c)). Stress is also an important initiator in rubbers subjected to mechanical deformation (Section 1.6.2(d)) and the technologically important phenomena of "fatigue" and "ozone cracking" [63,64] are associated with this stress-induced peroxidation. Recent evidence also suggests that stress may accelerate the rate of photooxidation of polyolefins [65].

It also seems likely by analogy with polymers that increased radical formation in stressed muscles may in part be a mechanochemical process although little research has been carried out to examine this possibility. However, Symons et al. have shown that peroxyl radicals are formed during bone fracture [66,67] and they attribute this to the breakage of collagen strands embedded in the bone. It seems likely that analogous physical processes may occur in severely fatigued muscles. Mechanooxidation seems to be a potentially important area for future peroxidation studies *in vivo*.

1.4 Termination
In most peroxidations, reactions 6(a) and 6(b) are rate determining in the radical chain process. Consequently, at ambient pressures, the main termination process is reaction (16). However, at low oxygen concentrations, and particularly in the case of hydrocarbons that give relatively stable alkenyl radicals termination through cabon-

centred radicals, reactions (17) and (18) may play a significant role even at ambient oxygen pressures [1,68]:

$$2\,R'_2C(H)OO\cdot \rightarrow R'_2C{=}O + R'_2CHOH + O_2 \qquad (16)$$

$$2R\cdot \rightarrow R\text{-}R \qquad (17)$$

$$R\cdot + ROO\cdot \rightarrow ROOR \qquad (18)$$

Tertiary alkylperoxyl radicals are much more stable than *secondary* since they cannot undergo facile disproportionation, reaction (16). Dialkyl peroxides are major products of the bimolecular reactions of *tert*-alkylperoxyl, shown typically for cumylperoxyl ($R_cOO\cdot$) in Scheme 1.6 [69].

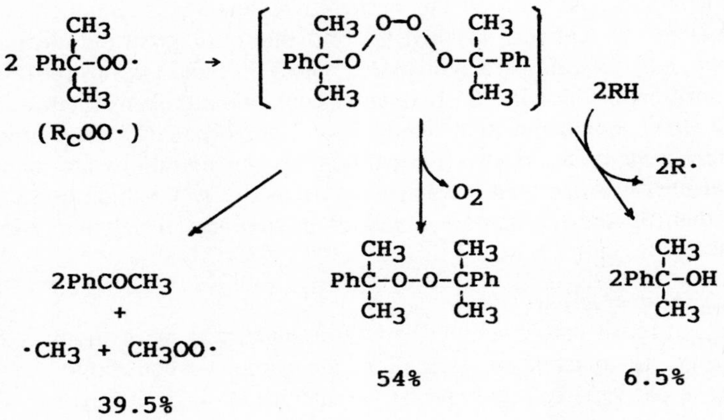
2 PhC(CH3)2-OO·
(RcOO·)
2RH
2R·
O2
2PhCOCH3
+
·CH3 + CH3OO·
39.5%
PhC(CH3)2-O-O-C(CH3)2Ph
54%
2PhC(CH3)2-OH
6.5%

Scheme 1.6 Products of α-cumylperoxyl termination [69]

The addition of a small molar proportion of tetralin to cumene dramatically reduces the peroxidation rate of the latter, and Russell [70] found that the rate of reaction of $R_cOO\cdot$ with tetralinperoxyl ($R_tOO\cdot$) was over a hundred times faster than the rate of reaction of $R_cOO\cdot$ with itself:

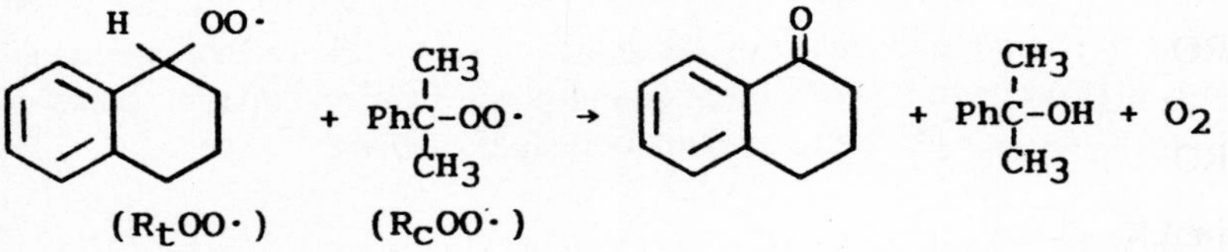
H OO·
+ PhC(CH3)2-OO·
(RtOO·)
(RcOO·)
O
+ PhC(CH3)2-OH + O2

(19)

As little as 2 mole % of tetralin in cumene results in 100% cross-termination by reaction (19) [70]. It is clear then that the rate of termination may dominate the overall rate of peroxidation and this was seen to be of considerable importance in the peroxidation of branched-chain polyolefins (Section 1.2).

The equations describing the steady state rates of oxidation of moderately reactive hydrocarbons under ambient oxygen pressures and of highly reactive hydrocarbons under conditions of reduced oxygen pressure are (i) and (ii) respectively [2]:

$$-\frac{d[O_2]}{dt} = k_6 k_{16}^{-1/2} r_i^{1/2} [RH] \qquad (i)$$

$$-\frac{d[O_2]}{dt} = k_5 k_{17}^{-1/2} r_1^{1/2} [O_2] \qquad (ii)$$

where r_i is the rate of intitiation and the k subscripts indicate the equations in the text. In the case of very oxidisable olefins, e.g. 2,6-dimethylheptan-2,5-diene, V, the rate of reaction 6(a) approaches that of reaction 5. Consequently, termination through the delocalised radical, VI, supervenes to relatively high oxygen pressures [71,72].

$$(CH_3)_2C=CHCH_2CH=C(CH_3)_2 \xrightarrow{X\cdot} (CH_3)_2\overset{\bullet}{C}\text{==}CH\text{==}CH\text{==}CH\text{==}C(CH_3)_2 \text{ VI} \qquad (20)$$
$$V \qquad\qquad + XH$$

The presence of a small amount of a relatively stable carbon radical such as VI or triphenylmethyl may determine the overall rate of oxidation. Thus, "stable" carbon-centred free radicals are effective retarders for the oxidation of many substrates at low oxygen pressures, due to reactions 17 and 18 [73]. It should be noted that this is not inhibition in the accepted sense of the word, since carbon-centred radicals always react with oxygen with continuation of the kinetic chain and this is more correctly called retarded cooxidation. It will be seen later (Section 1.6.2) that the extended conjugated polyene system produced in PVC during thermal degradation readily scavenges alkylperoxyl to give a delocalised polyene radical which can sacrificially co-oxidise with the main substrate. The same mechanism is probably involved in the "antioxidant" mechanism of ß-carotene and this will be discussed in Chapter 5.

1.5 Products of Peroxidation
1.5.1 *Hydroperoxides and their decomposition products*
The major route to the formation of the end products of peroxidation is through the hydroperoxide. This can occur by unimolecular thermolysis and photolysis (reaction 21) or by reduction to give an alkoxyl radical (reaction 22):

$$ROOH \xrightarrow{h\nu,\Delta} RO\cdot + \cdot OH \qquad (21)$$

$$ROOH \xrightarrow{e} RO\cdot + {}^-OH \qquad (22)$$

By far the most effective reducing agents in reaction (22) are the lower valency transition metal ions, notably Fe^{2+} and Cu^+ which are the major source of prooxidant species in both technological and biological systems.

The alkoxyl radical is highly reactive although not as reactive as the hydroxyl radical formed in the analogous Fenton reaction (reaction 15) and it can undergo further reduction, oxidation (or disproportionation), alkyl elimination and addition to a carbon-carbon double bonds [74].

Oxidation of C-H bonds by alkoxyl initiates the radical chain (reaction 23) and is the primary reason why hydroperoxides and transition meal ions are ubiquitous prooxidants in technological media:

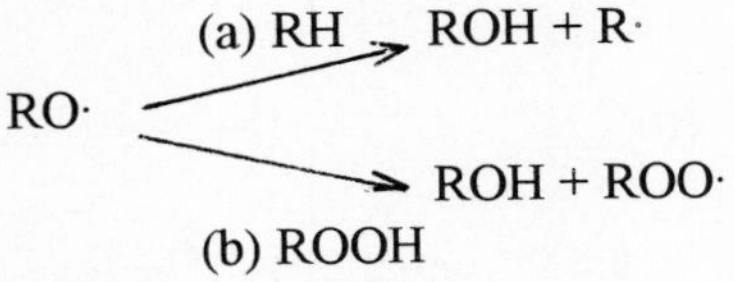

$$RO\cdot \begin{array}{l} \text{(a) } RH \rightarrow ROH + R\cdot \\ \text{(b) } ROOH \rightarrow ROH + ROO\cdot \end{array} \quad (23)$$

Reaction 24(a) is a termination step when it occurs in a molecular "cage" reaction with hydroxyl when there is a hydrogen on the a carbon atom:

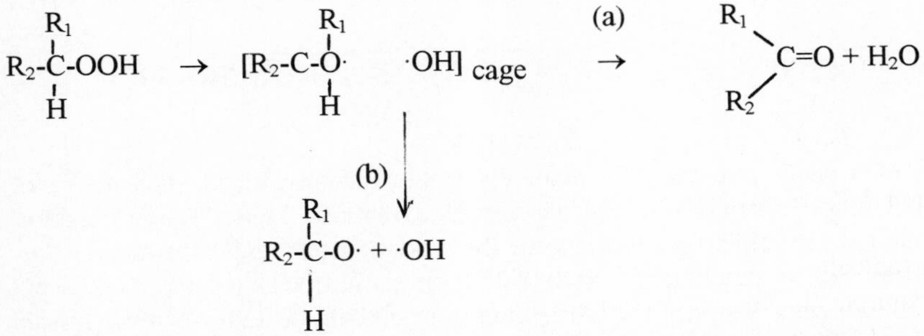

$$\underset{H}{\overset{R_1}{R_2-\overset{|}{C}-OOH}} \rightarrow [\underset{H}{\overset{R_1}{R_2-\overset{|}{C}-O\cdot}} \quad \cdot OH]_{cage} \xrightarrow{(a)} \underset{R_2}{\overset{R_1}{\diagdown}} C=O + H_2O$$

$$\downarrow (b)$$

$$\underset{H}{\overset{R_1}{R_2-\overset{|}{C}-O\cdot}} + \cdot OH \quad (24)$$

However a small proportion of the radicals escape from the "cage" even in the case of *secondary* hydroperoxides. Fragmentation of alkoxyl, reaction (25), is the predominent process occurring with *tertiary* alkoxyl. This is a propagation process which in the case of polymers also leads to molar mass reduction:

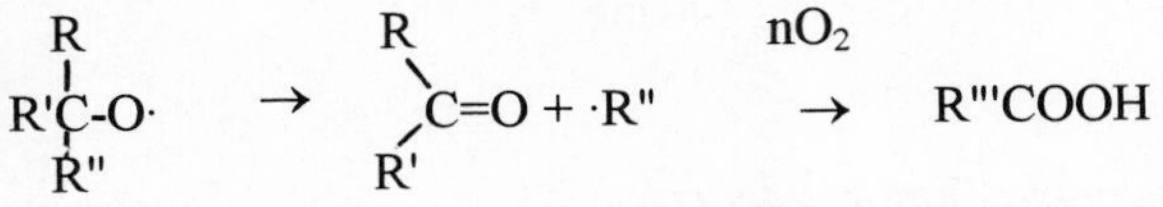

$$\underset{R''}{\overset{R}{R'\overset{|}{C}-O\cdot}} \rightarrow \underset{R'}{\overset{R}{\diagdown}} C=O + \cdot R'' \xrightarrow{nO_2} R'''COOH \quad (25)$$

The products are carbonyl compounds: initially aldehydes and ketones and ultimately carboxylic acids. Reactions (23a) and (25) are in competition and, whichever, predominates depends on the oxidisability (by hydrogen abstraction) of the substrate and the stability of the eliminated carbon-based radical ($\cdot R''$) [74]. Reaction (26) is also a propagation reaction which leads to molecular enlargement by ether formation and is the primary reason for the "drying" of paints and the cross-linking of unsaturated polymers during "ageing":

$$RO\cdot + >C=C< \rightarrow ROC-C< \quad (26)$$

Competition between α,β C-C bond scission (reaction 25) and molecular enlargement (reaction 26) depends primarily on temperature and on the nature of the unsaturation in the olefin. A 1,1-dialkylethylene is much more suceptible to alkoxyl attack than a 1,2-dialkylethylene [75].

As already been noted, many vinyl monomers react with alkylperoxyl under ambient conditions by reaction (6b). Styrene, if not effectively stabilised by the addition of an inhibitor, readily forms a 1:1 copolymer with oxygen at ambient oxygen pressures [76], Scheme 1.7. At low oxygen pressures, oxygen is a retarder of styrene polymerisation because the alkylperoxyl radical is less reactive toward styrene than the styryl radical and a minor by-product are styrene oxide, benzaldehyde and formaldehyde.

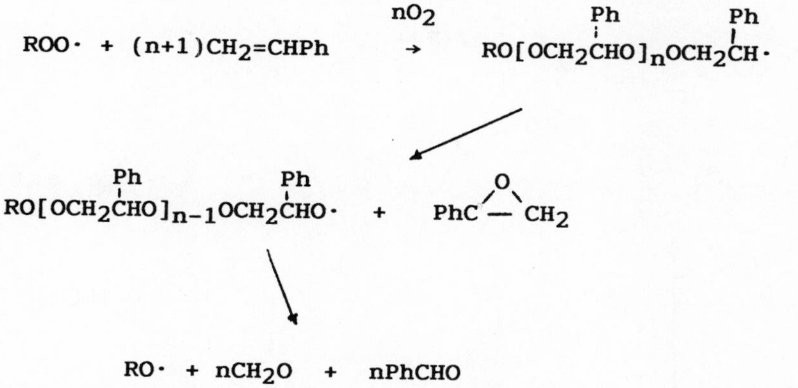

$$ROO\cdot + (n+1)CH_2=CHPh \xrightarrow{nO_2} RO[OCH_2\overset{Ph}{C}HO]_nOCH_2\overset{Ph}{C}H\cdot$$

$$RO[OCH_2\overset{Ph}{C}HO]_{n-1}OCH_2\overset{Ph}{C}HO\cdot \; + \; PhC\overset{O}{\underset{}{\diagdown}}CH_2$$

$$RO\cdot + nCH_2O + nPhCHO$$

Scheme 1.7: Products formed in the copolymerisation of styrene with oxygen [76]

Alkylperoxyl radicals can also, under favourable conditions, epoxidise olefins in good yield. An example is the epoxidation of β-di-*iso*-butene in 50% yield during peroxidation [77] (see Scheme 1.8). The driving force for this reaction is the formation of the relatively stable *tert*-alkyl radical, Scheme 1.8, reaction (a).

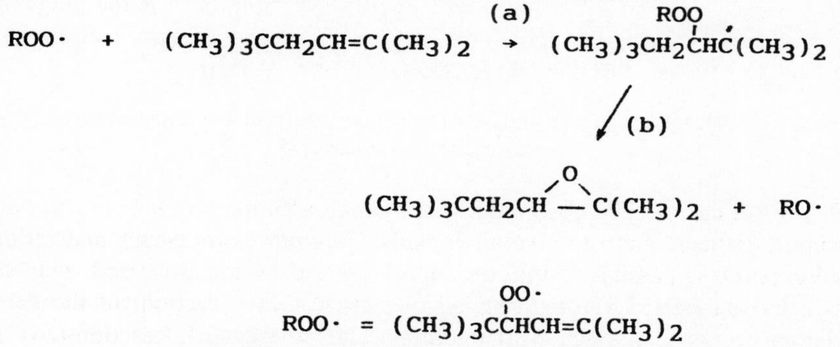

$$ROO\cdot \; + \; (CH_3)_3CCH_2CH=C(CH_3)_2 \; \overset{(a)}{\rightarrow} \; (CH_3)_3CH_2\overset{ROO}{C}HC(CH_3)_2$$

$$\downarrow (b)$$

$$(CH_3)_3CCH_2CH\overset{O}{\underset{}{\diagdown}}C(CH_3)_2 \; + \; RO\cdot$$

$$ROO\cdot \; = \; (CH_3)_3C\overset{OO\cdot}{C}HCH=C(CH_3)_2$$

Scheme 1.8: Peroxidation of di-*iso*-butene [77]

Many biological molecules contain methylene-interrupted dienes which on peroxidation undergo a double bond isomerisation to give potentially polymerisable 1,3-dienes. Molecular enlargement does not occur to any appreciable extent until about one mole of oxygen has been absorbed, Scheme 1.9 [78].

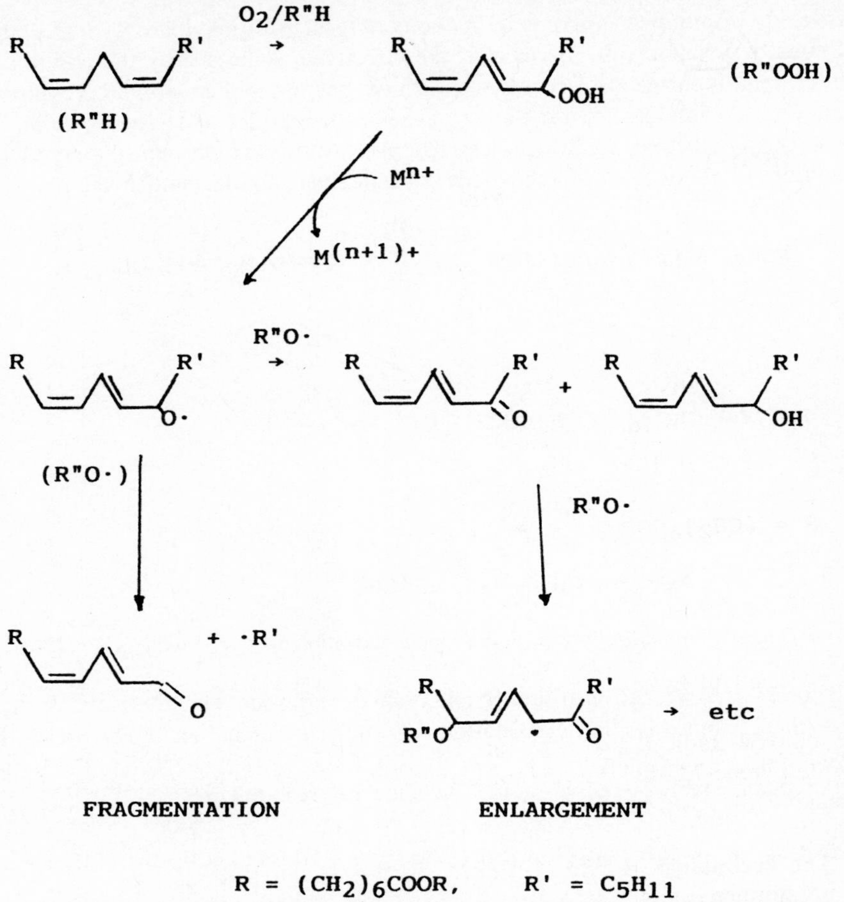

Scheme 1.9: Molecular reduction and enlargement during metal ion catalysed peroxidation of polyunsaturated esters [78]

Marnett [79,80] has recently observed intramolecular addition of alkoxyl to an adjacent double bond (Scheme 1.10) to give an epoxide. This process is preferred to hydrogen abstraction and it is possible to trap the carbon-centred radical by means of a "stable" phenoxyl derived from 2,4,6-tri-*tert*-butyl phenol. It seems likely that intramolecular alkoxylation occurs in parallel with intermolecular enlargement but the latter is the reason for the formation of highly cross-linked polymers during the "drying" of paint (see Section 1.6.1) and it seems likely that it also plays a part in foam and plaque formation during LDL oxidation in atheroslerosis (Chapter 2).

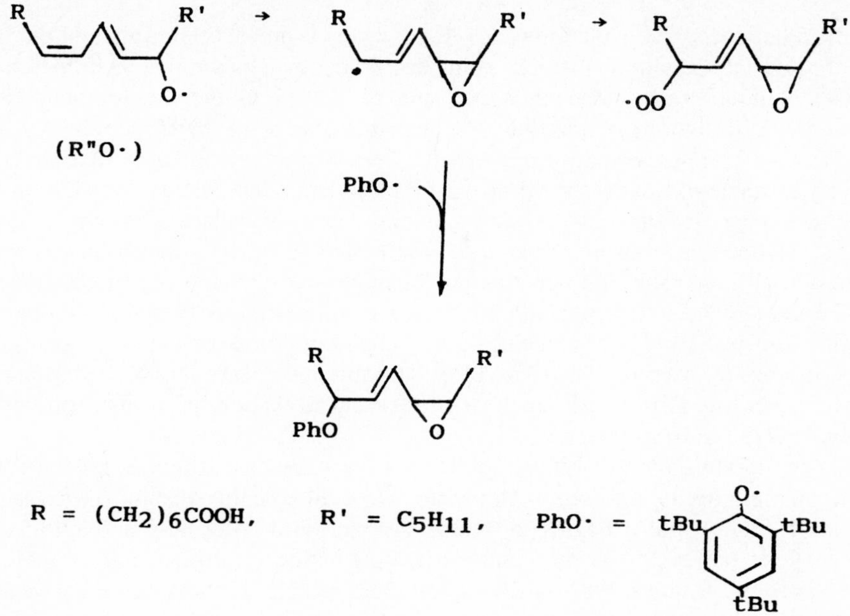

Scheme 1.10 Epoxide formation during metal ion catalysed lipid peroxidation [79, 80]

The peroxyl radical produced in Scheme 1.10 is also capable of epoxidising benzo[a]pyrene-7,8-diol by the mechanism described above for di-*iso*-butene [81] (Chapter 2).

1.6 The Technological Effects of Peroxidation
Polyunsaturated esters from natural products form the basis of two major industries: the edible fats and oils processing industries and the paint and varnish industries. They both use as raw materials the triglycerides of the fatty acids. However, whereas the surface coatings industry *utilises* oxidation to give cross-linked films, the food industry is primarily concerned with *avoiding* oxidation of the same materials during processing and use. The latter topic will be discussed in Chapter 2. Similarly the polymer manufacturing industries (polydienes, polyolefins, polyvinyl chloride, polystyrene, polyamides, polyurethanes, etc.) go to considerable lengths to prevent oxidative changes to their products both during manufacture and in service.

1.6.1 *Surface coatings by peroxidation of polyunsaturated fatty esters*
The surface coating industries utilise the poyunsaturated components of plant oils. Linseed oil, consisting primarily of glyceryl trilinoleate, undergoes rapid hydroperoxidation followed by polymerisation ("drying") particularly in the presence of a small amount of a cobalt carboxylate by the mechanism discussed in the previous Section (see Scheme 1.9). During the initial stages of "drying", the conjugated peroxy ester readily copolymerises with oxygen in the same way as styrene and indene (Scheme 1.3). However peroxidic cross-links are unstable and break down to give alkoxyl radicals. Ultimately ether and carbon-carbon bonds are formed [82] giving a tough

cross-linked polymer. The oxidation of 1,4-dienic esters is autoaccelerating, and the rate passes through a maximum which coincides with the maximum hydroperoxide concentration at an early stage after exposure of a film to the environment [82]. Subsequently, both hydroperoxides and conjugated double bonds decay rapidly and the "dry" film absorbs relatively little oxygen. It is really quite remarkable that such an oxidatively unstable precursor should give a film that can resist further degradation for long periods even in the outdoor environment. However, chain scission, to give aldehydes, alcohols, carboxylic acids and even carbon dioxide and water, occurs concurrently with cross-linking and this process continues through the lifetime of the film and eventually becomes predominant. This is manifested initially as loss of gloss of the paint film followed by cracking and "chalking" and, eventually, complete breakdown due to loss of elasticity [83]. Unsaturated esters based on aromatic carboxylic acids (the alkyds) are much more resistant to deterioration than the purely lipid-based polyunsaturated glycerides

The autoaccelerating nature of paint "drying" is normally an advantage, but for some applications rapid drying *ab initio* is desirable. Tung oil contains 80% of α-eleostearic acid, a cis-1,3-diene and it begins to oxidatively cross-link after only a fraction of a mole of oxygen has been absorbed and no hydroperoxide is formed until 1 mole of oxygen has been absorbed [78]. α-Eleostearic acid which contains three conjugated double bonds oxidises five time faster than linoleic esters but the trans β-eleostearic esters oxidise at only half the rate of the α-eleostearates [34]. Because of their tendancy to fast gellation during application, 1,3-dienes are difficult to use alone but they are frequently added to accelerate the drying of the 1,4-dienes.

1.6.2 *Thermal oxidation of polymers*
Polymeric materials vary remarkably in their resistance to oxidation. This is a consequence not only of the inherent peroxidisability of the molecular structure, but also of the way in which they are formulated with antioxidants and stabilisers to reduce both the initation and propagation reactions important to the radical chain mechanism. Polymer stabilisation will be discussed in Chapters 3 and 4, but the remarkable abilty of antioxidant technology to adapt inherently oxidisable molecules, for example the highly unsaturated poly(1,3-dienes) to the demands of the modern automotive industry, is quite outstanding. Even today, the detailed chemistry involved in the ability of a complex combination of antioxidant synergists and vulcanising ingredients to combat the effects of high temperatures and dynamic stress for long periods of time in service is not fully understood. But the durability of the automotive tyre has been steadily improved, mainly by empirical experimentation, to match the ever more severe conditions of modern transport.

The polyunsaturated elastomers, cis-poly(isoprene), (cis-PI), cis-poly(butadiene), (cis-PB) and styrene-butadiene rubber (SBR) are the basis of modern tyre technology and are at the same time among the most oxidisable of all polymers. Not only is the hydrocarbon structure itself highly susceptible to peroxidation, but the process of "vulcanisation" (polysulphide cross-linking) further increases the sensitivity of the tyre to thermo-oxidation by facile radical formation (see Scheme 1.11) [63,84,85]. At the same time, sulphur cross-linking and the presence of a substantial proportion of carbon black increases the resistance of rubber to "fatigue" (mechanooxidation). Sulphur acts as a source of polymer reactive thiyl radicals which facilitate restructurisation of the rubber network under stress [63,64].

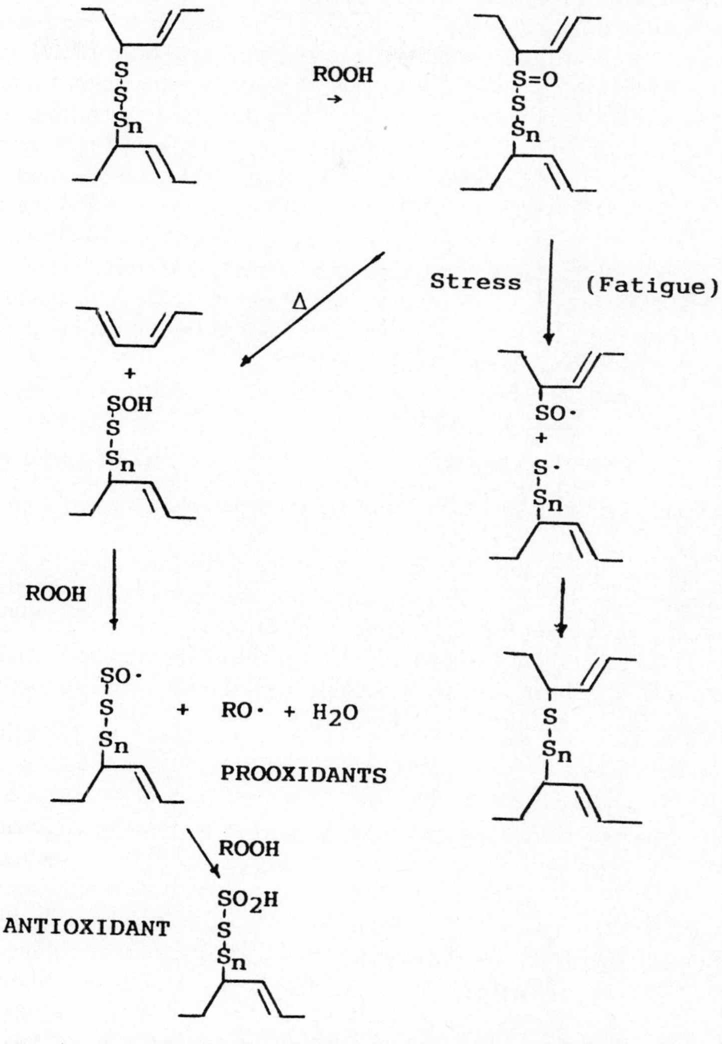

Scheme 1.11 Oxidative reactions of polysulphides in vulcanised rubber [63, 84, 85]

The hydrocarbon back-bone of the cis rubbers peroxidises with chain scission via the allylic hydroperoxide. This reaction predominates in cis-polyisoprene but, in the case of polybutadiene and its copolymers, cross-linking also occurs (see Scheme 1.12) through peroxyl addition to reactive vinyl groups. At ambient temperatures, peroxy gels are the primary products in unvulcanised rubbers, and in vulcanised rubbers at elevated temperatures there is an increase in modulus due to the formation of stable ether cross-links in the fabricated product during service [86] (see Scheme 1.12). In commercial practice, a blend of cis-PI and the PB-based elastomers is sometimes used to minimise changes in mechanical properties due to the above oxidative changes.

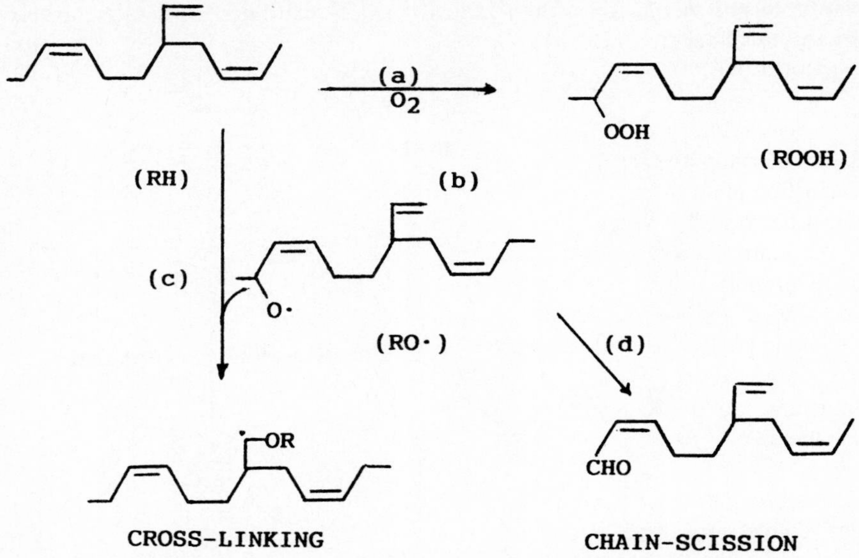

Scheme 1.12: Chain scission and cross-linking in polybutadiene during peroxidation

At the opposite end of the stability scale lie the fully fluorinated polymers of which polyfluoroethylene (PTFE) is the most important commercial example. PTFE in its pure form contains no carbon-hydrogen bond or olefin group and cannot therefore undergo reaction 5. Between these two extremes lie the major thermoplastic and fibre-forming polymers, all of which contain accessible C-H bonds. The approximate order of increasing oxidative stability is as follows [87]:

$$-CH_2\underset{PP}{\overset{CH_3}{CH}}- \;<\; -\underset{PE}{CH_2CH_2}- \;<\; -CH_2\underset{PS}{\overset{Ph}{CH}}- \;<\; -\underset{PA}{CH_2NHCOCH_2}- \;<\; -\underset{PET}{CH_2OCOPh}-$$

$$<\; -CH_2\underset{PVC}{\overset{Cl}{CH}}- \;<\; -CH_2\underset{\underset{PMMA}{COOCH_3}}{\overset{CH_3}{C}}- \;<\; -CH_2\underset{PAN}{\overset{CN}{CH}}- \;<<\; -\underset{PTFE}{CF_2CF_2}$$

It should be noted that most commercial polymers cannot be properly described by the formal structures shown above. There are a number of reasons for this.

(a) Structural defects

As manufactured, polymers normally contain minor structural defects, notably olefinic groups and chain branches which have a profound effect on their peroxidisability. Thus polymethylene made by decomposition of diazomethane is much more resistant to oxidation than commercial polyethylenes. The latter, like polypropylene and PVC, normally contains vinyl, vinylidine and vinylene unsaturation which is the locus of initial peroxidation, leading to sensitisation of the saturated components of the polymer [88-91] during environmental exposure (Section 1.6.3).

(b) Heterogeneity

Some thermoplastics (eg PS, PVC, etc) are composite materials containing a proportion of an unsaturated rubber, notably poly-BD, to improve impact resistance and it has been found that these have a profound sensitising effect on the oxidation of the main substrate [92,93]. Thus "high impact" polystyrene (HIPS) contains a small proportion of grafted polybutadiene which is mich less stable to oxidation than "pure" polystyrene [30,31]. The graft copolymer of acrylonitrile, styrene and butadiene (ABS), which contains a higher proportion of poybutadiene than HIPS is even more susceptible to peroxidation [32] and the addition of polyunstaurated polymers can have a catastrophic effect on the durability of otherwise durable polymers.

(c) Mechanooxidation during processing

The conversion of polymer raw materials to industrial or domestic products involves heating oxygen saturated pellets to high temperatures in a screw extruder. The viscous liquid is thus subjected to severe shear deformation which leads to mechanical scission of the macromolecule with the formation of alkyl radicals at the ends of the broken chains. In the presence of even very minor concentrations of oxygen, hydroperoxides are formed at the end of the broken chains which subsequently thermolyse to give highly reactive "oxyl" free radicals. This is the most important initiating mechanism in otherwise oxidatively stable polymers and requires the addition of "processing stabilisers" (mechanoantioxidants) at this stage in the life of the polymer. The chemical consequences of macroalkyl radical formation is shown typically for polyolefins in Scheme 1.13.

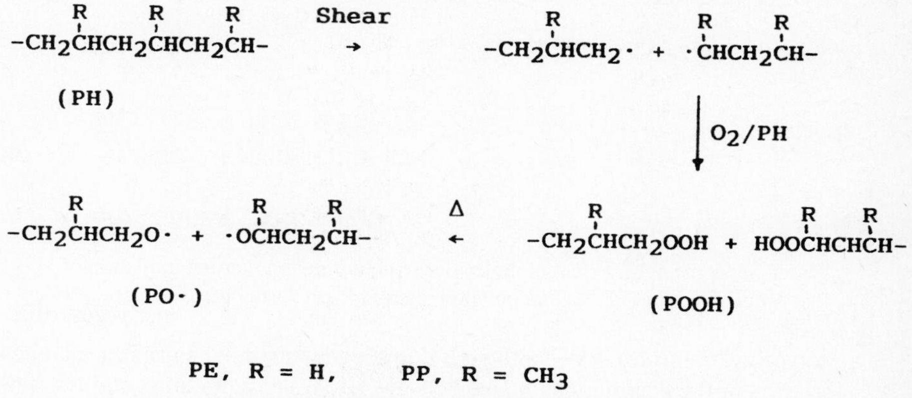

$$-CH_2\overset{R}{C}HCH_2\overset{R}{C}HCH_2\overset{R}{C}H- \xrightarrow{Shear} -CH_2\overset{R}{C}HCH_2\cdot + \cdot\overset{R}{C}HCH_2\overset{R}{C}H-$$

(PH)

$$\downarrow O_2/PH$$

$$-CH_2\overset{R}{C}HCH_2O\cdot + \cdot O\overset{R}{C}HCH_2\overset{R}{C}H- \xleftarrow{\Delta} -CH_2\overset{R}{C}HCH_2OOH + HOO\overset{R}{C}HCH\overset{R}{C}H-$$

(PO·)   (POOH)

PE, R = H,   PP, R = CH_3

Scheme 1.13: Mechanooxidation of polyolefins

Mechanochemical scission occurs with all thermoplastic poymers during processing and subsequent reactions of "oxyl" radicals in the polymer may lead to either chain scission or cross-linking, depending on the chemical structure of the polymer and the oxygen pressure in the mixer [62,94]. Thus polyethylene with its substantially unbranched hydrocarbon chain appears to cross-link with molecular enlargement whereas polypropylene undergoes predominent chain scission with reduction in molar mass (see Scheme 1.14) [95]. In practice, both chain scission and cross-linking occur together and that which predominates depends on the conditions (notably temperature and oxygen concentration). However, both are disadvantageous not only to the manufacturing process itself but also during subsequent service due to the presence in the polymer of hydroperoxides which sensitise the fabricated product to photooxidation. Rheological changes and hence durability in service can be minimised by the incorporation of antioxidants (processing stabilisers) which deactivate the initially formed mechano-radicals (Chapter 3).

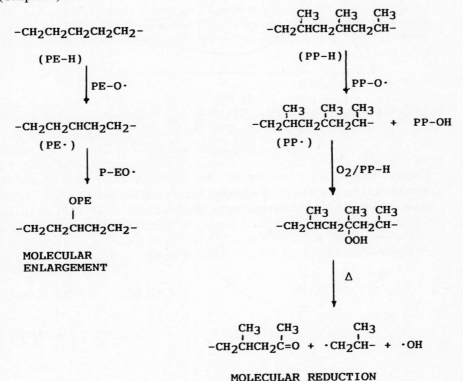

Scheme 1.14: Effect of molecular structure on the thermal oxidation of polyethylene (PE-H) and polypropylene (PP-H) during processing [62, 95]

Poly(vinyl chloride), PVC, although it undergoes a similar initial mechanochemical scission of the polymer chain (see Scheme 1.15), shows a more complex subsequent behaviour in which a proportion of the mechanoradicals undergo hydrogen chloride elimination with the formation of conjugated polyunsaturation along the polymer chain in an "unzipping" free radical chain reaction during processing [94] and recycling [96]. At the same time, some macroalkyl radicals also react with oxygen present in the system to give hydroperoxides [97] (Fig. 1.1) which are unstable under these conditions.

Sec. 1.6] Technological Effects of Peroxidation

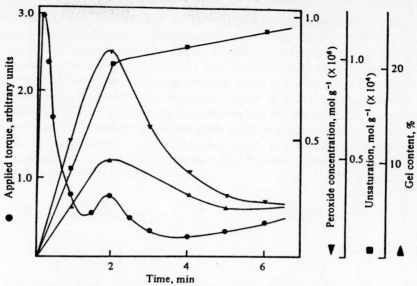

Fig. 1.1 Effect of applied torque (●) on PVC during high temperature processing in an internal mixer: olefinic unsaturation (■), peroxides (▼) and cross-linked polymer (gel) (▲). Reproduced with permission from Polymer Degradation and Stabilisation, N.Grassie and G.Scott, Cambridge University Press, 1985, p.107.

Cooray and Scott have shown that hydroperoxides react rapidly and stoichiometrically with a molar excess of hydrogen chloride even at 50°C [98,99] to give "oxyl" radicals and chlorine atoms, Scheme 1.15.

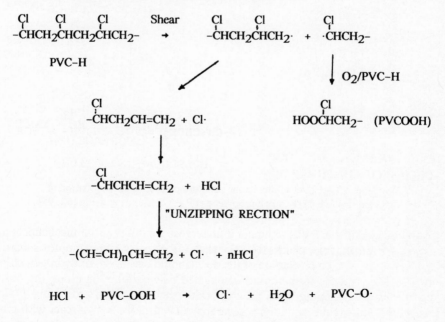

Scheme 1.15: Mechanodegradation of PVC during processing

Both radical species are highly reactive and readily hydrogen abstract from methylene groups in the PVC chain thus initiating the further radical chain elimination of hydrogen chloride. The polyenyl radical so produced can readily scavenge PVC peroxyl radicals (ROO·). This is essentially a cross-linking reaction and the viscosity of the polymer is found to increase as "peroxy gel" builds up in the polymer [97] (Fig. 1.1). The peroxy gel is, however, unstable at the temperature of processing and the peroxide cross-links are gradually replaced by more stable ether linkages through a similar attack of alkoxyl radicals on the polyconjugated unsaturation. Molecular reduction proceeds in parallel with molecular enlargement by main-chain carbon-carbon bond scission, Scheme 1.16, and after 20 minutes processing at 210°C the average molar mass decreases by almost an order of magnitude.

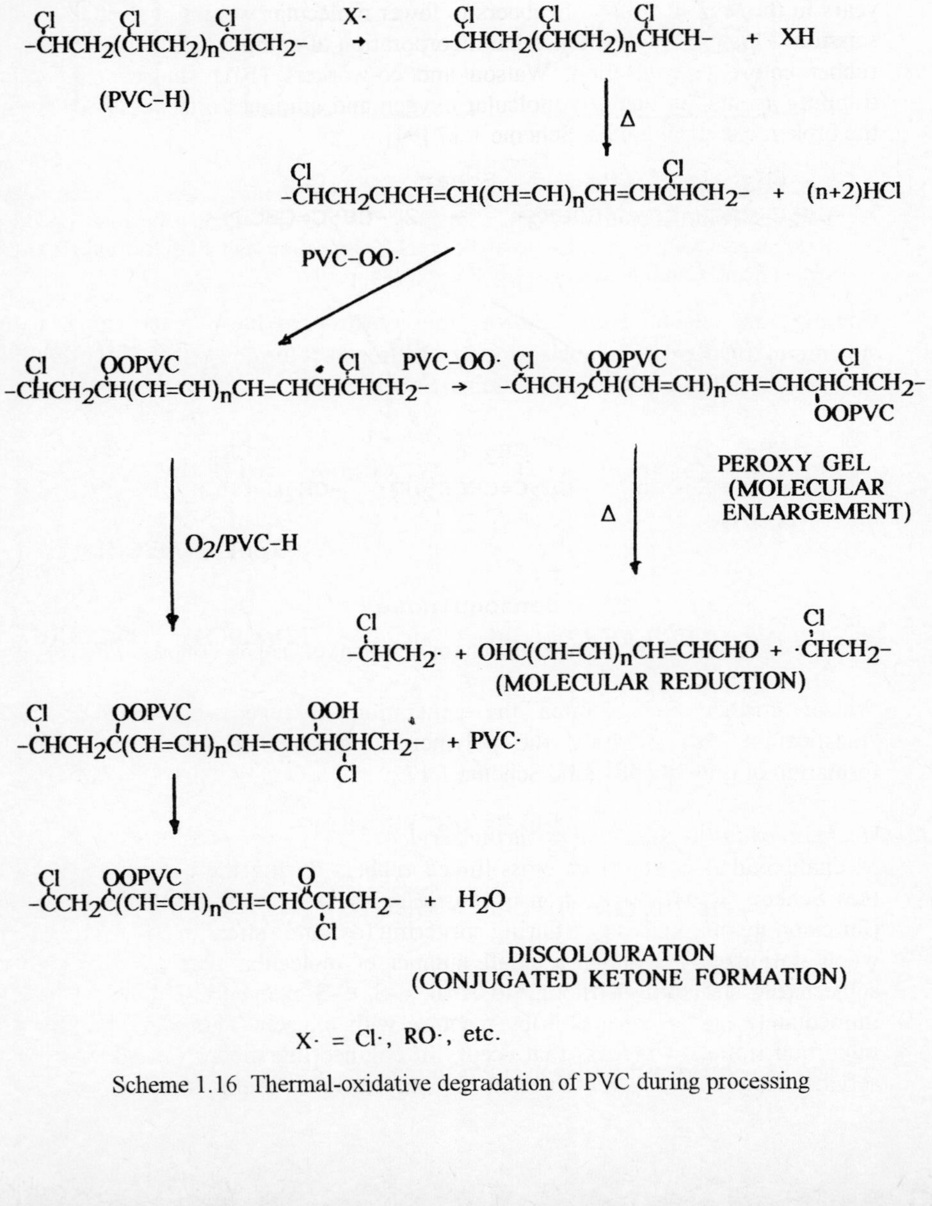

$$\begin{array}{c}\text{Cl} \quad\quad \text{Cl} \quad\quad \text{Cl} \\ -\text{CHCH}_2(\text{CHCH}_2)_n\text{CHCH}_2- \end{array} \xrightarrow{\text{X·}} \begin{array}{c}\text{Cl} \quad\quad \text{Cl} \quad\quad \text{Cl} \\ -\text{CHCH}_2(\text{CHCH}_2)_n\text{CHCH}- \end{array} + \text{XH}$$

(PVC–H)

$$\downarrow \Delta$$

$$\begin{array}{c}\text{Cl} \quad\quad\quad\quad\quad\quad\quad\quad\quad\quad\quad\quad \text{Cl}\\ -\text{CHCH}_2\text{CHCH}=\text{CH}(\text{CH}=\text{CH})_n\text{CH}=\text{CHCHCH}_2- \end{array} + (n+2)\text{HCl}$$

$$\swarrow \text{PVC–OO·}$$

$$\begin{array}{c}\text{Cl} \quad \text{OOPVC} \quad\quad\quad\quad\quad\quad\quad\quad\quad \text{Cl}\\ -\text{CHCH}_2\text{CH}(\text{CH}=\text{CH})_n\text{CH}=\text{CHCHCHCH}_2- \end{array} \xrightarrow{\text{·Cl} \;\; \text{PVC–OO·}} \begin{array}{c}\text{Cl} \quad \text{OOPVC} \quad\quad\quad\quad\quad\quad\quad \text{Cl}\\ -\text{CHCH}_2\text{CH}(\text{CH}=\text{CH})_n\text{CH}=\text{CHCHCHCH}_2- \\ \quad\quad\quad\quad\quad\quad\quad\quad\quad\quad\quad\quad \text{OOPVC} \end{array}$$

PEROXY GEL
(MOLECULAR ENLARGEMENT)

$$\downarrow \Delta$$

$$\text{O}_2/\text{PVC–H} \downarrow$$

$$\begin{array}{c}\text{Cl}\\ -\text{CHCH}_2\cdot \end{array} + \text{OHC}(\text{CH}=\text{CH})_n\text{CH}=\text{CHCHO} + \begin{array}{c}\text{Cl}\\ \cdot\text{CHCH}_2- \end{array}$$
(MOLECULAR REDUCTION)

$$\begin{array}{c}\text{Cl} \quad \text{OOPVC} \quad\quad\quad\quad \text{OOH}\\ -\text{CHCH}_2\text{C}(\text{CH}=\text{CH})_n\text{CH}=\text{CHCHCHCH}_2- \\ \quad\quad\quad\quad\quad\quad\quad\quad\quad\quad\quad \text{Cl} \end{array} + \text{PVC·}$$

$$\downarrow$$

$$\begin{array}{c}\text{Cl} \quad \text{OOPVC} \quad\quad\quad\quad\quad \text{O}\\ -\text{CCH}_2\text{C}(\text{CH}=\text{CH})_n\text{CH}=\text{CHCCHCH}_2- \\ \quad\quad\quad\quad\quad\quad\quad\quad\quad\quad \text{Cl} \end{array} + \text{H}_2\text{O}$$

DISCOLOURATION
(CONJUGATED KETONE FORMATION)

X· = Cl·, RO·, etc.

Scheme 1.16 Thermal-oxidative degradation of PVC during processing

In practice, PVC is not normally subjected to such prolonged high temperature treatment and much more important to the polymer technologist is the yellow-brown discolouration which develops even during the early stages of mechanooxidation. This is due in part to the extended polyenic unsaturation, but this is not in itself sufficient to account for the discoloration observed. Conjugated carbonyl groups are also formed at a relatively early stage during processing [100] and these are much more intensely coloured than polyenic conjugation alone. Conjugated carbonyls result from the alternative breakdown of pendant peroxides along the polymer chain (Scheme 1.16) and are effective traps for peroxyl and alkyl radicals and particularly under photooxidative conditions: this is an important antioxidant process (Section 1.6.3).

Mechanooxidation is unique to macromolecules and has been used for over 150 years in the "mastication" of rubbers to lower molecular weight by oxidative chain scission. This process facilitates the incorporation of compounding ingredients into rubber before vulcanisation. Watson and co-workers [101] showed that radical trapping agents, particularly molcular oxygen and quinones effectively "stabilised" the broken chemical bonds, Scheme 1.17 [94].

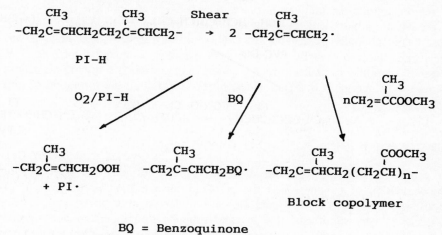

Scheme 1.17 Mechanooxidation (mastication) of cis-poly(isoprene)

Watson further demonstrated the generation of macroalkyl radicals during "mastication" by initiating the polymerisation of vinyl monomers with the formation of graft copolymers, Scheme 1.17.

(d) Mechanooxidation of polymers during service

Mechanooxidation occurs in cross-linked rubbers during the process of "fatigue" (see Scheme 1.11). Application of tensile or shear stresses to vulcanised rubbers (for example in a rubber tyre during cornering (dynamic stress or in a tyre side-wall when stationary) results in a small number of molecular segments having been subjected to strains high enough to break S-S, C-S or even C-C bonds which are immediately made irreversible by reaction with oxygen. This is by far the most important initiation process that occurs in engineering rubber components during service. However rubbers subjected to compression undergo the phenomenon of "compression set" due to the re-organisation of polymer chains broken by mechanooxidation in a new configuration.

Rubbers subjected to ozone in the absence of stress do not show physical evidence of ozone attack although the chemistry shown in Scheme 1.4 is known to occur. Under conditions of static stress on the other hand, unsaturated rubbers undergo rapid cracking in the presence of only parts per billion of ozone. The chemistry occurring is the same as in the absence of stress, but the effect of the stress is to initiate micro-crack formation with restructurisation of the zwitterion scission products [56,57].

1.6.3 *Photooxidation of polymers*

In general, polymers peroxidise and undergo physical deterioration much more rapidly in the presence of sunlight than they do thermally at the same temperature. The peroxidation chain reaction (reactions 5 and 6) occurs in light as well as in the dark. The main difference lies in the initiating processes discussed in Section 1.3.3 and a great deal of attention has been devoted to identifying the chemical species responsible for initiating photooxidation in polymers [102]. The most damaging wavelengths present in sunlight reaching the earth are those between 295 nm (the cut-off wavelength of the earth's atmosphere) and 350 nm. These frequencies have the ability to break weak chemical bonds, particularly the O-O bond in hydroperoxides which, as was seen above, are almost universally present to some extent in commercial polymers. UV light above 300 nm also promotes carbonyl compounds to excited states which may in turn pass on their energy to hydroperoxides, further facilitating the formation of initiating radicals [103]:

$$>C=O \xrightarrow{h\nu} [>C=O]^* \xrightarrow{ROOH} >C=O + RO\cdot + \cdot OH \qquad (27)$$

In general a linear relationship has been shown to exist between the rate of photooxidation of polymers and the peroxide concentration in the polymer before exposure [90,91,102,104,105]; and organo-soluble transition metal ions, particularly iron play a very significant role in catalysing the formation and decomposition of hydroperoxides since the metal is maintained in its most reactive reduced state by photolysis:

$$FeL_3 \xrightarrow{h\nu} L\cdot + FeL_2 \xrightarrow{ROOH} RO\cdot + HOFeL_2 \xrightarrow{LH} FeL_3 + H_2O \qquad (28)$$

$$\downarrow RH$$

$$LH + R\cdot$$

In polymers, L is normally a carboxylate group since these are formed by photooxidation, but when L is a sulphur ligand, FeL_3 may be an antioxidant which when destroyed by light inverts to give a photoprooxidant. This chemistry is now used to control plastics waste from commercial sources and will be discussed in Chapter 4.

PVC is particularly sensitive to the presence of hydroperoxides formed during the processing operation [105,106]. It was seen in Scheme 1.15 that hydroperoxides are formed as a result of mechanochemical scission of the PVC chain in the presence of oxygen. At the same time, olefinic unsaturation and conjugated carbonyl groups are also formed in PVC during processing and that these are the cause of subsequent thermal instability with the formation of a yellow and ultimately dark brown discolouration. However during photooxidation the rate of carbonyl formation is rapidly autoretarded and on extended photooxidation the colour bleaches with disruption of the conjugation and eventual disintegration of the polymer to lower molecular weight fragments. The fading of the yellow colour, known as "photo-bleaching" [107] also occurs in the dark (see Scheme 1.16). The removal of "oxyl" radicals by the conjugated carbonyl structure with the formation of relatively stable delocalised polyeneoxyl radicals (Scheme 1.18) is an effective retardation process which, together with partial screening of the polymer from incident light leads to autoretardation of photoperoxidation. This process can be repeated many times in the case of an extended polyconjugated sequence but the peroxidation products are ultimately unstable in the presence of light and break down with chain scission and reduction of polymer molecular weight [108]. Carbon dioxide, carbon monoxide and water are formed in the later stages of photooxidation [108] and this is consistent with the accumulation of adjacent peroxidic groups at the extremities of the conjugation.

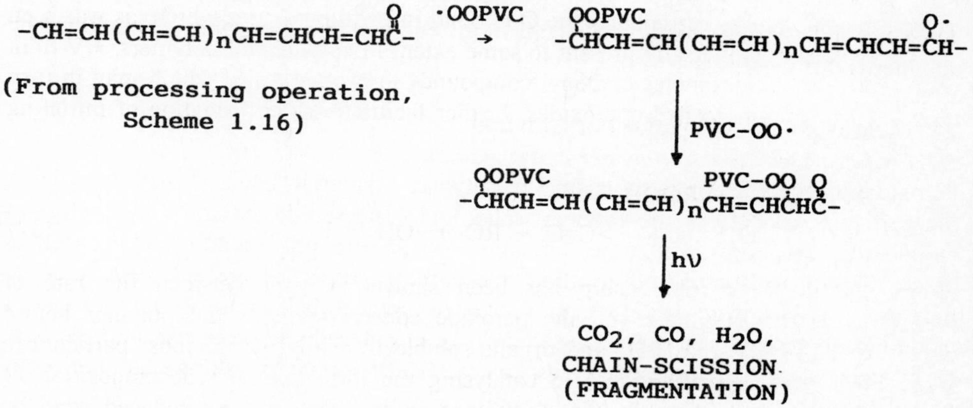

$$-CH=CH(CH=CH)_nCH=CHCH=CH\overset{O}{\underset{H}{C}}- \quad \xrightarrow{\cdot OOPVC} \quad \overset{OOPVC}{-\overset{|}{C}HCH=CH(CH=CH)_nCH=CHCH=\overset{O\cdot}{\overset{|}{C}H}-}$$

(From processing operation, Scheme 1.16)

$$\downarrow PVC-OO\cdot$$

$$\overset{OOPVC}{-\overset{|}{C}HCH=CH(CH=CH)_nCH=CH\overset{PVC-OO}{\underset{|}{C}H}\overset{O}{\underset{H}{C}}-}$$

$$\downarrow h\nu$$

$$CO_2, \ CO, \ H_2O,$$
$$CHAIN\text{-}SCISSION$$
$$(FRAGMENTATION)$$

Scheme 1.18 Photoantioxidant role of conjugated carbonyl formed in PVC during processing

Macroalkyl hydroperoxides as noted in Section 1.3.3 are important indigenous photo-sensitisers for polymers. However, some dyestuffs and pigments have also been shown to produce ROS (e.g. $^1O_2$, O_2^-, H_2O_2 and $\cdot OH$) by energy transfer to ground state oxygen (Section 1.3.3). The mechanisms of photsensitisation in polymers has been discussed in detail by Rabek [109] and by Allen [110,111] and the reader is referred to these reviews for further infomation.

1.6.4 *The effect of polymer morphology in polymer degradation*
Commercial polymers are generally not homogeneous materials. The polyolefins, polyamides and polyesters all contain crystalline regions embedded in amorphous polymer. The two domains behave very differently to oxidation [112]. Normally, the crystallites are impervious to oxygen and therefore much more resistant to peroxidation and chain scission begins in the more oxygen-accessible "tie-bonds" between the

crystallites, resulting in the process of chemi-crystallisation and physical disintegration of the polymer at a relatively low degree of oxidation [112-114]. The rate of physical deterioration of semi-crystalline polymers therefore has much more to do with morphology than with rate of oxidation. However, there is a related benefit for polymer stabilisation arising from morphological effects; namely that organic antioxidants and stabilisers are also exclusively soluble in the amorphous domains and are thus concentrated where they are most required (Chapter 4).

Polymers above their glass transition temperature (e.g. polystyrene) are much more resistant to peroxidation than their rate of oxidation in solution might suggest. There are several contributory reasons for this. The first is that oxygen is considerably less soluble in polymers than it is in mobile organic liquids of similar chemical composition [115]. One result of this is that the ratio of alkyl to alkylperoxyl radicals is greater than it is in low molecular weight solvents so that cross termination by reactions 17 and 18 which have low activation energies become much more frequent. Since the rate of oxygen diffusion is also considerably reduced in polymers compared with mobile liquid substrates [113], even highly peroxidisable rubbers may remain unchanged at the interior of thick sections over many years. Below the glass transition temperature macromolecules lack mobility and it is more difficult for the bond angles in the free radicals formed in reaction 6 to assume the optimal conformation for electron delocalisation. Consequently the activation energy for hydrogen abstraction by alkylperoxyl is increased.

1.7 Measurement of Polymer Peroxidation
1.7.1 *Characterisation of polymer deterioration*
Peroxidation precedes property change in polymeric materials,but it is the latter that is the particular concern of the polymer technologist,and this is measured by change in engineering properties with time. Of these, tensile strength, modulus, elongation at break, impact resistance and dynamic mechanical properties are relatively easy to measure [3]. However, in order to relate mechanical behaviour to the effects of oxidation, there has been increasing interest in recent years in characterising the very early stages of peroxidation, which, as was seen in the previous section, may not occur homogeneously throughout the polymer bulk.

Hydroperoxide formation generally rises to a maximum and then decays at a relatively early stage in peroxidation, whereas the major stages of oxygen absorption and associated changes in mechanical properties are accompanied by the formation of hydroperoxide breakdown products as evidenced by absorbance in the infra-red at 1715 cm^{-1} ketone), 1740 cm^{-1} (ester), 1710 cm^{-1} (carboxylic acid), 1730 cm^{-1} (aldehyde), 1763 cm^{-1} (perester) [116,117]. Some individual absorbances can be separated by FTIR but more frequently the measurement of the area under the carbonyl envelope is used to give a semiquantitative estimate of peroxide break-down products. Hydroperoxides frequently exist in polymers in two forms: isolated (non-hydrogen bonded) hydroperoxides which give a sharp absorbance in the infra-red at 3500 cm^{-1} and which can be used semi-quantitatively to measure the initial stages of peroxidation in polyethylene [95]; and hydrogen-bonded hydroperoxide which is the major species at higher concentration and in polymers such as polypropylene which gives "clusters" of vicinal hydroperoxide groups (Scheme 1.4) absorbing in a broad IR band centred at 3350 cm^{-1}. The latter are hydrogen bonded and consequently the absorption is too broad to be a good measure of hydroperoxide formation. However, hydroperoxides can also be readily measured by chemical methods [88,95,97], and recently a sensitive FTIR method has also been developed involving "derivitisation" of the hydroperoxide. This involves the quantitative conversion of hydroperoxide to nitrate by reaction with NO at low temperatures (-78 to -20) followed by FTIR measurement of nitrate [118].

Chemiluminescence (sometimes called "oxyluminescence") provides a very sensitive method for determining the rate of peroxidation of polymers due to the ability of modern photodetection techniques to detect extremely low luminescence emissions (10^{-8} to 10^{-10} lumens). Chemiluminescence occurs from triplet carbonyl which is excited by the decomposition of hydroperoxides [119,120] or possibly through termination of peroxyl radicals [120]. Chemiluminescence intensity from polymers in inert atmospheres appears to correlate well with hydroperoxide concentration in the polymer [121] and provides an additional method of following the induction time (τ) to peroxidation.

1.7.2 *Accelerated testing*
In many applications of polymers, particularly for durable products used in construction or automotive engineering, the lifetime has to be tens of years and the problem that faces the polymer technologist is to predict the useful lifetime of the product. For this reason some means must be found of accelerating the chemical processes occurring by intensifying one or more of the environmental factors which contribute to oxidative degradation. Many such tests have been developed, initially in the lubricating oil, surface coating and rubber industries [122] but more recently in the plastics and fibre industries [123]. They generally involve higher than ambient temperatures to accelerate thermal oxidation [59,122] and more intense sources of UV light to exacerbate "weathering" of polymers [83,124,125].

Both accelerating procedures involve pitfalls which are not evident at first sight. Although thermal degradation of polymers may obey the Arrhenius relationship over the relatively small temperature intervals involved in air oven ageing tests, these are generally remote from ambient temperatures, and very few long term evaluations have been carried out to relate accelerated heat-ageing tests to ambient conditions. Some spectacular misuses of thermal methods for the measurement of polymer durability have been reported by Gugumus [123]. The measurement of oxidation induction time (OIT) by differential thermal analysis (DTA) or thermogravimetric analysis (TG) are particularly simple techniques which for this reason has attracted a great deal of support among plastics technologists. These procedures are normally carried out at very high temperatures (up to 200°C) and in the case of polypropylene no correlation was found between this test at 190°C and an air oven ageing test at 150°C. This is not surprising since in the first case the polymer is a liquid and in the second case a solid. However, the differences are much more fundamental than this, since the performance of antioxidants and stabilisers under conditions of use depends crucially on the physical behaviour of the additives in the polymer. In an air stream, loss of antioxidant from the polymer may be rate controlling and this increases with increasing temperature, increasing air rate over the surface of the polymer, and decreasing the thickness of the sample [126] (Chapter 4). DTA-OIT may be a useful tool in quality control since comparison of a stabilised and an unstabilised sample of polymer will certainly show a difference, but is quite valueless as a predictor of long-term performance. Oven ageing tests are more useful, and their usefulness increases the closer are the test conditions to those experienced in practice. A detailed understanding of the physico-chemical factors governing the behaviour of antioxidants and stabilisers undergoing thermal oxidation is essential in the study of durable polymers, and although accelerated tests must continue to be used, the interpretation of results must be modulated by the developing understanding of how antioxidants interact both chemically and physically with polymers. This will be discussed further in Chapter 4.

The position is somewhat better in the correlation of accelerated UV testing with outdoor weathering. In this case similar temperatures are involved and the validity of an accelerated test as a predictor of outdoor durability depends fundamentally on matching the distribution of the incident wavelengths to that of sunlight. This has been done very successfully with modern xenon arc weatherometers [123]. However, attempts to computer model photooxidation in order to predict polymer lifetimes using rate constants for the fundamental chemical reactions involved [127] have not been signally successful. Many other factors have to be taken into account. These include other environment influences such as water (which can remove light stabilisers and antioxidants), environmental pollutants, notably O_3, SO_2 and NO_x, which photosensitise polymers to peroxidation [87] and even physical stress [65,128]. These parameters can in principle all be incorporated in modern UV weatherometers but the possibility of including every relevant parameter in a single test for the purpose of lifetime prediction still seems a long way off and may be unachievable due to the vagaries of the service environment. However correlations between accelerated and outdoor tests are statistically satisfactory and the prediction of polymer durability in the outdoor environment from accelerated weathering tests is now reasonably satisfactory.

REFERENCES

1. L. Bateman, *Quart. Rep.*, 8, 147 (1954).
2. J.L. Bolland, *Quart. Rep.*, 3, 1 (1949).
3. G. Scott, *Atmosperic Oxidation and Antioxidants*, Elsevier, Amsterdam, 1965
4. *Atmospheric Oxidation and Antioxidants*, Vols. I-III, Ed. G. Scott, Elsevier, Amsterdam, 1993.
5. G.P. Armstrong, R.H. Hall and D.C. Quin, *J. Chem. Soc.*, 666 (1950).
6. A. Robertson and W.A. Waters, *J. Chem. Soc.*, 1574 (1948).
7. G. Scott, *Atmospheric Oxidation and Antioxidants*, Elsevier, Amsterdam, 1965, p.127 et seq.
8. E.A. Oberright, S.J. Leonardi and A.P. Kozacik in *Additives in Lubricants*, ACS Symposium, Div. Pet. Chem., Atlantic City, 1958, p.115.
9. J.D. Holdsworth, G. Scott and D. Williams, *J. Chem. Soc.*, 4692-99 (1964).
10. K.J. Humphris and G. Scott, *J. Chem. Soc., Perkin Trans. II*, 831-5 (1973).
11. K.J. Humphris and G. Scott, *J. Chem. Soc., Perkin Trans. II*, 617-20 (1974).
12. C. Armstrong, M.J. Husbands and G. Scott, *Europ. Polym. J.*, **15**, 241-8 (1979).
13. M.J. Husbands and G. Scott, *Europ. Polym. J.*, **15**, 249-53 (1979).
14. M.J. Husbands and G. Scott, *Europ. Polym. J.*, **15**, 879-87 (1979).
15. S. Al-Malaika and G. Scott, *Europ. Polym. J.* **16**, 503-9 (1980).
16. B.B. Cooray and G. Scott, *Europ. Polym. J.* **16**, 169-77 (1980).
17. G. Scott, *Atmospheric Oxidation and Antioxidants*, Elsevier, Amsterdam, 1965, pp. 86-8.
18. R.H. Hall in *Basic Organic Chemistry, Part 5, Industrial Products*, Eds. J.M. Tedder, A. Nechvatal and A.H. Jubb, Wiley, New York, 1975, p.133.
19. G. Scott, *Atmospheric Oxidation and Antioxidants*, Elsevier, Ansterdam, 1965, pp.24-31.
20. G.A. Russell, *J. Am. Chem. Soc.*, **78**, 1035, 1041 (1956).
21. G.A. Russell, *J. Am. Chem. Soc.*, **78**, 1047 (1956).
22. S. Al-Malaika and G. Scott in *Degradation and Stabilisation of Polyolefins*, Ed. N. S. Allen, App. Sci. Pub., London, 1983, Chapter 6.
23. D.J. Carlsson and D.M. Wiles, *J. Macromol. Sci.*, Rev. Macromol. Chem., *C14*,(1), 65 (1976).
24. D.J. Carlsson, A. Garton and D.M. Wiles in *Developments in Polymer Stabilisation-1*, Ed. G. Scott, App. Sci. Pub., London, 1979, Chapter 7.
25. J.A. Howard, *Adv. Free Rad. Chem.*, **4**, 49 (1973).
26. C. Decker, F.R. Mayo and H. Richardson, *J. Polym. Sci.* Chem. Ed., *11*, 2879 (1973).
27. C. Decker and F.R. Mayo, *J. Polym. Sci.*, Chem. Ed., *11*, 2847 (1973).
28. C. Sadrmohaghegh and G. Scott, *Polym. Deg. Stab.*, **3**, 469 (1981-2).
29. G. Scott in *Atmospheric Oxidation and Antioxidants*, Vol. II, Ed. G. Scott, Elsevier, Amsterdam, 1993, pp. 387 et seq.
30. A. Ghaffar, A. Scott and G. Scott, *Europ. Polym. J.*, **11**, 271-5 (1975).
31. A. Ghaffar, A. Scott and G. Scott, *Europ. Polym. J.*, **12**, 615-20 (1976).
32. M. Ghaemy and G. Scott, *Polym. Deg. Stab.*, **3**, 233-42 (1981).
33. G. Scott and M. Tahan, *Europ. Polym. J.*, **13**, 997-1005 (1977).
34. G. Scott, *Atmospheric Oxidation and Antioxidants*, Elsevier, Amsterdam, 1965,

34. G. Scott, *Atmospheric Oxidation and Antioxidants*, Elsevier, Amsterdam, 1965, p.67 et seq.
35. A. Bromberg and K. A. Muszket, *J. Am. Chem. Soc.*, **91**, 2860 (1969).
36. A. Charlesby, *Atomic Radiation and Polymers*, Pergamon Press, Oxford, 1960; A. Chapiro, *Radiation Chemistry of Polymeric Systems*, Interscience, 1962.
37. P. Wardman in *Atmospheric Oxidation and Antioxidants*, Vol. III, Ed. G.Scott, Elsevier, Amsterdam, 1993, Chapter 4.
38. D.J. Carlsson in *Atmospheric Oxidation and Antioxidants*, Vol. II, Ed. G.Scott, Elsevier, Amsterdam, 1993, Chapter 11.
39. T.S. Dunn, E.E. Williams and J.L. Williams, *Radiat. Phys. Chem.*, **12**, 287 (1982).
40. J.L. Williams, T.S. Dunn and V.T. Stannett, *Radiat. Phys. Chem.*, **12**, 291 (1982).
41. J.F. Rabek, *Mechanisms of Photophysical Processes and Photochemical Reactions in Polymers*, Wiley, New York, 1987, Chapters 7 and 14..
42. C.S. Foote, *Acc. Chem. Res.*, **1**, 104 (1968).
43. D.R. Kearns, *Chem. Rev.*, **71**, 395 (1971).
44. *Singlet Oxygen, Reactions with Organic Compounds and Polymers*, Eds. B. Rånby and J. F. Rabek, Wiley, New York, 1978.
45. G. Scott in *Atmospheric Oxidation and Antioxidants*, Vol. I, Ed. G. Scott, Elsevier, Amsterdam, 1993, Chapter 3.
46. K. Gollnick, *Singlet Oxygen*, Eds. B. Rånby and J. F. Rabek, Wiley, New York, 1978, Chapter 10.
47. K.E. Russell and J.K.S. Wan, *Macromol.* **6**, 669 (1973).
48. H.C. Ng and J.E. Guillet in *Mechanisms of Photophysical Processes and Photochemical Reactions in Polymers*, Eds. B. Rånby and J. F. Rabek, Wiley, New York, 1978, p.278.
49. J.F. Rabek, Y.J. Shur and B. Rånby in *Singlet Oxygen*, Eds. B. Rånby and J.F. Rabek, Wiley, Chichester, 1978, Chapter 26.
50. G. Scott in *Singlet Oxygen*, Eds. B. Rånby and J.F. Rabek, Wiley, Chichester, 1978, Chapter 23.
51. D. Bellus, *Singlet Oxygen*, Eds. B. Rånby and J.F. Rabek, Wiley, 1978, Chapter 9.
52. R.W. Murray in *Polymer Stabilisation*, Ed. W. L. Hawkins, Wiley Interscience, New York, 1972, Chapter 5.
53. S.D. Razumovskii and G.E. Zaikov in *Developments in Polymer Stabilisation-6*, Ed. G. Scott, App. Sci. Pub., London, 1983, Chapter 7.
54. R. Criegee, Ber. **88**, 1878 (1955).
55. P.S. Bailey, *Chem. Rev.*, **58**, 925 (1958).
56. G. Scott, *Atmospheric Oxidation and Antioxidants*, Elsevier, Amsterdam, 1965, pp.477-91.
57. R.P. Lattimer, R.W. Layer and C.K. Rhee in *Atmospheric Oxidation and Antioxidants*, Vol. II, Ed. G. Scott, Elsevier, Amsterdam, 1993, Chapter 7.
58. B. Halliwell and J.M.C. Gutteridge, *Free Radicals in Biology and Medicine, Second Edition*, Clarendon Press, Oxford, 1989, p. 12.
59. T. Colclough in *Atmospheric Oxidation and Antioxidants*, Vol. II, Ed. G. Scott. Elsevier, Amsterdam, 1993, Chapter 1.
60. G. Scott, *Atmospheric Oxidation and Antioxidants*, Elsevier, Amsterdam, 1965, p. 402.

61. J. Sohma in *Developments in Polymer Degradation-2*, Ed. N. Grassie, Applied Science Publishers, London, 1979, Chapter 4.
62. G. Scott in *Atmospheric Oxidation and Antioxidants*, Vol. II, Ed. G. Scott, Elsevier, Amsterdam, 1993, Chapter 3.
63. G. Scott, *Rubb. Chem. Tech.*, **58**, 269-83 (1985).
64. G. Scott, *J. Int. Rubb. Res.*, **5(3)**, 163 -77 (1990).
65. W.K. Busfield and P. Taba, *Polym. Deg. Stab.*, **51**, 185-96 (1996).
66. M.C.R. Symons, *Free Rad. Biol. Med.*, **20**, 831-35 (1996).
67. R. Partridge, M.C.R. Symons and L.J. Wyatt, *J. Chem. Soc. Farad. Trans.*, **89**, 1285-6 (1993).
68. G.A. Russell, *J. Chem. Ed.*, **36**, 111 (1969).
69. H.S. Blanchard, *J. Am. Chem. Soc.*, **81,** 4548 (1959).
70. G.A. Russell, *J. Am. Chem. Soc.*, **77**, 4583 (1955).
71. L. Bateman and A.L. Morris, *Trans. Farad. Soc.*, **49**, 1026 (1953).
72. G. Scott, *Atmospheric Oxidation and Antioxidants*, Elsevier, Amsterdam, 1965, pp. 99 et seq.; G. Scott in *Atmospheric Oxidation and Antioxidants*, Vol. I, Ed. G. Scott, Elsevier, Amsterdam, 1993, p. 11 et seq.
73. G.A. Russell, *J. Am. Chem. Soc.*, **78**, 1047 (1956).
74. G. Scott, *Atmospheric Oxidation and Antioxidants*, Elsevier, Amsterdam, 1965, pp. 44-51.
75. G. Scott, *Atmospheric Oxidation and Antioxidants*, Elsevier, Amsterdam, 1965, p. 413.
76. A.A. Miller and F.R. Mayo, *J. Am. Chem. Soc.*, **78**, 1017 (1956); J.L. Bolland and G. Gee, *Trans. Farad. Soc.*, **42**, 236 (1950).
77. G.H. Twigg, *Chem. Eng. Sci., Proc. Conf. Oxidat. Pocesses*, Suppl. **3**, 5 (1954).
78. G. Scott, *Atmospheric Oxidation and Antioxidants*, Elsevier, Amsterdam, 1965, pp.19-23.
79. L.J. Marnett, *Chem Res. Toxic.*, **6**, 413 (1993).
80. L.J. Marnett and A.L. Wilcox in *Free Radicals and Oxidative Stress*, Eds. C. Rice-Evans, B. Halliwell and G.G. Lunt, Biochem. Soc. Symp., **61**, 65-72 (1995).
81. L.J. Marnett in *Free Radicals in Biology*. Ed. W.A. Pryor, Academic Press, New York, VI, 1984, 63.
82. G. Scott, *Atmospheric Oxidation and Antioxidants*, Elsevier, Amsterdam, 1965, pp. 368-79.
83. *Film Formation, Properties and Deterioration*, Ed. C.R. Bragdon, Interscience, New York, 1958.
84. C. Armstrong, F.A.A. Ingham, J.G. Pimblott, G. Scott and J.E. Stuckey, *Proc. Int. Rubb. Conf.* 1972, F2.
85. G. Scott, *Mechanisms of Reactions of Sulphur Compounds*, Ed. N. Kharasch, **4**, 99-110 (1969).
86. G. Scott, *Atmospheric Oxidation and Antioxidants*, Elsevier, Amsterdam, 1965, pp. 412-26.
87. N. Grassie and G. Scott, *Polymer Degradation and Stabilisation*, Cambridge University Press, Cambridge, 1985.
88. M.U. Amin, G. Scott and L.M.K. Tillekeratne, *Europ. Polym. J.*, **11**, 85-9 (1975).
89. G. Scott, *ACS Symp.* **25**, 340-66 (1976).

90. K.B. Chakraborty and G. Scott, *Polymer*, **18**, 98-99 (1977).
91. G. Scott in *Developments in Polymer Degradation-1*, Ed. N. Grassie, Applied Science Publishers, London, 1977, Chapter 7.
92. G. Scott in *Developments in Polymer Stabilisation-1*, Ed. G. Scott, Applied Science Publishers, London, 1979, Chapter 9.
93. C. Sadrmohaghegh, G. Scott and E. Setudeh, *Polym. Plast. Technol. Eng.*, **24**, 149-88 (1985).
94. G. Scott, *Polym. Deg. Stab.*, **48**, 315-24 (1995).
95. K.B. Chakraborty and G. Scott, *Europ. Polym. J.*, *13*, 731 (1977)
96. G. Scott in *Recycling of PVC*, Ed. F.P. LaMantia, ChemTec Pub., 1996, Chapter 1.
97. G. Scott, M. Tahan and J. Vyvoda, *Eur. Polym. J.*, **14**, 377-83 (1978).
98. B.B. Cooray and G. Scott, *Chem. & Ind.*, 741-2 (1979).
99. B.B. Cooray and G. Scott, *Eur. Polym. J.*, **16**, 169-77 (1980).
100. K.S. Minsker, M.I. Abdullin, S.V. Kolesov and G.E. Zaikov in *Developments in Polymer Stabilisation*, Ed. G. Scott, App. Sci. Pub., London, 1983, Chapter 5.
101. W.F. Watson, *Trans. I.R.I.* **29**, 32 (1953).
102. G. Scott in *Atmospheric Oxidation and Antioxidants*, Vol. II, Ed. G. Scott, Elsevier, Amsterdam, 1993, Chapter 8.
103. S-K.L. Li and J.E. Guillet, *Macromol.*, **17**, 41 (1984).
104. K. B. Chakraborty and G. Scott, *Eur. Polym. J.* **13**, 731-37 (1977).
105. B.B. Cooray and G. Scott in *Developments in Polymer Stabilisation-2*, Ed. G. Scott, App. Sci. Pub, London, 1980, Chapter 2.
106. G. Scott, M. Tahan and J. Vyvoda, *Eur. Polym. J.*, **14**, 1021-26 (1978).
107. D. Braun and R.F. Bender, *Eur. Polym. J.*, Suppl., 269 (1969)
108. G. Scott and M. Tahan, *Eur. Polym. J.*, **11**, 535-9 (1975).
109. *Mechanisms of Photophysical Processes and Photochemical Reactions in Polymers*, J.F. Rabek, Wiley, New York, 1987, p. 554 et seq.
110. N.S. Allen in *Degradation and Stabilisation of Polyolefins*, Ed. N.S. Allen, App. Sci. Pub., London, 1993, Chapter 8.
111. N.S. Allen, *Polym. Deg. Stab.*, **44**, 357-74 (1994).
112. W.L. Hawkins, W. Matreyek and F.H. Winslow, *J. Polym. Sci.* **41**, 1 (1959).
113. N.S. Billingham in *Atmospheric Oxidation and Antioxidants*, Vol. II, Ed. G. Scott, Elsevier, Amsterdam, 1993, Chapter 4.
114. P. Vink in *Degradation and Stabilisation of Polyolefins*, Ed. N.S. Allen, App. Sci. Pub., London, 1993, pp. 225-37.
115. E.T. Denisov in *Developments in Polymer Stabilisation-5*, Ed. G. Scott, App. Sci. Pub., London, 1982, p. 30.
116. S. Al-Malaika in *Atmospheric Oxidation and Antioxidants,* Vol. I, Ed. G. Scott, Elsevier, Amsterdam, 1993, Chapter 3.
117. S.S. Stivala, J. Kimura and S.M. Gabbay, *Degradation and Stabilisation of Polyolefins*, Ed. N.S. Allen, App. Sci. Pub., London, 1983, p.131.
118. D.J. Carlsson, R. Brousseau, C. Zhang and D.M. Wiles, *Am. Chem. Soc. Symp. Ser.*, **364**, 376 (1988).
119. N.C. Billingham and G.A. George, *J. Polym. Sci., B Polym. Phys.*, 257 (1990).
120. E.M.Y. Quinga and G.D. Mendenhall, *J. Am. Chem. Soc.* **105**, 6250 (1983).
121. N.C. Billingham, *ACS Polymer Preprints*, **34 (2)**, 237-8 (1993).

References

122. G. Scott, *Atmospheric Oxidation and Antioxidants*, Elsevier, Amsterdam, 1965, Chapter 6.
123. F. Gugumus in *Developments in Polymer Stabilisation-8*, Ed. G. Scott, Elsevier App. Sci. Pub., London, 1987, Chapter 6.
124. A. Davies and D. Sims, *Weathering of Polymers*, App. Sci. Pub., London, 1983.
125. J.F. Rabek, *Photostabilization of Polymers, Principles and Applications*, Elsevier App. Sci., 1990, pp.486-503.
126. G. Scott in *Atmospheric Oxidation and Antioxidants*, Vol.II, Elsevier, Amsterdam, 1993, pp.279-282.
127. A.C. Sommersall and J.E. Guillet, *ACS Symp. Ser.* **280**, 211 (1985).
128. N.A. Rapoport and G.E. Zaikov in *Developments in Polymer Degradation-6*, App. Sci. Pub., London, 1985, p.207.

2

The Biological Effects of Peroxidation

2.1 Causes of Peroxidation in Biological Substrates

Although the free radical chemistry of autoxidation has been studied in technological substrates since the 1930s it is only during the past two decades that biochemists have realised the profound significance of free radical peroxidation *in vivo* and particularly its involvement in many diseases. It is now clear that polyunsaturated fatty acids (PUFA) in the biological cell peroxidise in the same way as the polyunsaturated esters in edible oils during storage or in paint films during "drying". PUFAs do not normally peroxidise in normal cells due to the exceptional efficiency of the biological antioxidants which are in a state of continual dynamic replenishment. However, in cells subjected to severe oxidative stress, the antioxidant defences may become depleted and it is now accepted that many common diseases are associated with lipid peroxidation and low antioxidant status.

It is not always clear whether disease is the cause or the effect of peroxidation. Poli and co-workers have proposed [1] that three criteria should be applied in determining whether reactive oxygen species (ROS) are involved in the initiation or in the biochemical development of a disease. These are:

a) The chronological characterisation of free radical excess at the beginning of and during the development of the disease.
b) The existence of a direct statistical correlation between free radical stress and pathological events.
c) Demonstrable beneficial effects of antioxidants at each stage in the initiation and development of the disease.

The last has so far proved to be by far the most useful diagnostic criterion. Antioxidant intervention in disease will be discussed in Chapter 5 but it should be noted that an inverse relationship is frequently observed between antioxidant status (that is the overall levels and balance of antioxidants in the serum) and the incidence of diseases such as cardiovascular disease and cancer [2-4].

Although destructive peroxidation does not normally occur in healthy cells with a balanced complement of antioxidants, there is an increasing body of evidence to suggest that controlled peroxidation plays a positive role in cell division [5,6]. However, the sudden generation of ROS can dramatically upset this balance with an overload of the antioxidant defence system. Animal tissues contain a variety of enzymes that are capable of reducing ground state dioxygen to the radical-ion superoxide in a "radical burst" [7]. Superoxide ($O_2 \cdot -$), in spite of its name, is relatively unreactive but it is in equilibrium with its cognate protonated form, which is much more reactive toward PUFA (Table 2.1) [8].

$$O_2 \cdot - + H^+ \rightarrow \cdot OOH \tag{1}$$

pK_a for the above equilibrium is 4.75 which means that at pH 7-7.5 less than 1% of the reduced oxygen species is in the form of the more reactive hydroperoxyl [9]. $O_2 \cdot -$ is actually a mild reducing agent with a redox potential of -0.33 V and is capable of regenerating Fe^{2+} from Fe^{3+}, reaction (4) [10]. However the further reduction products of $O_2 \cdot -$ (Chapter 1) and in particular the hydroxyl radical, $\cdot OH$, formed by the one electron reduction of hydrogen peroxide, e.g. reaction (2) is a much more powerful oxidising agent (Table 2.1).

$$H_2O_2 + Fe^{2+} \rightarrow OH^- + \cdot OH + Fe^{3+} \tag{2}$$
$$H_2O_2 + Fe^{3+} \rightarrow H^+ + \cdot OOH + Fe^{2+} \tag{3}$$
$$O_2 \cdot - + Fe^{3+} \rightarrow O_2 + Fe^{2+} \tag{4}$$

Table 2.1 Reactivity of reactive oxygen species toward allyl compounds, $M^{-1}s^{-1}$ [10]

Reactive Oxygen Species		 bis-allyl	 mono-allyl
Hydroxyl	HO·	10^9	10^9
Alkoxyl	LO·	10^6	10^6
Hydroperoxyl	HOO·	10^2	10^2
Lipid peroxyl	LOO·	10^2	10^2
Superoxide	$O_2 \cdot -$	0	low
Hydrogen peroxide	H_2O_2	0	low
Lipid hydroperoxide	LOOH	0	low
Singlet oxygen	$^1O_2$	0	10^6
Ozone	O_3	low	10^6

The hydroxyl radical is by far the most reactive of all biologically generated reactive oxygen species. Its redox potential (+1.9V) is very much higher than superoxide and most of the other ROS formed by peroxidation (e.g. LOO·, LO, etc.) lie between these values [9]. It should be noted that hydrogen peroxide *in the absence of reducing agents* is not very reactive, and indeed H_2O_2 may actually deactivate the hydroxyl radical due to its reduction to the more stable hydroperoxyl [9]:

$$H_2O_2 + \cdot OH \rightarrow H_2O + HOO \cdot \tag{5}$$

The lifetimes of ROS in organic substrates is inversely related to their reactivity. Table 2.2, due to Pryor [11], shows that free radical lifetimes vary by ten orders of magnitude from the highly reactive hydroxyl to the relatively unreactive alkylperoxyl.

Table 2.2 Typical lifetimes of radicals involved in peroxidation [11]

Radical species	Conditions	$t_{½}$, s
HO·	1M linoleate	10^{-9}
LO·	100 mM linoleate	10^{-6}
LOO·	1 mM linoleate	10
L·	0.2 mM O_2	10^{-8}

There are two important consequences of this variation in radical reactivity:

a) The greater the reactivity of "oxyl" radicals, the lower is their selectivity toward C-X bonds. Thus hydroxyl reacts indiscriminately with most C-X bonds (where X is H, O, N, S etc., including DNA) and adds to double bonds at diffusion controlled rates. Peroxyl radicals by contrast preferentially attack the most labile C-H bond in the molecule and do not normally cause the breakage of hetero bonds. Alkoxyls behave more like hydroxyl, but are much less reactive and in general less mobile.

b) Hydroxyl and alkoxyl radicals have very short "action distances" and do not travel more than a few nanometers from their site of generation. Peroxyl radicals (including superoxide) can "act at a distance" and may by migration cause damage at sites remote from their origin [9].

2.2 Products of Lipid Peroxidation

The lipids of the cell membrane are known to be the primary site of peroxidation in the cell. They are the structural components of the membrane and are arranged in a bilayer radial structure on the outside of the cell [8]. The bilayer consists of phosphoglycerides (I), acyl glycerols (II) and cholesterol (IIIa) and its polyunsaturated fatty acid esters (IIIb) together with randomly distributed protein:

R_1COOCH_2
|
R_2COOCH O
| ||
CH_2OPOR_3
|
O^-

I

R_1COOCH_2
|
R_2COOCH
|
R_3COOCH_2

II

III(a), R = H
III(b), R = C_{16}-C_{18}-acyl

A substantial proportion (up to 30%) of the fatty acids are polyunsaturated and hence highly peroxidisable. Cholesterol (IIIa), on the other hand, is relatively resistant to peroxidation, since it lacks the highly reactive methylene group found in the doubly activated 1,4-dienes of the polyunsaturates. In contrast, the polyunsaturated fatty esters of cholesterol (e.g. III, R = linoleate) readily undergo co-oxidation and co-polymerisation with the associated phospholipids and triglycerides in low density lipoprotein (LDL) [12].

The free radical oxidation chemistry involved in the transformation of arachidonic acid (IV) to prostaglandin (V) and related polyunsaturated fatty acids has been studied in some detail in recent years because of its occurrence in a wide range of cells and the ability of the alkylperoxide intermediates to carry out the co-oxidation of a wide range of organic substrates catalysed by the enzyme prostaglandin H synthase, PGHS (Scheme 2.1) [13]. The surprising feature of this chemistry is that the essential radical steps are identical to those occurring *in vitro* [14], in spite of the fact that they proceed under strict enzymatic control in the cell. Both the formation of endoperoxide (cyclooxygenase) and the formation of hydroperoxide (peroxidase) in Scheme 2.1 are controlled by PGHS which both produces and uses a hydroperoxide produced in the classical radical-chain reaction in which alkyl and alkylperoxyl radicals alternate (see Chapter 1).

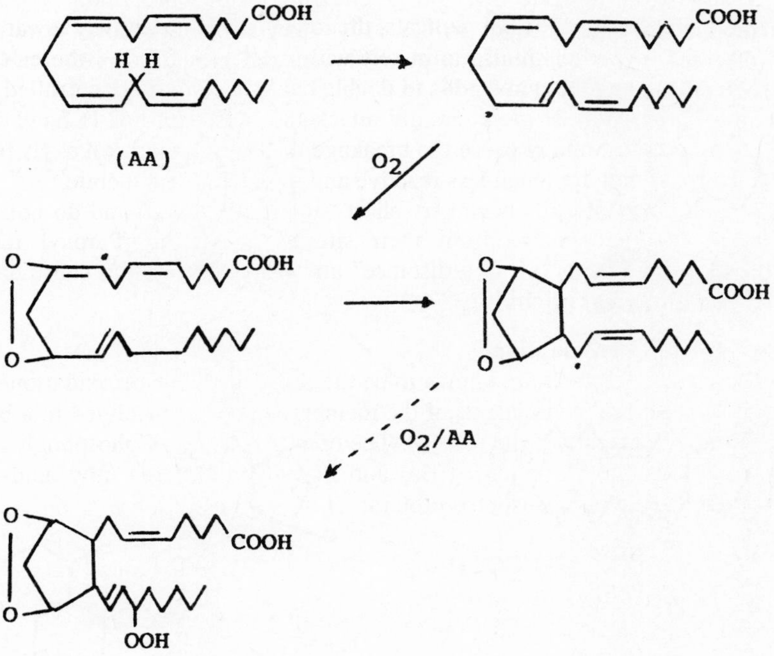

Endoperoxide

Scheme 2.1 Initial steps in the arachidonic acid (AA) oxidation cascade [13]

Evidence for the intermediacy of carbon-centred radicals has been found in the inhibition of the arachidonic acid oxidation "cascade" by antioxidant spin traps such as 2-methyl-2-nitrosopropane, MNP, which deactivates the intermediate carbon radical, L·, formed by hydrogen abstraction from C_{13} [15,16]:

$$L· + t\text{-BuN}=O \rightarrow \underset{L}{\overset{t\text{-Bu}}{>}}N\text{-O·} \tag{6}$$

MNP

Many phenolic antioxidants also inhibit this process.

The peroxidation of linolenic acid follows a very similar course to arachidonic acid both *in vivo* and *in vitro* [17]. Both cyclooxidation and hydroperoxidation occur in sequence, indicating that the basic radical chemistry occurring is not dependent on the bio-initiation step. The volatile end products of lipid metabolism are readily identified by gas chromatography coupled with mass spectrometry (GC-MS). They are all products of β-scission of the alkoxyl radical, the homolytic breakdown product of the primarily formed hydroperoxides. The more important of these transformation products are used to measure peroxidation in animals and include pentane, which can be monitored in the breath and malondialdehyde (MDA). The mechanism of their formation is shown in Scheme 2.2 for arachidonic acid; but similar α-scission breakdown products are general for PUFA [17]. They are universally found to be elevated in diseases involving increased peroxidation and provide a useful means of measuring such effects.

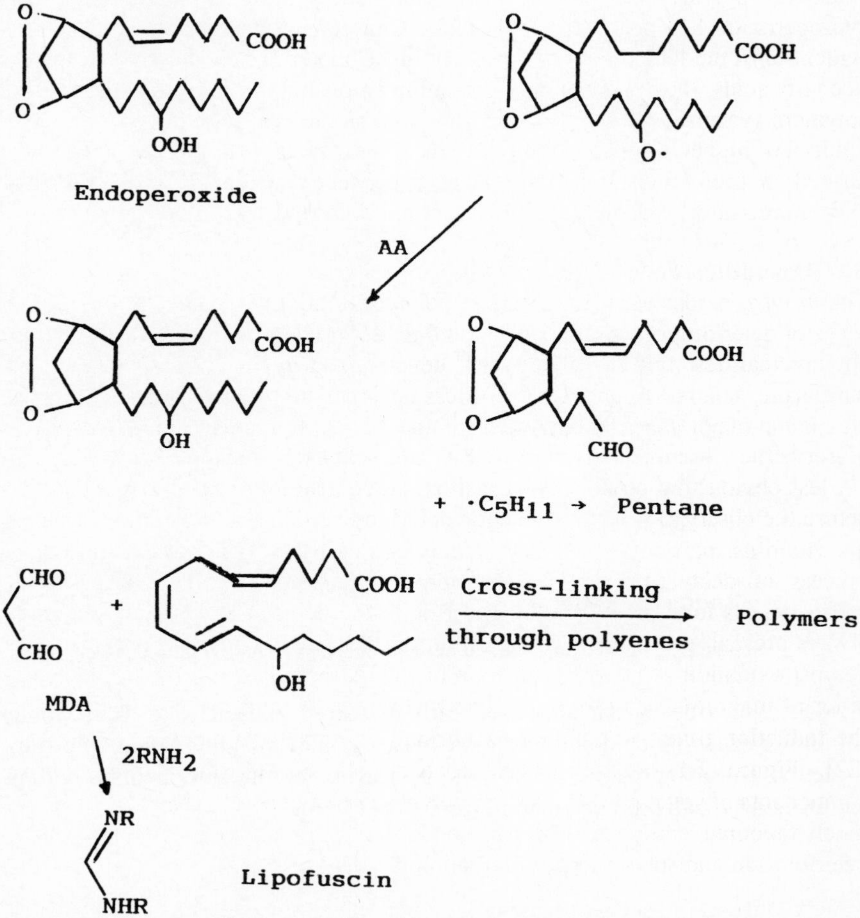

Scheme 2.2 Production of malondialdehyde, pentane and lipofuscin from arachidonic acid [13]

An equally important group of products formed in lipid oxidation are the oligomeric and polymeric products which are formed through the conjugated diene hydroperoxides.

It has long been recognised [18] that unsaturated fatty acids containing conjugated double bonds copolymerise with oxygen to give oligomeric and, in some cases, high molecular weight polyperoxides and polyethers (Chapter 1). Thus, methyl octadeca-9,11-dienoate copolymerises with oxygen in high yield to give oligomers of the fatty acid with oxygen (VI(a), VI(b)) [19]:

$$[-\overset{R}{C}HCHOO-]_n \quad \text{and} \quad [-\overset{R}{C}HCH=CH\overset{R'}{C}HOO-]_n \quad R = (CH_2)_5CH_3, \; R' = (CH_2)_7COOCH_3$$
$$\text{CH=CHR'} \hspace{5cm} n \text{ (average)} < 3$$

$$\text{VIa} \hspace{4cm} \text{VIb}$$

The 1,4-dienes rapidly hydroperoxidise to the conjugated hydroperoxides and co-oxidative polymerisation does not commence until a substantial amount of hydroperoxide has been formed [18,20]. Consequently molecular enlargement is not evident until the later stages of peroxidation (Chapter 1, Scheme 1.9). In the case of the free fatty acids, the degree of polymerisation is low but the same chemistry leads to high polymers (varnish, plaque, etc.) in the case of the triglyceryl esters, I and II of the lipids. At higher temperatures or in the presence of metal ions, ether linkages are formed in preference to peroxy linkages between 1,3-dienic fatty acids; and at low oxygen pressures, carbon-carbon linkages are preferred.

2.3 Rancidification of Fats and Oils

The *in vitro* oxidation of the esters of polyunsaturated fatty acids (PUFAs) is the major cause of deterioration of edible oils and fats [21]. Oxidative rancidification, which is the obvious manifestation of "off-flavour" development in the "visible" fats such as butter, margarine, salad and cooking oils or less obviously in the deterioration of meat, cheese, etc., is the major cause of food wastage and has led to intensive efforts over the years to preserve the "essential fatty acid" (EFA) components by means of added antioxidants.

A less obvious but none the less real oxidative deterioration of fatty foods occurs even before the observation of organoleptic deterioration. This is the removal by oxidation of the vitamins, particularly A,C, E, β-carotene and folic acid which are used up in the process of delaying oxidative deterioration (Chapter 5). This has two important consequences for human health. The first is the depletion of the antioxidants that are always present, particularly in the unsaturated vegetable oils whose need for protection against oxidation as the concentration of PUFA increases. Methyl linoleate oxidises an order of magnitude more rapidly than methyl oleate, and just 1% of linoleate removes the induction time to peroxidation normally observed with oleate ($\cong$ 800 h at 20°C) [22]. Figure 2.1, adapted from Kochar [21] relates the peroxidisability of the components of vegetable oils to their polyunsaturate content. This shows clearly that the much favoured components of the modern diet are in fact the main cause of their deterioration and subsequent peroxidation *in vivo*.

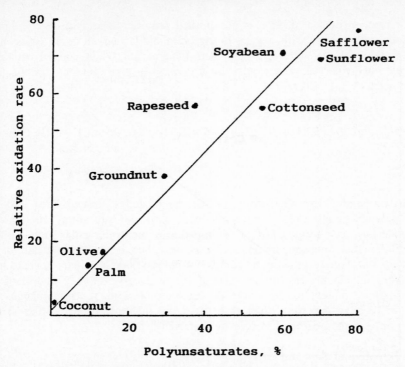

Fig. 2.1 Effect of the polyunsaturated components of edible oils on their peroxidisability

It will be seen in Chapter 3 that, in order to control oxidation in readily peroxidisable substrates, the concentration of antioxidants must be increased pro-rata to the rate of initiation. Interestingly, evolution has compensated for the increasing oxidisability of polyunsaturated oils by increasing the concentration of the lipid soluble antioxidants, notably vitamin E. Figure 2.2 relates vitamin E concentration in vegetable oils to the concentration of polyunsaturates in the oils (adapted from standard food composition tables [23]). The saturated and monoenic fats have been excluded from this correlation since they do not contain the 1,4-dienic structure and are therefore relatively unreactive to oxidation in the absence of an initiator (see Chapter 1).

The relationship shown between the concentrations of VE and PUFA in different oils is very significant and ensures that the highly peroxidisable polyunsaturated oils have increased protection from peroxidative destruction in the plant. However, in spite of nature's providence, the concentration of vitamin E in commercial oils is insufficient to compensate *in vivo* for the level of initiation provided by the PUFA (Chapter 6). This is because a very significant proportion of naturally occurring vitamin E (up to 40%) is destroyed during storage and subsequent processing of the oils for food use. It will be seen in Chapter 5 that this has important consequences for the nutrient value of margarines and spreads based on processed polyunsaturated vegetable oils since the removal of the naturally occurring antioxidants leaves the polyunsaturated lipids in the cells inadequately protected.

may be utilised in the production of ROS, notably superoxide and hydrogen peroxide [26]. Although this figure has been disputed [27], there is little doubt that mitochondria are a major source of ROS *in vivo*. Their production is increased in response to insult by toxic chemicals [28] and the ageing process has been associated with an increase in ROS production [25]. However, the most commonly attributed source of ROS is by the "oxidative burst" of phagocyte cells (Section 2.7) which occurs during inflammation.

Duthie lists the following among the diseases and clinical conditions which have been shown to be associated with elevated lipid peroxidation [29]: adult respiratory distress syndrome, ageing, Alzheimer's disease, angina, rheumatoid arthritis, atherosclerosis, cancers (some), cataract, dermatitis (some), diabetes mellitus, emphysema, favism, hyperoxia, hypertension, infertility (some), iron overload, Keshan's syndrome, ischaemia/ reperfusion injury, malaria, malignant hyperthermia, Parkinson's disease, radiation injury, retinopathy and muscular dystrophy. In addition, some voluntary actions of humans also lead to increase in lipid peroxidation. These include alcohol abuse, smoking, overconsumption of polyunsaturated fats, excess exercise and overexposure to sunlight. Some sources of oxidative stress have received considerable publicity, (e.g. smoking and cancer, sunburn and melanoma, ionising radiation and cancer, etc.), but some are still relatively unknown to the medical profession although they offer prospects for prevention that were unheard of until a few years ago. Particularly important is the relationship between dietary intake of polyunsaturated fats and diseases associated with peroxidation. Before considering the evidence for the involvement of oxidative processes in disease, it is necessary to briefly review the techniques that are currently available for detecting and quantifying the chemical effects of peroxidation in the lipids and hydroxyl radical damage to DNA.

2.4.1 *Recognition and measurement of biological peroxidation*
It is not easy to identify, still less to quantify, the radical intermediates of the peroxidation chain reaction or of the radical initiation processes that precede it. Great progress has been made in increasing the sensitivity of electron spin resonance (ESR) equipment and in the development of "continuous flow" [30,31] and "stopped flow" [32] techniques, but detailed kinetics of radical processes within the cell remain undefined. The problem is compounded in biological system by the difficulty of bringing the sample to the magnetic probe under conditions that realistically reflect the chemistry occurring *in vivo*. Thornally has commented that the biological application of ESR using these techniques "would be limited to studies of metalloproteins and semiquinones, with all other investigations so far removed from physiological setting that the validity of the model would be questionable" [33]. However, Thornally also points out that the situation has been transformed by the discovery and application of spin-trapping procedures, particularly by Janzen and co-workers [34]. The essential principle is that short-lived radicals, particularly carbon-centred radicals which cannot be studied by conventional ESR, are converted by reaction with the spin-trap (ST) to long-lived spin adducts (R-ST·) whose hyperfine splitting is characteristic of the radical from which it was derived:

$$R\cdot + ST \rightarrow R\text{-}ST\cdot \qquad (7)$$

This procedure can be readily carried out under physiological conditions *in vitro* and is being increasingly used *in vivo* by injection of spin-traps into laboratory animals.

Interestingly, no acute toxicity problems have been observed with nitroxyl-generating spin-traps [35-38] and indeed, as was seen in Chapter 1, "stable" aminoxyl radicals are effective antioxidants at relatively low oxygen concentrations. It will be shown in Chapter 3 that they are frequently catalytic antioxidants under these conditions.

The most commonly used spin-traps are 2-methyl-2-nitroso propane, MNP (see equation 6), α-phenyl-N-*t*-butylnitrone (PBN) and 5,5-dimethyl-1-pyrroline-N-oxide (DMPO):

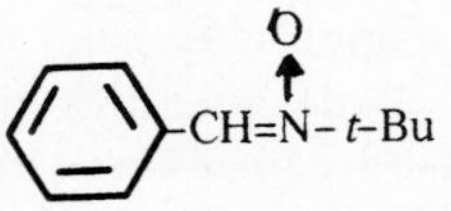

PBN

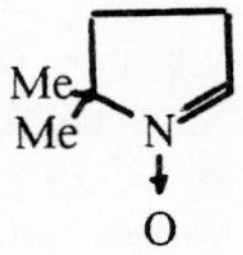

DMPO

All the important radicals involved in biological peroxidation have been identified by the use of one or other of the above spin-traps (Table 2.4) [33,39], and although there is sometimes doubt as to whether the initially formed radical or a secondary radical is being formed *in vivo* (e.g. in the case of peroxyl radicals), there is no doubt at all that free radicals are involved at every stage of biological peroxidation. Many antioxidants give rise to "stable" free radicals which can be identified and quantified by ESR and this will be discussed in Chapter 3.

Table 2.4 ESR Identification of biologically important free radicals [33]

Radical	Spin-Trap	g-Value	Hyperfine parameters
$O_2^{\cdot -}$	DMPO	2.0061	$\alpha_N=14.3G, \alpha^\beta_H = 14.3G, \alpha^\beta_H=11.7G, \alpha^\gamma_H=1.25G$
	PBN	2.0057	$\alpha_N=14.28, \alpha^\beta_H=2.25G, \alpha^\gamma_H=1.25G$
HO·	DMPO	2.0050	$\alpha_N=14.9G, \alpha^\beta_H=14.9G$
	PBN	2.0057	$\alpha_N=15.3G, \alpha^\beta_H=2.8G$
Cysteinyl-S·	DMPO	2.0047	$\alpha_N=15.3G, \alpha^\beta_H=17.26G$
	PBN	-	$\alpha_N=15.7G, \alpha^\beta_H=3.4G$
	MNP	2.0065	$\alpha_N=18.4G$
Glutathionyl-S·	DMPO	-	[RH1]$\alpha_N=15.4G, \alpha^\beta_H=16.2G$
LOO·	DMPO	-	$\alpha_N=14.8G, \alpha^\beta_H=12.6G$
LO·	DMPO	-	$\alpha_N=12.84G, \alpha^\beta=H6.48, \alpha^\gamma_H=1.68G$

Quantification of the stable oxidation products of lipids is of considerable relevance in following the progress of diseases involving peroxidation. It was seen above that the peroxidation of the lipids is characterised by the formation of hydroperoxides and their breakdown products; notably aldehydes and in particular malondialdehyde, conjugated dienes resulting from oxidative conjugation of 1,4-dienic fatty acids and low molecular weight hydrocarbons (ethane and pentane).

By far the most common method of monitoring lipid peroxidation involves the measurement of malondialdehyde by its reaction with thiobarbituric acid (TBA):

$$\text{TBA} + \text{CH}_2(\text{CHO})_2 \rightarrow \text{product} + 2\text{H}_2\text{O} \tag{8}$$

The pink coloured product absorbs in the region of 532 nm and the intensity of this absorption is used quantitatively after calibration with pure TBA. The reaction is carried out under acidic conditions, and to be quantitative all peroxides present must be decomposed to aldehydes; and Halliwell and co-workers [40,41] have pointed out that iron salts are also required to complete this process. However, iron also catalyses further peroxidation, and antioxidants are sometimes used to inhibit this process [42]. Since thiobarbituric acid is not highly specific for malondialdehyde, the results of this measurement are generally expressed as the measurement of thiobarbituric acid reactive substances: "TBARS".

Fluorescent Schiff's bases formed by reaction of aldehydes (including MDA) with amines can also be used as a very sensitive measure of the formation of carbonyl compounds by peroxidation [43]. These are also formed naturally in the "age pigments" and can be used as an endogenous measure of oxidation due to ageing (Section 2.10)

Direct measurement of hydroperoxides is also being increasingly used to monitor peroxidation of lipids. This can be done conveniently by measuring the fluorescence of the conjugated hydroperoxides in hexane extracts of plasma [44].

It was seen earlier that ethane and pentane are minor products of the oxidation of the polyunsaturated lipids. They are readily measurable in the breath by GLC and represent a complementary method of measuring peroxidation [45,46].

There is considerable interest in the possible involvement of hydroxyl derived from hydrogen peroxide by Fenton chemistry in direct attack on DNA and this has led to the development of techniques for the identification and quantification of DNA scission products which are generally associated with hydroxylation of DNA bases [47], some of which have themselves been shown to have mutagenic activity [48,49]. A number of markers of DNA hydroxylation have been identified and quantified by GC/MS after derivatisation. These include 5-hydroxycytosine, 5-(hydroxymethyl)-uracyl, 5,6-dihydroxyuracyl, 4,6-diamino-5-formamidopyrimidine, 2-hydroxyadenine, 8-hydroxy-adenine and 8-hydroxyguanine [50]. The chemistry underlying the formation and quantification of these products has been comprehensively reviewed by Breen and Murphy [51].

2.5 Atherosclerosis

The evidence for an association between polyunsaturated fatty acids in plasma and atherosclerosis is very convincing both from epidemiological studies [2-4,52] and from *in vivo* studies [53-55] which show the take-up by the arterial wall of oxidised low density lipoprotein (LDL), a cholesterol fraction containing a high proportion of linoleate in the phospholipid layer. Mitchinson [55] has given a simple account of the progress of the disease which begins with the recognition of "damaged" LDL by macrophages in the endothelium, progressing to visible lesions or "fatty streaks" and finally leading on to foam cells and necrosis of the macrophages with arterial restriction.

The aetiology of atherosclerosis is much less clear and is still the subject of considerable controversy. The historical development of the "cholesterol theory" has been reviewed by Addis and Warner [24]. The involvement of dietary cholesterol as the primary cause of atherosclerosis appears to have an element of folk-lore attached to it and recent evidence suggests that pure cholesterol is not atherogenic, even in sensitive animals [56] and the response of humans to cholesterol is weak [57,58]. Mitchinson and co-workers [59,60] have shown that although the linoleate (CL) and arachidinate (CA) esters of cholesterol lead to the formation inside the macrophages of ceroid rings, cholesterol itself and cholesterol oleate (CO) which is relatively inert to peroxidation (see Chapter 1) did not lead to ceroid accumulation at all. Furthermore, toxic oxidation products including cholest-5-en-3β,7β-diol were formed from CL and CA but not from CO and the former were inhibited by α-tocopherol [61].

Peng and co-workers [62] showed that some cholesterol oxidation products are reactive toward the endothelium and lead to balloons and craters in rabbits. Jacobson and co-workers [63] have called for a reinterpretation of cholesterol feeding trials and of human epidemiological data that does not take into account the cholesterol oxidation products in food. Cholesterol itself is not readily peroxidisable in the absence of an initiator since it lacks a highly activated methylene group (the allylic methylenes in cholesterol have similar reactivity to those in oleic acid). However, it is normally esterified in the cell with linoleic acid [54] which is much more reactive toward oxygen. Theoretical considerations, then, together with epidemiological and experimental studies [24,64] (Chapter 6) suggest that the key event in the initiation of atherosclerosis is the peroxidation of the polyunsaturated components of LDL. This may occur either in the food itself or subsequently in the body, but peroxidation precedes the formation of cholesterol oxidation products. This is supported by the above observations of Mitchinson and his co-workers [59,60] on the crucial importance of the esterifying fatty acid.

Antioxidant status is reduced in patients with CHD [65]. Individuals with angina have low plasma concentrations of the antioxidant vitamins C,E and β-carotene. Furthermore, LDL oxidation is delayed if both vitamins C and E or the flavonoid antioxidants are present [66]. A detailed discussion of how antioxidants protect against CHD will be deferred until Chapter 6, but epidemiological studies of CHD incidence in European populations indicate that an appreciable intake of total fats is compatible with a low rate of CHD provided that the supply of a-tocopherol is high [65]. Gey and his co-workers [3] have concluded that it is important to achieve an optimal ratio of the antioxidant vitamins to the polyunsaturated oils and fats in the food supply in order to minimise diseases associated with peroxidation.

The sequence of chemical reactions leading to the atherogenic effects of oxidised cholesterol in lipids can now be tentatively proposed. The intermediate peroxyl radical formed in the polyunsaturated LDL is almost certainly the source of cholesterol epoxides and hydroperoxides [67-69] and is also involved in the cooxidative polymerisation with the associated polyunsaturated glycerides by the mechanism discussed in Chapter 1 (see Schemes 1.9 and 1.10) leading to molecular enlargement. The most likely oxidative reactions involved in cholesterol oxidation are summarised in Scheme 2.4 and this occurs concomitantly with oligomerisation. It follows inevitably that interception of the initially formed alkylperoxyl radical should inhibit all the subsequent pathological effects of oxidised lipids, including cholesterol oxidation products.

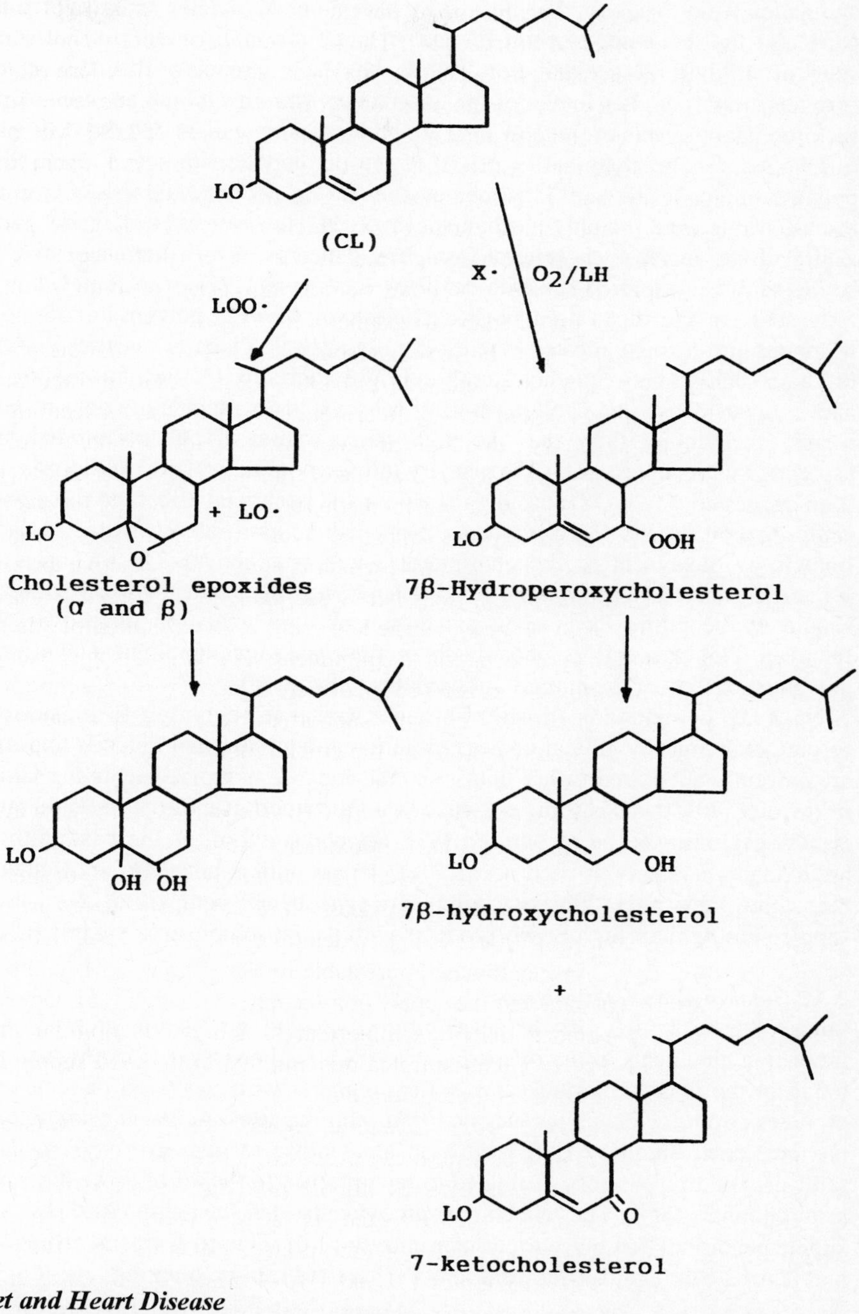

(CL)

LOO·

X· O_2/LH

Cholesterol epoxides
(α and β)   + LO·

7β-Hydroperoxycholesterol

7β-hydroxycholesterol

+

7-ketocholesterol

2.5.1 Diet and Heart Disease

There is now unequivocal evidence that the incidence of cardiovascular disease is affected by diet and in particular by fat intake. At present about 40% of energy intake in the British diet is provided by fats and it is recommended by the UK Department of Health [70] that this should be reduced to 35% in order to reduce CVD. Heart disease has been related to fat intake in many other countries, but there are interesting anomalies which suggest that high fat intake alone is not the determining factor. In particular the incidence of heart disease is much lower in Mediterranean countries in spite of a high fat consumption [2]. It has been proposed that this is due to a commensurately higher intake of the antioxidant vitamins which are able to neutralise the prooxidant effects of the polyunsaturated fats which will be discussed in more detail in Chapter 6. The chemical corollary of the peroxidation theory is that reduction of polyunsaturates in the total fats imbibed should also decrease the risk of cardiovascular disease if this were the only mechanism involved. However, the "classical" view is that saturated fats increase cholesterol formation which is in turn the cause of CVD. This connection was explored in some detail by Keys et al. [71], who found that saturated fatty acids in the diet raised plasma cholesterol whereas polyunsaturates lowered it. Monounsaturates (e.g. oleate) were essentially neutral. There is certainly good evidence of an association between cholesterol and CVD mortality [72-78], but as was discussed above, the evidence that cholesterol itself causes CVD is much more open to question. It seems much more likely that the high concentration of polyunsaturated fatty acids associated with cholesterol is the primary initiation process in atherosclerosis, and it has been suggested [60,76-78] that the reason for the beneficial effects of the monenic fatty acid, olive oil, in the "Mediterranean diet" may be associated with the replacement of linoleic by oleic acid in the cholesterol esters (see above). On the basis of recent evidence, Addis and co-workers [24,57] have suggested that food technology research should in the future focus upon prevention of lipid oxidation instead of cholesterol removal. This message is very much in tune with epidemiological evidence on the preventive effects of the natural antioxidants (Chapter 6).

The LDL peroxidation theory of atherosclerosis is beginning to impact on dietary recommendations by official bodies. Although polyunsaturated fish oils and margarines are currently advocated rather than monounsaturates as replacements for saturated fats in the diet [70], there is some evidence that the importance of the antioxidant vitamins is now beginning to be recognised [82]. Supplementation of the basic fatty foods by antioxidants has not yet been adopted as official policy, but there are indications that the more forward thinking food manufacturing companies are increasingly supplementing their high PUFA products with the fat soluble antioxidants (Chapter 5).

2.5.2 Smoking and heart disease

Smokers have a high risk of developing CVD [83] due to the presence of radical-producing chemicals in the inhaled smoke [84]. Higher levels of conjugated dienes are found in the plasma of smokers and this, together with increased pentane exhalation, provides strong evidence for elevated lipid peroxidation. Aqueous extracts of cigarette tar have been found by ESR readily to autooxidise to give hydroxyl and superoxide radicals, and this has been attributed to the presence in the tar of *ortho* and *para* benzo-semiquinones [85]. The effects of prooxidants can be suppressed by vitamin E supplementation [86] but normal concentrations of vitamin E in the alveola fluid [87] and plasma carotenoids concentrations [88] are reduced by smoking.

2.6 Cancer

The progress of cancer is characterised by three phases; initiation, promotion and progression [89]. The first of these is believed to be primarily by modification of cell DNA by a carcinogen which is taken into the body from an external source. This frequently involves the oxidative modification of a procarcinogen (see Section 2.6.2) and DNA which can also be disrupted by hydroxyl radicals. Radiotherapy is based upon the formation of hydroxyl radicals by radiolysis of water (reaction 9) in the environment of cancer cells in such a concentration as to kill the cell [90]:

$$H_2O \xrightarrow{\sim\sim\sim} H\cdot + \cdot OH \qquad (9)$$

Hydroxyl radicals from hydrogen peroxide formed by ascorbic acid in the presence of 3-amino-1,2,4-triazole (ATA), a catalase inhibitor, are also cytotoxic to tumour cells [91] and it has been proposed that site-specific prooxidant effects of catalase could be an important way forward in cancer chemotherapy [91,92].

The physiological effects of hydroxyl radical generation are "sensitised" by molecular oxygen ($\approx$ 3 times), implying peroxidation of cell components. However, the high reactivity of the hydroxyl radical (See Tables 2.1 and 2.2) requires that the radical has to be produced close to the cell nucleus to be damaging to DNA, for example by site specific Fenton reaction [93-95]. It has been suggested that DNA-bound copper is particularly implicated in this process.

There is recent evidence to suggest that some diseases involving chronic inflammation are associated with increased incidence of cancers. Thus, cystic fibrosis which is associated with a high level of lipid peroxidation in the lungs leads to increased damage to DNA and higher incidence of cancer [96]. This kind of evidence, coupled with an ever increasing body of evidence for the formation of oxidative scission products of DNA produced by Fenton chemistry occurring spontaneously or under the influence of exogenous chemical agents [51], suggests that the hydroxyl radical is intimately involved in carcinogenesis by DNA modification. Evidence for the involvement of peroxidation intermediates in tumour promotion and progression is even stronger [97]; both peroxides and hydroperoxides have been shown to be tumour promoters [98,99]. However, the most convincing evidence comes from observation of inhibition of tumour growth by antioxidants [100,101] and this will be discussed in detail in Chapter 6.

2.6.1 Diet and Cancer

It has been recognised for some time from population studies and from experiments in animals [102-105] that a high fat intake is associated with a higher incidence of certain types of cancer notably mammary and colon cancers [106]. Animal studies have also shown that polyunsaturates are more effective cancer promoters in the presence of procarcinogens than saturated fats (Fig. 2.3), and the evidence implicates the polyunsaturated components of fats as the primary source of hydroperoxides.

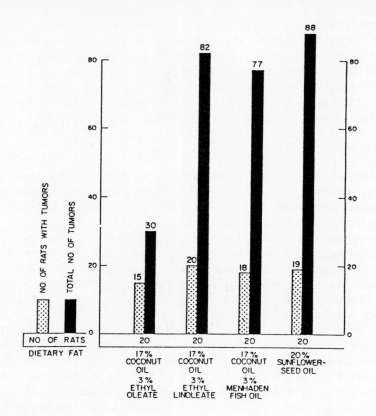

Fig. 2.3 Effects of dietary fat on development of mammary tumours induced in female Sprague-Dawley rats. A single oral dose of DMBA was given at 50 days after birth with the commencement of the indicated diet and the rats were autopsied four months after commencement of diet. (Reproduced with permission from K. K. Carroll in *Cancer and the Environment*, Eds. H. B. Demopoulos and M. A. Mehlman, Pathotos. Pub. Inc., Illinois, USA, 1980, p. 257)

It has been pointed out by Newberne and McConnel [107] that the successful promotion of polyunsaturates as "preventives" of cardiovascular disease by the (American Heart Association) has resulted in an increase in PUFA in the diet from 7% in 1909 to 15% in 1980 in spite of concerns that excessive ingestion of PUFA is potentially toxic [108]. Both total fats and polyunsaturates play a part in the initiation and promotion of cancer, but it is the latter that lead to the rapid depletion of the antioxidant defences [109]. Carrol and Hopkins [110] have shown that only 15% of ethyl linoleate or of Menhaden fish oil in coconut oil fed to rats increases tumour development almost to that of sunflower seed oil, whereas oleic acid had no such effect (Fig. 2.3).

From a theoretical point of view this is entirely consistent with the principles of peroxidation outlined in Chapter 1 where it was observed above that very small amounts of polyunsaturated compounds increase the peroxidisability of more stable substrates out of proportion to their concentration. Recent evidence supporting these conclusions has been comprehensively reviewed by Welsch [111].

In spite of the above consensus on the effects of polyunsaturates in general there is increasing evidence that some specific polyunsaturates can inhibit tumourogenic processes in animals [111]. Most interest has centred on the chemoprevention of mammary gland tumours in animals fed diets rich in fish oils, and in particular those containing appreciable concentrations of the ω-3 long chain unsaturated fatty acids, of which the most abundant are 5,8,11,14,17-eicosapentaenoic acid, EPA (C20:5) and 4,7,10,13,16,19-docosahexenaenoic acid, DHA (C22:6). These are more effective than the shorter chain ω-3 fatty acids, for example, α-linolenic acid (C18:3) [112]. Thus, in mice inoculated with human breast carcinomas, growth was greatest in mice fed high levels of corn oil, intermediate in mice fed high levels of butter and lowest in mice fed a high fish oil (Menhaden oil) diet [113-115]; and in mice fed a high fat diet in which corn oil (CO) to fish oil (FO) ratio was varied (18%CO/5%FO to 5%CO/18FO%), carcinoma growth and metastasis were significantly suppressed in mice fed the highest amount of fish oil compared with those fed the highest amount of corn oil [116].

Although it is not yet clear why different polyunsaturated fatty acids should behave so differently in the promotion and suppression of cancer, it is evident that the products of peroxidation rather than the PUFAs themselves are responsible for the lysis of human breast carcinoma cells *in vitro*. In particular, the effectiveness of different fatty acids in lysing breast carcinoma cells correlates with the intracellular concentration of TBARS produced [117]. Furthermore, both the endogenous antioxidants (uric acid, SOD and GSP) and exogenous antioxidants (vitamins A, E and BHA) suppressed the rate of cytolysis in cultured breast carcinoma cells [118,119] and reduced the amount of TBARS in tumour tissues in mice fed fish oils [113,114], whereas iron promoted the effect [119]. A number of peroxidation products of the polyunsaturate fatty acids have been proposed and in some cases evaluated in an attempt to identify both tumour promoting and tumour suppression. These include the hydroperoxides themselves and their transformation products formed in prostaglandin-type chemistry [111]. There is good evidence that the 13-hydroperoxy- and 13-hydroxylinoleic acids can promote DNA synthesis [120] and cell growth, and a variety of hydroperoxides, including hydrogen peroxide itself, have been shown to cause cell proliferation [5,6,121]. The cytotoxic effects of the n-3 poly-unsaturates have not so far been fully explained, but Chajès et al. have proposed [122] that LDL containing peroxidised PUFA are taken up by human mammary tumour cells which lack oestrogen receptors leading to lysis. Pre-treatment of the LDL with vitamin E counteracted this effect. It should be noted however that minor concentrations of fish oil may simply promote peroxidation of other PUFAs, and the major peroxidation breakdown products will be derived from the latter and not from ω-3 acids.

The desirability of an increased intake of vitamin E in high PUFA diets was first noted by Horwitt [123], but this has been made explicit by subsequent epidemiological studies which have emphasised the requirement for an optimum ratio of the vitamin antioxidants (particularly C, E and β-carotene) to polyunsaturated fats as is found in Mediterranean diets where the cancer incidence is lower [124]. Two unsaturated fatty acids (C18:2(n-6) and C20:4(n-6)) are essential for weight gain in animals, and others (e.g. C18:3(n-3)) are desirable [125]. However, the small amounts required can probably be compensated for by a relatively minor increase in the antioxidant vitamins in the diet (Chapter 4).

Addis has commented that the popular addiction to the vegetable oil-based polyunsaturates is unlikely to be easily reversed [24] because of the high profile publicity campaign initiated by national and international bodies which advocate without discrimination increase of polyunsaturates in the diet. This in turn is powered by the food industry which has seen it as a commercial opportunity. However, there is some evidence of a change in emphasis. For example, although the World Cancer Research Fund still emphasises the "beneficial" effects of replacing saturated fats by polyunsaturated fats in the diet [126], this is balanced by a more scientifically based recommendation to increase vitamin E intake from whole grain cereals and pulses, ß-carotene from "yellow-orange" fruits and vegetables (carrots, sweet potatoes, peaches, apricots, oranges, cantaloupe melons and bananas) and dark green vegetables (spinach, spring greens and broccoli). The latter category is also recommended for vitamin C, together with fresh fruits (citrus fruits, strawberries, etc.). Sources of the antioxidant vitamins will be discussed further in Chapter 5.

2.6.2 Chemical Carcinogens

Although the evidence discussed above suggests that tumour promotion occurs due to peroxides in cells containing elevated polyunsaturates, there is also little doubt that this is primarily a result of activation in the body of fat-soluble but, in their unmodified forms, physiologically inactive chemicals (procarcinogens) to carcinogens [127-129]. Perhaps the most studied of the procarcinogens are the polycyclic aromatic hydrocarbons and in particular 3,4-benzpyrene (BP) and 7,12-dimethyl-benz[a]anthracene (DMBA). The transformation of these compounds occurs in the liver, the first stage being catalysed by cytochrome P450. [130,131]. The sequence is shown typically for BP in Scheme 2.5. The second epoxidation stage involves alkylperoxyl radical addition to an aromatic double bond [132]. It is inhibited *in vivo* by a variety of antioxidants [131], although McCay et al have found [127] that the beneficial effects of antioxidants are at least partially neutralised in a high PUFA diet. Frenkel et al. have found that hydrogen peroxide is produced from DMBA in topical treatment of mice and this was associated with damage to DNA as evidenced by the formation of oxidised bases (e.g. 8-OHdGuo) in the skin [133]. This behaviour is similar to that of the phorbol esters and is associated with inflammation of the epidermis.

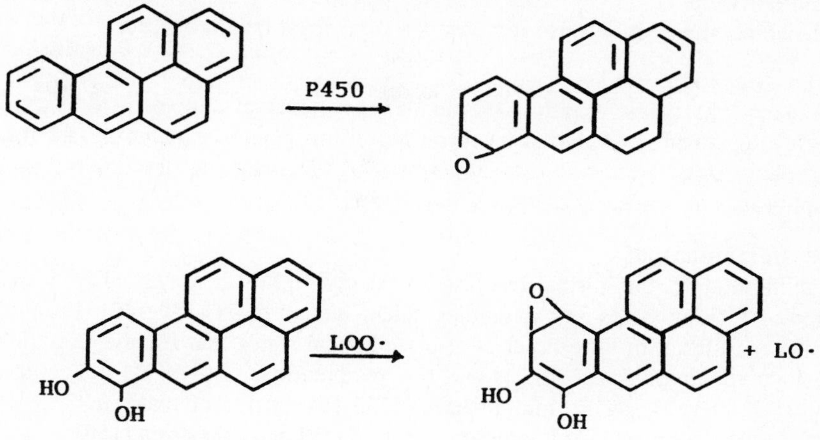

Scheme 2.5 Epoxidation of Benzo[α]pyrene [130,132]

Many other procarcinogens require an oxidative environment for activation to the active carcinogen [131]. A recent example of procarcinogen activation because of its relevance to the environmental problem of air pollution from petrol engines and cigarette smoke is benzene. Chronic exposure causes aplastic anaemia and leukaemia. Benzene is converted to 1,2,4-trihydroxy benzene (THB) by cytochrome P450 oxidation in the liver, and it accumulates in the bone marrow [134] where further oxidation occurs and ROS are generated [134,135]. The hydroquinones/quinones are believed to redox cycle with the formation of superoxide and its further reduction products [136] (see Scheme 2.6).

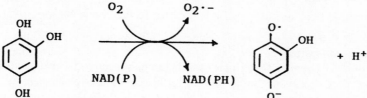

Scheme 2.6 Redox cycling prooxidant action of 1,2,4-trihydroxybenzene [136]

2.6.3 Smoking and Cancer

The incidence of cancer of the lung and of the larynx are both strongly correlated with the number of cigarettes smoked daily [137]. Correspondingly, the cessation of smoking leads to reduction of risk relative to that of non-smokers [137]. As was noted in Section 2.5.2 cigarette tars are a potent source of prooxidants and there seems to be little doubt that lipid peroxidation initiated by tars in cigarette smoke is the major cause of cancer of the lung in smokers, and there is some evidence that this may be due to ROS formation by redox cycling of semiquinones which have become covalently bound to DNA [138]. As was seen above, benzene is a major source of semiquinones and the concentration in the expired air from smokers is an order of magnitude higher than it is from non-smokers [137]. Since a single cigarette emits 12-480 µg of benzene, passively inhaled cigarette smoke also represents a significant health hazard.

2.6.4 Alcohol abuse
Smokers who drink heavily have an increased risk of cancer of the larynx [119]. A highly significant correlation has been found for pentane exhalation and liver cirrhosis for alcoholics. This was associated with a decrease in arachidonic acid in their membranes indicating preferential oxidation [139].

2.7 Inflammation
Inflammation is the normal response of the cell to hostile invasion by microorganisms. Phagocytosis involves an oxygen dependent radical burst [140-142]. Inflammation also is the manifestation of a number of pathological states that involve peroxidation. They include rheumatoid arthritis [143-145], pancreatitis, [146-149], ischaema-reperfusion injury [150-152], severe malnutrition, [153,154], cerebral malaria [155,156], asthma [157], cystic fibrosis [158], pneumoconiosis [159] and asbestosis [160].

The formation and reactions of reactive oxygen species at inflammatory sites has been reviewed by Halliwell and Gutteridge [161] and by Bendich [142] and a number of publications have been devoted to the biochemistry and pathology of inflammation [162-164]. The mechanisms of generation of ROS are complex since oxygen radicals are produced at different sites in the defence mechanism (e.g. by macrophages, polymorphonuclear leukocytes (PMNs), lymphocytes, etc. [165,166]. Superoxide by reduction of molecular oxygen is generally the primary product of these processes generated by redox enzymes such as xanthine-xanthine oxidase, hypoxanthine-xanthine [166], the flavoprotein cytochrome-b-NADPH [167] and cytochrome P450 metalloenzyme systems [167]. Peroxidases are also important in the oxidative defence mechanism, and prostaglandin synthase (PGHS), the key enzyme in the arachidonic acid cascade (Scheme 2.2), is particularly involved in the oxidation of xenobiotics [13,166,169]. A detailed discussion of ROS generation in biological systems is outside the scope of this book and the reader is directed to the above authoritative reviews for further information.

2.7.1 Rheumatoid arthritis
Synovial fluid, a polymer based upon the heteropolysaccharide hyaluronic acid is the lubricating fluid between the joints. In arthritis, oxidative damage is caused by phagocytic cells entering the joint with consequent high ROS activity leading to lipid peroxidation and the formation of low molar mass oligosaccharides, causing subsequent damage to bone and cartilage [167,170]. The presence of elevated levels of lipid peroxidation products has been observed in the breath of RA patients as measured by increased pentane formation [171] and by the presence in the synovial fluid of diene conjugation and TBA reactive substances (TBARS) [172,173]. In both cases the extent of hydroperoxide breakdown products correlates with the severity of the disease. Iron frequently plays a part in radical formation in synovial fluid [40] and again there is a correlation between free iron (detectable by bleomycin) and the extent of the damage [173]. There is an increase in ferritin in the synovium [174,175], probably resulting from intermittent haemorrhage and a corresponding decrease in haemoglobin in the serum. As will be seen later (Chapter 6, Section 1.7.1) the vitamin antioxidants have some activity in decreasing the effect of the Fenton reaction. However, iron complexing proteins, notably lactoferrin and transferrin, are much more effective inhibitors of iron

catalysed peroxide decomposition [176], and desferrioxamine, a naturally occurring iron complexing agent was found to be beneficial in restoring haemoglobin concentrations in human subjects [177,178].

Physical factors are also implicated in synovial fluid degradation. During exercise, the pressure in normal joints is close to ambient. In rheumatoid joints by contrast, the resting pressure is elevated and on exercise it may rise to levels that exceed the capillary perfusion pressure [170,173]. The resulting ischaemia results in the build-up due to hypoxia of reducing enzymes in the xanthine-xanthine oxidase or NADPH in the neutrophils within the inflamed joint. This in turn leads to increased ROS generation as the oxygen in the blood supply returns to normal after exercise and is an example of reperfusion injury (see below):

2.7.2 Hypoxia-reperfusion injury

The example quoted above illustrates a general principle which has been observed in other diseases. It has been recognised for some time that interruption of blood flow to the heart (ischaemia) causes tissue damage, but further damage is also caused when the blood supply is restored (reperfusion) [179,180]. McCord suggested [181] that the latter effect was due to the build-up of hypoxanthine which on re-oxygenation converts ground state oxygen to superoxide in the presence of xanthine oxidase:

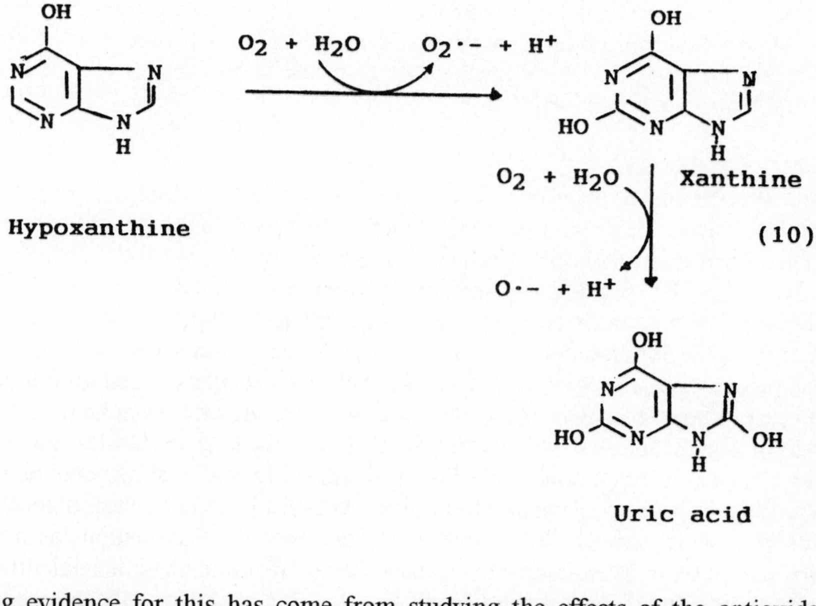

Convincing evidence for this has come from studying the effects of the antioxidant enzymes in animals. Thus Jolly et al. [182] found that a combination of superoxide dismutase and catalase protected the dog heart from reperfusion injury. Similar effects have been observed after reperfusion of the rat kidney [183,184] and the canine pancreas [184].

The significance of oxygen radicals in ischaemia-reperfusion has considerable importance in organ transplant surgery. There is clearly little that can be done physically to avoid reperfusion and considerable emphasis has fallen upon the prevention of the effects of reperfusion by the use of antioxidants [185-187] (Chapter 6).

The brain is particularly sensitive to oxidative damage. This appears to be due to the high level of polyunsaturated fatty acids in the brain and the relatively low concentrations of endogenous antioxidants, notably catalase and glutathione peroxidase [188]. Ascorbic acid also appears to be concentrated in the grey and white matter of the central nervous system [189]. This is certainly protective under normal conditions but in the case of damage to the brain, for example after head injury [190] or cerebral ischaemia [191], bleeding may lead to an increase in "free" iron concentration, forming an effective prooxidant combination by reducing Fe^{3+} to damaging Fe^{2+} due to participation in the Fenton reaction.

2.7.3 Pancreatitis

Acute pancreatitis has been graphically described by Braganza as the gland being "cannibalised by its own prematurely activated enzymes" [148]. Inflammation of the pancreas is initiated in a variety of situations, some arising from physiological phenomena within the pancreatic duct (e.g. gallstones) and some incited by xenobiotics of which the most important is alcohol [148,192], but many chemicals and drugs are also known to be involved [193]. Cytochrome P450 is particularly implicated in both acute and chronic pancreatitis in the generation of ROS [148,192,194]. Very convincing evidence for the involvement of radicals in pancreatitis has come from the identification of lipid peroxidation products, notably conjugated diene polyunsaturated fatty acids in duodenal bile [195], but more particularly from the therapeutic effects of antioxidants [196] (Chapter 6).

2.7.4 Cystic fibrosis

Cystic fibrosis results from a defective gene which leads to pulmonary infection and diminished pancreatic function [158]. Poor fat absorption leads to deficiencies of the fat soluble antioxidant vitamins, notably α-tocopherol, β-carotene and selenium. Infection of the lungs by bacteria results in a vigorous inflammatory response causing bronchospasm, increased secretions and other changes that lead to a vicious cycle of infection and inflammation [197,198]. Following stimulation, neutrophils and macrophages in the bronchalveola lavage reduce molecular oxygen to superoxide and hydrogen peroxide which normally constitute an effective mechanism for killing invading organisms [140,141]. However, they also damage the host tissue and Brown and Kelly [44] have reported that plasma malondialdehyde is elevated in about 40% of cystic fibrosis patients, comparable to that reported for patients with diabetes mellitus [199] and acute hepatic failure [200]. Other markers of oxidation were somewhat equivocal [200]. Antioxidant status is also changed in patients with cystic fibrosis [40], but not always in the direction expected. Vitamin E is on the low side but within normal limits. However, ascorbic acid, uric acid and thiol groups were substantially elevated compared with controls. Direct measurement of total radical-trapping antioxidant potential (TRAP) [201] indicated that measured chain-breaking antioxidant activity of plasma of CF patients was considerably lower than that expected on the basis of the concentrations of the antioxidants present, suggesting that under the conditions encountered one or more of the components is acting as a prooxidant. Ascorbic acid which is known to be a potential prooxidant in the presence of iron [40] (Chapter 5) may be implicated. However, it should be noted that the TRAP assay provides no indication of other types of antioxidant present (e.g. peroxide decomposers and metal deactivators) and therefore provides only a partial description of the antioxidant defences. This will be discussed in Chapters 3 and 5.

It has been observed that the incidence of cancer is increased in CF sufferers [202], suggesting that there may be increased oxidative damage to DNA by the free radicals produced by pulmonary infection and inflammation. Consistent with this, increased concentrations of guanine bases were observed in the urine of CF patients [203].

2.7.5 Disorders of prematurity

Premature infants suffer from a number of diseases associated with peroxidation [204,205]. The most important of these are chronic lung disease, retinopathy and intraventricular haemorrhage. A unifying hypothesis to account for capillary damage and haemorrhage is that increased radical generation occurs due to hyperventilation of newborn premature infants, and it is now recognised that only the minimum amount of supplemental oxygen necessary to keep the infants alive should be used.

This problem is exacerbated by an imbalance of the antioxidant vitamins in the premature infant. Plasma vitamin E concentrations are low in both full-term and pre-term infants but, whereas the former replete vitamin E stores to that of the adult population within one week, premature infants have low vitamin E levels up to eight weeks after birth .[206]. By contrast, vitamin C concentrations in the plasma of premature infants is higher than that found in adults and Powers and co-workers [207] have shown a clear correlation between the degree of immaturity and vitamin C concentration. Caeruloplasmin which oxidises and complexes iron [208], thus removing it from involvement in the Fenton reaction, is also depleted in the plasma of preterm infants compared with full-term infants or adults. However, ascorbic acid competes with this process and acts as an antagonist for caeruloplasmin by reducing Fe^{3+} to Fe^{2+} [209-211].

2.7.6 Adult respiratory distress syndrome

Adult humans are much more resistant to the effects of elevated oxygen concentrations than are premature children and small animals. Humans can breathe pure oxygen for several hours without discomfort, but pure oxygen at pressures greater than 2 atmospheres is acutely toxic to the nervous system, causing convulsions [212]. Adult respiratory distress syndrome (ARDS) can also occur in humans at ambient oxygen pressures and is manifested as acute respiratory failure due to pulmonary oedema. ARDS generally results from shock, notably tissue damage due to burns or massive infections [213].

Evidence for radical formation in ARDS has come primarily from experiments in animals. Vitamin E deficient rats are very susceptible to pulmonary oxygen toxicity, but injection of rats exposed to pure oxygen with lyposomes containing SOD and catalase substantially increased survival times. SOD or catalase alone had much smaller protective effects [214]. This is what would be expected In view of the synergism between SOD and Cat (Chapter 4).

2.7.7 Inflammatory bowel disease

Inflammatory bowel disease (IBD) is the collective term for Crohn's disease and ulcerative colitis [215]. Ascorbic acid [216] and glutathione [217] have been shown to be present in reduced concentration in the mucosa, and it is suggested that this results from the presence of metal ion peroxidation catalysts in the faeces [218]. Indeed, as was seen above, ascorbic acid can play a prooxidant role in the presence of transition metal ions and it seems likely that the removal of transition metal ions by increased fibre in the diet could play a significant role in preventing IBD.

2.7.8 Disorders of severe malnutrition

There is increasing evidence that, in parts of the world where diet is restricted to a limited number of staple components, a deficiency may develop in the antioxidant defence mechanisms, particularly of children. Typical diseases which are thought to arise from malnutrition are kwashiorkor [154,155] and cerebral malaria [156,157]. Golden and Ramdath have pointed out [155] that malnutrition is not the same as hunger with its associated low energy intake. Malnourished children may have an adequate even though monotonous energy intake. It is rather the lack of variety and the absence of the more important micronutrients in the diet that leads generally to anorexia leading to the necessity for forced feeding a primary treatment.

Kwashiorkor is characterised by oedema, severe "fatty liver" and changes in the colour of the hair and skin [155]. The plasma levels of vitamin E are markedly reduced in children with kwashiorkor [155,219] as is the concentration of the retinoid antioxidants [220,221]. Selenium required for glutathione peroxidase (GPx) is also below normal levels in children with kwashiorkor [222-224]. These deficiencies are probably partially caused by poor nutrition but also result from severe bacterial attack [225]. Finally there is evidence of iron overload in the plasma of Kwashorkor children [155], thus increasing radical damage by Fenton chemistry. Evidence for this was found in high transferrin saturation, high plasma levels of ferritin and excess hepatic iron.

2.8 Iron Overload

2.8.1 Idiopathic haemochromatosis

Haemochromatosis is a relatively rare inherited condition in which more dietary iron than usual is absorbed by the gut, leading to liver damage, weakness, skin pigmentation and diabetes (sometimes called "bronze diabetes"). In severe cases it may lead to cardiac malfunction [226,227]. A secondary effect of haemochromatosis is arthritis [228] which is known to be associated with excess iron in the synovial fluid. Increased lipid peroxidation has been observed in iron overloaded spleens [229] and similar effects have been seen in animals [40]. The disease, if diagnosed in the early stages, can be readily controlled by venesection and the ferritin level is maintained at the lower end of the "normal" range. Gutteridge and co-workers have shown [230] that total plasma iron greater than 40μmol/l leads to bleomycin detectable iron and peroxidation of phospholipids which does not occur in normal plasma.

2.8.2 Thalassaemia

Another disease of iron overload is found in children suffering from thalassaemia which is treated by repeated blood transfusion. The effect is to overload the system with iron and haemochromatosis results [231], leading to liver damage and cardiac malfunction. There is at present no completely effective treatment of iron overload in thalassaemia since venesection, the most efficient way of eliminating excess iron is ruled out. Treatment by injection with desferrioxamine the most widely used metal chelating agent [232] is only partially successful due to the large amounts required to reach the iron storage pools (Section 5.5.4).

2.9 Environmental Damage
2.9.1 Age-related cataract
The term age-related cataract is used to differentiate the accumulation of protein damage which results in opacification of the lens with time from similar effects due to metabolic disorders. The eye contains a number of photosensitisers of which the aromatic ketone, kynurenine and its derivatives are probably the most important [233]. Aromatic ketones are well known triplet carbonyl photosensitisers above 300 nm (Chapter 1), and cataract is thus the effect of life-long exposure of the eye to the photochemical effects of incident light in the presence of sensitisers.

Kyneurenine

The lens consists of 98% of proteins which gradually becomes cross-linked due to UV sensitised oxidation of free thiols to disulphides [234]. In this situation it may seem surprising that the lens of the eye can remain clear into old age. Of course in many instances it does not and the lens has to be removed and replaced by a replaceable man-made substitute. It will be seen in Chapter 6 that the healthy eye contains a rich combination of endogenous and exogenous antioxidants and a nutritious diet provides the appropriate synergistic combinations to counteract the repeated photooxidative insult in the outdoor environment.

2.9.2 UV skin damage
The UV-A (320-380 nm) component of sunlight has been strongly implicated in the consequences of exposure to sunlight [235]. Morliere et al. [236] have analysed the relative contribution of UV-A and the shorter (and hence more energetic) wavelengths, UV-B (280-320 nm). These authors have derived an activation spectrum for TBARS formation from skin fibroblasts on irradiation at different wavelengths, and find that the TBARS/quantum falls almost linearly by two orders of magnitude with wavelength increase from 275 nm to 425 nm. In spite of this differential in photochemical potency, they concluded that wavelengths below 310 nm make a relatively minor contribution to photosensitising effectiveness. There appear to be two contributory reasons for the relative ineffectiveness of the shortest wavelengths of the sun's spectrum. The first is the relatively small UV-B doses actually falling on the body (only a few hundred Jm^{-2}); and the second is that the endogenous sensitisers (notably the porphorins, flavins, pterins, etc. which absorb primarily in the 400 nm region) appear to be the predominant UV sensitisers in skin. The haematoporphorins in particular have been shown [237] to generate singlet oxygen and UV-A radiation reduces Fe^{3+} associated with ferritin [238]. It is well known from polymer chemistry that this represents a general method of producing "Fenton active" Fe^{2+} and indeed this chemistry is the basis of a process for commercial "time-controlled" photodegradation of plastics [239] in which the initiating process is the formation of Fe^{2+} from an antioxidant Fe[III] sulphur complex (Section 4.1.3). Many naturally occurring iron chelates behave in this way:

$$\text{FeL}_3 \xrightarrow{h\nu} \text{L}\cdot + \text{FeL}_2 \xrightarrow{\text{ROOH}} \text{RO}\cdot + \text{L}_2\text{FeOH} \xrightarrow{\text{LH}} \text{L}_3\text{Fe} + \text{H}_2\text{O} \quad (11)$$

R = H or alkyl

The most common biological ligands must be carboxylic acids. Citrate has been shown to be an effective photosensitiser in the presence of hydrogen peroxide [240], and Vile and Tirrell have shown that hydrogen peroxide is also formed during the UV-A irradiation of skin cells [235].

2.9.3 *Effects of ionising radiation*

Ionising radiation leads to the radiolysis of water yielding hydrogen atoms, hydroxyl radicals and hydrated electrons:

$$\text{H}_2\text{O} \rightarrow e^-_{aq} + \text{H}\cdot + \cdot\text{OH} \quad (12)$$

Secondary products formed in smaller molar yield in the absence of substrates capable of being attacked by the primary free radicals are dimerisation products molecular hydrogen and hydrogen peroxide. Direct interaction of high energy radiation may occur with most biological molecules leading to bond scission followed by reaction with oxygen similar to the radiolysis of polymers. This in turn leads to subsequent peroxidation (Chapter 1). However, due to the preponderance of water in biological systems, damage by hydroxyl radicals is at least as important as direct radiolysis and in some circumstance it may be more important. Breen and Murphy [51] have carried out an extensive investigation of the reactions of oxyl radicals with DNA which illuminate the chemistry governing the initiation of mutagenesis. Wardman has suggested that about one third of the damage to DNA which leads to strand breakage and mutation is due to "direct hit" [242] and the remainder due to "clusters" of radicals produced by hydroxyl radicals which appear to be more effective in double strand breakage than individual hydroxyl radicals produced by chemical reaction from hydrogen peroxide [243].

Oxygen enhances the damaging effect of radiation on cells two to threefold [242] but isolated DNA is not sensitised by oxygen and may even be protected. This is an important effect in radiotherapy since it permits a higher degree of cancer cell killing than can be achieved in the absence of air, but it is also important in radioprotection because although it is difficult if not impossible to inhibit the primary processes of radiolysis in biological systems, it is easier to inhibit the subsequent oxidative chemistry [244]. Wardman has noted [242] that the oxygen concentration in many normal tissues is around 50 µmol dm^{-3} (partial pressure $\cong$ 40 torr) and that this is sufficient to give maximal radiosensitisation of mammalian cells *in vitro*. Considering this from the standpoint of prevention, it seems likely that under normal physiological conditions, oxygen-centred rather than carbon- or nitrogen-centred radicals will be the main targets for inhibition.

Chromosome-damaging (clastogenic) effects have been observed in the plasma of humans many years after they were subjected to ionising radiation [245] and it has been shown that "clastogenic factors" (CFs) stimulate superoxide production by neutrophils and monocytes [246], and this can be prevented by superoxide dismutase [247]. This opens up interesting prospects for the role of antioxidants (Chapter 6).

2.9.4 Lung damage

The lung is particularly susceptible to damage from environmental pollutants. These range from the gaseous products of fuel combustion, of which the most important are ozone (O_3), oxides of nitrogen (NO_x), oxides of sulphur (SO_2, SO_3) and aerosol particulates derived from the combustion of diesel fuels.. Smoke from cigarettes contains oxides of nitrogen as well as semi-quinones which generate superoxide and hydroxyl radicals during autooxidation [248,249]. In addition to these widely distributed pollutants there are some highly specific causes of lung diseases arising from industrial processes. These include asbestos, coal dust and the dipyridyl herbicides widely use in agriculture.

Quite apart from man's activities, naturally occurring radioactivity, for example from radon, first identified in uranium mines and known to be the cause of cancer, has been recognised to be present at very significant levels in houses [250] in areas where there are rock fissures (e.g. granite and limestone). Recent research has suggested that radon is more important as a causative factor in the induction of myeloid leukaemia [251] than as a cause of lung cancer [252]. The main "antidote" to radon is correct ventilation of buildings since the outdoor concentration is normally very much lower.

The most widespread specific lung disease which is either caused by or exacerbated by environmental pollutants is asthma. Engine exhausts are a source of NO_x and O_3 and both are also found generally in the environment as a result of photochemical reactions in the atmosphere catalysed by the hydrocarbon products of the petrochemical industries [253-255]. The basic chemical reactions leading to ozone are:

$$\cdot NO_2 \xrightarrow{h\nu} \cdot NO + O\cdot \qquad (13)$$

$$O_2 + O\cdot \rightarrow O_3 \qquad (14)$$

Ozone reacts extremely rapidly with compounds containing double bonds to give peroxidic compounds (Chapter 1) and Pryor et al. have observed that (as in the case of the motor car tyre), ozone is almost certainly consumed as it contacts the surface of the air/lung interface [256]. However, as in the case of the polyunsaturated elastomers, damage is not limited to the surface layers and the primarily produced peroxidic species are stable enough to penetrate further into the lung tissues where they mediate more long-term oxidative damage [257]. Ozone is thus a very potent ROS in biological systems and has been shown to react with PUFA to give free radicals [258] and to stimulate protein synthesis in the lung [259].

NO_x also reacts with double bonds (Section 1.2.3) and has been shown to initiate lipid peroxidation [260]. One of the products of this reaction, HNO_2, has been shown to be an antioxidant in technological systems by virtue of its ability to reduce peroxyl radicals and in some systems this is an antioxidant process (Chapter 3):

$$O=NOH + ROO\cdot \rightarrow NO_2 + ROOH \qquad (15)$$

Sulphur oxides are antioxidants in the absence of light, since they have the ability to destroy hydroperoxides catalytically as do the derived sulphites, sulphinic esters, etc. The latter are also effective CB-D antioxidants (Chapter 4). However, in some circumstances they may reduce oxygen to superoxide [261,262]:

$$\text{HOSO}^- \underset{H^+}{\overset{}{\longrightarrow}} \begin{array}{l} \text{(a) ROO·} \longrightarrow \overset{O}{\underset{\|}{\text{HOSO·}}} + \text{ROOH} \\ \\ \text{(b) O}_2 \longrightarrow \overset{O}{\underset{\|}{\text{HOSO·}}} + \text{O}_2\text{·}^- \end{array} \qquad (16)$$

The triplet state of SO_2 which is readily formed in sunlight is also a powerful prooxidant which can readily hydrogen abstract from labile C-H bonds.

Until recently, diesel engines were thought to have environmental advantages over petrol engines in that they did not produce lead in the effluent, they have low CO and hydrocarbon emissions and NO_x emissions are comparable to cars fitted with catalytic converters [262]. However, they produce much higher levels of aerosol soot particles than petrol engines and these have been shown to be the cause of asthma in railway workers [263] and to be carcinogenic in animals [264]. Both adsorbed NO_x [265,266] and adsorbed SO_2 [261] have been shown to be involved in cytotoxic effects and Hippeli et al. [262] have proposed that reaction (16(b)) is catalysed by redox functional groups in the diesel particles. These have not so far been unequivocally identified but quinones are implicated [267,268] and it is already clear that not all carbon particles are equally redox active and hence equally capable of generating ROS. This behaviour is typical of carbon particles formed in combustion processes. Typical structures contain oxidising (e.g. quinone) and reducing (e.g. phenolic) components which vary in a complementary way depending on the environment. An important consequence of the effects of prooxidant loaded diesel particles in the lung is to deplete essential antioxidants, notably ascorbic acid, uric acid and a-tocopherol [269,270]. Redox cycling of quinones/hydroquinones has also been proposed by Zang et al. [271] to explain the free radical reactivity of cigarette smoke.

It has been known since the 1970s that asbestos causes mesothelioma, a form of cancer which has been the cause of many deaths in recent years. The disease appears to be associated with silicate fibres of critical length which are phagocytosed by pulmonary macrophages with associated radical formation [272,273]. Asbestos fibres also contain iron either as part of the structure ($Na_2O \cdot Fe_2O_3 \cdot 3FeO \cdot SiO_2$) or as a surface contaminant and it has been suggested [274] that free iron catalyses the formation of hydroxyl radicals by the Fenton reaction. Kinnula et al [276] have shown that hydrogen peroxide is involved in mesotheloima and that added antioxidants, notably catalase and glutathione peroxidase have reduced cell damage, although these workers did not relate radical formation to the presence of iron.

A very similar explanation has been proposed recently by Dalal et al. [275] to account for coal worker pneumoconiosis. Coal dusts gave rise to hydroxyl in the presence of hydrogen peroxide and by spin trapping the ·OH concentration was shown to be related to the iron content and the rate of lipid peroxidation was found to be increased in the same order as ·OH generation. Very significantly catalase and the iron chelator desferrioxamine both effectively inhibited ·OH formation.

2.9.5 Physical exercise
Oxygen uptake by the body increases several-fold during physical exercise. Most of this is utilised in energy production in the mitochondria particularly in the muscles.

However between 2% and 5% of the oxygen may be converted to reactive oxygen species (ROS) via superoxide reduction. Davies et al. [277] showed by ESR that free radical concentrations increased in the leg muscles of rats subjected to treadmill running to exhaustion and this has since been confirmed by other ESR studies [278,279]. Increased radical concentrations could be a direct result of mechanochemical scission of macromolecules in the muscles similar to that observed in polymers under conditions of mechanical stress or "fatigue" (Chapter 1, Section 1.6.2(c) and Chapter 4, Section 3.5.1(d)) and which has also been seen to occur in the ultrasonic (mechanochemical) degradation of DNA [280]. In polymers (including DNA) this is an initiating process for peroxidation.

The detection and quantification of ESR signals is not unequivocal evidence of peroxidative damage but this has been obtained *in vivo* by the detection of increased ethane and pentane in the breath and of MDA in the tissues of human subjects after severe exercise [277,281-283]. The elevated formation of 8OHdG which indicates oxidative damage to DNA has also been observed in marathon competitors [283].

The effect of exercise on the endogenous antioxidants will be discussed in Chapter 6 but it should be noted here that the depletion of GSH with concomitant increase in GSSG is a valuable marker of intense endurance exercise in both man [284] and animals [2282], and the ratio [GSSG]/[GSH] has been found to increase in proportion to the severity of the exercise under conditions of prolonged exercise but not in severe short-term exercise. There is little doubt then that oxidative stress is considerably increased in mammals during severe exercise.

2.10 Ageing

Ageing is not strictly a disease, although it is associated with an increase in disease (e.g. cardiovascular disease, cancer, Alzheimer's Disease, Parkinson's Disease, etc.) and may in part be a result of disease. During the past four hundred years the average life-span of human beings has increased from about 40 years to 75 [285,286]. The primary reason for this has been the increasing control by medical science of life-threatening diseases and the reduction of other non-medical environmental hazards. Over this same period, however, there has been almost no increase in maximum life-span (life-span potential, LSP); the overall effect has been rather to push the average age of the population upward without affecting the age at which the oldest die.

The evidence seems to suggest that ageing is at least in part genetically controlled [286-288] and many biologists and some biochemists believe that genetic control is absolute. There are two rather different versions of this theory:

a) There is a genetically controlled limit to cell growth and hence a predestined cell death. This is sometimes called the "death gene" theory [286].
b) There is a genetically controlled reduction in the body's oxidative defence mechanisms with ageing due to the programmed reduction in the synthesis of the endogenous antioxidants or the reduced absorption of the exogenous antioxidant vitamins [288,289].

It has been reported that "super fruit flies" can be selectively bred which live twice as long as normal members of the species. They may achieve this by producing an unusually active version of superoxide dismutase (SOD) [290] which catalyses the conversion of superoxide to hydrogen peroxide and in combination with the ubiquitous antioxidant, catalase, SOD is a powerful antioxidant (see Chapter 5). Genetic manipulation to improve antioxidant defences is clearly a possibility but is outside the scope of this book, the primary purpose of which is to examine the possibility of developing improved antioxidant systems based on an understanding of how cells are destroyed by free radicals and how antioxidants act.

It was seen in previous sections that reactive oxygen species (ROS) produced by reduction of molecular oxygen are involved in the detoxification of invading organisms and chemicals. But they also initiate lipid peroxidation in healthy cells and are known to be involved in the diseases of old age, notably cancer, Alzheimer's disease, Parkinson's disease and rheumatoid arthritis [29]. The natural antioxidant defences in healthy individuals are generally adequate to neutralise "stray" radicals, but these tend to decrease in concentration with age so that the level of oxidation products in the body also increases with age.

2.10.1 Chemical evidence for peroxidation during ageing

Some of the more important oxidation products of polyunsaturated fatty acids are now recognised to correlate with the irreversible changes associated with ageing that occur in human cells [288]. Products that have been identified include malondialdehyde (MDA), pentane and ethane which are the common oxidation products of 1,4-dienes containing three or more double bonds (see Section 2.2). MDA is believed to be a major source of lipofuscin (see Scheme 2.2), dark coloured products formed by reaction with protein amino groups in increasing quantity in many organs in the body with age but in particular in the nerve cells, in the brain and in the heart [288].

The rate of lipofuscin formation in different species correlates generally with the life-span of the species. Thus, the rate of lipofuscin formation in the hearts of dogs is about five times faster than that of humans and roughly corresponds to the difference in their life-spans [291]. Lipofuscin formation has been shown to be associated with activity. Human postural muscles have lower amounts than muscles involved in movement [292]. Houseflies deprived of their wings live approximately 2.5 time longer than normal flies [293,294]. These observations suggest that peroxidation is having a detrimental effect during exercise; and lipid peroxidation products (e.g. pentane) have been found to be elevated in the breath of exercised humans [295], and these were found to be reduced by dietary supplementation with vitamin E. Although MDA in exercised rodents was elevated, training has been found to increase glutathione peroxidase and reductase, catalase and SOD in exercised rats [296] even although vitamin E levels are depleted [297].

Peroxidative stress leads to physical changes in the cell membrane due to chemical cross-linking [298]. This involves not only the dimerisation of the lipids themselves through the polymerisation processes known for many years in the "drying" of polyunsaturated oils (see Chapter 1), but is also due to the cross-linking of proteins in the process of lipofuscin formation and the oxidation of protein thiols to disulphides. These processes all intensify with age but are reduced by life-long dietary restriction and are independent of cholesterol levels [298-300].

2.10.2 Metabolic rate, peroxidation and ageing

There is a correlation between the metabolic rate of animals and the rate of oxidative damage as measured by the formation of oxidation products (MDA, paraffins and lipofuscin). Cutler has shown that the specific metabolic rate of a species, (SMR), defined as the rate of oxygen consumption per unit body weight, is inversely related to the life span potential (LSP) of the species [286]. This is shown for the mammals in Fig. 2.4.

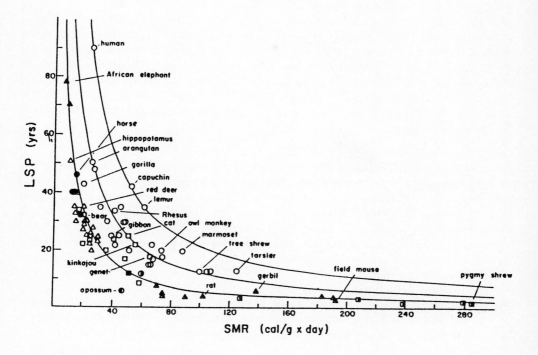

Fig. 2.4 Life-span potential (LSP) of animals as a function of specific metabolic rate (SMR). o - primates; △ artiodactyla; □ carnivora; ● perissodactyla; ▲ proboscida; ■ hydrocoidea; ◐ logomorpha; ▲ rodentia; ◨ insectivora; ◌ marupilia. (Reproduced with permission from R. G. Cutler, *Free Radicals in Biology*, Vol. VI, Ed. W. A. Pryor, Academic Press, New York, 1984, Chapter 11).

The life-span energy potential (LSP x SMR) is about 920 kJ/g for most mammals, but humans have a much higher value (3850 kJ/g), suggesting that humans are more resistant to the ageing effects of oxygen. Cutler and co-workers found that the concentrations of several antioxidants in the livers of primates and rodents are strongly correlated with life-span energy potential (LEP) [301]. This is shown typically for SOD in figure 2.5. A similar relationship with LEP has also been found for uric acid, a powerful water-soluble antioxidant in the primates and a positive but non-linear relationship for β-carotene. α-Tocopherol concentrations in the plasma of mammals also showed a good correlation but ascorbic acid in mammalian tissues did not. This may be a consequence of an imbalance in the antioxidant defences since vitamin C is known to be an effective prooxidant in the presence of transition metal ions (Chapter 5). Cellular glutathione, a peroxidolytic antioxidant, has also been shown to be reduced with age [302].

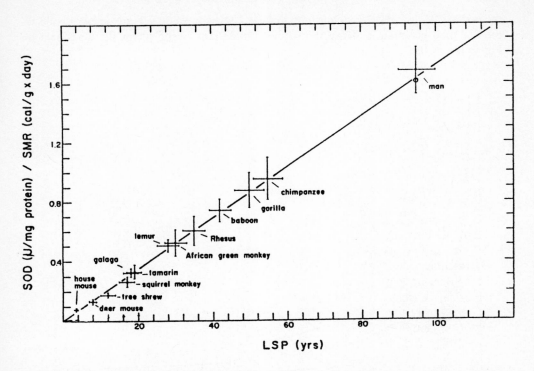

Fig. 2.5 Relationship between life-span energy potential (LEP) of animals and the concentration of SOD in the liver. (Reproduced with permission from R.G. Cutler, *Free Radicals in Biology*, Vol. VI, Ed. W. A. Pryor, Academic Press, New York, 1984, Chapter 11).

REFERENCES

1. G.Poli, F. Biasi, R. Carini and E. Chiarpotto in *Free Radicals in the Environment, Medicine and Toxicology*, Eds. H.Nohl, H.Esterbauer and C.Rice-Evans, Richelieu Press, London, 1994, p. 493 et seq.
2. K.F. Gey, G.B. Brubacher and H.B. Stähelin, *Am. J. Clin. Nutr.*, **45**, 1368 (1987).
3. K.F. Gey, U.K. Moser, P. Jordon, H.B. Stähelin, M. Eichholzer and E. Lüdin, *Am. J. Clin. Nutr.*, **57** (suppl)., 787S-97S (1993).
4. K.F. Gey, *Brit. Med. Bull.*, **49**, No.3, 679 (1993)
5. D.G. Pobedimskij and E.B. Burlakova in *Atmospheric Oxidation and Antioxidants*, Vol III, Ed. G. Scott, Elsevier, Amsterdam, 1993, Chapter 9.
6. R.H. Burdon, *Free Radicals Biol. & Med.*, **18**, 775 (1995)
7. P.G. Winyard, D. Perrett, G. Harris and D.R. Blake in *Biochemistry of Inflammation*, Eds. J.T. Whicher and S.W. Evans, Kluwer Academic Publishers, 1992, Chapter 6.
8. E. Niki in *Atmospheric Oxidation and Antioxidants*, Vol.III, Ed. G. Scott, Elsevier, Amsterdam, 1993, p.5.
9. D.C. Borg in *Oxygen Free Radicals in Tissue Damage*, Eds. M. Tarr and F. Samson, Birkhäuser, Boston, 1993, Chapter 2.
10. E. Niki in *Atmospheric Oxidation and Antioxidants*, Vol.III, Ed. G. Scott, Elsevier, Amsterdam, 1993, p.7.
11. W.A. Pryor, *Ann. Rev. Physiol.* **48**, 657-667 (1986).
12. B. Frei, R. Stocker and B.N. Ames, *Proc. Nat. Acad. Sci.*, quoted in Ref. 8, p.22.
13. G.A. Reed in *Atmospheric Oxidation and Antioxidants*, Vol. III, Ed. G. Scott, Elsevier, Amsterdam, 1993, Chapter 7.
14. N. Morisali, H. Sprecher, G.E. Milo and D.G. Cornwell, *Lipids*, **17**, 893 (1982)
15. R.P. Mason, B. Kalyanaraman, B.E. Tainer and T.E. Eling, *J.Biol. Chem.*, **255**, 5019-5022 (1980).
16. J. Schreiber, T.E. Eling and R.P. Mason, *Arch. Biochem. Biophys.*, **249**, 126-136 (1986).
17. H.W. Gardner in *Free Radicals: Chemistry, Pathology and Medicine*, Eds. C. Rice-Evans and T. Dormandy, Richelieu Press, London, 1988, 75 et seq.
18. G. Scott, *Atmospheric Oxidation and Antioxidants*, Elsevier, 1965, p.24 et seq.
19. W. Kern and A.R. Heinz, *Makromol. Chem.*, **16**, 81 (1985).
20. G. Scott, *Atmospheric Oxidation and Antioxidants*, Elsevier, Amsterdam, 1965 p.374.
21. S.P. Kochhar in *Atmospheric Oxidation and Antioxidants*, Vol.II, Ed. G. Scott, Elsevier, Amsterdam, 1993, Chapter 2.
22. G.Scott, *Atmospheric Oxidation and Antioxidants*, Elsevier, Amsterdam, 1965, p.359.
23. McCance and Widdowson's *The Composition of Foods*, 5th edition, Eds. B. Holland, A.A. Welch, I.D. Unwin, D.H. Buss, A.A. Paul and D.A.T. Southgate, Royal Soc. Chem., 1994.
24. P.B. Addis and G.J. Warner in *Free Radicals and Food Additives*, Eds. O.I. Aruoma and B. Halliwell, Taylor and Francis Ltd., 1991
25. M.K. Shigenaga and B.N. Ames in *Natural Antioxidants in Human Health and Disease*, Ed. B. Frei, Academic Press, San Diego, 1994, Chapter 3.

26. A. Boveris, N. Oshino and B. Chance, *Biochemical J.*, **128**, 617-630 (1972).
27. J.P. Kehrer and C.V. Smith in *Natural Antioxidants in Human Health and Disease*, Ed. B. Frei, Academic Press, San Diego, 1994, Chapter 2.
28. A. Boveris and B. Chance, *Biochem. J.*, **134**, 707-716 (1973).
29. G.G. Duthie, *Europ. J. Clin. Nutrit.*, **47**, 759-764 (1993).
30. N.T. Dixon and R.O.C. Norman, *Nature*, **196**, 891-92 (1962).
31. I. Yamazaki in *Free Radicals in Biology*, Vol. III, Ed. W.A. Pryor, Academic Press, New York, 1977, pp.183-218.
32. N. Klimes, C. Lassmann and B. Ebert, *J. Magn. Res.*, **37**, 53-59 (1980).
33. P.J. Thornalley in *Atmospheric Oxidation and Antioxidants*, Vol. III, Ed. G. Scott, Elsevier, Amsterdam, 1993, Chapter 2.
34. E.G. Janzen and B.J. Blackburn, *J. Am. Chem. Soc.*, **91**, 4481-4490 (1969).
35. J.V. Bannister, P. Bellavite, M.C. Serra, P.J. Thornally and F. Rossi, *FEBS Lett.*, **145**, 323-6 (1981).
36. H.A. O. Hill and P.J. Thornally, *FEBS Lett.*, **125**, 235-8 (1981).
37. C.M. Arroyo, J.H. Kramer, B.F. Dickens and W.B. Weglicki, *FEBS Lett.*, **221**, 101-4 (1987).
38. K.R. Maples, S.J. Jordan and R.P. Mason, *Mol. Pharmacol.*, **33**, 344-50 (1988).
39. G. Beuttner, *Free Rad. Biol. Med.*, **3**, 259-303 (1987).
40. B. Halliwell and J.M.C. Gutteridge, *The Importance of Free Radicals and Catalytic Metal Ions in Human Disease, Molecular Aspects of Medicine*, **8**, 89-193 (1985).
41. S.F. Wong, B. Halliwell, R. Richmond and W.R. Skowroneck, *J. Inorg. Biochem.*, **14**, 127-34 (1981).
42. T. Asakawa and S. Matsushita, *Lipids*, **14**, 401-6 (1979).
43. A.L. Tappel in *Free Radicals in Biology*, Vol. 4 Ed. W.A. Pryor, Academic Press, New York, 1979, pp. 1-85.
44. R.K. Brown and F.J. Kelly, *Pediatric Res.*, **36**, 487-93 (1994).
45. G.D. Lawrence and G. Cohen, *Anal. Biochem.*, **122**, 283-90 (1982).
46. A.L. Tappel and C.J. Dillard, *Fed. Proc.*, **40**, 174-8 (1981).
47. M. Dizdaroglu, *Mutat. Res.*, **275**, 331-342 (1992).
48. S.S. Wallace, *Br. J. Cancer*, **55** (**suppl. VIII**), 111-25 (1987).
49. A.P. Grollman and M. Moriya, *Trends Gen.*, **9**, 246-9 (1993).
50. M. Dizdaroglu, *Methods Enzymol.*, **234**, 3-16 (1994).
51. A.P. Breen and J.A. Murphy, *Free Rad. Biol. Med.*, **18**, 1033-77 (1995).
52. K.F. Gey, *Bibl. Nutr. Dieta.* **37**, 53 (1986); R.A. Riemersma, D.A. Wood, C.C.A. Macintyre, R.A. Elton, K.F. Gey and M.F. Oliver, *Lancet*, **337**, 1, (1991).
53. S.R.J. Maxwell, *J. Clin. Pharm. and Therapeutics*, **18**, 85 (1993).
54. S. Parsarathy in *Free Radicals in the Environment, Medicine and Toxicology*, Eds. H. Nohl, H.Esterbauer and C.Rice-Evans, Richelieu Press, London, 1994, p.163-179.
55. M.J. Mitchinson, *BJPC*, **48**, 149 (1994).
56. C.B. Taylor, S.K. Peng, N.T. Werthessen, P. Tham and K.T. Lee, *Am. J. Clin. Nutr.*, **32**, 40-57 (1979).
57. P.B. Addis and S.W. Park in *Food Toxicology: A Perspective on the Relative Risks*, 1989, p.297-330.
58. J. Edington, M. Geekie, R. Carter, L. Benfield, K. Ficher, M. Ball and J. Mann, *Brit. Med. J.*, **294**, 333-6 (1987)

References

59. R.Y. Ball, K.L.H. Carpenter, J.H. Enright, S.L. Hartley and M.J. Mitcheson, *Brit. J. Exp. Path.*, **68**, 427 (1987).
60. K.L.H. Carpenter, J.A. Ballantine, B. Fussel, J.H. Enright and M.J. Mitcheson, *Atherosclerosis*, **83**, 217 (1990).
61. V.C. Reid, C.E. Brabbs and M.J. Mitchinson, *Atherosclerosis*, **92**, 251 (1992).
62. S.K. Peng, C.B. Taylor, J.C. Hill and R.J. Morin, *Atherosclerosis*, **54**, 121-33.
63. M.S. Jacobson, M.G. Price, A.E. Shamoo and F.P. Heald, *Atherosclerosis*, **57**, 209-17 (1985).
64. J.K. Donnelly and D.S. Robinson, *Free Radical Research*, **22**, 147-176 (1995).
65. M.C. Bellizzi, M.F. Franklin, G.G. Duthie and W.P.T. James, *Europ. J. Clin. Nutr.*, in press, by kind permission of the authors.
66. G.G. Duthie, J.A.G. Beattie, J.R. Arthur, M. Franklin, P.C. Morrice and W.P.T. James, *App. Nutr. Investig.*, **10**, 313-6 (1994).
67. *Free Radical Substitution Reactions*, Eds. K.U. Ingold and B.P. Roberts, Wiley Interscience, New York, 1970, pps. 148-153.
68. A. Padwa and L. Brodsky, *Tetrahedron Lett.*, 1045-1048 (1873).
69. H. Hart and P.B. Lavrick, *J. Org. Chem.*, **39**, 1793-4 (1974).
70. *Nutritional Aspects of Cardiovascular Disease*, Department of Health, 1994, HMSO, London.
71. A. Keys, J.T. Anderson and F. Grande, *Lancet*, 959-56 (1957); ibid, Metabolism, **14**, 747-87 (1965); D.M. Hegsted, R.B. McGandy, M.L. Myers and F.J. Stare, *Am. J. Clin. Nutr.*, **17**, 281-95.
72. M.J. Martin, S.B. Hulley, W.S. Browner, L.H. Kuller and D. Wentworth, *Lancet*, **2**, 933-6 (1986).
73. V. Goldbourt, E. Holtzmann and H.N. Neufeld, *Brit. Med. J.*, **290**, 1239-43 (1985).
74. D. Reed, K. Yano and A. Kagan, *Am. J. Med.*, **80**, 871-8.
75. G. Rose and M. Shipley, *Brit. Med. J.*, **293**, 306-7 (1986).
76. S.J. Pocock, A.G. Shaper and A.N. Phillips, *Brit. Med. J.*, **298**, 998-1002.
77. A. Keys, *Circulation*, **41** (4 Suppl.), 11-211 (1970)
78. T.L. Robertson, H. Katio and T. Gordon, *Am. J. Cardiol.*, **39**, 244-9 (1977).
79. E.M. Berry, S. Eisenberg, D. Haratz, Y. Friedlander, Y. Norman, N.A. Kaufman and Y. Stein, *Am. J. Clin. Nutr.*, **53**, 899-907 (1991).
80. P. Reaven, S. Parthasarathy, B.J. Grassie, et al., *Am. J. Clin. Nutr.*, **54**, 701-6 (1991).
81. M. Abbey, G.B. Belling, M. Noakes, F. Hirata and P.J. Nestel, *Am. J. Clin. Nutr.*, **57**, 391-8 (1993).
82. *The Scottish Diet*, Report of a Working Party to the Chief Medical Officer for Scotland, The Scottish Office, December 1993.
83. A. Shaper, *Coronary Heart Disease: Risk and Reasons*, Current Medical Literature Ltd., 1989.
84. D.F. Church and W.A. Pryor, *Environm. Health Perspect.*, **64**, 111 (1985).
85. L.Y. Zhang, K. Stone and W.A. Pryor, *Free Rad. Biol. Med.*, **19**, 161-7 (1995).
86. G.G. Duthie, J.R. Arthur and W.P.T. James, *Am. J. Clin. Nutr.*, **53**, 1061S (1991).
87. E.R. Pacht, H. Kaseki, J.R. Mohamed, D.G. Cornwell and W.B. Davis, *J. Clin. Invest.* **77**, 786 (198.).
88. C.K. Chow, R.R. Thacker, C. Changchit, R.B. Brigges, S.R. Rehm, J. Humble and J. Thurber, *J. Am. Coll. Nutr.*, **5**, 305 (1986).

89. T.W. Kensler, N.E. Davidson and K.Z. Guyton in *Atmospheric Oxidation and Antioxidants*, Vol. III, Ed. G. Scott, Elsevier, 1993, Chapter 12.
90. P. Wardman in *Atmospheric Oxidation and Antioxidants*, Vol. III, Ed. G. Scott, Elsevier, Amsterdam 1993, Chapter 4.
91. L. Benade, T. Howard and D. Burk, *Oncology*, **23**, 33-43 (1969).
92. G. Block and R. Schwartz in *Natural Antioxidants in Human Health and Disease*, Ed. B. Frei, Academic Press, 1994, Chapter 11.
93. M. Dizdaroglu, G. Rao, B. Halliwell and E. Gajewski, *Arch. Biochem. Biophys.*, **285**, 317-324 (1991).
94. B. Halliwell, *FASEB J.*, **1**, 358-364 (1987).
95. A. Samuni, J. Aronovitch, D. Godimger, M. Chevion and G. Czapski, *Europ. J. Biochem.*, **137**, 119-24 (1983).
96. G.M. Makrigiorgos, E. Bump, C. Huang, J. Baranonowska-Kortylewicz and A.I. Kassis, *Free Rad. Biol. Med.*, **18**, 669-78 (1995).
97. H.B. Demopoulos, D.D. Pietronigro, E.S. Flamm and M.L. Seligman in *Cancer and the Environment*, Eds. H.B. Demopoulos and M.A. Mehlman, Pathotox, 1980, p.273
98. T.J. Slaga, A.J.P. Klein-Szanto, L.L. Triplett, L.P. Yotti and J.E. Trosko, *Science*, **213**, 1023-5 (1981).
99. J.F. O'Connell, A.J.P. Klein-Szanto, D.M. DiGiovanni, J.W. Freis and T.J. Slaga, *Cancer Res.*, **46**, 2863-5 (1986).
100. L.W. Wattenberg, *J. Nat. Cancer Inst.*, **48**, 1425 (1972)
101. L.W. Wattenberg, *J. Nat. Cancer Inst.*, **70**, 11 (1978).
102. K.K. Carrol and H.T. Khor, *Prog. Biochem. Pharmacol.* **10**, 308 (1975).
103. W.M. King, D.M. Bailey, D.D. Gibson, J.V. Pitha and P.B. McCay, *J. Nat. Cancer Inst.*, **63**, 657 (1979).
104. E.L. Wynder in *Cancer and the Environment*, Eds. H.B. Demopoulos and M.A. Mehlman, Pathotox, 1980, p.171-192.
105. A. Tannenbaum, *Cancer Res.*, **2**, 468-75 (1942).
106. K.K. Carrol in *Cancer and the Environment*, Eds. H.B. Demopoulos and M.A. Mehlman, Pathotox, 1980, p. 254.
107. P.M. Newberne and J.F. O'Connell in *Cancer and the Environment*, Eds. H.B. Demopoulos and M.A Mehlman, Pathotox, 1980, p.333.
108. E.R. Pinckney, *Am. Heart J.*, **85**, 723-6 (1973).
109. S.N. Gershoff, *J. Am. Vet. Med. Assoc.*, **166**, 455 (1975).
110. K.K.Carroll and G.J. Hopkins in *Advances in Medical Oncology, Research and Education*, Ed. B.W. Fox, Pergamon Press, Oxford, Vol. 5, Basis for Cancer Therapy 1, 1979, p.221-8.
111. C.W. Welsch, *Free Rad. Biol. Med.*, **18**, 757-73 (1995).
112. I.J. Tinsley, J.A. Schmitz and D.A. Pierce, *Cancer Res.* **41**, 1460-5 (1981).
113. M.J. Gonzalez, R.A. Schemmel, J.L. Gray, L. Dugan, L.G. Sheffield and C.W. Welsch, *Carcinogenesis*, **12**, 1231-5 (1991).
114. M.J. Gonzalez, R.A. Schemmel, L. Dugan, J.I. Gray and C.W. Welsch, *Lipids*, **18**, 827-32 (1993).
115. C.W. Welsch, C.S. Oakley, C.C. Chang and M.A. Welsch, *Nutr. Cancer*, **20**, 119-27 (1993).
116. D.P. Rose, H.A. Hatale, J.M. Connolly and J. Rayburn, *Cancer Res.*, **53**, 4686-90 (1993).

117. M. Begin, G. Ellis and D.F. Horrobin, *J. Nat. Cancer Inst.*, **80**, 188-94 (1988).
118. U.N. Das, *Cancer Lett.*, **56**, 235-43 (1991).
119. S. Takeda, D.F. Horrobin, M. Manku, P.G. Sim, G. Ellis and V. Simmons, *Anticancer Res.* **12**, 329-34 (1992).
120. A.W. Bull, N.D. Nigro, W.A. Golombieski, J.D. Krissman and L.J. Marnett, *Cancer Res.*, **44**, 4924-8 (1984).
121. R.H. Burdon, V. Gill and C. Rice-Evans, *Free Radicals Res. Comm.*, **7**, 149-59 (1989).
122. V. Chajès, M. Mahon and G.M. Kostner, *Free Rad. Biol. Med.*, **20**, 113-20 (1996).
123. M.K. Horwitt, *Am. J. Clin. Nutr.*, **27**, 1182 (1974).
124. K.F. Gey in *Free Radicals in the Environment, Medicine and Toxicology*, Eds. H. Nohl, H. Esterbauer and C. Rice-Evans, Richelieu Press, London, 1994, p.181
125. D.G. Cornwell and N. Morisaki in *Free Radicals in Biology*, Ed. W. Pryor, Academic Press, New York, 1984, Chapter 4.
126. *Dietary Guidelines to know your cancer risk*, World Cancer Research Fund, 1994.
127. P.B. McCay, M. King, L.F. Rikans and J.V. Pitha in *Cancer and the Environment*, Eds. H.B. Demopoulos and M.A. Mehlman, Pathitox, 1980, p.451.
128. A.H. Conney, *Cancer Res.* **42**, 4875 (1982).
129. E.C. Cavialieri and E.G. Rogan in *Free Radicals in Biology*, Ed. W. Pryor, Academic Press, Chapter 10.
130. G.A. Reed in *Atmospheric Oxidation and Antioxidants*, Ed. G. Scott, Elsevier Sci. Pub., New York, 1993, p.198.
131. T.W. Kensler, N.E. Davidson and K.Z. Guyton in *Atmospheric Oxidation and Antioxidants*, Ed. G. Scott, Elsevier Sci. Pub., London, 1993, p.333.
132. L.J. Marnett, J.T. Johnson and M.J. Bienkowski, *FEBS Lett.*, **106**, 13-16 (1979).
113. M.T. Smith, J.W. Yager, D.A. Eastmond, *Environ. Health Perspect.*, **82**, 23 (1989).
133. K. Frenkel, L. Wei and H. Wei, *Free Rad. Biol. Med.*, **19**, 373-80 (1995)
134. W.F. Greenlee, J.D. Sun and J.S. Bus, *Toxicol. App. Pharm.*, **59**, 187 (1989).
135. R.D. Irons, *J. Toxicol. Environ. Health*, **16**, 673 (1985).
136. L. Zhang, M.T. Smith, B. Bandy, S.J. Tamaki and A.J. Davison in *Free Radicals in the Environment, Medicine and Toxicology*, Eds. H. Nohl, H. Esterbauer and C. Rice-Evans, Richelieu Press, London, 1994, p.521.
137. E.L. Wynder in *Cancer and the Environment*, Eds. H.B. Demopoulos and M.A. Mehlman, Pathitox, 1980, 171
138. W.A. Pryor, K. Uehara and D.F. Church in *Oxygen Radicals in Chemistry and Biology*, Eds. W. Bors, M. Saran and D. Tact, de Gruyter, New York, 1984, p. 193.
139. M.R. Clemens, H. Einsele, C. Ladner and H.D. Waller in *Free Radicals in Chemistry, Pathology and Medicine*, Eds. C. Rice-Evans and T. Dormandy, Richelieu Press, London, 1988, p.393.
140. B.M. Barbior, *N. Engl. J. Med.*, **298**, 659-68 (1978).
141. B.M. Barbior, *N. Engl. J. Med.*, **298**, 721-5 (1978).
142. A. Bendich in *Natural Antioxidants in Human Health and Disease*, Ed. B. Frei, Acad. Press, San Diego, Chapter 15.
143. J.M.C. Gutteridge and B. Halliwell in *Atmospheric Oxidation and Antioxidants*, Ed. G. Scott, Elsevier, Amsterdam, Vol III, 1993, Chap. 3.

144. S.E. Edmonds, D.R. Blake, C.R. Morris and P.G. Winyard, *J. Rheumatol.* **20**, (Suppl., 37), 26 (1993).
145. V.R. Winrow, P.G. Winyard, C.J. Morris and D.R. Blake, *Brit. Med. Bull.*, **49**, 506, 522 (1993).
146. H. Sanfey, N.G. Sarr, N.G. Bulkley and J.H. Cameron, *Acta Physiol. Scandanavica*, **Suppl., 548**, 109 (1986).
147. H. Sanfey in *The Pathogenesis of Pancreatitis*, Ed. J.M. Braganza, Manchester University, 1991, Chap. 5,
148. J.M. Braganza in *Free Radicals, Chemistry, Pathology and Medicine*, Eds. C. Rice-Evans and T. Dormandy, Richelieu Press, London, 1988, p.357.
149. J.M. Braganza in *Recent Advances in Gastroenterology* 6, Ed. R. Pounder, Churchill Livingstone, 1986, pp.251-280.
150. K.A. Thompson and M.L. Hess, *Prg. Cardiovasc. Dis.*, **28**, 449 (1986).
151. J.M. McCord, *N.Eng. J. Med.*, **312**, 159 (1985).
152. R. Kloner, K. Przyklenk and P. Whittaker, *Circulation*, **80**, 1115 (1989).
153. M.H.N. Golden and D. Ramdath, *Proc. Nutr. Soc.*, **46**, 53 (1987).
154. M.H.N. Golden, D.D. Ramdath and B.E. Golden in *Trace Elements, Micronutrients and Free Radicals*, Ed. J.E. Dreosti, Humana Press, 1991, Ch.9.
155. D.I. Thurnham, S. Koottathep and D.A. Adelekan in *Free Radicals, Chemistry, Pathology and Medicine*, Eds. C. Rice-Evans and T. Dormandy, Richelieu Press, London, 1988, p.161-185.
156. N.H. Hunt, C.M. Thumwood, I.A. Clark and W.B. Cowden in *Free Radicals, Chemistry, Pathology and Medicine*, Richelieu Press, London, 1988, p.405-414.
157. J.S. Eiseric, A. van der Vliet, B. Halliwell and C.E. Cross, *Biochem. Soc. Trans.*, **23**, 2383 (1995).
158. R.K. Brown and F.J. Kelly, *Thorax*, **49**, 738-42 (1994).
159. N.S. Dalal, J. Newman, D. Pack, S. Leonard and V. Vallyathan, *Free Rad. Biol. Med.*, **18**, 11-20.
160. V.L. Kinnula, K.O. Raivio, K. Linnainmaa, A. Ekman and M. Klockars, *Free Rad. Biol. Med.*, **18**, 391-99 (1995).
161. B. Halliwell and J.M.C. Gutteridge, *Free Radicals in Biology and Medicine, 2nd Ed.*, Clarendon Press, Oxford, 1989.
162. J.I. Gallin, I.M. Goldstein and R. Snyderman, *Inflammation: Basic Principles and Clinical Correlates*, Raven Press, New York, 1992.
163. *Biochemistry of Inflammation*, Eds. J.T. Whicher and S.W. Evans, Kluwer Academic Pub., 1992.
164. *The Respiratory Burst and its Physiological Significance*, Eds. A.J. Sbarra and R.R. Strauss, Plenum Press, New York, 1988.
165. B. Halliwell and J.M.C. Gutteridge, *Biochem. J.*, **219**, 1-14 (1984).
166. I. Ginsburg and R. Kohen, *Free Radicals Res.* **22**, 489-517 (1995).
167. V.R. Winrow, P.G. Winyard, C.J. Morris and D.R. Blake, *Brit. Med. Bull.*, **49**, 506-522 (1993).
168. L.L. Ingraham and D.L. Meyer, *Biochemistry of Dioxygen*, Plenum Pub. Corp., 1985, Chapter 15.
169. D. Metodiewa and H.B. Dunford in *Atmospheric Oxidation and Antioxidants*, Vol. III, Ed. G. Scott, Elsevier, Amsterdam, 1993, Chapter 11.

References

170. S.E. Edmonds, D.R. Blake, C.J. Morris and P.G. Winyard, *J. Rheumatol.*, **20**, Suppl. 37, 26-31 (1993).
171. A.L. Tappel and F.W. Summerfield, *Arch. Bochem. Biophys.*, **233**, 408-16 (1984).
172. Z. Zhang, A.J. Farrel, D.R. Blake, K. Chadwick and P.G. Winyard, *FEBS Letts*, **321**, 274-8 (1993)
173. J. Lunec, S.P. Halloran, A.G. White and T.L. Dormandy, *J. Rheumatol.*, **8**, 233-45, (1981).
174. D.A. Rowley, J.M.C. Gutteridge, D. Blake and B. Halliwell, *Clin. Sci.*, **66**, 691-5 (1984).
175. D.R. Blake, N.D. Hall, P.A. Bacon, P.A. Dieppe, B. Halliwell and J.M.C. Gutteridge, *Lancet*, **ii**, 1142-44 (1981).
176. D.R. Blake, P.J. Gallagher, A.R. Potter, M.J. Bell and P.A. Bacon, *Arthritis Rheum.*, **27**, 495-501 (1984).
177. B. Halliwell, J.M.C. Gutteridge and D. Blake, *Phil. Trans. Roy. Soc.*, **B311**, 659-71 (1985).
178. N. Giordano, A. Fioravanti, S. Sanscasciani, R. Marcolongo and C. Borghi, *Brit. Med. J.*, **289**, 961-2 (1984).
179. D.J. Hearse, *J. Mol. Cell. Cardiology*, **9**, 605-16 (1977).
180. J.M. McCord, *New Engl. J. Med.*, **312,** 159-63 (1984?1985) (check)
181. J.M. McCord, *Fed. Proc.*, **46**, 2402 (1987).
182. S.R. Jolly, W.J. Kane, M.B. Baillie, G.D. Abrahams and B.R. Lucchesi, *Circ. Res.*, **54**, 277-85 (1984).
183. M.S. Paller, J.R. Hoidal and T.F. Ferris, *J. Clin. Invest.*, **74**, 1156-64 (1984).
184. H. Sanfrey, G.B. Bulkley and J.L. Cameron, *Ann. Surg.*, **200**, 405-413 (1984).
185. A.S. Casale, G.B. Bulkley, B.H. Bulkley, J.T. Flaherty, V.L. Gott and T.J. Gardner, *Surg. Forum*, **34**, 313-6 (1983).
186. G. Littarru, M. Battino, S.A. Santini and A. Mordente in *Free Radicals in the Environment, Medicine and Toxicology*, Eds. H. Nohl, H. Esterbauer and C. Rice-Evans, Richelieu Press, London, 1994, pps. 258 et seq.
187. B. Kalyanaraman, E.A. Konorev, J. Joseph and J.E. Baker, ibid, p.313-326.
188. G. Cohen in *Handbook of Neurochemistry*, Vol. 4, 2nd edition, Ed. A. Laijtha, Plenum Press, New York, 1983, pp. 315-330.
189. R. Spector and J. Eells, *Fed. Proc.*, **43**, 196-200 (1984).
190. S.S. Panter, S.M. Sadrzadeh, P.E. Hallaway, J. Haines, V.E. Anderson and J.W. Eaton, *J. Exp. Med.*, **161**, 748-54 (1985)
191. B.D. Watson, R. Busto, W.J. Goldberg, M. Santiso, S. Yoshida and M.D. Ginsberg, *J. Neurochem.*, **42**, 268-74 (1984).
192. J.M. Braganza in *The Pathogenesis of Pancreatitis*, Ed. J.M. Braganza, Manchester University Press, 1991. Chapter 6.
193. T.J-K. Toung, M. Sendak, R.J. Traystman, M.C. Rogers, G.B. Bulkley and J.C. Cameron, *Circulation*, **74**, 11-91 (1986).
194. B. Houston in *The Pathogenesis of Pancreatitis*, Ed. J.M. Baganza, Manchester University Press, 1991, Chapter 8.
195. J.M. Braganza in *The Pathogenesis of Pancreatitis*, Ed. J.M. Braganza, Manchester University Press, 1991, Chapter 6.
196. J.M. Braganza in *The Pathogenesis of Pancreatitis*, Ed. J.M. Braganza, Manchester University Press, 1991, Chapter 13.

197. M.W. Konstan and M. Berger in *Cystic Fibrosis*, Ed. P.B. Davis, Marcel Dekker, New York, 1993, pps. 219-75.
198. P.B. Davis in *Cystic Fibrosis*, Ed. P.B. Davis, Marcel Dekker, New York, 1993, 193-218.
199. Y. Sato, N. Hotta, N. Saxamoto, N. Ohnishi and K. Yagi, *Biochem. Med.*, **21**, 104-7 (1979).
200. K. Yagi, *Chem. Phys. Lipids*, **45**, 337-51 (1987).
201. S.C. Langley, R.K. Brown and F.J. Kelly, *Pediatric Res.*, **33**, 247-50 (1993).
202. J.P. Neglia, C.L. Wielinski and W. Warwick, *J. Pediatr.*, **119**, 764-7 (1991).
203. R.K. Brown, A. McBurney, J. Lunec and F.J. Kelly, *Free Rad. Biol. Med.*, **18**, 801-806 (1995).
204. J.R. Wispe and R.J. Roberts, *Clin. Perinatol.*, **14**, 651-1240 (1987).
205. F.J. Felly, *Brit. Med. Bull.*, **49**, 668-78 (1993).
206. F.J. Kelly, W. Rodgers, J. Handel, S. Smith and M.A. Hall, *Brit. J. Nuitr.*, **63**, 631-8 (1990).
207. K.M. Silvers, A.T. Gibson and H.J. Powers, *Arch. Dis. Child*, **71**, F40-F44, (1994).
208. J.C.M. Gutteridge and B. Halliwell in *Atmospheric Oxidation and Antioxidants*, Vol. III, Ed. G. Scott, Elsevier, Amsterdam, 1993, Chapter 3.
209. J.M.C. Gutteridge, *Clinical Sci.*, **81**, 413-7 (1991).
210. H.J. Powers, A. Loban, K. Silvers and A.T. Gibson, *Free Radicals Res.*, **22**, 57-65 (1995).
211. H.M. Berger, S. Mumby and J.C. Gutteridge, *Free Radicals Res.*, **22**, 555-9 (1995).
212. J.D. Ballentine, *Pathology of Oxygen Toxicity*, Academic Press, New York, 1982.
213. B. Wolach, T.D. Coates, T.E. Hhugli, R.L. Baener and L.A. Boxer, *J. Lab. Clin. Med.*, **103**, 284-93 (1984).
214. J.F. Turrens, J.D. Crapo and B.A. Freeman, *J. Clin. Invest.*, **73**, 87-95 (1984).
215. *Inflammatory Bowel Disease: Current Status and Future Approach*, Ed. R.P. MacDermott, Elsevier, Amsterdam, 1988.
216. G.D. Buffinton and W.F. Doe, *Free Radicals Res.*, **22**, 131-143 (1995).
217. B.H. Lauterburg, M.E. Blizer, M.E. Rowedder and R.W. Inauen in *Inflammatory Bowel Disease: Current Status and Future Approach*, Ed. R. P. MacDermott, Elsevier, Amsterdam, 1988, pps 273-7.
218. C.F. Babbs, *Free Rad. Biol. Med.*, **13**, 169-181 (1992).
219. L. Charley, J. Foreman, D. Ramdath, F. Bennett, B. Golden and M. Golden, *West Ind. Med. J.*, **34 (suppl.)**, 62-3 (1985).
220. D S. McLaren in *Calorie Deficiencies and Protein Deficiencies*, Eds. R.A. McCance and E.M. Widdowson, Churchill, London, 1968, pp. 191-99.
221. F.R. Smith, D.S. Goodman, M.S. Zaklama, M.K. Gagr, S. El Maraghy and V.N. Patwardhan, *Am. J. Clin. Nutr.*, **26**, 973-87 (1973).
222. R.F. Burk, W.N. Pearso, R.P. Wood and F. Viteri, *Am. J.Clin. Nutr.*, **20**, 723 33, (1967).
223. R.J. Levine and R.E. Olson, *Proc. Soc. Exp. Biol. Med.*, **134**, 1030-4 (1970).
224. C. Murphey, B. Golden, D. Ramdath and H.N. Golden in *Trace Element Metabolism in Man and Animals*, Eds. L.S. Hurley, C.L. Keen, B. Lonnerdal and R.B. Rucker, Plenum, New York, 1987, pp.11-12.

225. C. Christie, G.T. Hoskins and D.E. MacFarlane, *West Ind. Med. J.*, **34(suppl.)**, 47 (1985).
226. A. Jacobs, *Semin. Hematol.*, **14**, 89-113 (1977).
227. G.D. McLaran, W.A. Muir and R.W. Kellermeyer, *CRC Crit. Rev. Clin. Lab. Sci.*, **19**, 205-66 (1983).
228. H.R. Schumacher, *Arthritis Rheum.*, **25**, 1460-7 (1982)
229. A.D. Heys and T.L. Dormandy, *Clin. Sci.*, **60**, 295-301 (1981).
230. J.M.C. Gutteridge, D.A. Rowley, E. Griffiths and B. Halliwell, *Clin. Sci.*, **68**, 463-7 (1985).
231. C. Hershko, G. Graham, G.W. Bates and E.A. Rachmliewitz, *Brit. J. Haematol.*, **40**, 255-63 (1978)
232. B. Modell, E.A. Letsky, D.M. Flynn, R. Peto and D.J. Weatherall, *Brit. Med. J.*, **284**, 1081-4 (1982).
233. K.J. Reszka, P. Bilski, C.F. Chignell and J. Dillon, *Free Rad. Biol. Med.*, **20**, 23-34 (1996).
234. E.R. Berman, *Lens Biochemistry of the Eye*, Plenum, New York, 1991.
235. G.F. Vile and R.M. Tyrrell, *Free Rad. Biol. Med.*, **18**, 721-30 (1995).
236. P. Morliere, A. Moysan and I. Tirache, *Free Rad. Biol. Med.*, **19**, 365-71 (1995).
237. J.P. Thomas and A.W. Girotti, *Photochem. Photobiol.*, **47**, 79S (1988).
238. M. Aubailly, R. Santus and S. Salmon, *Photchem. Photobiol.*, **54**, 769-73 (1991).
239. G. Scott in *Degradable Polymers: Principles and Applications*, Eds. G. Scott and D. Gilead, Chapman & Hall, London, 1995, Chapter 9.
240. G. Scott, *J. App. Polym. Sci. Symp*, **55**, 3-14 (1994).
241. J. Van der Zee, B.B.H. Krootjes, C.F. Chignell, T.M.A.R. Dubbelman and J. Stevenink, *Free Rad. Biol. Med.*, **14**, 105-13 (1993).
242. P. Wardman in *Atmospheric Oxidation and Antioxidants*, Vol. III, Ed. G. Scott, Elsevier, Amsterdam, 1993, Chapter 4.
243. J.F. Ward, *Radiat. Res.*, **86**, 185-95 (1981).
244. Perspectives in Radioprotection, Eds. J.F. Weiss and M.G. Simic, *Pharmacol. Ther.*, **39**, 1-414.
245. G.S. Pant and N. Kamada, *Hiroshima Med. J. Sci.*, **26**, 149-54 (1977).
246. I. Emerit in *Free Radicals, Lipoproteins and Membrane Lipids*, Eds. A. Crastes de Paulet, et al., 1990, pp.99-104.
247. I. Emerit, R. Arutyunyan, N. Oganesian, A. Levy, L. Cernjavsky, T. Sarkisian, A. Pogossian and K. Asrian, *Free Rad. Biol. Med.*, **18**, 985-91 (1995).
248. W.A. Pryor, *Br. Cancer J.*, **55**, Suppl. VIII, 19-23 (1987).
249. L-Y. Zhang, K. Stone and W.A. Pryor, *Free Rad. Biol. Med.*, **19**, 161-7, (1995).
250. D.L. Henshaw, *Contemp. Physics*, **34**, 31-48 (1993).
251. D.L. Henshaw, J.P. Eatough and R.B. Richardson, *Lancet*, **335**, 1008-12, (1990).
252. B.L. Cohen, *Health Physics*, **68**, 157-74 (1995).
253. *Air Quality Criteria for Photochemical Oxidants*, U.S. Natl. Air Pollution Admin., Publ. No. AP-63, US Government Printing Office, Washington DC, 1970.
254. T. Goodish, *Air Quality*, Lewis Pub., 1991.
255. S.E. Manahan, *Environmental Chemistry*, Lewis Pub., 1991.
256. W.A Pryor, G.L. Squadrito and M. Friedman, *Free Rad. Biol. Med.*, **19**, 935-41 (1995).

257. S.D. Langford, A. Bidani and E.M. Postlethwaite, *Toxicol. Appl. Pharmacol.*, **132**, 122-30 (1995).
258. W.A. Pryor, D.G. Prier and D.F. Church, *Environm. Res.*, **24**, 42-52 (1981).
259. F.J. Kelly and S. Birch in *Free Radicals in the Environment, Medicine and Toxicology*, Eds. H. Nohl, H. Esterbauer and C. Rice-Evans, Richelieu Press, London, 1994, pp.393-408.
260. W.A. Pryor and J.W. Lightsey, *Science*, **214**, 435-7 (1981).
261. S.C. Hippeli and E.F. Elstner, *Z. Naturforsch.*, **44c**, 514-23 (1989).
262. S. Hippeli, B. Blaurock, A.V. Preen and E.F. Elstner in *Free Radicals in the Environment, Medicine and Toxicology*, Eds. H. Nohl, H. Esterbauer and C. Rice-Evans, Richelieu Press, London, 1994, pp.375-92.
263. J.F. Wade and L.S. Newman, *J. Occ. Med.*, **35**, 149-54 (1993).
264. R.O. McClellan, *Ann. Rev. Pharmacol. Toxicol.*, **27**, 279-300 (1987).
265. T. Handa, T. Yamauchi, M. Ohnishi, Y. Hisamatsu and T. Ishii, *Environm. Int.*, **9**, 335-41 (1983).
266. T.R. Henderson, A.P. Li, R.E. Royer and C.R. Clark, *Environ. Mutagen.*, **3**, 211-20 (1981).
267. Y. Kumagai, J. Taira and M. Sagai, *Free Rad. Biol. Med.*, **18**, 365-71 (1995).
268. D. Schuetzle, *Environ. Health Perspect.*, **47**, 65-80 (1983).
269. J.P. Eiserich, A. Van der Vliet, B. Halliwell and C. Cross, *Biochem. Sci. Trans.*, **23**, 238S (1995).
270. B. Blaurock, S. Hippeli, N. Metz and E.F. Elstner, *Arch. Toxicol.*, **66**, 681-87 (1992).
271. L-Y. Zhong, K. Stone and W.A. Pryor, *Free Rad. Biol. Med.*, **19**, 161-7 (1995).
272. L.A. Goodlich and A.B. Kane. *Cancer Res.*, **50**, 5153-63 (1992).
273. D.W. Kemp, P. Graceffa, W. Pryor and S. Weitzman, *Free Rad. Biol. Med.*, **12**, 293-15 (1992).
274. M. Gulumian and J.A. Van Wyk, *Chem. Biol. Interact.* **62**, 89 (1987).
275. W.S. Dalal, J.N. Newman, D. Pack, S. Leonard and V. Vallyathan, *Free Rad. Biol. Med.*, **18**, 11-20 (1995).
276. V.L. Kinnula, K.O. Raivio, K. Linnainmaa, A. Ekman and M. Klockars, *Free Rad. Biol. Med.*, **18**, 391-9 (1995).
277. K.J.A. Davies, T.A. Quintanilhla, G.A. Brooks and L. Packer, *Biophys. Res. Comm.*, **107**, 1198-1205 (1982).
278. M.J. Jackson, R.H.T. Edwards and M.C.R. Symons, *Biochem. Biophys. Acta* **847**, 185-90 (1985).
279. C.T. Kumar, V.K. Reddy, M. Prasad, K. Thyagaraju and P. Reddanna, *Mol. Cell Biochem.*, **111**, 109-15 (1992).
280. A.F. Fuciarelli, E.C. Sisk, R.M. Thomas and D.M. Miller, *Free Rad. Biol. Med.*, **18**, 231-8 (1995).
281. M.M. Kanter, L.A. Nolte and J.O. Holloszy, *J. App. Physiol.*, **74**, 965-9 (1993).
282. L.L. Ji, R.G. Fu and E. Mitchell, *J. Appl. Physiol.*, **73**, 1854-9 (1992).
283. H.M. Allessio, *Med. Sci. Sports Exer.* **25**, 218-24 (1993).
284. K. Gohil, C. Viguie, W.C. Stanley, G.A. Brooks and L. Packer, *J. Appl. Physiol.*, **64**, 115-9 (1988).
285. D. Harman, *Free Radicals in Biology*, Ed. W.A. Pryor, Vol. V, Academic Press, New York, 1982, Chap. 8.

286. R.G. Cutler, *Free Radicals in Biology*, Ed. W.A. Pryor, Vol. VI, Academic Press, New York, 1984, Chap. 11.
287. R.J. Melhorn and G. Cole, *Adv. Free Radical Biol. & Med.*, Vol. 1, Pergamon Press, 1985, p.165.
288. R.S. Sohal and R.G. Allen, *Adv. Free Radical Biol. & Med.*, Vol.2, Pergamon Press, 1986, p. 117.
289. C.L.D. Taranto, L.M. Viale, F. Beneduce and C.D.V. Blanko, *Free Rad. Biol. Med.*, **20**, 483-88 (1996).
290. *Evolutionary Biology of Ageing*, M.J.Rose, Oxford University Press, 1991
291. J.F. Munnel and R. Getty, *J. Geront.*, **23**, 154-8 (1968).
292. W. Kny, *Virchows Arch. Pathol. Anat. Physiol.*, **229**, 468-78 (1937).
293. R.S. Sohal and H. Donato, *Exp. Geront.*, **34**, 489-96 (1979).
294. R.S. Sohal, *Exp. Geront.*, **16**, 347-55 (1981).
295. C.J. Dilllard, R.E. Litov, W.M. Savin, E.E. Dumelin and A.L. Tappel, *J. Appl. Physiol.*, **45**, 927-32 (1978).
296. A.T. Quintanilha and L. Packer in *Biology of Vitamin E*, Ciba Foundation Symp. No 101, London, 1983, pp. 56-69.
297. K.M. Aikawa, A.T. Quintanilha, B.O. Lumen, G.A. Brooks and L. Packer, *Biosci. Rep.*, **4**, 253-7 (1984).
298. M. Choe, C. Jackson and B.P. Yu, *Free Rad. Biol. Med.*, **18**, 977-84 (1995).
299. G.E. Dobretsov, T.B. Borschevskaya, V.A. Petrov and Y.A. Vladimirov, *FEBS Lett.* **84**, 125-8 (1977).
300. S.Y. Yang and B.P. Yu in *Nutrition in the Aged*, Ed. R.R. Watson, CRC Press, 1993, pp.113-31.
301. J.M. Tolmasoff, T. Ono and R.G. Cutler, *Proc. Natl. Acad. Sci.*, USA, **77**, 2777 (1980).
302. J-H. Choi and B.P. Yu, *Free Rad. Biol. Med.*, **18**, 133-9 (1995).

3

Chain-breaking Antioxidants

The study of antioxidants emerged almost independently in several different chemical technologies. The early theories of Moureu and Dufraisse [1] are now only of historical interest since they were developed before the free radical theory of peroxidation had been proposed. Nevertheless, the experimental observations of these pioneers, particularly on the role of peroxides and their control during peroxidation, anticipated later investigations although they were largely ignored in the following decades [2]. Systematic investigations of antioxidant activity in petroleum technology began with the work of Lowry and his co-workers [3] and was elaborated by Bolland and co-workers at the British Rubber Producers Research Association [4]. The importance of these studies cannot be over-emphasised since they were firmly based on the chemistry of radical-chain peroxidation of "model" chemicals discussed in Chapter 1. Semi-empirical studies of phenolic and amine antioxidant activity in the inhibition of peroxide gel formation in petroleum [5], provided detailed structure-action relationships which were later formalised by relating antioxidant activity to electronic and steric features in the chain-breaking (CB) antioxidants [5,6].

Early empirical discoveries in rubber stabilisation were later extended to the polyolefins whose development as technological products was made possible by the discovery of synergistic (co-operative) interactions of different classes of antioxidant [7]. However, the credit for the discovery of antioxidant synergism must go to food scientists and in particular to H.S. Olcott and H.A. Mattill [8]. The principle of synergism was of considerable practical importance at the time, since it made possible the stabilisation of foodstuffs by combinations of antioxidants found in different kinds of food. Perhaps even more significantly, these early studies anticipated the later awareness of the importance of combinations of naturally occurring biological antioxidants acting in concert *in vivo* and this will be discussed in Chapter 5.

3.1 What are Antioxidants?

Each industry has developed its own terminology to describe the inhibition or retardation of peroxidation. In rubber technology the term **"antidegradant"** is widely used to describe the protection of rubber in engineering applications, particularly in tyre technology. There is some justification for this in the disparate nature of oxidation phenomena in rubber products. For example rubber undergoes observably different physical deterioration phenomena to which the terms "ageing", "perishing", "fatiguing", "stress-cracking", "resinification" and "ozone cracking" were used before a scientific understanding of peroxidation had developed [2]. Many of these terms are still used by rubber technologists. However, all these technological phenomena are manifestations of peroxidation and for this reason the term antioxidant rather than antidegradant is the preferred scientific description of the inhibition of oxidative degradation. This does not preclude the use of more functionally descriptive terms such as "antiozonant" or "antifatigue agent" but in this book the rather vague term "antidegradant" will be avoided.

Plastics technologists use the term "**stabiliser**" in much the same way as the rubber technologist uses "antidegradant". Typical examples are processing stabilisers (polyolefins), heat stabilisers (PVC, polyolefins), UV stabilisers (all polymers). These terms again obscure the fact that all stabilisers function at least in part by inhibiting peroxidation by one or more of the mechanisms that will be described in the following sections and for clarity, the more scientific terms for example, mechanoantioxidants, thermal antioxidants, photoantioxidants will be used whenever the underlying mechanism has been clarified.

In biology the term "antioxidant" is generally not used in a mechanistic sense and tends to encompass all phenomena which prevent or delay the *effects* of oxidation of biological substrates. It includes for example agents which promote repair of cell constituents or modulate cell-cell interactions [9-11]. For example, Deprenyl is a selective amine oxidase inhibitor used in the early treatment of Parkinson's disease [12] and it is sometimes called an antioxidant although it does not contain a recognised antioxidant group. Its protective role is as a very specific enzyme inhibitor. Its action may be compared conceptually to the function of "excited state quenchers" such as manganese salts which deactivate reactive centres in the surface of TiO_2, a white pigment for polymers (see Chapter 4). Mn ions are normally powerful prooxidants and their protective effect is very specific to this type of pigment. These effects, although of great biological or technological significance are site specific and mechanistically atypical, and consequently lie outside the scope of this book which is concerned with the mechanisms involved in the reactions of oxygen with organic molecules and the ways in which antioxidants inhibit these processes.

The peroxidation mechanism is summarised in Scheme 3.1 and the points at which antioxidants can interfere with these processes are indicated by hatched lines. Antioxidants fall into two mechanistic groups [5,6]: those which interrupt the radical chain reaction, the "chain-breaking" (CB) antioxidants, and those which inhibit or retard the formation of free radicals from their unstable precursors and in particular the hydroperoxides. This is the "preventive" antioxidant mechanism [5].

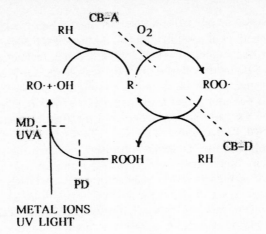

ANTIOXIDANT MECHANISMS

CB-A   CHAIN-BREAKING ACCEPTOR
CB-D   CHAIN-BREAKING DONOR
PD   PEROXIDE DECOMPOSER
MD   METAL DEACTIVATOR
UVA   UV ABSORBER

Scheme 3.1 Mechanisms of antioxidant action

In this chapter the principle chemical reactions involved in the chain-breaking antioxidants will be outlined. These can be further sub-divided into chain-breaking hydrogen or electron donors (CB-D) to peroxyl or hydroxyl radicals, and hydrogen or electron acceptors (CB-A) from carbon-centred radicals [6,13]. The first class comprises the well known phenol and arylamine antioxidants and the second the "stable" phenoxyl and nitroxyl radicals and quinonoid compounds. These processes are summarised in Scheme 3.2.

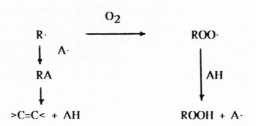

CHAIN-BREAKING ACCEPTOR   CHAIN-BREAKING DONOR
(CB-A)   (CB-D)

Scheme 3.2 Chain-breaking mechanisms of antioxidant action

The preventive antioxidants will be discussed in Chapter 4, together with "synergism" and "antagonism" which involve interactions between antioxidants when used in combination. Synergism may be simply defined as an effect which is greater than that expected on the basis of the additive effects of two antioxidants and antagonism as an effect which is less than the additive effect [7]. Sub-sets of synergistic antioxidants are described by the terms "homosynergists" and "heterosynergists" in which the synergistic behaviour occurs between antioxidants acting by the same or different mechanisms respectively. Antioxidants containing more than one antioxidant function in the same molecule are autosynergistic [5,7].

3.2 The Chain-breaking Donor Mechanism

Table 3.1 lists some of the main sub-groups of chain-breaking donor (CB-D) antioxidants, many of which are used in substantial quantities in lubricating oils and polymers.

Table 3.1 Chain-breaking Donor Antioxidants

KEY REACTIONS:
$ROO\cdot + AH \rightarrow ROOH + A\cdot$
$2A\cdot \rightarrow A-A$
$A\cdot + ROO\cdot \rightarrow AOOR$

CHEMICAL CLASS: HINDERED PHENOLS

EXAMPLES		CODE	USE
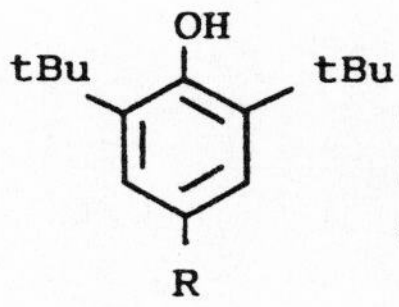 (a) BHT, R=CH_3		I	P,O
(b) 1076, R=$CH_2CH_2COOC_{18}H_{37}$			P
(c) 1010, R=$(-CH_2CH_2COOCH_2)_4C$			P
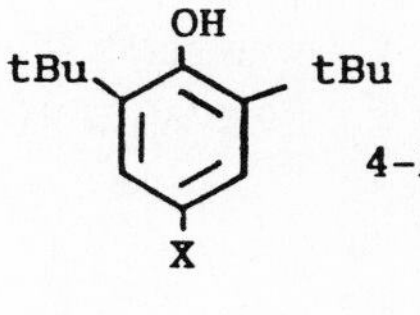 4-Activated hindered phenols		II	
X = OH			O
X = OCH_3			O
X = NHR (R = alkyl, acyl)			P
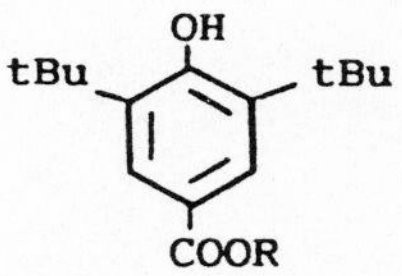 4-Carboxy hindered phenols		III	
(a) R = long-chain alkyl			P
(b) 2,4-di-*tert*-butyl phenyl			P
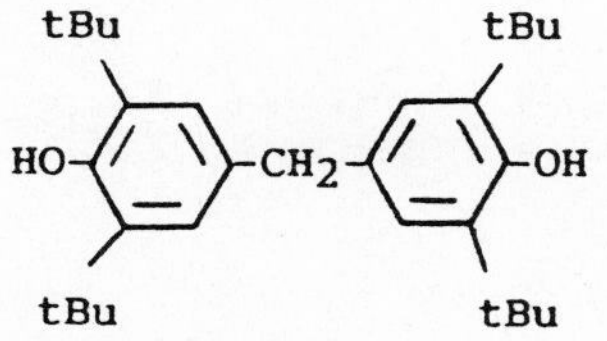 4-methylene-bis phenol		IV	P

P = Plastics, O = Oils, greases, etc

Table 3.1 (cont)

CHEMICAL CLASS: PARTIALLY HINDERED PHENOLS

EXAMPLES

Structure	Name	CODE	USE
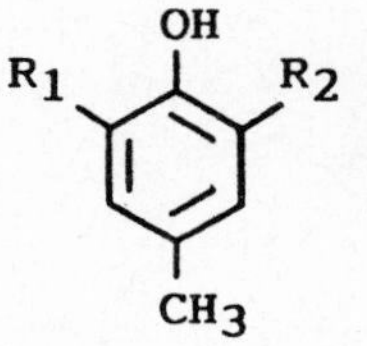 2,4-substituted phenol with R_1, R_2, CH_3	Tertiary alkylated phenols, R_1 = C_8–C_{12} α-branched alkyl, R_2 = CH_3 or R_1	V	P
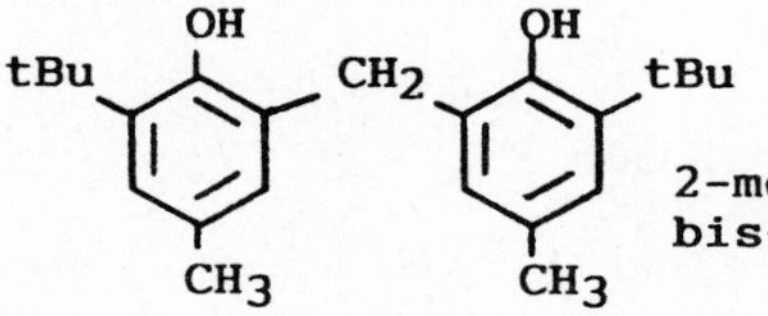 bis-phenol linked by CH_2, with tBu and CH_3 substituents	2-methylene-bis-phenols	VI	R
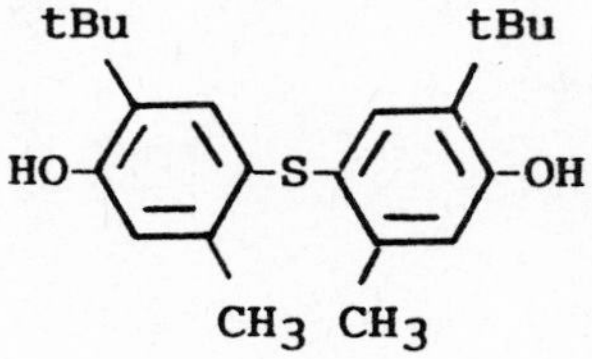 bis-phenol linked by S, with tBu and CH_3 substituents	4-bis-phenol sulphides	VII	P
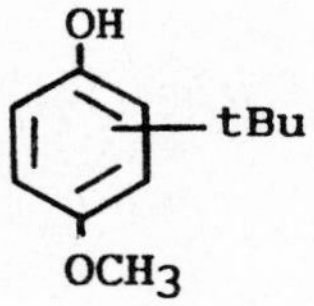 phenol with tBu and OCH_3	BHA	VIII	R,O

CHEMICAL CLASS: AROMATIC AMINES

Structure	Name	CODE	USE
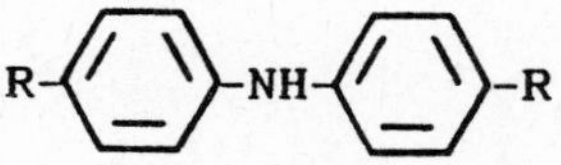 R–C$_6$H$_4$–NH–C$_6$H$_4$–R	OD, R = *tert*-Oct	IX	R,O
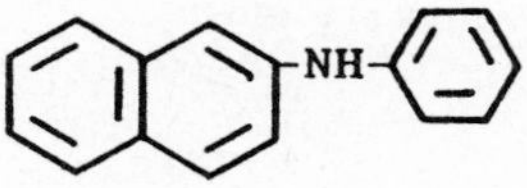 naphthyl–NH–phenyl	PβN	X	R,O
Ar–NH–C$_6$H$_4$–NH–Ar	DPPD, Ar = phenyl	XI	R
	DNPPD, Ar = naphthyl	XII	R,O

P = Plastics, R = Rubbers, O = Oils

Table 3.1 (cont)

CHEMICAL CLASS: AROMATIC AMINES

Structure	Name	CODE	USE
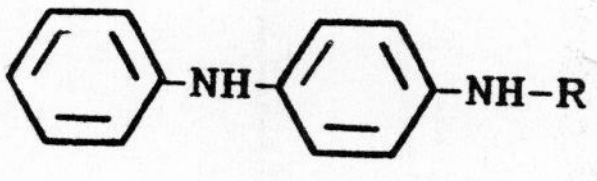 Ph-NH-C₆H₄-NH-R	IPPD, R = *iso*-Pr	XIII	R
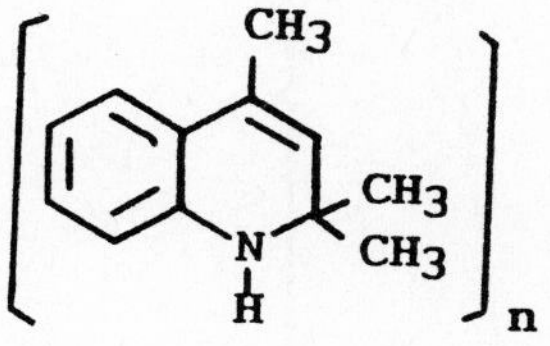 [2,2,4-trimethyl-1,2-dihydroquinoline]ₙ	DMDHQ	XIV	R,O
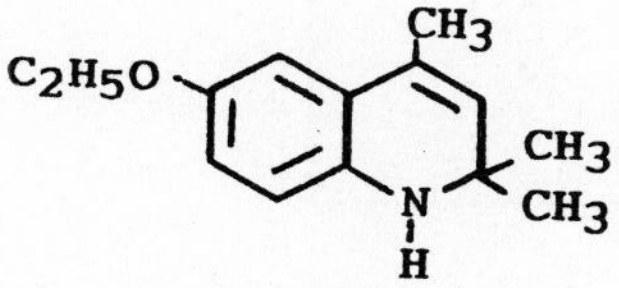 6-ethoxy-2,2,4-trimethyl-1,2-dihydroquinoline	Ethoxyquin	XV	R

R = Rubbers, O = Oils

The rate at which phenols and amines transfer a hydrogen to peroxyl radicals (reaction 1) was extensively studied in the 1940s and 1950s in a variety of model substrates [5] and is still used, mainly by biochemists, as a measure of "radical trapping" activity:

$$ROO\cdot + AH \xrightarrow{k_7} ROOH + A\cdot \qquad (1)$$

k_7 can be measured by a number of kinetic techniques involving the measurement of some function of peroxidation, for example, oxygen absorption, hydroperoxide formation, chemiluminescence, etc. during the induction time, τ (Fig. 3.1). This is normally carried out in pure model compounds in the presence of an alkyl radical generator [5] whose rate of homolysis is known. k_7 can then be used to assign a relative order of antioxidant activity to substituted phenols. Table 3.2, column D, shows the results of a typical evaluation of the relative CB-D effectiveness of 2,4,6-trialkyl phenols in tetralin relative to BHT ($R_1 = R_2 = t$-Bu, $R_3 = $ Me) which is given the arbitrary rating 100.

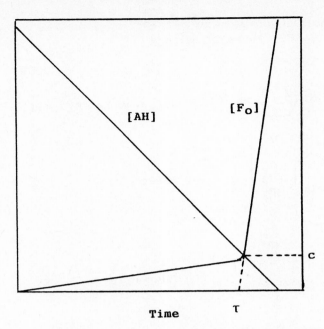

Fig. 3.1 Idealised description of initiated, CB-D inhibited peroxidation. [F_o] is a function of peroxidation (e.g. oxygen absorption, hydroperoxide formation, physical property change, etc.). [AH] = total primary and secondary antioxidant concentrations, τ = induction (inhibition) time and c = critical antioxidant concentration.

Phenol itself is relatively ineffective as a CB antioxidant, and electron attracting groups (e.g. NO_2 and COOH) decrease it even further. 4-Hydroxy and 4-alkoxy groups increase CB-D activity as do 4-methyl and to a lesser extent, 4-*tert*-butyl. 4-Methyl and 4-*tert*-butyl are less effective than their 2-isomers and the effect of more than one alkyl group is additive. However, in trialkyl phenols, optimal activity is obtained with methyl in the 4-position rather than *tert*-butyl. Partially hindered phenols are more effective than fully hindered phenols.

Table 3.2 Antioxidant Activities Relative to BHT of 2,4,6-trialkyl phenols [5]

R_1	R_2	R_3	Oil*	Lard*	Petroleum*	Tetralin+
H	H	H	0	–	2	4
Me	H	H	0	–	16	47
tBu	H	H	–	11	24	53
H	H	Me	0	0	10	15
H	H	NO_2	–	–	–	2.6
H	H	COOH	–	–	–	2.6
H	H	OH	–	–	–	280
H	H	OnBu	–	–	–	112
H	H	tBu	–	–	6	23
Me	Me	H	–	–	34	56
Me	H	Me	0	–	51	55
tBu	H	Me	36	36	96	84
Me	Me	Me	0	51	118	154
tBu	Me	Me	65	–	170	167
tBu	tBu	Me	100	100	100	100
tBu	tBu	tBu	36	75	84	72

* By measurement of induction time, τ. In petroleum, τ was an average of four independant studies [5].
+ By measurement of k_7.

However, model compound studies rarely give information very relevant to the performance of antioxidants in technological systems. Thus for example, in the saturated polyolefins, which are normally tested at relatively high temperatures (>100°C) in a flowing air stream, fully hindered and relatively involatile phenols (e.g. Table 3.1; I(b), I(c) and IV) are more effective than lower molecular weight antioxidants containing the same antioxidant function (e.g. BHT, Table 3.1, I(a)). In rubbers, partially hindered phenols (e.g. Table 3.1, V-VII) are more effective than the fully hindered related phenols. It is noteworthy that *tert*-alkyl phenols are unknown in biological substrates and high effectiveness is achieved in the tocopherols by a combination of methyl substitution and activation by the 4-chroman oxygen (Chapter 5).

3.2.1 *Structure-activity relationships in CB-D antioxidants*

Before discussing the reasons for antioxidant activity, it is important to recognise that although the primary function of CB antioxidants is to interrupt the kinetic chain reaction by removing alkylperoxyl radicals (reaction 1) they normally give rise to peroxides as co-products (Scheme 3.3). The latter can subsequently break down, particularly in the presence of UV light or at elevated temperatures, to give "oxyl" radicals thus re-initiating the chain reaction. Hydroperoxides are much less stable than dialkyl peroxides due to their susceptibility to induced decomposition, particularly by reduced transition metal ions, but even dialkyl peroxides undergo homolysis at under conditions commonly encountered in lubricating oils and polymers.

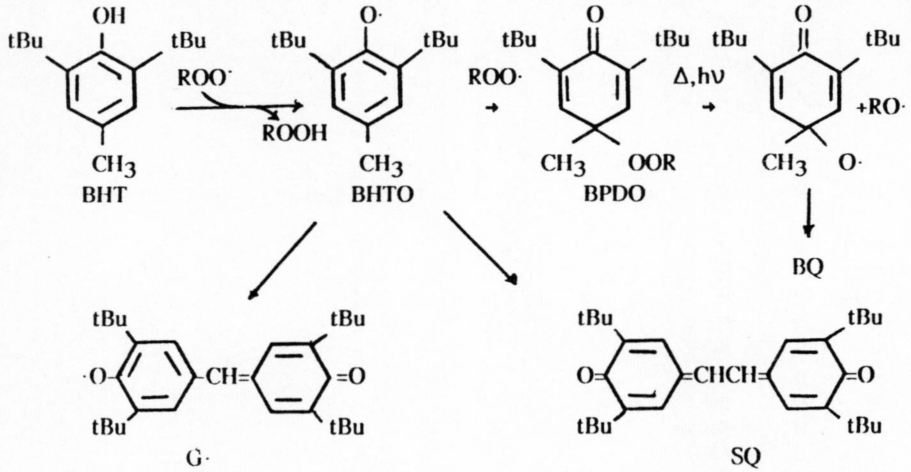

Scheme 3.3 Oxidative transformation products of BHT

Consequently, the effectiveness of phenolic antioxidants is prejudiced, and in many practical situations they have to be used in combination with antioxidants that act by complementary preventive mechanisms (see Chapter 4). The most important of these are the metal deactivators or the ionic peroxides decomposer, both of which inhibit the formation of free radicals from peroxidic intermediates.

k_7, in an alkyl radical initiated system, measures *only* peroxyl trapping (CB-D) activity Although it is widely used by biochemists to measure "total radical trapping antioxidant potential" (TRAP) in biological systems [14-18], it cannot by its very nature assess the contribution of preventive mechanisms, notably the peroxidolytic (PD) and metal deactivating (MD) processes (Scheme 3.1). It was at one time widely used in the assessment of the technological effectiveness of the chain-breaking (CB-D) antioxidants, and there is a satisfactory correlation between τ in petroleum and k_7 in tetralin for a variety of phenolic antioxidants (see Table 3.2). In practice, these single function anti-oxidants are now rarely used alone and the TRAP assay does not extend to antioxidants acting by complementary mechanisms. It does have utility in measuring the potential of CB-D antioxidants in body fluids or of comparing the activity of CB-D antioxidants. However, it does not measure "total antioxidant potential" and, if used uncritically, it may be quite misleading in multi-antioxidant biological systems which almost always involve synergism between the same or complementary classes of antioxidant.

A second but more theoretical limitation of the TRAP assay is that many CB-D antioxidants give rise to secondary oxidation products with antioxidant activity equal to or in some cases superior to the starting antioxidant. Pospíšil has comprehensively reviewed the formation of oligomeric products from both phenol [19] and amine [19,20-22] antioxidants. Thus, for example, diarylamines give dimerisation products with CB-D antioxidant activity essentially similar to that of the original amine. A typical example of the dimerisation of the simplest hindered phenol, BHT, is shown in Scheme

3.3. The antioxidant activity of the resulting quinones will be discussed in Section 3.1.2. The dehydro-dimerisation of typical paraphenylene diamines (PPDs) which are widely used as antioxidants in rubbers [20-22] is shown in Scheme 3.4.

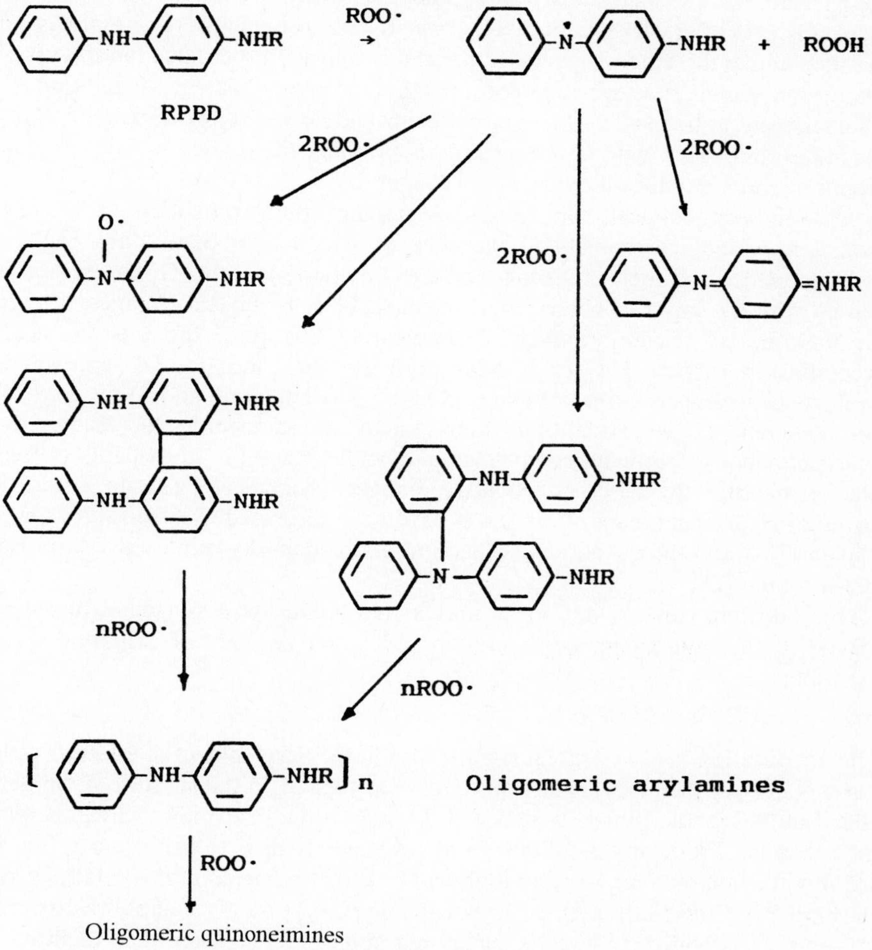

Scheme 3.4 Oxidation of PPD Antioxidants

The oligomeric products of phenols and aromatic amines are often more effective than their precursors as CB-D antioxidants. In technological systems, this results primarily from the lower volatility of the derived antioxidants, but it is also due to the formation of new oxygenated products such as quinones, quinoneimines, stable aryloxyls and aminoxyls [6] acting by the complementary CB-A mechanism which synergise with the original antioxidant. As will be seen below, under the appropriate conditions these antioxidants may also act catalytically with stoichiometric inhibition coefficients (f) many times higher than the antioxidants from which they were derived.

A very convenient method of monitoring total antioxidant potential is by measurement of the induction time, τ, to product formation in an uninitiated or preferably in a hydroperoxide initiated peroxidation. (Fig. 3.1). This provides a measure of combined

CB and preventive antioxidant activities of single antioxidants and their combinations. This technique is now widely used in technological and model media and a variety of oxidation functions (F_o) can be used. They include the formation of hydroperoxides and their breakdown products aldehydes and ketones which are readily monitored spectroscopically. τ is of greater practical relevance to both technologists and biologists than is k_7, since unlike the latter it generally correlates well with the induction time preceding change in technological or pathological parameters in real systems. In addition it allows an assessment to be made of the relevant contributions of radical trapping and peroxide decomposition. This will be illustrated in mechanistic studies of several important groups of peroxidolytic antioxidants in Chapter 4.

In both technological and biological systems, the termination of the chemical induction period corresponds to the time at which antioxidants are depleted and hydroperoxides and their breakdown products increase rapidly. This in turn correlates closely with the onset of observable technological or biological changes. For example the build-up of viscous products in lubricating oils [23], the loss of mechanical properties in rubbers [24] or plastics [25] and the formation of toxic breakdown products of hydroperoxides in biological systems (Chapters 1 and 5) all coincide with the depletion of antioxidants in the system. Consequently the measurement of hydroperoxides or carbonyl compounds and/or the decay of antioxidants is frequently used to monitor the onset of oxidative damage. Analytical techniques used for the quantitative determination of oxidation have been discussed by Foster [26] and by Al-Malaika [27] and the quantitative aspects of antioxidant depletion have been reviewed by Munteanu [28].

The induction time, τ, due to an antioxidant acting by a sacrificial mechanism, is related to the antioxidant concentration [AH] and the rate of initiation, r_i, by the equation:

$$\tau = 2[AH]/r_i \qquad (i)$$

This relationship is of very great significance in the development of effective stabilising systems in both chemical technology and biology since, in the absence of indigenous or added initiators, the initiating step will depend on the formation of hydroperoxides in the substrate. From this it follows from equation (i) that, to achieve a given level of stability, the antioxidant concentration will be directly related to the oxidisability of the substrate [5]. Thus in the case of the polydienic rubbers the level of antioxidant required to achieve durability (2-4%) is about ten times higher than that required in the essentially saturated polyethylene.

Transition state theory (see reaction 2) provides an explanation for the electronic and delocalisation effects of substituents in phenolic antioxidants. The activating effects of substituents in the arylamines are broadly similar.

Transition State

(2)

(i) Inductive effects

The effect of an electron releasing group, Y, is to facilitate the transfer of an electron to the electrophilic peroxyl radical thus reducing the transition state energy [29], whereas electron attracting groups have the opposite effect. As might be expected, the activity of CB-D antioxidants is related to critical oxidation potential [30] but there is a limit to the effectiveness of phenols with very low oxidation potentials, notably the unhindered polyalkyl phenols, due to the increasing probability of direct attack by molecular oxygen [5,6]:

$$O_2 + XPhOH \rightarrow \cdot OOH + XPhO\cdot \qquad (3)$$

(ii) Electron delocalisation

Groups in the 2, 4 and 6 positions which delocalise the unpaired electron in the derived phenoxyl or aminyl radicals (e.g. phenyl, alkoxyl, alkylamino, allyl, methyl, etc.) increase antioxidant activity [5]. Figure 3.2 shows the spectrum of the primary phenoxyl radical formed by hydrogen abstraction from hindered phenol BHT (Scheme 3.1) which was unambiguously identified by electron paramagnetic resonance (subsequently renamed electron spin resonance) over 30 years ago. The major quartet splitting indicates the interaction of the three methyl hydrogens with the unpaired electron formally on oxygen. 6.3% of the unpaired spin density is associated with the methyl by hyperconjugation [31].

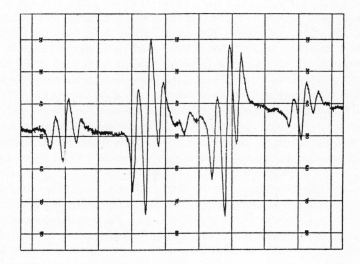

Fig. 3.2 Primary aryloxyl radical formed by oxidation of BHT
(Reproduced with permission from J.K. Becconsall, S. Clough and G. Scott, *Trans. Farad. Soc.*, **56**, 459 (1963)).

α,β- unsaturated substituents and aryl groups similarly increase antioxidant activity. Hydroxy, alkoxy, amino, alkylamino and sulphide are particularly effective in delocalising unpaired electrons in the aromatic ring. This depends on overlap of the unpaired electron with the ρ-type unpaired electrons of the oxygen or nitrogen in the 2 or 4 position in the aromatic ring.

Aromatic amines are in general more effective than phenols even though they do not normally contain ring alkyl substituents. The ranking order of aromatic amine antioxidants in rubbers is: diphenylamines < phenylnaphthyamine < dialkyl-*p*-phenylenediamines < diaryl-*p*-phenylene diamines. k_7 for the ρ-phenylene-diamines is almost three orders of magnitude higher than that of the N-alkyl anilines and two orders of magnitude higher than diarylamines [32]. However, k_7 is probably not the most important measure of antioxidant activity for this class of antioxidant, since it will be seen below that the derived oxidation products play a very important role as CB-A antioxidants in the protection of polymers during mechanical service. 4-alkylamino groups have a similar activating effect on the CB-D activity of phenols and, in both cases, this is associated with very low oxidation-reduction potentials [5].

Bis-phenol sulphides are widely used as antioxidants in polymers, and ESR evidence suggests [33] that delocalisation of the unpaired electron on sulphur contributes to their activity. However, there is also strong evidence [34,37] that the sulphide group provides an additional antioxidant function by hydroperoxide decomposition (Chapter 4).

(iii) Steric effects

Branched-chain alkyl groups (and in particular *tert*-alkyl) in the 2 and 6 positions of phenols increase antioxidant activity by increasing the stability of the phenoxyl radical by steric hindrance [5]. Thus, unhindered phenoxyls derived from benzene without ortho substituents have half-lives measured in microseconds, and cannot normally be observed by ESR except by dynamic methods in which radicals are formed at the same rate at which they decay. Hindered aryloxyls, e.g. BHTO by contrast have half-lives of several seconds. Tkác [33] has shown by ESR that not only does the stability of the aryloxyl increase with increasing 2,6 steric hindrance, but that the activation energy of hydrogen abstraction by alkylperoxyl decreases. The effect of steric hindrance in the 4-position is to decrease the ability of the phenoxyl to capture a second alkylperoxyl (Scheme 3.3) thus decreasing the normal stoichiometry of peroxyl radical capture (f) from 2 to 1;

Although phenols and amines represent by far the major chemical classes of CB-D antioxidants, other chemical species have been shown to have chain-breaking activity. Of particular significance in lubricating oils and polymers are sulphur compounds. However, in contrast to the inhibitive role of thiols in biological systems, simple thiols have not been found to be immediately effective in technological systems, and indeed they are chain transfer agents in polymerisation and redox initiators with hydroperoxides in rubbers [38]. It seems likely that the activity of sulphur compounds in polymer stabilisation is due to the formation of further oxidation product which act as hydroperoxide decomposers (Chapter 4).

3.2.2 Physical aspects of antioxidant effectiveness in polymers

It was noted above that oxidative dimerisation products of phenols and aromatic amines are generally more effective in technological systems than the parent antioxidants due to reduction in the volatility of the higher molecular weight products. During the development of polypropylene in the 1950s it was found that some of the common antioxidants (e.g. BHT) developed for oils and foodstuffs were almost without activity in this polymer, which is normally evaluated in an oven ageing test at temperatures between 100 and 140°C. This observation led to a search for less volatile antioxidants and to the development of higher molecular weight analogues of BHT of which 1076, I(b), and 1010, I(c), are typical examples [39,40]. The higher efficiency achieved by increasing the size of antioxidant molecules was found to be due entirely to the persistence of the antioxidant in the polymer matrix [39,40] and this in turn resulted from the reduced volatility from the surface of the polymer and reduced diffusion rate within the polymer matrix [41]. The antioxidant activity of the homologous series I(b), measured at the same molar concentration in decalin solution at 140°C by oxygen absorption in which the group R is varied from $-CH_2CH_2COOCH_3$ to $-CH_2CH_2COOC_{18}H_{37}$ showed that change in the size of the molecule had almost no effect on the intrinsic molar activity (D_c) of the antioxidant in solution [39,40] (Table 3.3). In polypropylene at 140°C in a flowing air stream (PP_o), the position was quite different and the same chemical structure was found to be virtually ineffective unless the half-life of the antioxidant in the polymer exceeded the expected optimal induction time. Under these conditions, the antioxidant performance is dominated by the volatility of the molecule from the polymer. A more complex behaviour is observed in polypropylene in a closed system (PP_c). In this case volatility is no longer important and antioxidant activity correlates closely with hydrocarbon solubility.

Table 3.3 Molar Functional Group Activities of 2,6-*tert*-butyl Phenols (I) (at 140°C, 2×10^{-3} mol/100g) [39]

Antioxidant, I	Solubility in hexane at 25°C (g/100g)	Half-life in N_2 stream (h at 140°C)	Induction time (hours)		
			D_c	PP_c	PP_o
R = CH_3 (BHT)	100	0.1	150	140	2
R = $CH_2CH_2COOCH_3$	32	0.28	25	95	2
R = $CH_2CH_2COOC_6H_{13}$	∞	3.6	23	312	2
R = $CH_2CH_2COOC_{12}H_{25}$	∞	83.0	20	420	2
R = $CH_2CH_2COOC_{18}H_{37}$	64	660.0	20	200	165

D_c = oxygen absorption in decalin (closed system), PP_c = oxygen absorption in polypropylene (closed system), PP_o = embrittlement failure in polypropylene (airflow).

An extension of the above concept is to covalently attach antioxidants to the polymer chain and this will be discussed in Chapter 4.

3.3 The Chain-breaking Acceptor Mechanism

It has long been known that electron acceptors, particularly quinones, are effective inhibitors for free radical polymerisation [42], and Szwarc showed [43] that the affinity of quinones for methyl radicals was decreased by electron releasing groups and increased by electron attracting groups. Although it was recognised by Milas [44] in early antioxidant investigations that some oxidising agents, including quinones, were antioxidants, it was not fully understood at that time how they were able to inhibit peroxidation. Watson and co-workers showed that benzoquinone is an efficient trap for macroalkyl radicals formed during "mastication" of rubber [45,46], a process used in rubber technology to "plasticise" (i.e. to reduce the molecular weight, Chapter 1, Scheme 1.17). Mechano-radical formation is thus a consequence of homolytic scission of the polymer chain by mechanical stress [47] and oxidation of macroalkyl radicals by quinones to give olefins, a process described by Haber and Willstätter over sixty years ago [48] is now known to be very important in both polymer technology and in biology:

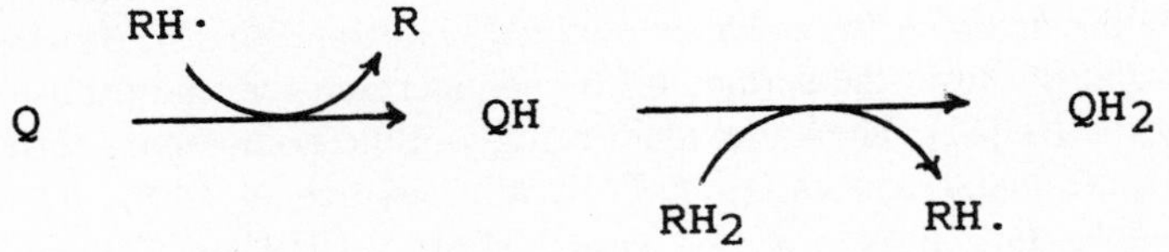

$$Q \xrightarrow{RH\cdot \quad R} QH \xrightarrow{\quad RH_2 \quad RH\cdot} QH_2 \qquad (4)$$

$$RH_2 = \begin{matrix} CH_2COOH \\ | \\ CH_2COOH \end{matrix} \qquad RH\cdot = \begin{matrix} \cdot CHCOOH \\ | \\ CH_2COOH \end{matrix} \qquad R = \begin{matrix} CHCOOH \\ \| \\ CHCOOH \end{matrix}$$

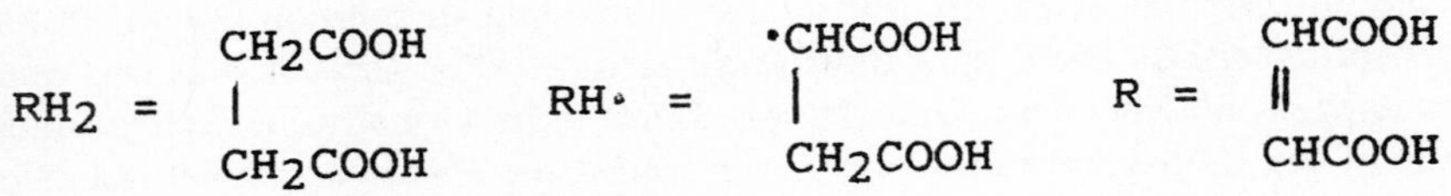

The same redox antioxidant process involving quinones, quinoneimines, aryloxyls and aminoxyls is believed to be involved in both the catalytic stabilisation of polymers during processing at high temperatures and in the prevention of "fatigue" in rubbers at ambient temperatures [49,50] (Section 3.5.1).

Typical chain-breaking acceptor antioxidants are listed in Table 3.4. Conditions favouring efficient cycling of redox couples derived from antioxidants have been outlined [47,51,52]. The following are the salient features:

a) Denisov has shown [51] that in polyolefins the solubility of oxygen is lower than in homologous liquid hydrocarbons and at 371K, $[R\cdot]/[ROO\cdot]$ is three orders of magnitude higher in than it is in liquid *iso*-pentane at the same temperature. Furthermore, except in very thin sections, there is a diffusion limitation on replenishment of oxygen in bulk polymers, thus favouring alkyl radical termination.
b) Redox cycling is also favoured when the rate of radical formation is high, for example during processing in a screw extruder or internal mixer, (Section 3.5.1(a)) or in photosensitised oxidations (Section 3.5.1(c)).
c) Redox cycling is similarly preferred in the case of resonance stabilised alkyl radicals. For example in partially degraded PVC carbon-centred polyenyl radicals are relatively stable and reaction with oxygen is reversible. Consequently hydrogen abstraction by "stable" phenoxyl or aminoxyl radicals can compete with hydroperoxide formation with extension of the conjugated system [52].

$$-CH_2(CH=CH)_nCH(OO)\cdot \begin{array}{c} \xrightarrow{PH} -CH_2(CH=CH)_n\overset{|}{C}HOOH \\ \xrightarrow{>N-O\cdot} -(CH=CH)_nCH=CH- \\ + >NOH + (O_2) \end{array} \quad (5)$$

In practice, more than one of these conditions is often simultaneously met in polymers. For example, conditions (a) and (b) obtain during high shear processing (all polymers), during photooxidation (hydrocarbon polymers) and (c) during fatigue (rubber products, e.g. tyres, from polyunsaturated rubbers).

Table 3.4 Chain-breaking Acceptor Antioxidants

KEY REACTIONS: R· + A· → RA
                                    RA → RH(C=C) + AH

CHEMICAL CLASS: "STABLE" PHENOXYLS

EXAMPLES		CODE	USE
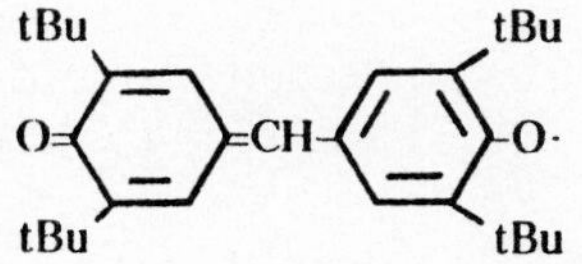 (tBu-substituted galvinoxyl structure)	Galvinoxy, G·	XVI	P

CHEMICAL CLASS: QUINONES AND SEMIQUINONES

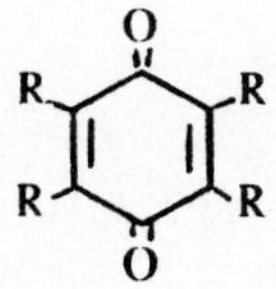 (R-substituted benzoquinone)	Benzoquinones, BQ	XVII	R
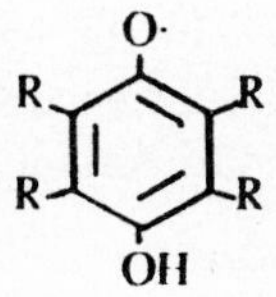 (R-substituted semibenzoquinone)	Semibenzoquinones, BQH·	XVIII	R
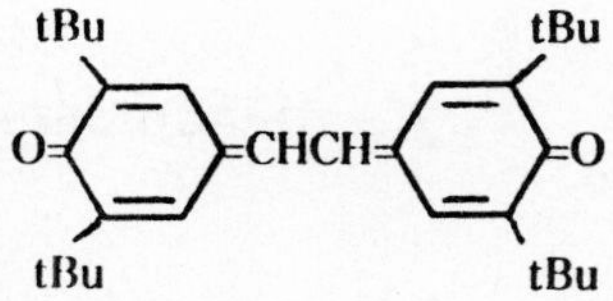 (tBu-substituted stilbenequinone)	Stilbenequinone, SQ	XIX	P

Table 3.4 (cont)

CHEMICAL CLASS: "STABLE" AMINOXYLS (NITROXYLS)

Structure	Name	Number	Notes
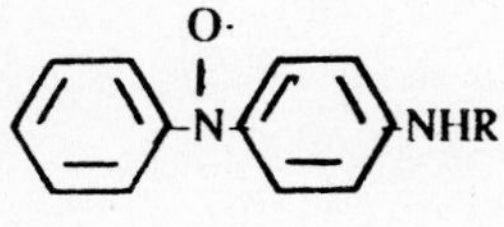 Ph-N(O·)-C₆H₄-NHR	Diphenylaminoxyl, DPAO	XX	O,R
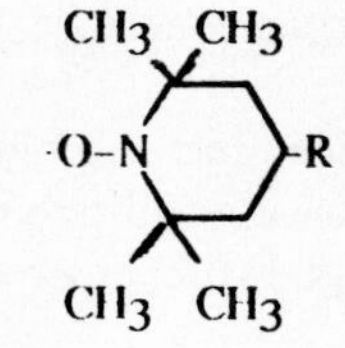 Tetramethylpiperidine N-oxyl with R substituent	Tetramethylpiperidinoxyl, TMPO	XXI	P

CHEMICAL CLASS: SPIN-TRAPS

KEY REACTIONS: R· + ST → RST·
RST· + A· → RSTA, etc

Structure	Name	Number	Notes
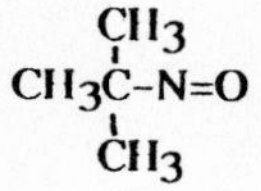 $(CH_3)_3C-N=O$	2-Methyl-2-nitrosopropane, MNP	XXII	P
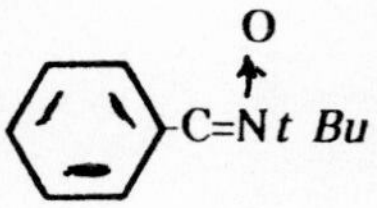 Ph-CH=N⁺(O⁻)-tBu	Phenyl-*tert*-butyl nitrone, PBN	XXIII	P

3.4 The Catalytic Chain-breaking Mechanism

Denisov was the first to propose that chain-breaking antioxidants could act catalytically [54]. During the inhibition of cyclohexanol peroxidation by α-naphthylamine it was found that the number of radicals scavenged per mole of inhibitor during the inhibition time in an initiated oxidation (f) was approximately 30 at 140°C. Similar effects were also observed with benzoquinone/quinhydrone ($f = 20$) and Cu^{2+}/Cu^+. In the latter inhibition process, it was concluded that f was virtually infinity. It was suggested that the intermediate hydroxyperoxyl radical was oxidised by the antioxidant to cyclohexanone:

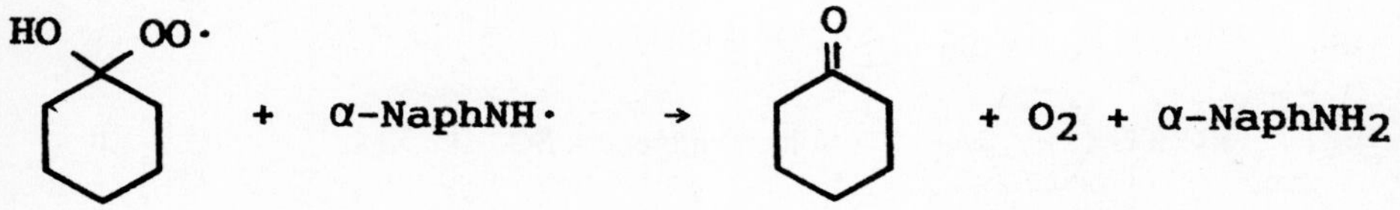

$$\text{HO-C}_6\text{H}_{10}\text{-OO·} + \alpha\text{-NaphNH·} \rightarrow \text{C}_6\text{H}_{10}\text{=O} + \text{O}_2 + \alpha\text{-NaphNH}_2 \quad (6)$$

The significance of the catalytic CB mechanism in hydrocarbons at elevated temperatures was later demonstrated by Berger and co-workers in mechanistic studies with model compounds of known structure [55]. They showed that aminoxyl radicals are continuously regenerated during oxidation at 130°C. Table 3.5 shows the stoichiometric inhibition coefficient, f, for diarylamines, dialkylamines and their oxidation products at 130°C. Alkoxy-, alkyl- and nitrodiphenylaminoxyls are effective catalytic inhibitors as are the parent alkoxy and alkyl diphenylamines but nitrodiphenylamine is not. Furthermore, the hydroxylamine formed by elimination from the trapped 4-ethoxydiphenylnitroxyl (Scheme 3.5) is also an effective catalytic antioxidant and, by oxidation by alkylperoxyl, recycles to regenerate nitroxyl. It appears that nitro-diphenylamine is not converted to the aminoxyl under the conditions of the oxidation.

Table 3.5 Stoichiometric Inhibition Coefficients (*f*) of Amines and their Oxidation Products in a paraffinic oil at 130°C [55]

Antioxidant	f
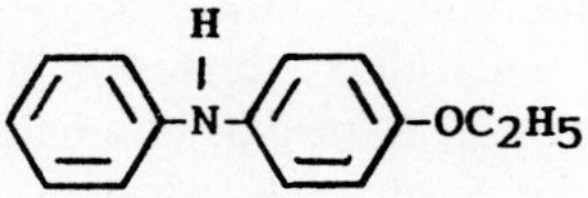 C$_6$H$_5$–NH–C$_6$H$_4$–OC$_2$H$_5$	41
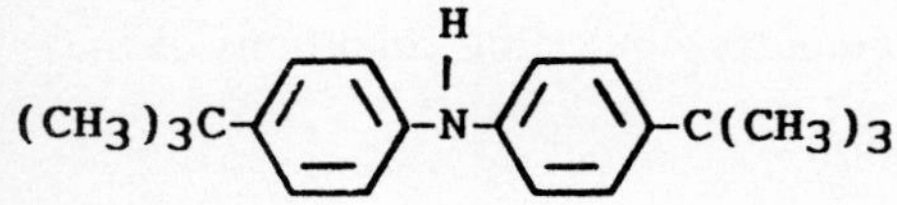 (CH$_3$)$_3$C–C$_6$H$_4$–NH–C$_6$H$_4$–C(CH$_3$)$_3$	52
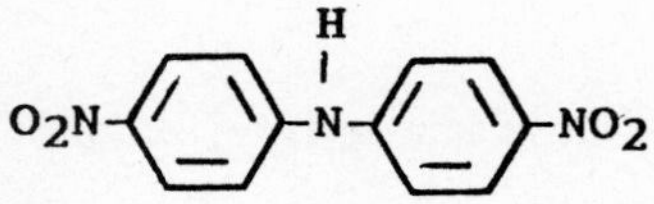 O$_2$N–C$_6$H$_4$–NH–C$_6$H$_4$–NO$_2$	0
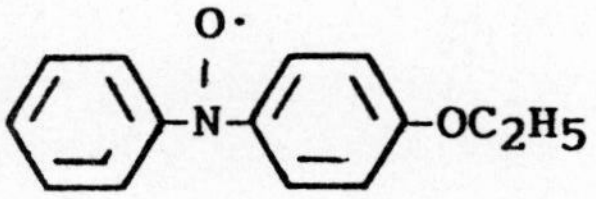 C$_6$H$_5$–N(O·)–C$_6$H$_4$–OC$_2$H$_5$	26
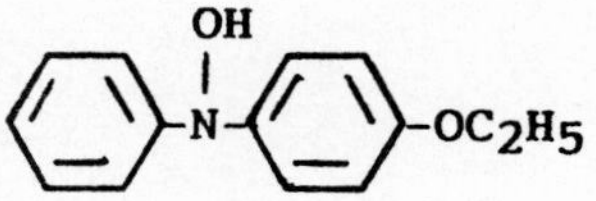 C$_6$H$_5$–N(OH)–C$_6$H$_4$–OC$_2$H$_5$	35
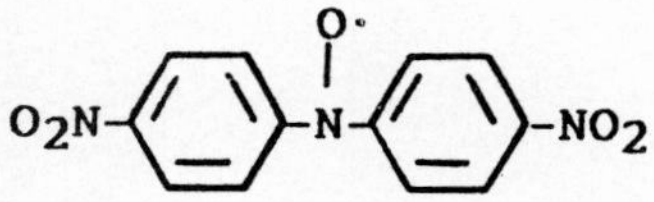 O$_2$N–C$_6$H$_4$–N(O·)–C$_6$H$_4$–NO$_2$	15
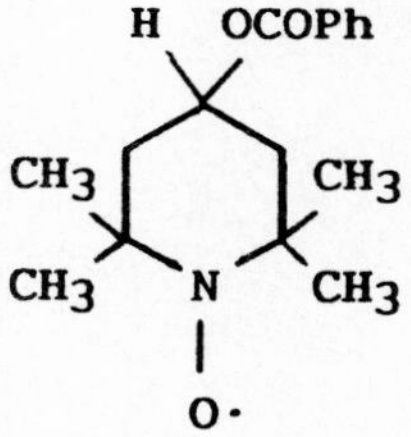 4-H,4-OCOPh-2,2,6,6-tetramethylpiperidine-N-oxyl	630

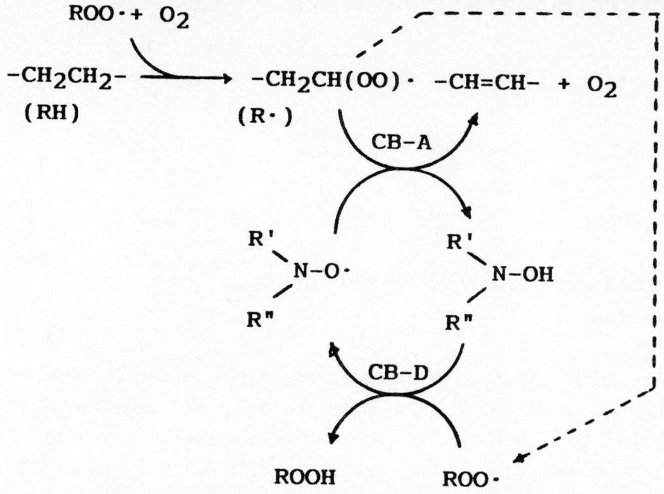

Scheme 3.5 Catalytic thermal antioxidant activity of aminoxyls

The diarylaminoxyls are not stable indefinitely under peroxidising conditions since the peroxyl radical can react in the aromatic ring (see Scheme 3.6). However the fully α-substituted dialky nitroxyls are very more stable and the stoichiometic inhibition factor f is an order of magnitude higher for the widely used 2,2',6,6'-tetramethylpiperidinoxyl light stabilisers which will be discussed in Section 3.5.1(c).

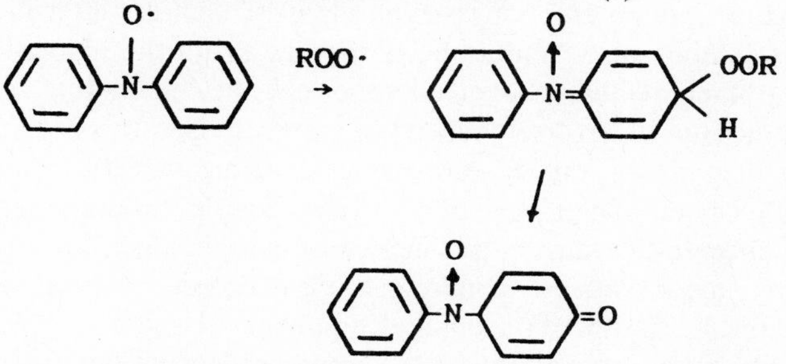

Scheme 3.6 Alternative reaction of Diphenylaminoxyl

The catalytic activity of nitroxyls as measured by f also varies with the temperature and with the chemical structure of the substrate [55]. The "hindered" piperidinoxyl is relatively ineffective in tetralin at 60°C in the absence of light, indicating that in this substrate, the CB-A step in Scheme 3.5 does not occur sufficiently rapidly at this temperature to allow the cycle to proceed.

Even at 130°C, straight-chain paraffins appear to participate less readily in the cycle than alkyl aromatics. This reflects the higher activation energy involved in hydrogen abstraction from an aliphatic hydrocarbon radical than from an aralkyl radical:

$$-CHCH=CHCH_2- \; + \; >NO\cdot \;\; \rightarrow \;\; -CH=CH-CH=CH- \; + \; >NOH \tag{7}$$

This will be reduced even further in the 2,4-dienyl radicals which are the predominant carbon-centred radical species found in biological systems.

Although the oxidation of carbon-centred radicals is frequently depicted as in reaction (7), it seems likely that at ambient oxygen pressures alkylperoxyl rather than alkyl reacts with nitroxyl (see Scheme 3.5) [56]. This reaction will be particularly favoured in the case of allylperoxyl or benzylperoxyl, due to electron delocalisation in the transition state:

$$\text{H}\cdots\text{ON}<$$
$$-\dot{\text{C}}\text{H}=\text{CH}=\text{CH}=\text{CH}\cdots\text{O}=\text{O}\cdot \rightarrow -\text{CH}=\text{CHCH}=\text{CH}- + >\text{NOH} + \text{O}_2 \qquad (8)$$

3.5 Applications of the Catalytic CB Process in Polymers
3.5.1 *Mechanoantioxidants*
(a) During processing

Many oxidising agents are effective CB-A antioxidants in polyolefins during processing at high temperature [57]. Under these conditions, three of the requirements for the operation of the catalytic CB-A/CB-D cycle are fulfilled, namely high rate of radical formation, relatively high temperature and a low oxygen concentration [58-63]. A standard procedure used to monitor the effects of processing on polymers involves rotor-mixing of the polymer at temperatures generally between 140° and 250°C. Although air is formally excluded from the mixing chamber, in practice a small amount is always present, dissolved in the polymer or trapped between the polymer particles. In the case of polypropylene, the molecular weight decreases with oxidative modification at the end of the broken chains but polyethylene, under the same conditions, undergoes cross-linking (Chapter 1). Antioxidants inhibit both processes and the effectiveness of processing stabilisers is monitored by measuring viscosity changes after processing for various times. This is done on a routine basis by measuring the amount of polymer extruded through a standard die in a given time and the "melt flow index" (MFI) monitored by this procedure is inversely related to the melt viscosity of the polymer. An alternative test for degradation during processing is to measure the change in MFI (ΔMFI) after a single or multiple passage of a polymer sample through a screw extruder [57]. The oxygen concentration during polymer processing is much lower than it is in liquid hydrocarbons under ambient conditions. Furthermore, radical generation is higher due to mechanical scission of the entangled polymer chains.

Under the above conditions, the quinonoid oxidation products of the hindered phenols (see Scheme 3.3) are much more effective mechanoantioxidants than the parent antioxidant, BHT, from which they are derived [47,53,57] (see Table 3.6).

Table 3.6 Equimolar effectiveness of quinonoid products derived from BHT as processing stabilisers [57]

Antioxidant[a]	ΔMFI[b]	
	4.5[c]	0.5[c]
BHT	100	270
G·	45	45
SQ	50	60
BQ	55	65

a For chemical structures see Scheme 3.3
b ΔMFI, percentage change in MFI on single passage through a screw extruder.
c Concentration of antioxidant, m mol kg^{-1}

Galvinoxyl, a "stable" aryloxyl radical was observed in an ESR study [53,58,62,63] to undergo continuous reversible reduction and re-oxidation under these conditions. The concentrations of galvinoxyl (G·, Table 3.4, XVI) and its cognate phenol, GH show complementary oscillation with the continuous formation of unsaturation in the polymer (see Fig. 3.3).

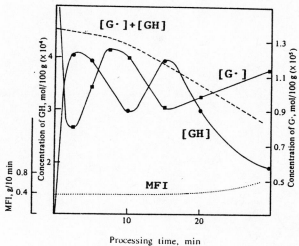

Fig. 3.3 Complementary oscillation of galvinoxyl [G·] and hydrogalvinoxyl [GH] concentrations during mechanooxidation of polypropylene during processing in an internal mixer at 200°C. MFI is the melt flow index of the polymer. Initial [G·] = 4.74 x 10^{-4} mol/100g. (Adapted with permission from R. Bagheri, K.B. Chakraborty and G. Scott, *Polym. Deg. Stab.*, **5**, 145 (1983).

The chemical interpretation of Fig. 3.3 is shown in Scheme 3.7. G· and GH together form a catalytic redox couple, removing on two radicals in each cycle of the chain reaction. This process is analogous to that observed by Berger et al with aminoxyls in liquid hydrocarbons (Section 3.5) and indeed the same oscillating phenomenon occurs when a piperidinoxyl radical, TMPO (Table 3.4, R=OH), is processed with polypropylene [59-61] or PVC [64].

Unsaturation was formed in the polymer and again no spin adducts were observed at 180°C. At ambient oxygen pressures, the oscillating process was rapidly quenched due to the destruction of the antioxidant radicals by reaction with peroxyl radicals at higher oxygen concentrations. This irreversible process occurs slowly even at the relatively low oxygen concentrations which obtain in commercial mixers and consequently the redox couple becomes much less effective as the oxygen concentration increases. For the same reason, f, which in the "closed" system was found to be approximately 50, decreased rapidly with increasing rate of oxygen diffusion into the system. The oscillating behaviour does however depend on the slow ingress of oxygen into the system [66] and is analogous to similar oscillations reported in some inorganic systems [65].

In the absence of antioxidant, the MFI of polypropylene begins to increase immediately but galvinoxyl and piperidinoxyls delay this change for up to 20 minutes at temperatures in the region of 200°C [58,63].

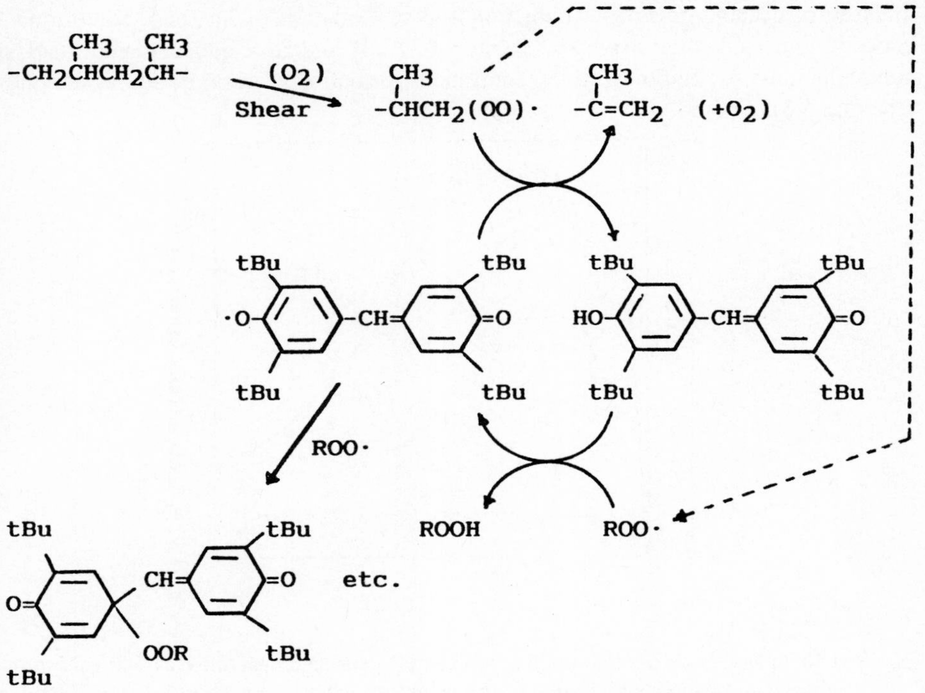

Scheme 3.7 Catalytic antioxidant activity of galvinoxyl (G·) during processing of polypropylene

Other oxidising agents behave similarly to the stable phenoxyls and aminoxyls in polypropylene during processing. These include Cu^{2+} (but not Fe^{3+}), I_2 and alkyl iodides (see Table 3.7). They are, however, of little practical utility in polymers due to their prooxidant behaviour under service conditions (see below).

Table 3.7 Catalytic Chain-breaking Antioxidants

GENERAL CATALYTIC MECHANISM

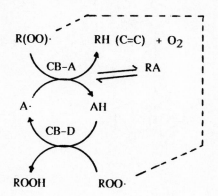

EXAMPLES	CONDITIONS	REDOX REACTION	
		A·	AH
Amines, aminoxyls R', R" = aryl R', R" = *tert*-alkyl	Δ, hυ	R'\N-O·/R"	R'\N-OH/R"
Phenols, phenoxyls	Δ	G·	GH
Benzoquinones, Stilbenequinones, etc	Δ	Q	QH·
Semiquinones	Δ	QH·	QH_2
Copper salts	Δ, hυ	Cu^{2+}	Cu^+
Iodine, alkyl iodides	Δ	I·	HI
Nitro compounds	hυ	O=N-O·	O=N-OH

One of the most effective processing stabilisers for polypropylene is α-tocopherol and from the standpoint of the packaging technologist, a major advantage of this naturally occurring antioxidant is its non-toxicity under conditions where it could "migrate" into foodstuffs [67]. Although α-tocopherol is not very effective in heat ageing (hot air oven) tests, this is of little practical concern and may even be an advantage in packaging where long term durability is an important cause of pollution due to the persistence of many items of packaging in the environment [68]. In the melt stability tests described above, α-tocopherol is more effective than conventional commercial processing stabilisers [67,69]. As in the case of BHT, the mechanoantioxidant activity of α-tocopherol (α-Toc-OH) is due, not to the parent phenol, but to the derived quinonoid products, notably the oxidative spirodimer (α-Toc-sd) and to a smaller extent, α-tocopheryl quinone (α-Toc-q). (see Scheme 3.8).

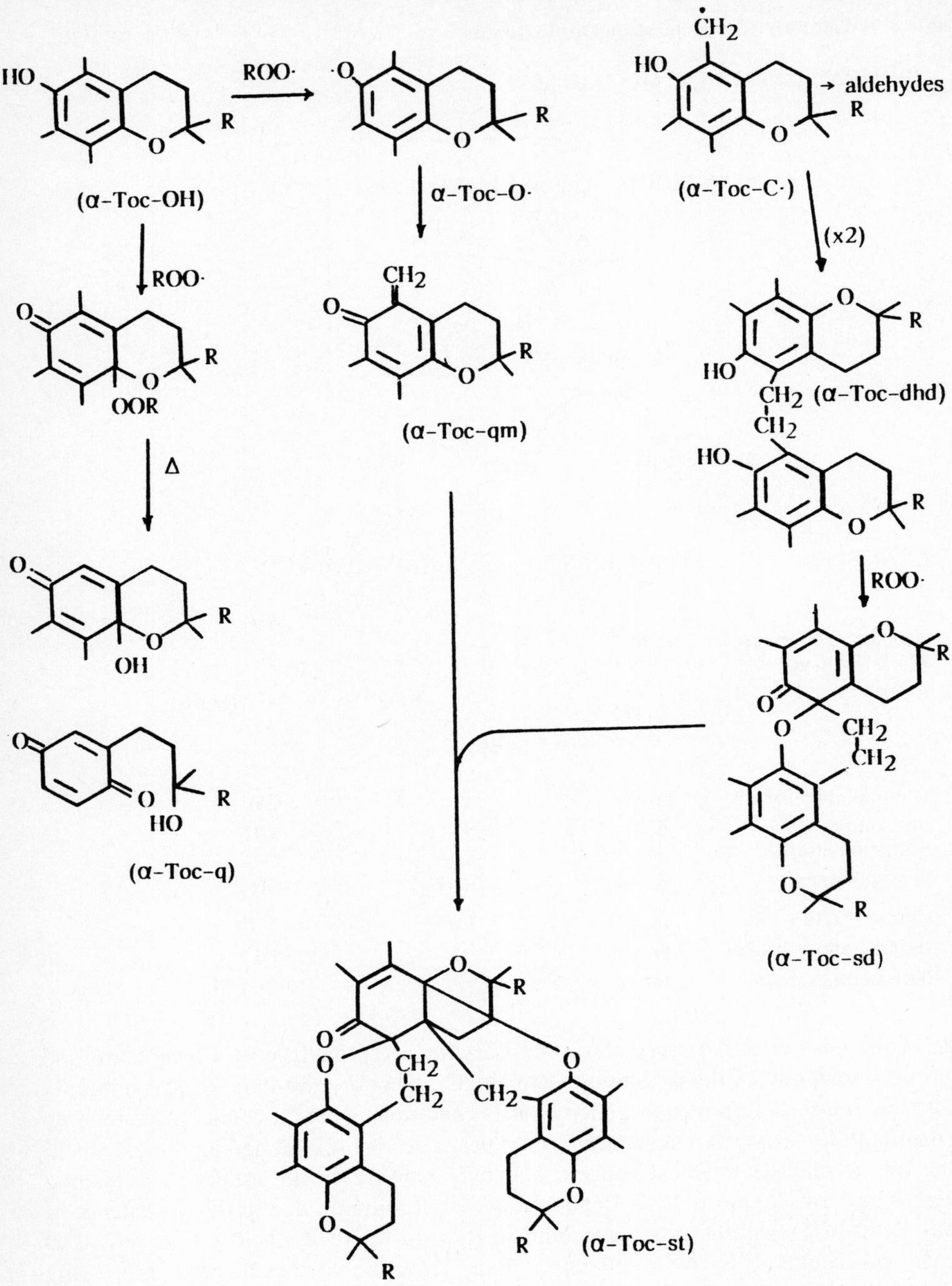

Scheme 3.8 Oxidation of α-tocopherol

These products are effective mechanoantioxidants in their own right and unlike α-tocopheryloxyl (α-Toc-O·), which does not appear to be reversibly reduced back to α-

Toc-OH by macroalkyl radicals in polymers, the quinones all form reversible catalytic antioxidants in combination with their cognate phenols. The quinonoid products, α-Toc-sd and α-Toq-q, are reduced back to the cognate phenols, α-tocopheryl dehydrodimer (α-Toc-dhd) and a-tocopheryl hydroquinone (α-Toc-hq) by mild reducing agents and it has been shown in solution studies [70] that α-Toc-sd is reduced to α-Toc-dhd even by α-Toc-OH. The proposed catalytic mechanism involving both Toc-sq and Toc-q in polypropylene is shown in Scheme 3.9 [67].

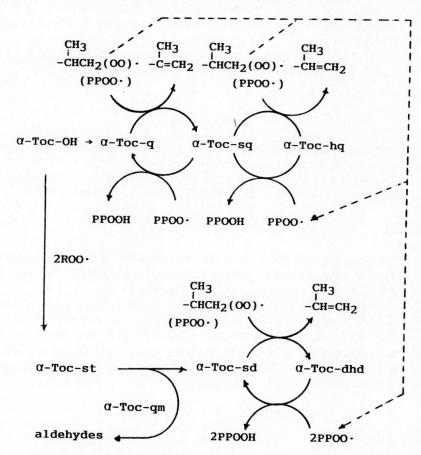

Scheme 3.9 Catalytic antioxidant activity of α-tocopheroquinone (α-Toc-q) and tocopherol *spiro*-dimer (α-Toc-sd) in polypropylene

It has recently been proposed [71] that the α-Toc-q/α-Toc-hq redox couple may also be involved in the antioxidant activity of α-Tocopherol *in vivo* in which a membrane-bound flavin enzyme acts as the reducing agent. However, at the relatively low oxygen concentrations found in many cells, reduction by polyconjugated allyl or allylperoxyl radicals cannot be ruled out:

$$\text{RCH}_2\text{CH}=\text{CHCH}=\text{CHCHR'} \xrightarrow[\text{(OO·)}]{\alpha\text{-Toc-q} \quad \alpha\text{-Toc-sq}} \text{RCH}=\text{CHCH}=\text{CHCH}=\text{CHR'} + \text{O}_2 \quad (9)$$

It was noted above that cupric stearate behaves in exactly the same way as the "stable" phenoxyls and aminoxyls in polypropylene during high temperature processing [72]. At lower temperatures, cupric ions react with alkyl radicals to give copper alkyls [73], but the latter are unstable at high temperatures or in the presence of light and the predominant reaction is under these conditions the observed reaction is olefin formation in the polymer backbone:

$$\text{Cu}^+\text{R} \xrightarrow{\Delta, h\nu} \text{Cu}^+ + >\text{C}=\text{C}< + \text{H}^+$$

$$\text{Cu}^{2+} \begin{array}{c} \xrightarrow{\text{R·}} \text{CB-A} \\ \xrightarrow{\text{ROOH}} \text{Cu}^+ + \text{ROO·} + \text{H}^+ \end{array} \quad (10)$$

Cu^+ is then readily reoxidised by peroxyl radicals in a CB-D reaction (see Table 3.7). However, cuprous ion decomposes hydroperoxides to alkoxyl radicals in a Fenton type reaction and this is the dominant reaction occurring in rubber and in lubricating oils at ambient oxygen pressures [74] where the CB-A/CB-D catalytic cycle is inefficient:

$$\text{Cu}^+ + \text{ROOH} \rightarrow \text{Cu}^+ + \text{RO·} + \text{OH}^- \quad (11)$$

In hydrocarbon polymers under photooxidative conditions where the rate of initiation is high and oxygen concentration is limited by diffusion, copper stearate behaves as a photoantioxidant [75] although it is not a thermal antioxidant.

Spin-traps are a potentially important source of aminoxyl radicals during polymer processing. Sohma demonstrated the mechanical scission of polyethylene at low temperatures by ESR. At 77K, comminution of polyethylene was found to give macroalkyl radicals at the end of the broken chain whereas γ-irradiation gives in-chain radicals [76]. However, even at 77K continued mechanical action causes disproportionation of the primary alkyl radical from polyethylene to give in-chain radicals:

$$\text{-CH}_2\text{CH}_2\text{·} + \text{-CH}_2\text{CH}_2\text{CH}_2\text{-} \xrightarrow{77K} \text{-CH}_2\text{CH}_3 + \text{-CH}_2\overset{\bullet}{\text{C}}\text{HCH}_2\text{-} \quad (12)$$

At 313K alkyl radicals are rapidly converted to alkylperoxyls which in turn decay. Spin trapping of radicals produced in polymers in solution by ultrasound have also failed to show the formation of the primary mechanochemical chain scission product. Thus Terabe et al. [77] showed that the main identified radicals spin trapped by pentamethylnitrosobenzene during ultrasonic degradation of polypropylene in benzene solution at 40°C were the in-chain isomers.

Scott and co-workers [13,78-82,206] applied the spin-trapping technique in polyolefins during high temperature processing. Both nitroso and nitrone spin-traps were found to give well characterised triplet spectra due to nitroxyl radicals [79]. The spin adducts were found to be highly effective mechanoantioxidants (see Table 3.8 and 3.9), stabilising the polymer viscosity for 15-30 minutes at 180°C with the formation of unsaturation in the macromolecule. The formation of hydroperoxides was retarded or even inhibited completely for more than 20 mins relative to a control without spin-trap. Only a fraction of the spin-traps was measured as nitroxyl radicals and the nitroxyl concentration rose to a maximum and then decayed. The mechanoantioxidant effect persisted long after the maximum radical concentration was attained, consistent with the proposed catalytic mechanism in Scheme 3.5.

Table 3.8 Nitroso spin-traps as catalytic mechanoantioxidants in polypropylene during processing

Nitroso Compound	Code	τ^*, min	g value
$(CH_3)_3C-N=O$	MNP	17	2.0059
$(CH_3)_3CCH_2C(CH_3)-N=O$	TMB	13	2.0065
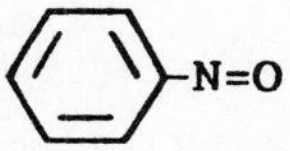 $C_6H_5-N=O$	NB	10	2.0071
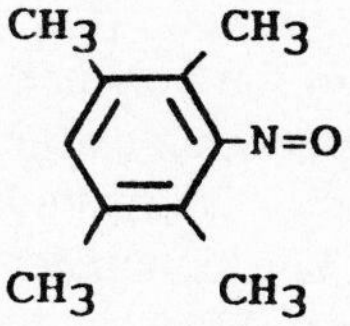 2,6-(CH$_3$)$_2$C$_6$H$_3$-N=O	NTMB	–	2.0062
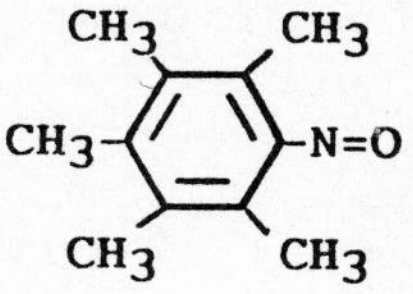 2,4,6-(CH$_3$)$_3$C$_6$H$_2$-N=O	NPMB	>20	2.0065
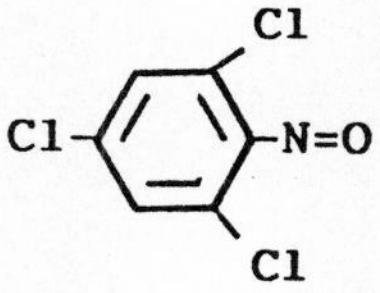 2,4,6-Cl$_3$C$_6$H$_2$-N=O	NTCB	–	–
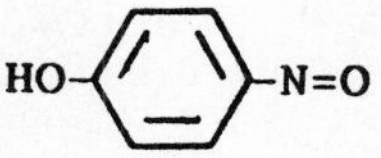 HO-C$_6$H$_4$-N=O	NHB	–	–
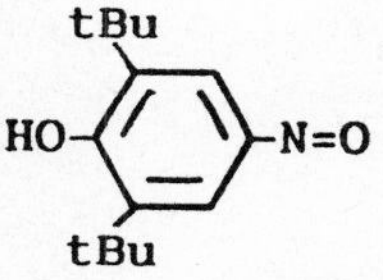 3,5-tBu$_2$-4-HO-C$_6$H$_2$-N=O	NBHB	–	–
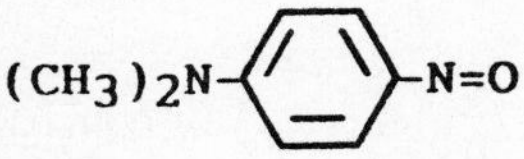 $(CH_3)_2N$-C$_6$H$_4$-N=O	NHN	–	–

* τ = Induction time to MFI increase at 180°C

Table 3.9 Aldonitrones as catalytic mechanoantioxidants in polypropylene during processing

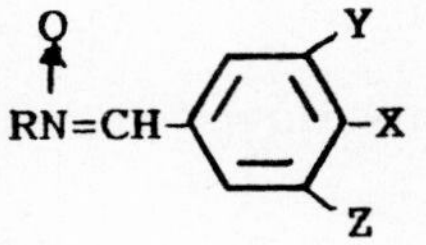

Aldonitrone				Code	MFI (10 min at 180°C, 10^{-3} mol/100g)
R	X	Y	Z		
Ph	OH	H	H	HDPN	0.28
Ph	NO_2	H	H	NDPN	0.37
Ph	OMe	H	H	MDPN	0.36
tBu	H	H	H	PBN	0.48
tBu	OH	H	H	HPBN	0.31
tBu	OH	Me	Me	MHPBN	0.27

A more detailed examination of the behaviour of 4-nitrosodiphenylamine (NDPA) in an internal mixer at 120°C in ethylene-propylene copolymer, showed [82] that approximately 50% of the spin-traps became chemically attached to the polymer after 10 mins processing in an internal mixer. 90% of the resulting product was extractable with an efficient solvent. However, only 67% of the original nitroso compound was present in the polymer as nitroxyl (Scheme 3.10, reaction (a)). The remainder was reduced to the corresponding alkyl hydroxylamine (reaction (b)) and free hydroxylamine by further thermal elimination with the formation of unsaturation in the polymer (reaction (c)).

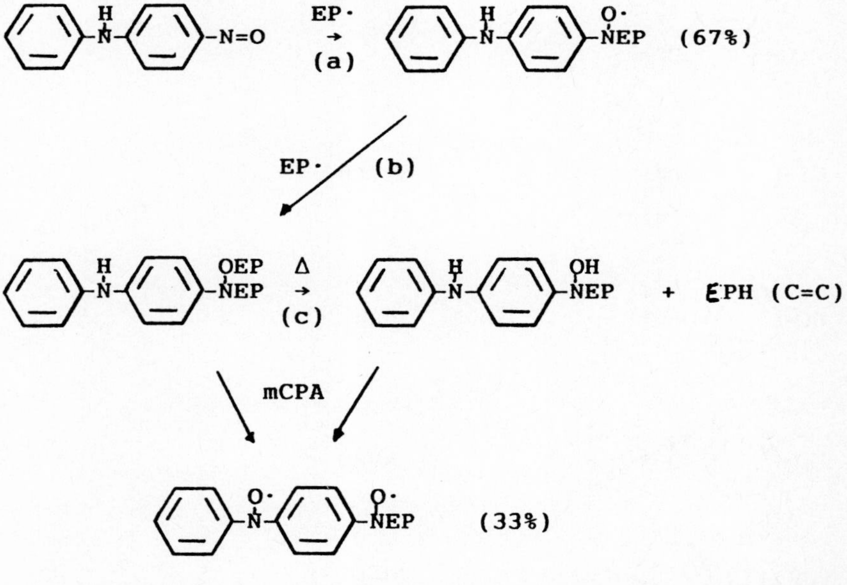

Scheme 3.10 Mechanochemical covalent attachment of 4-nitrosodiphenylamine to ethylene-propylene copolymer (EP-H) at 120°C

Both macroalkyl hydroxylamine and free hydroxylamine were shown to be readily oxidised to the corresponding aminoxyl by peracids in the cavity of the ESR spectrometer even at ambient temperatures. Peracids are co-products of mechano-oxidation by peroxidation of aldehydes and this provides an alternative mechanism for aminoxyl regeneration at lower temperatures. Other aromatic nitroso compounds behaved similarly. For example NPMB gave ~80% of polymer-bound adduct of which ~25% was nitroxyl and the remainder hydroxylamines [82]. Transient cross-linking of the polymer was observed with some nitroso compounds during processing [79,80]. This is consistent with the formation of thermally unstable macroalkyl hydroxylamine by reaction (b) in Scheme 3.10, giving rise to the free hydroxylamine which is a highly effective CB-D antioxidant. However. oxidation by peroxides and peracids during processing would also lead to regeneration of nitroxyl by a different mechanism [83].

Spin-trapping of mechanochemically generated macroalkyl radicals in vinyl polymers gives rise to relatively unstable aminoxyl radicals due to the presence of hydrogen on the α-carbon atom in the nitroxyl which can then disproportionate as shown in Scheme 3.11. These reactions proceed in parallel with the catalytic CB-A/CB-D cycle described in Scheme 3.5.

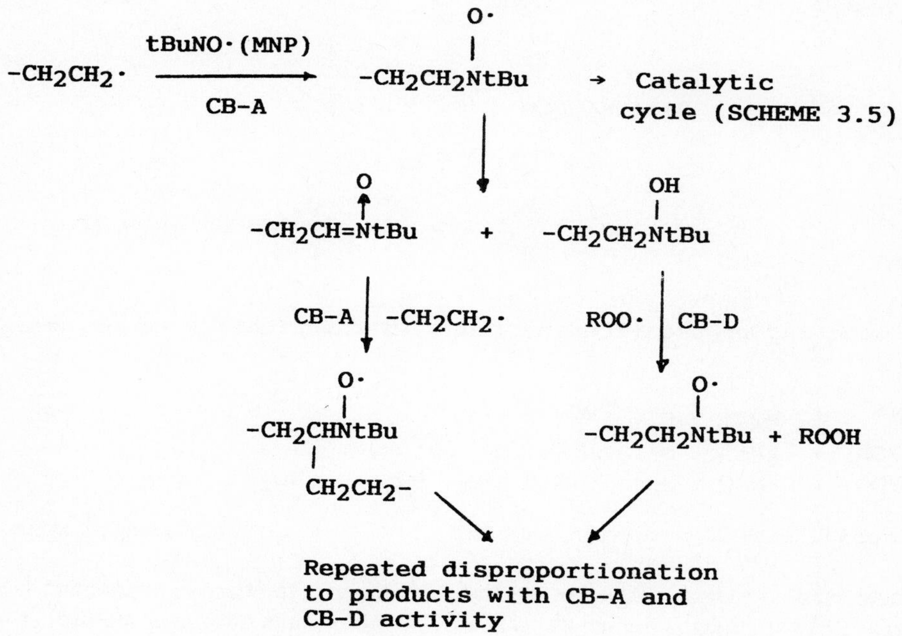

Scheme 3.11 Repeated radical trapping by MNP

The sustained removal of both mechanoalkyl and mechanoalkylperoxyl radicals is consistent with the remarkable mechanoantioxidant activity of spin-traps in polyolefins.

Aromatic spin-traps generally act as effective thermal antioxidants but poor photo-antioxidants whereas the reverse is true of aliphatic spin-traps. This will be discussed further below.

Nitrosamines have been shown [84] to be very effective mechanoantioxidants for polyolefins. Diaryl nitrosamines are used as vulcanisation retarders during cross-linking of rubber and they are known to generate nitric oxide (·N=O) which is a powerful alkyl radical trap [85]. N-nitrosodiphenylamine, NNDPA, is one of the most effective processing stabilisers for polypropylene so far reported and unlike its aliphatic analogue, dialkyl ntrosamine, NNDCA, it is effective even at relatively high oxygen concentrations:

NNDPA   NNDCA

Dialkyl nitrosamines cause extensive cross-linking of polypropylene, particularly during the early stages of processing, and it has been proposed, by analogy with the reactions of ·N=O in solution [86] that this is a result of the consecutive reactions of the initially formed nitric oxide with macroalkyl radicals in the polymer, Scheme 3.12.

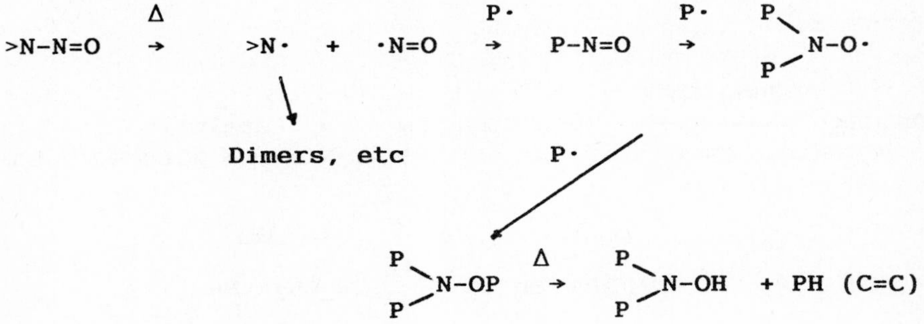

$$>N-N=O \xrightarrow{\Delta} >N\cdot + \cdot N=O \xrightarrow{P\cdot} P-N=O \xrightarrow{P\cdot} P_2N-O\cdot$$

$$\downarrow$$

Dimers, etc

$$P_2N-OP \xrightarrow{\Delta} P_2N-OH + PH\,(C=C)$$

Competing propagation reactions in the presence of oxygen;

$$P\cdot + O_2 \rightarrow POO\cdot$$
$$POO\cdot + >N\cdot \rightarrow >N-O\cdot + PO\cdot$$
$$POO\cdot + \cdot N=O \rightarrow POONO \rightarrow PO\cdot + NO_2$$

Scheme 3.12 Reactions of nitrosamines and derived nitric oxide in polymers during processing

Aromatic aminyl radicals formed in Scheme 3.12 dimerise to effective antioxidant (see Scheme 3.4) but aliphatic aminyl radicals are not antioxidants until oxidised further to aminoxyl. The evidence suggests then that two different types of aminoxyl are formed under these conditions; the first derived from the aminyl radical and the second from reactions of nitric oxide with macroradicals in the polymer, and the extent of these transformations depends strongly on competing prooxidant reactions which increase with increasing oxygen concentration in the system (see Scheme 3.12). It will be seen below that the aminoxyl products of the reaction which have been characterised by ESR [81] are very effective antifatigue agents in rubbers and photoantioxidants in polyolefins, and this chemistry is also highly relevant to the behaviour of ·N=O in biological systems.

(b) During fatiguing of rubbers

Fatigue was seen in Chapter 1 to be the mechanochemical oxidation of cross-linked rubbers through the formation of macroalkyl radicals under conditions of dynamic deformation at ambient temperatures [87]. The resulting peroxides initiate further oxidative degradation of the rubber with loss of mechanical properties. The effectiveness of antifatigue agents is normally measured as the time taken for breakage of a rubber strip subjected to cyclical extension and relaxation [88]. Quinones have been known for many years to give protection under these conditions [89], extending appreciably the time to crack formation and sample failure. Katbab et al. [56] found that 2,6-di-*tert*-butyl benzoquinone and 3,5,3',5' tetra-*(tert*-butyl) stilbene-4,4'-quinone (SQ) (Table 3.4, XVII, XVIX) which are oxidation products of BHT (Scheme 3.3) were significantly more effective than the latter as mechanoantioxidants for rubber during "fatiguing" at ambient temperatures. Quinones are thus able to oxidise allyl or more likely allylperoxyl radicals formed during the mechanooxidation of rubber. Scheme 3.13 outlines the mechanism proposed to account for the antioxidant activity of quinones in cis-polyisoprene under conditions of limited oxygen concentration and rapid mechanochemical generation of macroalkyl radicals and this is closely analogous to the action of quinones during polymer processing at higher temperatures (see Table 3.7). The semiquinones may either disproportionate to quinone and hydroquinone or participate further in redox reactions with allylperoxyl with which it may act either as a reducing agent or as an oxidising agent depending on the conditions (see Scheme 3.13).

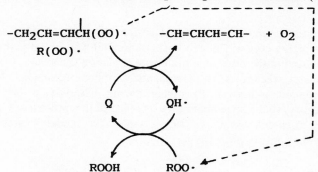

$2QH\cdot \rightarrow Q + QH_2$

$QH\cdot + -CH_2CH=CHCH(OO)\cdot \rightarrow QH_2 + -CH=CHCH=CH- + O_2$

Scheme 3.13 Antifatigue activity of quinones

Hindered phenols have almost no antifatigue activity in rubbers. Aromatic amines and in particular the 4-alkylaminodiphenylamines (e.g. Table 3.1, IPPD, XIII) by contrast are very widely used and represent a major group of large tonnage antioxidants used in tyre technology. Diarylamines give rise to two kinds of oxidation product with CB-A activity [90,91] , The first are the quinoneimines of varying molecular weight (see Scheme 3.4) [22] which behave similarly to the quinones and the second are the derived aminoxyls which are formed from the parent arylamines during the early stages of fatigue ageing (see for example Fig. 3.4). Nitroxyl concentration was found to rise to a maximum and then fall to a low stationary value in the rubber vulcanisate [92] during fatiguing at ambient temperatures.

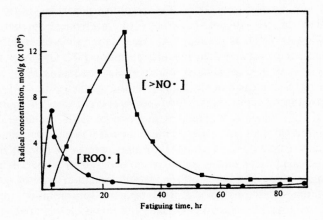

Fig. 3.4 Growth and decay of alkylperoxyl [ROO·] and aminoxyl [>NO·] concentrations during the flexing (mechanooxidation) of rubber containing 1g/100g of IPPD at ambient temperature. Reproduced with permission from A. Katbab and G. Scott, *Chem. Ind.*, 573 (1980).

The nitroxyl is partially reduced to hydroxylamine and partly to the parent amine during mechanooxidation even at ambient temperatures [85,93], and it is this facility to be oxidised and reduced in the process of inhibiting oxidation (Scheme 3.14) that makes the aromatic amines so effective as "antidegradants" in rubber technology. The catalytic activity of the aromatic aminoxyls in rubber during fatiguing is of considerable theoretical significance, since the temperatures involved in this process are considerably lower than those discussed earlier in this Chapter for model hydrocarbons and thermoplastic polymers during processing.

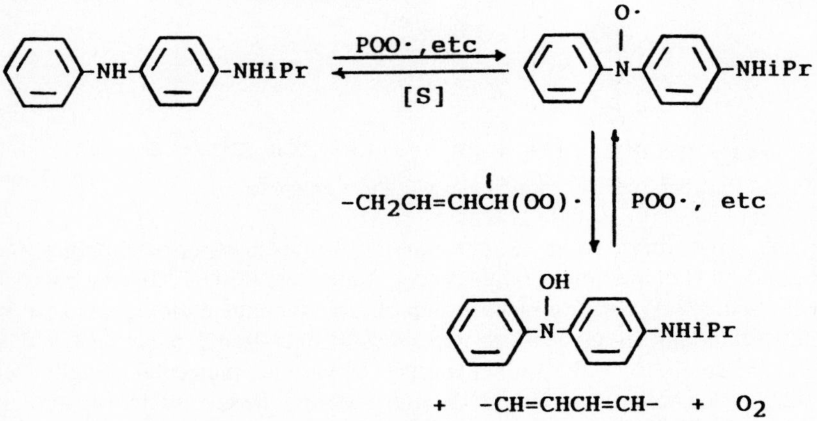

Scheme 3.14 Reversible oxidation and reduction of IPPD during fatiguing (mechanooxidation) of rubber

The driving force for the reaction of nitroxyl in Scheme 3.14 is the lability of the allylic C-H bond and the formation of conjugated double bonds in the rubber molecule (reaction 8). There is convincing evidence for the formation of conjugated dienes in the rubber by Scheme 3.14, suggesting that direct hydrogen abstraction plays the major role in the catalytic mechanism rather than the formation of an intermediate alkyl hydroxylamine [94,94]. However, irrespective of the detailed mechanism, it is clear that nitroxyl can redox cycle under relatively low temperature conditions in an unsaturated peroxidising substrate, deactivating many alkyl and peroxyl radicals in the process. This mechanism is borne out by the fact that the derived nitroxyls and hydroxylamines are at least as effective and in some cases more effective than the parent arylamines (Table 3.10). Moreover, the 2,2',6,6'-tetramethyl piperidines (TMP) are essentially inactive as antifatigue agents as are their derived oxidation products. Unlike the arylamines, TMP has no CB-D activity and although the nitroxyl may be formed, it is reduced back to the amine by sulphur compounds in the vulcanisate [85,93].

The N-nitrosamines derived from the arylamine antioxidants are also catalytic antioxidants are in every case more effective as antifatigue agents in vulcanised rubbers than the parent amine [85,95] (Table 3.10). Particularly surprising is the high mechano-antioxidant activity of N-nitrosodiphenylamine, NNDPA, (Table 3.10, R =R' = H, X = NO) which is more effective than the commercial antifatigue agent, IPPD (Table 3.10, R = H, R' = NiPr, X = H). NNDPA has been used for many years as a retarder for rubber vulcanisation and it seems likely that macromolecular nitroxyls derived from ·N=O discussed in the previous section make a major contribution to the antifatigue activity of the nitrosamines.

Table 3.10 Antifatigue (mechanoantioxidant) activity of diarylamines and derived products [85, 93]

Antioxidant		Fatigue life, h			
		X = H	X = O·	X = OH	X = NO
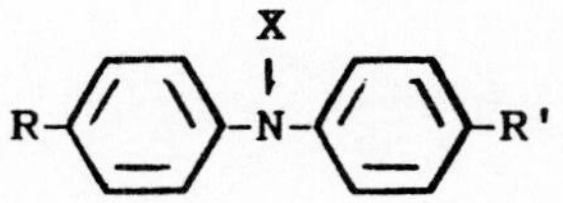 R	R'				
H	NiPr	274	337	–	400
MeO	MeO	207	411	56[a]	336
tOct	tOct	83	88	107	112
H	H	209	–	–	345
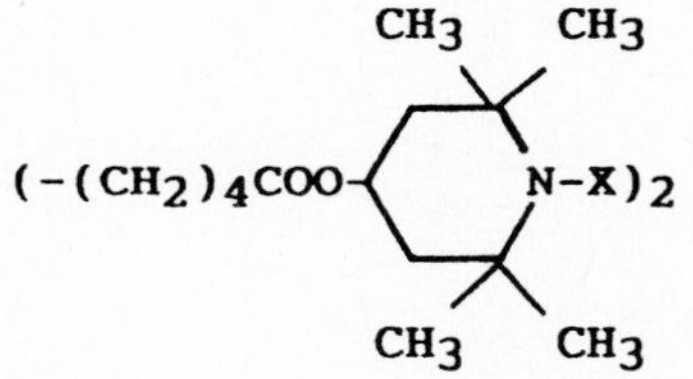 (-(CH$_2$)$_4$COO-[TMP])$_2$		25	30	39	–

[a] insoluble in rubber

C-nitroso aromatic compounds are also partially converted to nitroxyl and macroalkyl hydroxylamines during processing in EP rubbers (see Section (a)). The adduct formed from 4-nitrosodiphenylamine (NDPA) in EPR was found to be somewhat less effective than IPPD as an antifatigue agent at the same concentration in natural rubber (see Table 3.11) but it was more effective than IPPD after solvent extraction.

Table 3.11 Antifatigue activity of polymer-bound nitroso spin adducts in natural rubber sulphur vulcanisates (antioxidant conc.1g/100g) [82]

Antioxidant	Cycles to break ($\times 10^{-2}$)	
	Unextracted	Extracted
None	400	345
NDPA adduct	1150	500
NPMB adduct	860	380
IPPD additive	2200	315

The aminoxyl adduct from NPMB which contains no antioxidant function other than nitroxyl was moderately effective before extraction and as effective as IPPD after extraction. Antifatigue activity after extraction reflects the fact that the aminoxyls are partially chemically attached to the EPR "carrier". Mechanochemistry may thus be turned to advantage in producing more substantive antioxidants in the fabricated product.

Nitrones have some activity as antifatigue agents which appears to depend on the extent of mechanooxidation before vulcanisation [96]. 4-Hydroxy-α-phenyl nitrones are more effective than analogous nitrones without a hydroxy group in the aromatic ring (see Table 3.12) and, like the aromatic amines, they appear to function initially as CB-D antioxidants by hydrogen abstraction from phenol, although nitroxyl radicals are the only unpaired electron species that can be detected by ESR [96]. The proposed mechanism is shown in Scheme 3.15.

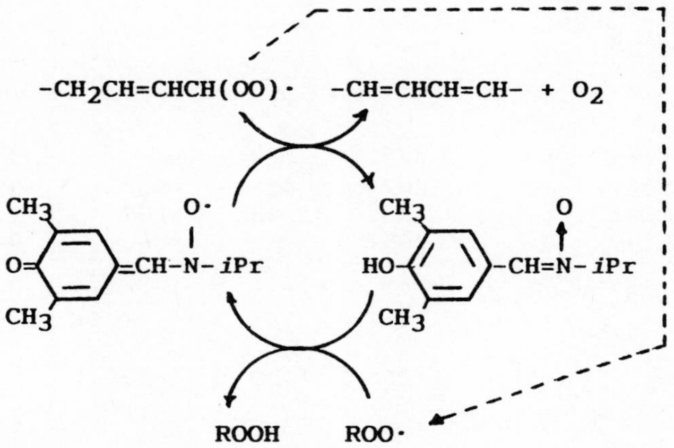

Scheme 3.15 Antifatigue mechanism of 4-Hydroxy-α-phenyl nitrones

It is clear from Table 3.12 that the mechanoantioxidant activity of α-phenyl nitrones is very sensitive to substituents in the phenyl ring but the normal rules governing hindered phenol activity do not apply. MHPBN, MHPPN, partially hindered phenols are more effective than the fully hindered BHPMN and BMHPPN but this may be a consequence of the physical behaviour (e.g. volatility) of the additives in the polymer since BMHPPN, a bis phenol, is more effective than its lower molecular weight homologue, BHPMN [96]. As might be anticipated, the thermal (air oven) antioxidant activity of this class of antioxidants is again broadly related to molecular weight.

Table 3.12 Aldonitrones as antifatigue agents in a sulphur vulcanisate [96]

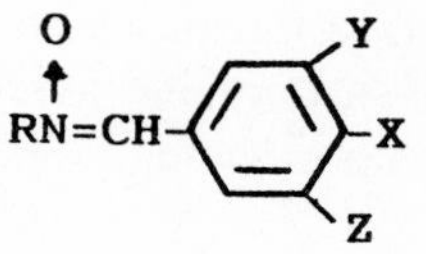

Aldonitrone				Code	Time to fatigue failure, hours
R	X	Y	Z		
tBu	H	H	H	PBN	26
tBu	OH	H	H	HPBN	30
tBu	OH	Me	Me	MHPBN	110
iPr	OH	Me	Me	MHPPN	130
Me	OH	Me	Me	MHPMN	80
$(CH_2)_2$	OH	tBu	tBu	BMHPPN	105
Me	OH	tBu	tBu	BHPMN	45
No additive					20
IPPD as additive					270

(c) Photoantioxidants

Until the mid-1970s the most effective light stabilisers were the "UV absorbers" (UVAs) which acted primarily by absorbing the damaging short wavelengths of the sun's spectrum. As will be seen later this is an oversimplification since many UV absorbers have CB-D activity [97] and they powerfully synergise with peroxide decomposers [98]. However, in their CB capacity, they act sacrificially and are relatively ineffective alone. A very significant discovery was that the 2,2,6,6-tetramethylpiperidines (TMP) are very effective light stabilisers, although they do not absorb significantly in the UV.

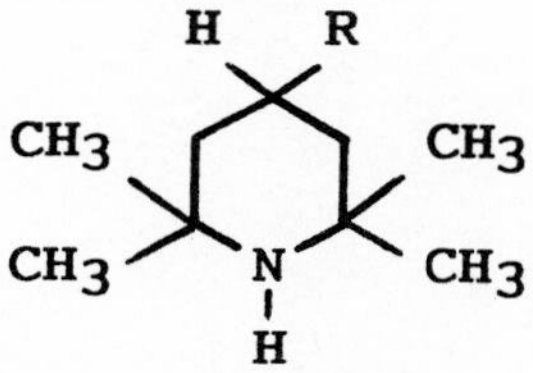

TMP 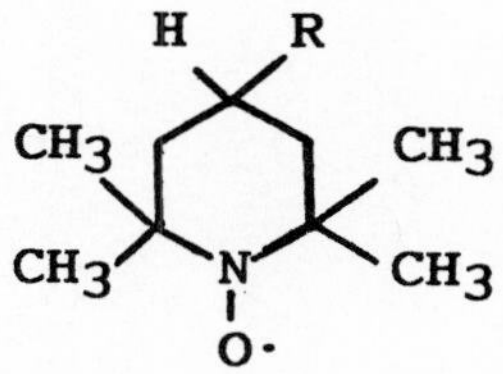   TMPO

Unlike the aromatic amines, aliphatic amines are not chain-breaking antioxidants under thermal conditions due to the high N-H bond strength [99]. They do however undergo redox reactions with hydroperoxides to give a mild prooxidant effect during processing with the formation of carbonyl compounds in the polymer (Scheme 3.16 [100]).

The high photoantioxidant activity of the TMPs which were popularly described as "hindered amine light stabilisers" (HALS) stimulated a great deal of scientific interest in the mechanism of their action. Explanations proposed for their activity were that they could quench singlet oxygen [101] or triplet carbonyl [102], or that they could complex prooxidant metal ions [103]. Although these mechanisms have not been completely rejected and may contribute to photoantioxidant activity in specific situations, they cannot by themselves account for the extraordinary efficiency of the HALS.

A key to the mechanism of the hindered piperidines was that during photooxidation in the polymer they were oxidised to the corresponding "stable" aminoxyl radicals [104] (Scheme 3.13). As noted above, aliphatic amines are not antioxidants and their conversion to nitroxyls involves radical generation. Thus during the initial stages of photooxidation HALS catalyse carbonyl formation in the polymer in the absence of chain-breaking antioxidants [105].

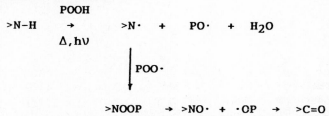

Scheme 3.16 Transformation of TMP (>N-H) during processing and photooxidation of polymers

During both thermal and photooxidation, aminoxyl concentration rises to a maximum and then falls to a lower stationary concentration. This is shown for the photooxidation of polypropylene in Fig. 3.5.

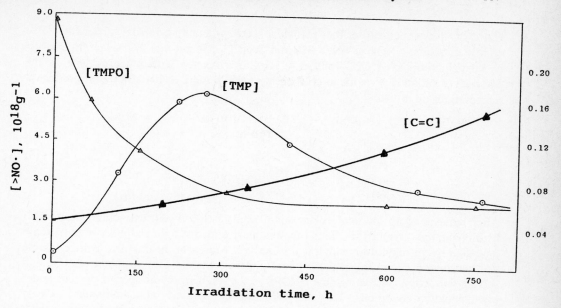

Fig. 3.5 Growth and/or decay of [>NO·] concentration from TMP and TMPO during photooxidation of polypropylene at ambient temperature. [C=C] indicates the formation of olefinic unsaturation in the polymer. Initial concentrations: [TMP] = 3 x 10^{-4} mol/100g, [TMPO] = 6 x 10^{-4} mol/100g. Adapted from *Polym. Deg. Stab.* **4**, 11 (1982) with permission.

Macroalkyl hydroxylamines (>NOP, Scheme 3.17)) are the main reservoirs for the nitroxyl radical in the polymer and when pre-formed in solution they regenerate nitroxyl on re-oxygenation [106]. There is evidence from the formation of unsaturation in polymers [105,107] that under photoxidative conditions elimination of hydroxylamine occurs in a reaction analogous to that observed thermally at higher temperatures (Scheme 3.17).

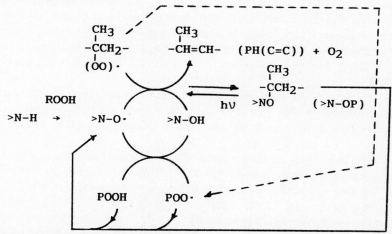

Scheme 3.17 Catalytic chain-breaking photoantioxidant mechanism of aminoxyls (>N-O·) in polypropylene

However, it was seen earlier that peroxidic species (notably peracids) in polymers also readily re-oxidise >NOP to >NO·. Hu and Scott [83] found that 2,2,6,6-tetramethyl-4-hydroxypiperidinoxyl (XXI, Table 3.4, R = OH) reacted with polypropylene in the presence of a dialkyl peroxide to give mainly >NOPP with >NOH as a minor product at temperatures up to 140°C:

$$PP-H + RO· \rightarrow ROH + PP· \xrightarrow[140°C]{>NO·} PPON< + (>NOH)$$

$$\downarrow R\overset{O}{C}OOH$$

$$PPC=O + >NO· \qquad (13)$$

After removal of all low molecular weight products by solvent extraction, macroalkyl hydroxylamine (>NOPP) was found to be a more effective light stabiliser in polypropylene than the nitroxyl from which it was derived. Moreover, macroalkyl hydroxylamine was substantially retained in the polymer under repeated cycles of photooxidation and solvent extraction, indicating that >NOPP is the main reservoir species present under photooxidative conditions. This results in much improved photoantioxidant activity compared with conventional HALS which is rapidly extracted from the polymer under the same conditions (Table 3.13).

Table 3.13 Comparison of >NOPP with related hindered amine (TMP) under conditions of alternating UV exposure (100 h) and acetone extraction (10 h)

HALS in PP film	UV/extraction cycles	Embrittlement time, h
0.4% >NOPP	8	800
0.4% TMP	1	<100

It has been found [97,105] that free hydroxylamines (>NOH) are more effective than the nitroxyls (>NO·), which are in turn more effective photoantioxidants than the parent HALS (>NH). However, under conditions where physical loss of low molecular weight additives is important, for example in domestic applications where polypropylene fibres may be exposed to solvent leaching or in outdoor film applications in greenhouses, etc., then macroalkyl hydroxylamines are more effective than lower molecular weight alternatives.

TMPs are among the more effective stabilisers against high energy radiation of polymers [108]. Again they appear to act again through the derived aminoxyls, which are highly effective scavengers for macroalkyl radicals formed by radiolysis of the polymer.

It was seen in Section (a) above that in-chain aminoxyl radicals can be formed by mechanochemical reactions of spin-traps with polymers. It might therefore be expected that the resulting macroalkyl nitroxyls could have photoantioxidant activity. This has been shown to be the case for the aliphatic C-nitroso spin adducts (e.g. from 2-methyl-2-nitrosopropane, MNP) which have similar activity to the piperidinoxyls. By contrast, spin-adducts containing aromatic groups [79] are almost devoid of activity due to the photo-destabilising effect of the aromatic ring. The effectiveness of MNP increases with increasing processing time, corresponding to an increase in [>NO·] in the polymer. The aromatic α-nitrones behave very similarly to the aromatic nitroso compounds and it was shown that the aromatic nitroxyls are similarly photo-unstable [80].

Nitrosamines derived from TMP, for example DNTMP, are much more effective photo-antioxidants on a molar basis than are the amines from which they are derived [512]:

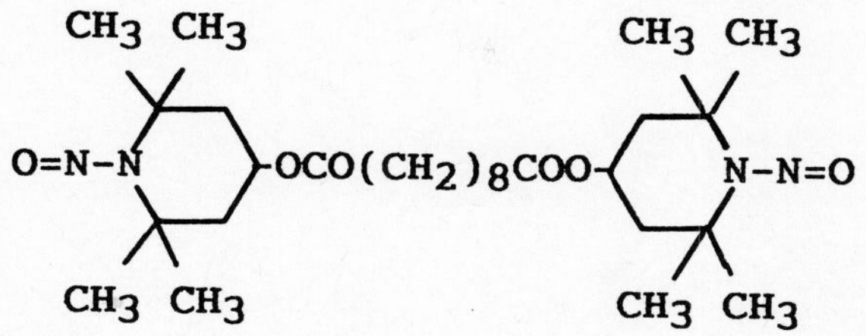

$$O=N-N\underset{CH_3\ CH_3}{\overset{CH_3\ CH_3}{\bigg\langle}}\!-OCO(CH_2)_8COO-\underset{CH_3\ CH_3}{\overset{CH_3\ CH_3}{\bigg\rangle}}\!N-N=O$$

As was noted above, two different types of aminoxyl radical are formed in this system and these differ considerably in stability depending on the substitution on the carbon α- to nitrogen. Nitroxyl radicals with an α-hydrogen, formed by reaction between ·NO and polyolefin macroradicals, readily undergo disproportionation to nitrones and hydroxylamines, both of which are themselves antioxidants; the first by the spin trapping (CB-A) mechanism and the second by the hydrogen donating (CB-D) mechanism as discussed previously (see Scheme 3.12). As in the case of other aminoxyl photoantioxidants, there is a steady increase in unsaturation in the polymer containing nitrosamines during the induction time [81], confirming the catalytic mechanism.

Unbranched dialkyl nitroxyls, hydroxylamines and nitrones constitute a potentially powerful class of catalytic CB-A/CB-D antioxidants and this principle has recently been exploited commercially in the development of N,N-dialkyl hydroxylamine antioxidants and light stabilisers (Irganox FX) which have been reported to act as shown in reaction 14 [109]:

$$(RCH_2CH_2)_2N\text{-}OH \xrightarrow{ROO\cdot} (RCH_2CH_2)_2N\text{-}O\cdot \rightarrow \begin{array}{c} \overset{\uparrow O}{RCH_2CH_2N=CHCH_2R}\ (CB\text{-}A) \\ + \\ (RCH_2CH_2)_2NOH\ \ (CB\text{-}D) \end{array} \quad (14)$$

Aliphatic nitro compounds are also very effective photoantioxidants for polyolefins, comparing with TMP in efficiency [110]. Low molecular weight additives such as 2-nitro-2-methylpropane are too volatile in the polymer to be commercially useful, but higher molecular weight nitroalkanes and in particular macromolecular nitroalkanes are highly effective [111]. Solvent extraction has relatively little effect on their activity of nitrated polypropylene (NPP) (Table 3.14).

Table 3.14 Photoantioxidant activity of nitroalkanes

R-NO$_2$	UV Embrittlement time, ν						
	2*	4*	5*	7*	10*	15*	20*
(CH$_3$)$_3$C-NO$_2$	550	-	800	-	400	260	260
(CH$_3$)$_3$CCH$_2$C(CH$_3$)$_2$-NO$_2$	690	-	800	-	1070	850	800
Ph-NO$_2$	200	-	250	-	300	300	300
PP-NO$_2$(NPP)(Unext)	290	430	-	580	-	-	-
(Ext)	280	410	-	565	-	-	-

* Concentration, 10^3 mol/100g

Unlike the aliphatic nitroso compounds, aliphatic nitro compounds are thermally stable but photolytically unstable and undergo two photolysis processes in parallel to give NO$_2$ and hyponitrous acid [112,113]:

$$RCH_2CH_2NO_2 \xrightarrow{h\nu} RCH=CH_2 + HONO \quad (15)$$
$$RCH_2CH_2NO_2 \xrightarrow{h\nu} RCH_2CH_2\cdot + NO_2$$

Detailed analysis of the products formed in polyolefins has shown that both NO$_2$ and HNO$_2$ are involved in the photoantioxidant activity in a catalytic mechanism (see Scheme 3.18) analogous to that discussed above for the >NO·/>NOH couple. Confirmation for this mechanism is the very rapid formation of unsaturation in the polymer and the more gradual accumulation of oxidation products, nitric acid and nitrite and nitrate esters. Nitric acid and nitrate esters are probably the major stable end-products and their formation irreversibly removes the photoantioxidant species from the polymer.

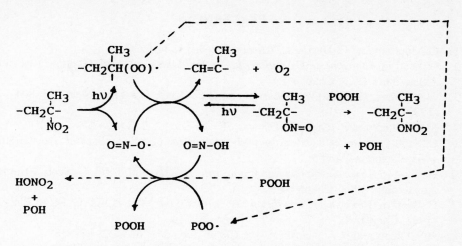

Scheme 3.18 Photoantioxidant mechanism of nitroalkyl compounds

A new light stabiliser for polypropylene based on the above principle has recently been introduced to the market by Akzo Nobel as Armosorb UV 101.

REFERENCES

1. C. Moureau and C. Dufraisse, *Chem. Rev.*, **3**, 113 (1926-7).
2. G. Scott in *Atmospheric Oxidation and Antioxidants*, Vol. I Ed. G. Scott, Elsevier, Amsterdam, 1993, Chapter 1.
3. C.D. Lowry, G. Egloff, J.C. Morrell and G.C. Dryer, *Ind. Eng. Chem.*, **25**, 804 (1933).
4. J.L.Bolland and P. ten Have, *Discuss. Farad. Soc.*, **2**, 252 (1947).
5. G. Scott, *Atmospheric Oxidation and Antioxidants* (1st Ed.), Elsevier, Amsterdam, 1965, Chapters 4 and 5.
6. G. Scott in *Atmospheric Oxidation and Antioxidants*, Vol. I, Ed. G. Scott, Elsevier, Amsterdam, 1993, Chapter 4.
7. G. Scott in *Atmospheric Oxidation and Antioxidants*, Vol. II, Ed. G. Scott, Elsevier, (1993, Chapter 9).
8. H.S. Olcott and H.A. Mattill, *J. Am. Chem. Soc.*, **58**, 1627, 2204 (1936).
9. B. Halliwell and J.M.C. Gutteridge, *Free Radicals in Biology and Medicine*, 2nd Ed., Clarendon Press, Oxford, 1989.
10. K.J.A. Davies in *Free Radicals and Oxidative Stress*, Eds. C. Rice-Evans, B. Halliwell and G.G. Lunt, *Biochem. Soc. Symp.*, **61**, Portland Press, 1995, pp.11-18.
11. B. Halliwell in *Free Radicals and Oxidative Stress*, Eds. C. Rice-Evans, B. Halliwell and G.G. Lunt, *Biochem. Soc. Symp.*, **61**, Portland Press, 1995, p.75.
12. J.W. Tetrud and J.W. Langston, *Science*, **243**, 519-22 (1989); The Parkinson Study Group, New England J. Med., **321**, 1364-71 (1989); **328**, 176-88 (1993).
13. G. Scott in *Developments in Polymer Stabilisation-7*, Ed. G. Scott, Applied Science Pub., 1984, London, Chapter 2.
14. G.W. Burton and K.U. Ingold, *J. Am. Chem. Soc.*, **103**, 6472 (1981).
15. G.W. Burton and K.U. Ingold, *Science*, **224**, 569-73 (1984).
16. D.D.M. Wayner, G.W. Burton, K.U. Ingold, L.R.C. Barkley and S.J. Locke, *Biochim. Biophys. Acta*, **924**, 408-19 (1987).
17. E. Niki, *Am. J. Nutr.*, **54**, 1119S-24S (1991).
18. B. Frei, *Am. J. Nutr.*, **54**, 1113S-18S (1991).
19. J. Pospíšil in *Developments in Polymer Stabilisation-1*, Ed. G. Scott, App. Sci. Pub., London, 1979, Chapter 1.
20. J. Pospíšil in *Developments in Polymer Stabilisation-7*, Ed. G. Scott, Elsevier App. Sci., London, 1984, Chapter 1.
21. J. Pospíšil, *Polym. Deg. Stab.*, **34**, 85-109 (1991).
22. J. Pospíšil, *Adv. Polym. Sci.*, **124**, 87-189 (1995).
23. T. Colclough in *Atmospheric Oxidation and Antioxidants*, Vol. II, Ed. G. Scott, Elsevier, Amsterdam, 1993, Chapter 1.
24. G. Scott, *Atmospheric Oxidation and Antioxidants*, Elsevier, 1965, Chapters 6-10.
25. F. Gugumus in *Developments in Polymer Stabilisation-8*, Ed. G. Scott, Elsevier App. Sci., London, 1987, Chapter 6.
26. G.N. Foster in *Oxidation Inhhibition in Organic Materials*, Eds. J. Pospíšil and P.P. Klemchuk, CRC Press, Vol. II, 1990, Chapter 7.
27. S. Al-Malaika in *Atmospheric Oxidation and Antioxidants*, Vol. I, Ed. G. Scott, Elsevier, Amsterdam, 1993, Chapter 2.

28. D. Munteanu in *Mechanisms of Polymer Degradation and Stabilisation*, Ed. G. Scott, Elsevier App. Sci., London, 1990, Chapter 7.
29. C.E. Boozer, G.S. Hammond, C.E. Hamilton and J.N. Sen, *J. Am. Chem. Soc.*, **77**, 3238 (1955).
30. L.F. Fieser, *J. Am. Chem. Soc.*, **52**, 9 (1930).
31. J.K. Becconsall, S. Clough and G. Scott, *Trans. Farad. Soc.*, **56**, 459-72 (1963).13.
32. A.F. Bickel and E.C. Kooyman, *J. Chem. Soc.*, 2217 (1957).
33. A.Tkáč in *Developments in Polymer Stabilisation-8*, Ed. G. Scott, Elsevier App. Sci., London, 1987, Chapter 3.
34. W.L. Hawkins and H.J. Sautter, *J. Polym. Sci.*, **A1**, 3499 (1969).
35. J.R. Shelton in *Developments in Polymer Stabilisation-4*, Ed. G. Scott, Applied Science Pub., London, 1982, Chapter 2.
36. G. Scott in *Developments in Polymer Stabilisation-6*, Ed. G. Scott, Applied Science Pub., London, 1982, Chapter 2.
37. S. Al-Malaika, K.B. Chakraborty and G. Scott in *Developments in Polymer Stabilisation-6*, Ed. G. Scott, App. Sci. Pub., London, 1983, Chapter 3.
38. G. Scott, *Atmospheric Oxidation and Antioxidants*, Elsevier, Amsterdam, 1965, pp. 392-400.
39. G. Scott in *Developments in Polymer Stabilisation-4*, Ed. G. Scott, Applied Science Pub., London, Chapter 6.
40. G. Scott in *Atmospheric Oxidation and Antioxidants*, Vol. II, Elsevier, Amsterdam, 1993, Chapter 5.
41. N.C. Billingham in *Atmospheric Oxidation and Antioxidants*, Vol. II, Ed. G. Scott, Elsevier, Amsterdam, 1993, Chapter 4 .
42. C. Walling, *Free Radicals in Solution*, Wiley, New York, 1957, p.166.
43. M. Szwarc and J.H. Binks, *Kekulé Symp.*, Chem. Soc. Pubn., 1958, p.262.
44. N.A. Milas, *Chem. Rev.*, **10**, 295 (1932).
45. M. Pike and W.F. Watson, *J. Polym. Sci.*, **9**, 229 (1952).
46. W.F. Watson, *Trans. I.R.I.*, **29**, 32 (1953).
47. G. Scott in *Atmospheric Oxidation and Antioxidants*, Vol.II, Elsevier, Amsterdam, 1993, Chapter 3.
48. F. Haber and R. Willstätter, *Ber.* **64**, 2844 (1931).
49. A. Ogunbanjo and G. Scott, *Europ. Polym. J.*, **21**, 541 (1985).
50. A.A. Katbab, A. Ogunbanjo and G. Scott, *Polym. Deg. and Stab.*, **12**, 333 (1985).
51. E.T. Denisov in *Developments in Polymer Stabilisation-5*, Ed. G. Scott, App. Sci. Pub., London, 1982, Chapter 2.
52. J.B. Adeniyi and G. Scott, *Polym. Deg. Stab.*, **17**, 117-29 (1987).
53. R. Bagheri, K.B. Chakraborty and G. Scott, *Chem. Ind.*, 865 (1980).
54. E.T. Denisov, *Izv., AN SSSR, Ser. Khim.*, 327 (1969).
55. H. Berger, T.A.B.M. Bolsman and D.M. Brouwer in *Developments in Polymer Stabilisation-6*, Ed. G. Scott, App. Sci. Pub., London, 1983, Chapter 1.
56. E.T. Denisov, *Russ. Chem. Revs*, **65** (1996).
57. T.J. Henman in *Developments in Polymer Stabilisation-1*, Ed. G. Scott, App. Sci. Pub., London, 1979, Chapter 2.
58. R. Bagheri, K.B. Chakraborty and G. Scott, *Polym. Deg. Stab.*, **5**, 145 (1983).
59. R. Bagheri, K.B. Chakraborty and G. Scott, *J. Polym. Sci.*, **22**, 1573-8 (1984).
60. S. Al-Malaika, E.O. Omikorede and G. Scott, *Polym. Commun.*, **27**, 173 (1986).

61. S. Al-Malaika, E.O. Omikorede and G. Scott, *J. App. Polym. Sci.*, **33**, 703 (1989).
62. G. Scott, *Makromol. Chem.*, Macromol. Symp., **27**, 1-23 (1989).
63. G. Scott in *Developments in Polymer Stabilisation-7*, Ed. G. Scott, Elsevier App. Sci., London, 1984, Chapter 2.
64. J.B. Adeniyi, S. Al-Malaika and G. Scott, *J. App. Polym. Sci.*, **32**, 6063-71 (1986).
65. I. Prigogine and R. LeFever, *J.Chem. Phys.*, **48**, 1695 (1968).
66. J. Li, PhD Thesis, *Chemical Modification of Polymers*, Aston University, 1988.
67. G. Scott in *Free Radicals and Oxidative Stress*, Eds. C. Rice-Evans, B. Halliwell and G.G. Lunt, *Biochem. Symp.*, **61**, Portland Press, London, 1995, pp.235-46.
68. *Degradable Polymers, Principles and Applications*, Eds. G. Scott and D. Gilead, Chapman & Hall, London, 1995.
69. S. Al-Malaika, Z-A. Lin and G. Scott, unpublished work.
70. W.A. Skinner and P. Aloupovic, *Science*, **140**, 803 (1963).
71. I. Kohar, M. Baca, C. Suarna, R. Stocker and P.T. Southwell-Keely, *Free Rad. Biol. Med.*, **19**, 197-207 (1995).
72. R. Bagheri, K.B. Chakraborty and G. Scott, *Polym. Deg. Stab.*, **11**, 1-7 (1985).
73. J.K. Kochi, *J. Am. Chem. Soc.*, **84**, 3271 (1962).
74. G. Scott, *Atmospheric Oxidation and Antioxidants*, Elsevier, Amsterdam, 1965, p. 260, 402.
75. F. Rasti and G. Scott, *Europ. Polym. J.*, **16**, 1153-8 (1980).
76. J. Sohma in *Developments in Polymer Degradation-2*, Ed., N. Grassie, App. Sci. Pub., London, 1979, Chapter 4.
77. S. Terabe, K. Kuruma and R. Konaka, *J. Chem. Soc.*, Perkin Trans II, 1252 (1972)
78. K.B. Chakraborty and G. Scott, *J. Polym. Sci.*, **22**, 553-8 (1984).
79. K.B. Chakraborty, G. Scott and H. Yaghmour, *J. App. Polym. Sci.*, **30**, 189-203 (1985).
80. K.B. Chakraborty, G. Scott and H. Yaghmour, *J. App. Polym. Sci.*, 3267-81 (1985).
81. K.B. Chakraborty, G. Scott and H. Yaghmour, *Polym. Deg. Stab.*, **10**, 221-35 (1995).
82. G. Scott, *Polym. Deg. Stab.*, **48**, 315-24 (1995).
83. X-J. Hu and G. Scott, *Polym. Deg. Stab.*, **52**, 301-4 (1996).
84. K.B. Chakraborty, G. Scott and H. Yaghmour, *Polym. Deg. Stab.*, **10**, 221-35.
85. H.S. Dweik and G. Scott, *Rubb. Chem. Tech.*, **57**, 908 (1984).
86. B.A. Gingrass and W.A. Waters, *J. Chem. Soc.*, 1920 (1954).
87. E.V. Reztsova, B.G. Lipkena and G.L. Slonimskii, *Zh. Fiz. Khim.*, **33**, 656 (1959).
88. G. Scott, *Atmospheric Oxidation and Antioxidants*, Elsevier, Amsterdam, 1965, Chapter 6.
89. G. Scott, *Atmospheric Oxidation and Antioxidants*, Elsevier, Amsterdam, 1965, p.469-475.
90. J. Pospišil in *Developments in Polymer Stabilisation-4*, Ed. G. Scott, App. Sci. Pub., London, 1979, Chapter 1.
91. G. Scott in *Developments in Polymer Stabilisation-4*, Ed. G. Scott, App. Sci. Pub., London, 1981, Chapter 2.
92. A.A. Katbab and G. Scott, *Chem. Ind.*, 573 (1980), *Europ. Polym. J.*, **17**, 559 (1981).
93. H.S. Dweik and G. Scott, *Rubb. Chem. Tech.*, **57**, 735 (1984).
94. L.P. Nethsinghe and G. Scott, *Europ. Polym. J.*, **20**, 213 (1984).

References

95. G. Scott, *Rubb. Chem. Tech.*, **58**, 269-83 (1985).
96. L.P. Nethsinghe and G. Scott, *Rubb. Chem. Tech.*, **57**, 779-91 (1984).
97. G. Scott in *Atmospheric Oxidation and Antioxidants*, Vol.II, Ed. G. Scott, Elsevier, Amsterdam, 1993, Chapter 8.
98. G. Scott in *Atmosheric Oxidation and Antioxidants*, Vol.II, Ed. G. Scott, Elsevier, Amsterdam, 1993, Chapter 9.
99. V.Ya. Shlyapintokh and V.B. Ivanov in *Developments in Polymer Stabilisation-5*, Ed. G. Scott, App. Sci. Pub., London, 1982, p.44.
100. K.B. Chakraborty and G. Scott, *Chem. Ind.*, 237 (1978).
101. V.B. Ivanov, V.Ya. Shlyapintokh, O.M. Khvostach, A.B. Shapiro and E.G. Rozantsev, *J. Photochem.*, **4**, 313 (1975)0.
102. N.S. Allen, J. Homer and J.F. McKellar, *Macromol. Chem.*, **179**, 1575 (1978).
103. S.P. Fairgreave and J.R. McCallum, *Polym. Deg. Stab.*, 8, 107 (1984).
104. V.Ya. Shlyapintokh, E.V. Bystrikzkaya, A.B. Shapiro, L.N. Smirnov and E.G. Rozantsev, Izv. Akad. *Nauk SSSR, Ser. Khim.*, 1915 (1973).
105. R. Bagheri, K.B. Chakraborty and G. Scott, *Polym. Deg. Stab.*, **4**, 1-16 (1982).
106. D.J. Carlsson, A. Garton and D.M. Wiles, *Developments in Polymer Stabilisation-1*, Ed. G. Scott, App. Sci. Pub., London, 1979, Chapter 7.
107. K.B. Chakraborty and G. Scott, *Polymer*, **21**, 252 (1980). (1985).
108. D.J. Carlsson in *Atmospheric Oxidation and Antioxidants*, Vol. II, Elsevier, Amsterdam, 1993, Chapter 11.
109. H. Zweifel, *11th Bratislava IUPAC/FECS Int. Conf. on Polymers, Thermal and Photoinduced Oxidation of Polymers and its Inhibition in the Upcoming 21st Century*, June 24-28, 1996, ML13.
110. S. Al-Malaika, T. Czechai, G. Scott and L.M.K. Tillekeratne, *Polym. Deg. Stab.*, **26**, 375-84 (1989).
111. G. Scott and S. Al-Malaika, *US Pat.*, 5,098,957 (1992).
112. R.E. Rebbert and N. Slagg, *Bul. Soc. Chim. Belg.*, **71**, 709 (1962).
113. R.H. Whitfield and D.I. Davies, *Polym. Photochem.*, **1**, 261-74 (1981)

4

Preventive Antioxidants, Synergism and Technological Performance

Hydroperoxides are the most important molecular products of the peroxidation chain reaction. Because of their instability and consequent radical generating capacity under a wide range of environmental conditions, the removal of hydroperoxides without the formation of free radicals constitutes a very important and complementary antioxidant mechanism to the CB processes discussed in the last Chapter. Peroxide decomposition by ionic peroxidolysis (PD) is therefore a ubiquitous preventive antioxidant mechanism. However, absorption (screening) of UV light (UVA) and deactivation of transition metal ions (MD) which inhibit peroxide photolysis or thermolysis are complementary to the CB and PD mechanisms (see Table 4.1), and since each antioxidant class operates in a different mechanism, the combined effect of additives from each class is synergistic. This will be discussed in detail in Section 4.4.

Table 4.1 Preventive antioxidants

(a) CATALYTIC HYDROPEROXIDE DECOMPOSERS (PD-C)

   CHEMICAL CLASS: SULPHUR COMPOUNDS

   KEY REACTIONS: $-S- \xrightarrow{ROOH} -\overset{O}{\underset{\|}{S}}- \xrightarrow{ROOH} -SO_2H \begin{matrix} \nearrow SO_x \\ \searrow -SO_3H \end{matrix}$

EXAMPLES		CODE	USE
$(ROCOCH_2CH_2)_2S$	DRTP	XXIV	P
$(R_2NCSS)_2M$	(a) ZnDRC, M=Zn	XXV	R, P
	(b) NiDRC, M=Ni		P
$(ROCSS)_2$	ZnRX, M=Zn	XXVI	R
$((RO)_2PSS)_2M$	(a) ZnDRP, M=Zn	XXVII	O
	(b) NiDRP, M=Ni		P
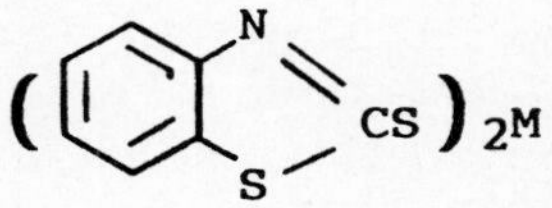 $\left(\underset{S}{\overset{N}{\bigcirc\!\!\!\!\!\bigcirc}}\!CS\right)_2 M$	(a) MBT, M=H	XXVIII	R
	(b) ZnMBT, M=Zn		R
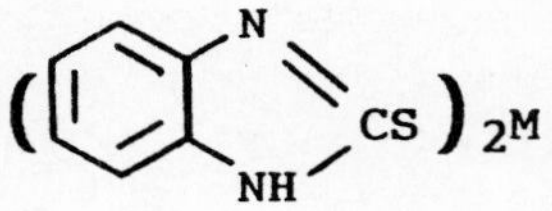 $\left(\underset{NH}{\overset{N}{\bigcirc\!\!\!\!\!\bigcirc}}\!CS\right)_2 M$	ZnMBI, M=Zn	XXIX	R

(b) STOICHIOMETRIC HYDROPEROXIDE DECOMPOSERS (PD-S)

   CHEMICAL CLASS: PHOSPHITE ESTERS

   KEY REACTION: $(RO)_3P + ROOH \rightarrow (RO)_3P=O + ROH$

EXAMPLE		CODE	USE
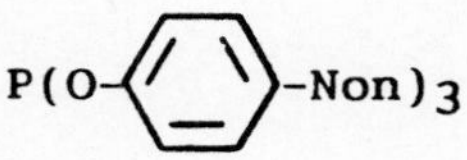 $P(O\text{-}C_6H_4\text{-}Non)_3$	TNPP	XXX	R, P

Table 4.1 (cont)

(c) METAL DEACTIVATORS (MD)

    CHEMICAL CLASS: PHENOLIC HYDRAZIDES

    KEY REACTIONS: CB-D, METAL CHELATION

EXAMPLE		CODE	USE
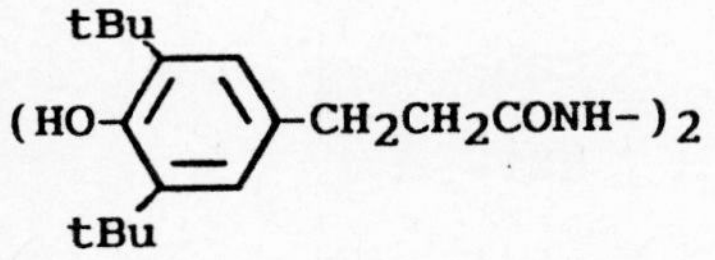 (HO–[tBu,tBu-C$_6$H$_2$]–CH$_2$CH$_2$CONH–)$_2$	1024MD	XXXI	P

(c) HYDROGEN CHLORIDE SCAVENGERS

    CHEMICAL CLASS: TIN CARBOXYLATES

    KEY REACTIONS: $R_2Sn(XR)_2 \xrightarrow{2HCl} R_2SnCl_2 + 2RXH$

		CODE	USE
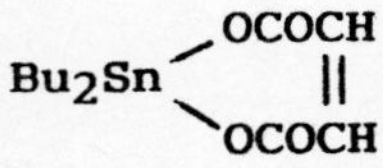 Bu$_2$Sn(OCOCH=OCOCH)	DBTM	XXXII	P
Oct$_2$Sn(SCH$_2$COOOct)$_2$	DOTG	XXXIII	P

(d) U.V. ABSORBERS (UVA)

    CHEMICAL CLASS: PHENOLS

    KEY REACTIONS: [keto-enol phototautomerism between X···H–O and X–H···O=C forms under $h\nu$ / $-\Delta$]

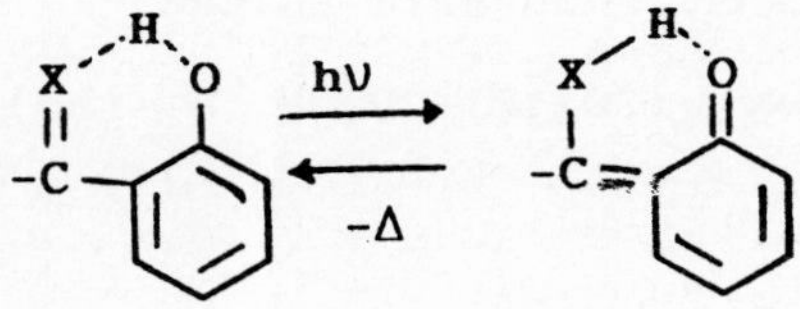

Table 4.1 (cont)

EXAMPLES		CODE	USE
(structure) HRBP		XXXIV	P
(structure) HRBT		XXXV	P
(structure) NiBOP		XXXVI	P

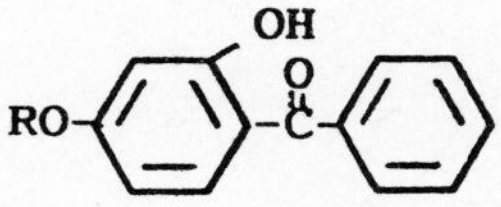

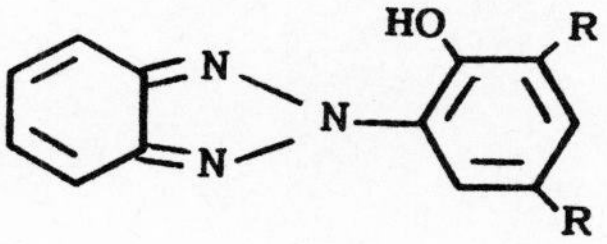

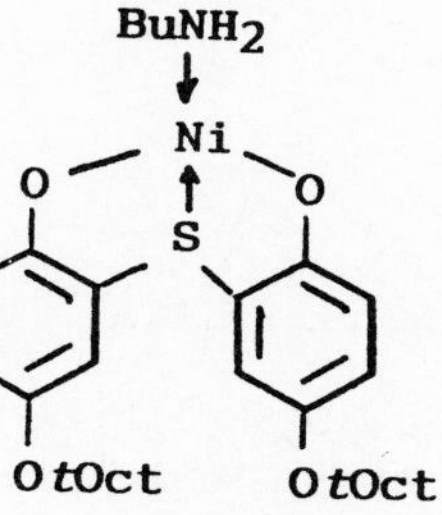

The practical performance of antioxidants and stabilisers frequently depends less on intrinsic antioxidant activity of the molecule than on its physical behaviour (solubility, volatility and diffusion rate) within the substrate, and the results of recent research to determine the optimal antioxidant structure will be reviewed in Section 4.5.

4.1 Peroxidolytic Mechanisms

Sulphur has held a central position in the stabilisation of lubricating oils and polymers for many years. Natural mineral oils contain indigenous sulphur compounds which have a major influence on the oxidative stability of "unpurified" oils [1,2], and in sulphur vulcanised rubbers the sulphur cross-links and accelerator transformation products play a dominant role in oxidative stabilisation [3]. The technological phenomena associated with oxidation have been reviewed in many publications [1-9] and in this section emphasis will be placed on the mechanisms involved in the antioxidant and prooxidant activity of peroxidolytic compounds.

4.1.1 *Aliphatic and aromatic sulphides*

An investigation of dialkyl sulphides related to the sulphur cross-link in rubber by Bateman and his co-workers [10-13] showed that simple alkyl monosulphides and disulphides act as antioxidants in model substrates and that activity was strongly dependent upon the chemical structure of the sulphide [11] (see Table 4.2). The straight chain dialkyl monosulphides were found to be much less effective than their α-branched dialkyl analogues, whereas in the dialkyl disulphides, the reverse was the case.

Table 4.2 Antioxidant activity of sulphides and their oxidation products in squalene [11]

Antioxidant	Concentration (M)	RR*
nBuSnBu	0.25	1.7
tBuStBu	0.25	256
Cyc$^+$StBu	0.25	150
MeCH=CHCHMeSMe	0.25	3.1
nBuSSnBu	0.25	121
tBuSStBu	0.25	1.6
PhSSPh	0.25	2.0
nBuSOnBu	0.01	1.0
tBuSOtBu	0.01	144
Cyc$^+ t$Bu	0.005	69
MeCH=CHCHMeSOMe	0.26	7
nBuSOSnBu	0.005	248
tBuSOStBu	0.004	140
PhSOSPh	0.01	68

* Retardation ratio
$^+$ Cyc = 2,3-cyclohexene

It was shown by Bateman et al [12] that the antioxidant activity of the aliphatic sulphides is associated with the thermal instability of the sulphoxides formed by oxidation with hydroperoxides. In subsequent kinetic studies Shelton and Davis [14] demonstrated that the homolysis product formed from the sulphoxide is itself thermally unstable and converts rapidly to thiolsulphinate in the absence of hydroperoxides and this in turn disproportionates to more stable products, reaction 1 [15]. Furthermore, there is a close relationship between the antioxidant effectiveness of the monosulphides and the first order rate constant for sulphoxide decomposition [16] (see Table 4.3).

$$(CH_3)_3CSC(CH_3)_3 \xrightarrow{ROOH} (CH_3)_3C\overset{O}{\underset{\|}{S}}C(CH_3)_3 \xrightarrow[\Delta]{k} (CH_3)_3CSOH + (CH_3)_2C=CH_2$$

$$\downarrow \times 2 \quad (1)$$

$$(CH_3)_3C\overset{O}{\underset{\|}{\overset{\|}{S}}}SC(CH_3)_3 \;+\; (CH_3)_3CSSC(CH_3)_3 \xleftarrow[\times 2]{\Delta} (CH_3)_3C\overset{O}{\underset{\|}{S}}SC(CH_3)_3 + H_2O$$

The above mechanism accords with the fact that sulphides inhibit autooxidation in an autoretarding mode [9,17]; that is, they are not initially antioxidants and may even be prooxidants [18], but they become antioxidants during autooxidation.

Table 4.3 First order rate constants for the decomposition of sulphoxides at 100°C [15]

$$R_1 \overset{O}{\underset{\|}{S}} R_2$$

R_1	R_2	$10^6 k\ (s^{-1})$	k_{rel}*
nPr	nPr	0.06	0.02
iPr	iPr	6.0	0.7
Me	tBu	6.3	1.0
iPr	tBu	46	4.4
tBu	tBu	1170	93
Ph	iPr	1.3	0.31
Ph	tBu	205	33
nhept	$(CH_2)_2COOEt$	390	280
$(CH_2)_2COOEt$	$(CH_2)_2COOEt$	850	300

*Corrected for the number of β-hydrogens

The long-chain thiodipropionate esters, DRTP (Table 4.1, XXIVI) are widely used as antioxidants in thermoplastic polymers and have also been permitted for use as antioxidants in fatty foods [19]. Mechanism studies of the thiodipropionate esters and their derived sulphoxides in model substrates [17,18] showed that they were catalysts for hydroperoxide decomposition. At low hydroperoxide to sulphoxide molar ratios, they show an initial prooxidant effect before being converted to effective antioxidants in a hydroperoxide initiated peroxidation [18,20] (see Fig. 4.1).

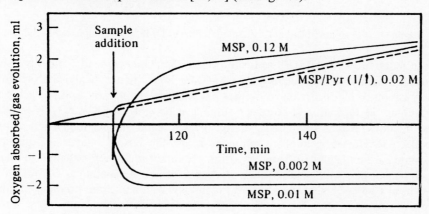

Fig. 4.1 Peroxidation of cumene initiated by cumene hydroperoxide (CHP. 0.1M) in the presence of methyl-β-sulphinoproprionate (MSP) and its pyridine salt. Numbers on curves are concentrations of MSP, M. (Reproduced with permission from C. Armstrong, M.J. Husbands and G. Scott, *Europ. Poly. J.*, **15**, 244 (1979)).

Armstrong and Scott [21] showed that the formation of sulphenic acid is reversible but becomes irreversible when rapidly removed by oxidising radicals (e.g. galvinoxyl or alkyperoxyl) or by hydroperoxides. However, a combination of hydroperoxide with dimethylsulphinyldiproprionate (DMSDP) at low molar ratios gave free radicals capable of initiating autooxidation and polymerisation and it was concluded that the derived sulphenic acid was responsible for this effect. This chemistry is summarised in Scheme 4.1:

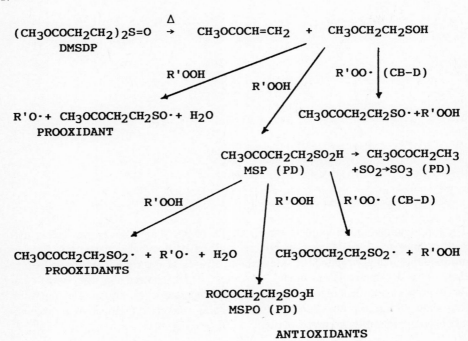

Scheme 4.1 Prooxidant and antioxidant activities of DMSDP decomposition products

At high molar [ROOH]/[DSDP] ratios, however, the sulphenic acid is rapidly oxidised to sulphinic and sulphonic acids [22] which are ionic (non-radical) catalysts for the decomposition of hydroperoxides. Further oxidation of the organic sulphur acids leads ultimately to the formation of inorganic sulphur acids, SO_2 (H_2SO_3), SO_3 (H_2SO_4) [22]. Bases, including excess sulphoxide, inhibit the acid catalysed decomposition of hydroperoxide [23]. However, like the sulphenic acids, the further oxidation products are prooxidants at low [ROOH]/[S] molar ratios. SO_2 and the further oxidation products of the sulphenic acid, MSP and MSPO, showed a similar behaviour to DRDSP [23]. It was seen in Chapter 1 that the ionic and homolytic decomposition products of cumene hydroperoxide are characteristically different (Scheme 1.1); and based on this chemistry, a diagnostic product analysis developed by Oberright [24] permitted a distinction to be made between a radical and ionic decomposition of cumene hydroperoxide by sulphur-containing antioxidants. At [CHP]/[SO_2] molar ratios greater than 1, the decomposition was found to be almost exclusively ionic but at lower ratios, radical reactions play a significant part. Hydroxyl and sulphinyl radicals have been observed by ESR during the reaction of hydroperoxides with SO_2 [25], and the competition between the two opposed processes is shown in Scheme 4.2 [18].

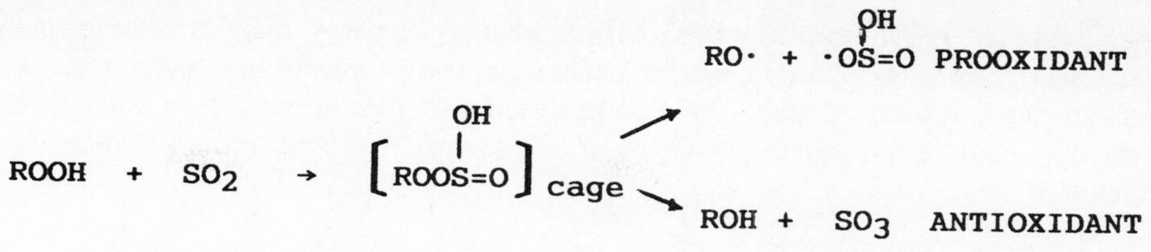

$$\text{ROOH} + \text{SO}_2 \rightarrow \left[\text{RO}\overset{\text{OH}}{\underset{|}{\text{OS=O}}} \right]_{\text{cage}} \nearrow \text{RO·} + \text{·}\overset{\text{OH}}{\underset{|}{\text{OS=O}}} \quad \text{PROOXIDANT}$$
$$\searrow \text{ROH} + \text{SO}_3 \quad \text{ANTIOXIDANT}$$

Scheme 4.2 Competitive prooxidant and antioxidant reactions of sulphur dioxide

It will be seen later that because of the latent tendency to prooxidant reactions, the thiodipropionate esters are almost always used in combination with a chain-breaking antioxidant in technological systems. Nevertheless, once the initial prooxidant stage has been passed, the thiodipropionate esters are very powerful antioxidants and in a detailed kinetic analysis of the decomposition of CHP by SO_3, MSP and MSPO, it was found that the first had the highest activity and the last the lowest [22] (see Fig. 4.2). However, the order of effectiveness may be quite different in lubricating oils or in polymers where SO_x may be readily lost by volatilisation and the organo-soluble organic acids, MSP and MSPO are probably the main peroxidoltic antioxidants under these conditions. Furthermore, the pyrolytic decomposition of MSP to SO_2 appears to take place only at elevated temperatures [22], so that sulphur oxides may play a major antioxidant role during processing of polymers but the sulphinic and sulphonic acids are likely to be more important at ambient temperatures.

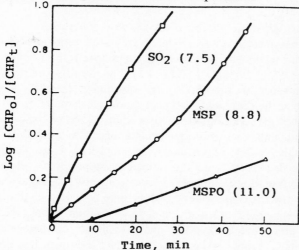

Fig. 4.2. Kinetics of the decomposition of CHP by sulphur compounds. (Numbers in parenthesis are concentrations of sulphur compounds, 10^4 mol l^{-1}, . [CHP] = 0.5 M). For structures of sulphur acids see Scheme 4.1. (Reproduced with permission from C. Armstrong, K.J. Humphris and G. Scott, *Europ. Polym. J.*, **15**, 241 (1979)).

The O-H bonds in the sulphenic acid and its further oxidation products, MSP and MSPO are extremely labile and all three have been found to be effective CB-D antioxidants in azo-initiated oxidations. The significance of this process in a fully oxygenated system is difficult to assess due to the competition between peroxyl radicals and hydroperoxides for the sulphur acids. It is probably less important than the PD mechanism but it may well synergise with the latter. However, there is some evidence that under conditions of rapid radical generation and relatively low oxygen pressure (e.g. during fatiguing of rubber [26] or processing of polymers [27]), a catalytic CB cycle may operate analogous to that described for aryloxyl and aminoxyl radicals in Chapter 3:

$$\underset{RS-OH}{(O)_n} \quad \overset{R'OO\cdot \quad R'OOH}{\underset{R'H(C=C) \quad R'\cdot}{\rightleftarrows}} \quad \underset{RS-O\cdot}{(O)_n} \qquad (2)$$

$$n = 0-2$$

Di-n-alkyldisulphides are in general more effective than the corresponding monosulphides as peroxidolytic antioxidants (see Table 4.2), and it has been found that the products of their oxidation, the thiolsulphinates, are thermolabile, rearranging to give predominantly stable products [16] (Reaction 1). However, in the presence of hydroperoxides they are rapidly oxidised to sulphur acids. Hawkins and Sautter [28] showed that diphenyl disulphides are readily oxidised to the corresponding thiolsulphinates and sulphur acids which are effective catalysts for hydroperoxide decomposition. Although aromatic sulphides are not very effective antioxidants alone, there is some evidence that the antioxidant efficiency of the long-established bis-phenol sulphides may be due to autosynergism between the sulphide (PD) and the phenol (CB-D) (see Section 4.4.3).

4.1.2 *Heterocyclic thiols, aliphatic dithioic acids and their derivatives*

A chemically diverse but mechanistically related group of antioxidants, having in common a thiol or metal thiolate group, finds application in a wide range of technological media [1,8,15,24,29,30,31,] (Table 4.1). The most important members of this class are mercaptobenzothiazole (MBT), its disulphide (MBTS) and its zinc complex (ZnMBT). When these are used as vulanisation accelerators or are produced during vulcanisation, they give vulcanisates with a high degree of oxidative stability [29,30]. A related compound, zinc mercaptotobenzimidazolate (ZnMBI) is also an effective synergist with CB antioxidants in rubbers. Zinc complexes of dialkyldithiocarbamic acids (MDRC), are also formed in "sulphurless" vulcanisation of rubber by using the corresponding disulphides as curing agents, and the analogous zinc dialkyl dithiophosphates (MDRP) are widely used as antioxidants in engine oils. As a class, this group of compounds effectively chelates transition metal ions and this constitutes a second preventive antioxidant mechanism in technological systems containing iron, manganese, copper and other trace metals. However, it is important to recognise that "metal deactivation" accounts for part but not the whole of the activity of metal dithiolates since the transition metal complexes are themselves very powerful peroxidolytic antioxidants.

(a) Mercaptobenzothiazoles and their metal complexes
Mercaptobenzothiazole, MBT and its zinc complex, ZnMBT (Table 4.1) both give rise to efficient catalysts for the decomposition of hydroperoxides. Husbands and Scott in a detailed study of their oxidation, coupled with the measurement of the activity as antioxidants of the oxidation products, have provided a relatively clear understanding of the mechanism of their action which is common to the more complex systems discussed

below [31,32]. The behaviour of both MBT and ZnMBT in a cumene hydroperoxide (CHP) initiated oxidation of cumene shows a strong dependence on the molar ratio of hydroperoxide to sulphur compound, HP/S (HP/S = [ROOH]/[S]) in the system [32]. At HP/S below 10 an autoretarding prooxidant effect was observed from the beginning, whereas at HP/S 30, complete inhibition associated with gas evolution was seen. Examination of the products formed by HPLC at different HP/MBT ratios (see Fig. 4.3) showed the transient formation of mercaptobenzothiazole disulphide (MBTS) and the formation of stable benzothiazole (BT).

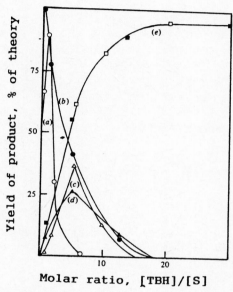

Fig. 4.3. Products formed during the oxidation of mercaptobenzothiazole (MBT) and its oxidation product mercaptobenzothiazole disulphide (MBTS) by *tert*-butyl hydroperoxide (TBH) with increasing molar ratio [TBH]/[S] where S is the sulphur compound. (a) MBTS (MBT), (b) MBTS (MBTS), (c) BT (MBT), (d) BT (MBTS), (e) BTSO (MBT, MBTS). Compounds in parenthesis are the starting materials. (Reproduced with permission from M.J. Husbands and G. Scott, *Europ. Polym. J.*, **15**, 241 (1979)).

At 62°C, benzothiazolesulphonic acid (BTSO) was the other major stable end product. The formation of BT and BTSO requires the intermediate formation of a sulphinic acid (BTS) which was too unstable to be isolated under these conditions but which is stable in the form of its zinc complex (see below). The reaction sequence is shown in Scheme 4.3. At 140°C, BTSO also loses SO_2 to give hydroxybenzothiazole (HBT).

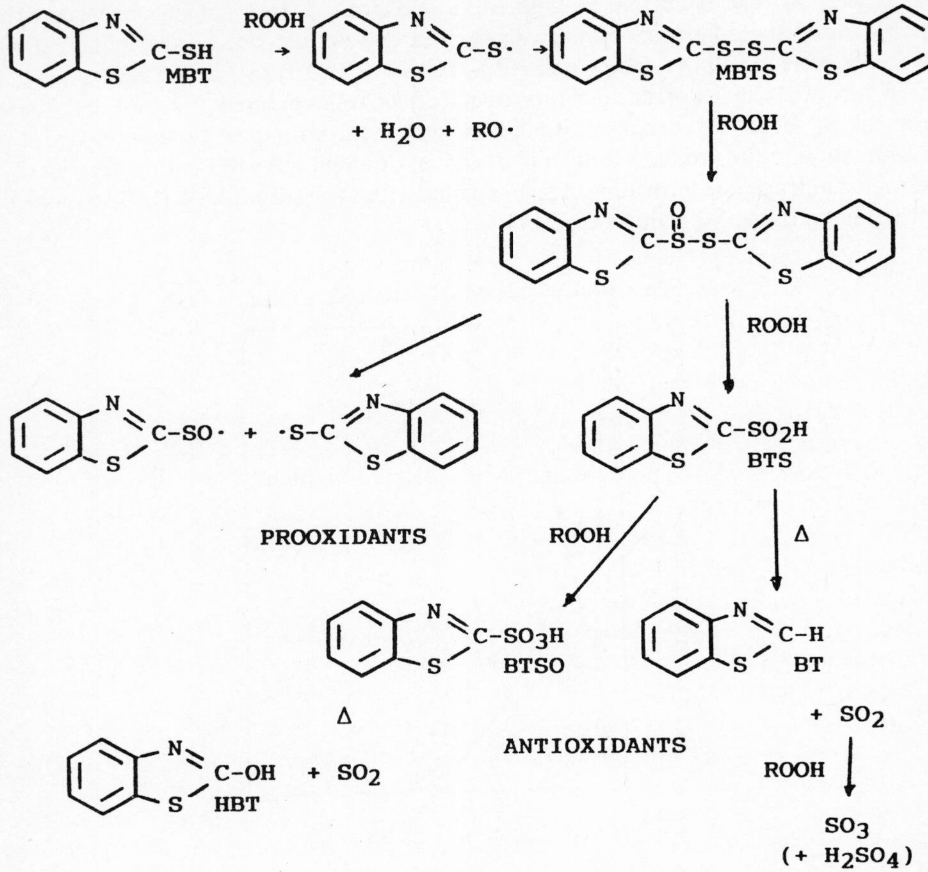

Scheme 4.3 Oxidation of mercaptobenzothiazole by hydroperoxide

BTSO is a very powerful CB-D antioxidant in cumene initiated by AIBN whereas MBT and MBTS have no such activity [32]. However, compounds that show CB-D activity also tend to be prooxidants in a hydroperoxide initiated system, and this may substantially neutralise their radical trapping activity when small amounts of hydroperoxides or hydrogen peroxide are already present from some prior source. This dual activity, summarised in Scheme 4.3, has in the past led to much contradiction in the published literature, since a slight change in concentration of hydroperoxide can transform the antioxidant from an apparent prooxidant to an antioxidant and vice versa. In the final analysis it is the catalytic peroxidolytic process which dominates in the later stages of the oxidation and which is responsible for reducing hydroperoxide concentrations to a low level.

MBTS is also a major transformation product from ZnMBT and is a PD antioxidant in its own right. However in this case an intermediate stable zinc benzothiazolyl sulphinate (ZnBTS) is also formed. This is an important reservoir for sulphur acids by reaction with hydroperoxides at higher temperatures.

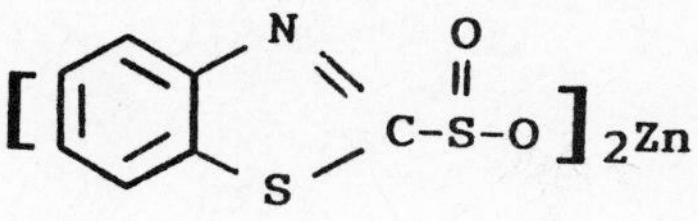

[structure: benzothiazole fused ring with N=C-S-O group, ×2, Zn] ZnBTS

As in the case of MBT, ZnMBT gave benzothiazole as one of the oxidation products indicating a similar series of reactions to that outlined in Scheme 4.3. A common feature of the peroxidolytic antioxidant activity of both MBT and ZnMBT is an ionic decomposition of hydroperoxides, preceded by a radical generating reaction which is transient at high HP/S ratios but may be dominant at low HP/S ratios. Consequently, MBT and MBI have both been used to chemically "plasticise" rubbers by inducing oxidation during processing [33].

Phenol and acetone are the main products of cumene hydroperoxide decomposition in the presence of MBT and ZnMBT at HP/S ratio 30 indicating an essentially peroxidolytic antioxidant activity due to the derived sulphur acids. However, it was found that the addition of a base (pyridine) exacerbated the prooxidant effects of ZnMBT and delayed the autoretardation stage. Removal of volatile acidic species from the head-space during oxygen absorption of paraffin oil at also 140°C markedly reduced the induction period associated with ZnMBT.

In summary, SO_3 (H_2SO_4) and BTSO are the ultimate stable peroxidolytic antioxidants formed through a series of hydroperoxide oxidations from both MBT and ZnMBT. Both involve prooxidant intermediates and the pattern of chemical reactions is similar to those observed in the oxidation of the thiodpropionate esters. Both processes have served as useful models to assist in the understanding of other more complicated systems discussed below.

A new class of peroxidolytic antioxidants for polyolefins has been reported based on the 4-alkyl-2-mercaptothiazoline structure [34-36]. Unlike the MBT analogues, these are powerful photoantioxidants, particularly when R is a solubilising alkyl chain. Although they initially form the disulphide analogous to MBTS, they subsequently undergo oxidation and rearrangement to give light stable, UV absorbing products which slowly produce sulphur acids (see Scheme 4.3a).

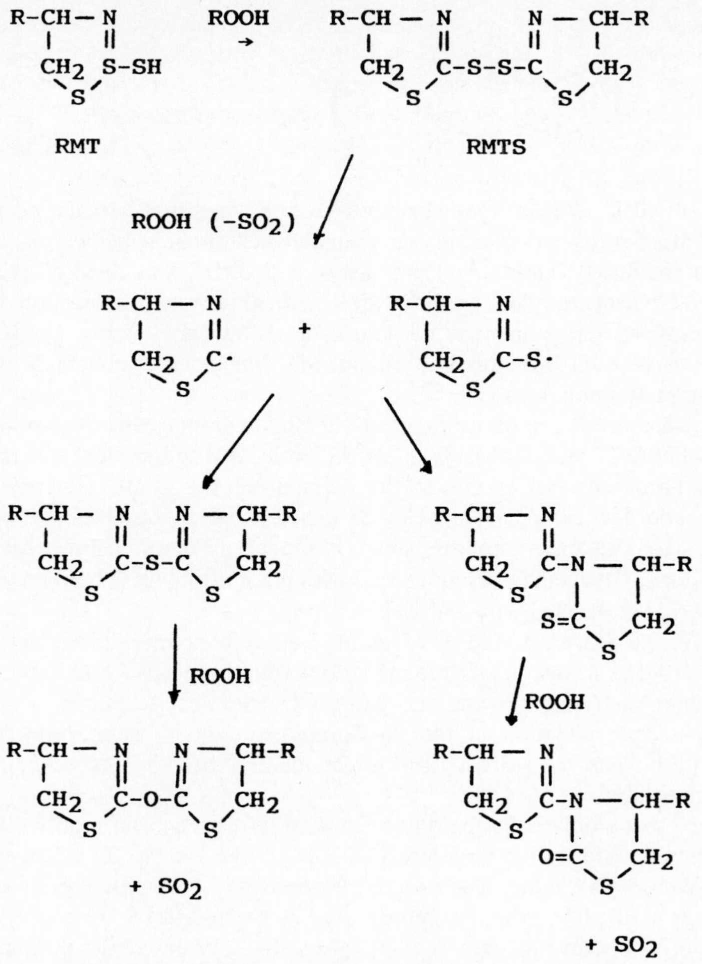

Scheme 4.4 Oxidative transformation of 4-alkyl-2-mercaptothiazolines (RMT) [34-36]

(b) Metal dithiolates

MDRC, MDRP and MRX (Table 4.1) show basic similarities to, but detailed differences from, ZnMBT [37]. They all give rise to sulphur acids with both prooxidant and antioxidant activity but ultimately they are all effective heterolytic hydroperoxide decomposers. The alkali metal dithiocarbamates are important copper chelating agents in aqueous solution and the oil soluble zinc complexes undergo metathesis with transition metals in organic solution thus performing the same function. This has been found to be important in rubber technology where zinc dialkyl dithiocarbamates (ZnDRC) transform copper salts from prooxidants to antioxidants.

The contribution to overall antioxidant activity from peroxyl trapping (CB-D) reactions has been the subject of considerable discussion [31,36,38-44]. Much of this debate results from the very different criteria used by different groups to determine antioxidant effectiveness. It was seen earlier that radical trapping using an alkyl radical generator frequently underestimates overall antioxidant activity since it gives no information about the preventive mechanism. In an early study of the mechanism of the dithiocarbamates and dithiophosphates, Scott and co-workers [17] used both an alkyl radical generator (azo-bis-*iso*-butyronitrile, AIBN) and a hydroperoxide (tetralin hydroperoxide, THP) to initiate the zinc diethyl dithiocarbamate inhibited oxidation of tetralin at **the same antioxidant concentration and the same temperature.** This comparison is shown in Figs. 4.4 and 4.5 and it is evident that ZnDEC behaves very differently in the two systems. With the azo initiator, it has almost no effect at concentrations at which it gives complete and sustained inhibition in the hydroperoxide initiated oxidation. A slight initial prooxidant effect is observed with the latter, similar to but smaller than that observed with ZnMBT. The structurally related zinc dinonyl-dithiophosphate under the same conditions gave a similar inhibition in THP initiated peroxidation, whereas a thiodipropionate ester gave the autoretarding behaviour discussed above. The main products obtained from the reaction of the dithiocarbamate with cumene hydroperoxide (CHP) were sulphur oxides from the dithiocarbamate and phenol and acetone from CHP, indicating an essentially non-radical hydroperoxide decomposition.

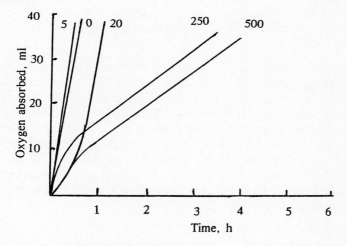

Fig. 4.4 Effect of ZnDEC on the peroxidation of tetralin initiated by azo-bis-*iso*-butyronitrile (AIBN) at 50°C. Numbers on curves are ZnDEC concentrations (10^3 g/100g). (Reproduced with permission from J.D. Holdsworth, G. Scott and D. Williams, *J. Chem. Soc.*, 4692 (1964)).

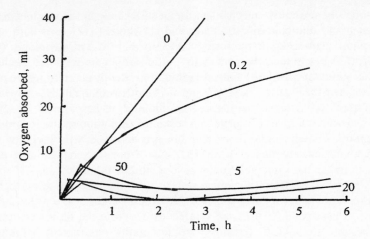

Fig. 4.5 Effect of ZnDEC on the peroxidation of tetralin initiated by tetralin hydroperoxide (THP) at 50°C. Numbers on curves are ZnDEC concentrations (10^3 g/100g). (Reproduced with permission from J.D. Holdsworth, G. Scott and D. Williams, *J. Chem. Soc.*, 4692 (1964).

Extensive investigations in recent years have confirmed the dominating importance of the peroxidolytic mechanism for the thermal antioxidant activity of the zinc dithiolates [31,36,45-49]. Many workers have shown a three step decomposition of hydroperoxides and this is shown typically under the same conditions for the photoantioxidants nickel dibutyldithiocarbamate (NiDBC), nickel dibutyldithiophosphate, (NiDBP) and a nickel dibutylxanthate (NiBX) in Fig. 4.6 [37]. The first rapid stage is stoichiometric and radical generating and, as in the case of MBT is favoured by low HP/S ratios. The second catalytic stage is ionic, catalysed by acidic oxidation products. In the region of HP/S = 10 there is a sharp change from a predominantly radical generating reaction to a predominantly ionic reaction.

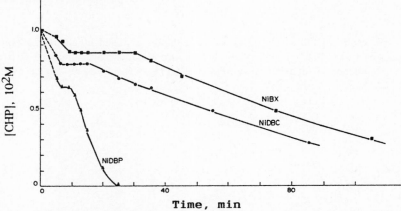

Fig. 4.6 Decomposition of CHP (10^{-2}M) in chlorobenzene at 110°C in the presence of nickel butylxanthate (NiBX), nickel dibutyldithiocarbamate (NiDBC) and nickel dibutyldithiophosphate (NiDBP), all at 2 x 10^{-4}M. (Reproduced with permission from S. Al-Malaika and G. Scott, *Europ. Polym. J.*, **16**, 504 (1980).

The formation of the peroxidolyic species from the transition metal dithiocarbamates proceeds through the thiocarbamoyl disulphide (DRDTS), see Scheme 4.5 [33]. This is formed by attack of both hydroperoxide and alkylperoxyl at the metal centre. For the latter reason, the nickel dithiocarbamates show some CB-D antioxidant activity [49], although this is weak compared with the peroxidolytic antioxidant activity of NiDRC in a hydroperoxide initiated system. The subsequent oxidative transformation products have not been examined in detail, but it appears likely by analogy with ZnMBT and MRDP (see below) that SO_2 and isothiocyanate are formed through the intermediate dialkydithiocarbamoyl sulphonic acid (DRDTA). Alternatively or in addition SO_2 may be formed through an intermediate unstable sulphinic acid.

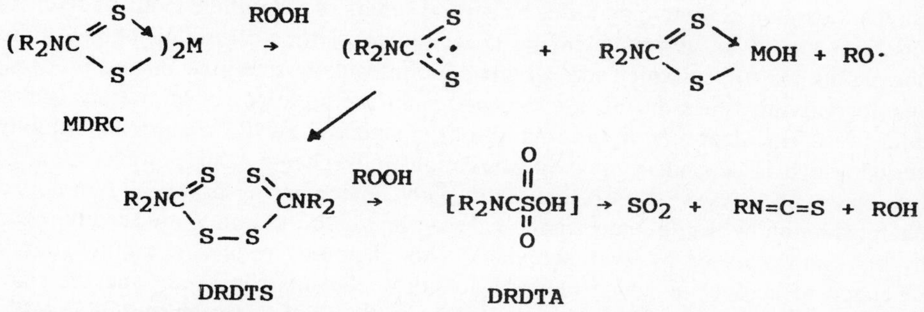

Scheme 4.5 Oxidative transformation of transition metal dialkyl dithiocarbamates (MDRC)

Zinc dithiocarbamates (ZnDRC), unlike ZnMBT and the transition metal dithiocarbamates (see below) do not appear to form disulphides [31,48]. They do however form the corresponding zinc thiopercarbamate ZnDRSO which, as in the case of ZnBTS, is the major source of sulphur acids, Scheme 4.6.

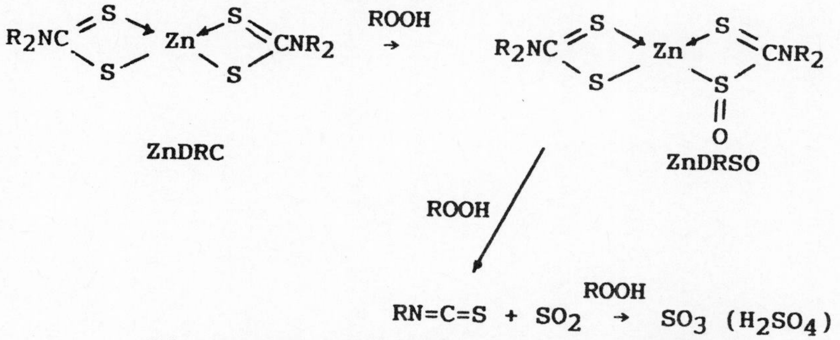

Scheme 4.6 Oxidation mechanism of zinc dithiocarbamates

The formation of the ultimate peroxide decomposer is inhibited by a typical chain-breaking antioxidant (BHT), confirming that this is a free radical process [47]. The same catalytic antioxidant species are formed irrespective of the intermediate and the addition of a base ($CaCO_3$, pyridine) also completely inhibits the peroxidolytic antioxidant activity [47]. The formation of isothiocyanate and SO_2 have been observed as products of this process [17].

It was noted earlier that the dithiolates are effective chelating agents for prooxidant transition metal ions. Thus the zinc dithiocarbamates undergo ready metathesis with iron, copper, cobalt, nickel and manganese salts in hydrocarbon polymers during processing to give the corresponding transition metal complexes which unlike most metal chelates are not simply inert but act as effective peroxidolytic antioxidants due to the presence of sulphur in the molecule. Thus, copper ions which are normally prooxidants in vulcanised rubbers become antioxidants in a TMTD "sulphurless" vulcanisates due to the formation of copper dimethyl dithiocarbamate (CuDMC) and the antioxidant activity lasts just as long as the sulphur ligand is present [50]. Once it is destroyed by the chemistry discussed above the activity of copper "inverts" to give its normal prooxidant effect. In polyolefins, the transition metal dithiocarbamates differ very considerably in their photostability. As a consequence, the duration of their photoantioxidant activity may differ by several orders of magnitude. This phenomenon has been turned to advantage in the development of time-controlled photostabilised polyolefins [51-58] in which the polymers are effectively stabilised during processing and for varying times during use depending on the additive concentrations and the intensity of incident UV light. Iron dithiocarbamates (FeDRC) cause an inhibition period which is dependent upon concentration (see Scheme 4.7). However once they have been photolysed, the liberated iron carboxylates act as powerful initiators for photooxidation by the "photo-Fenton" reaction in which the iron is alternately reduced by light and oxidised by hydroperoxides. This chemistry results in a quite dramatic inversion of activity in which both the length of the induction time and the rate of ultimate photooxidation of the polymer depend on the FeDRC concentration [51-57].

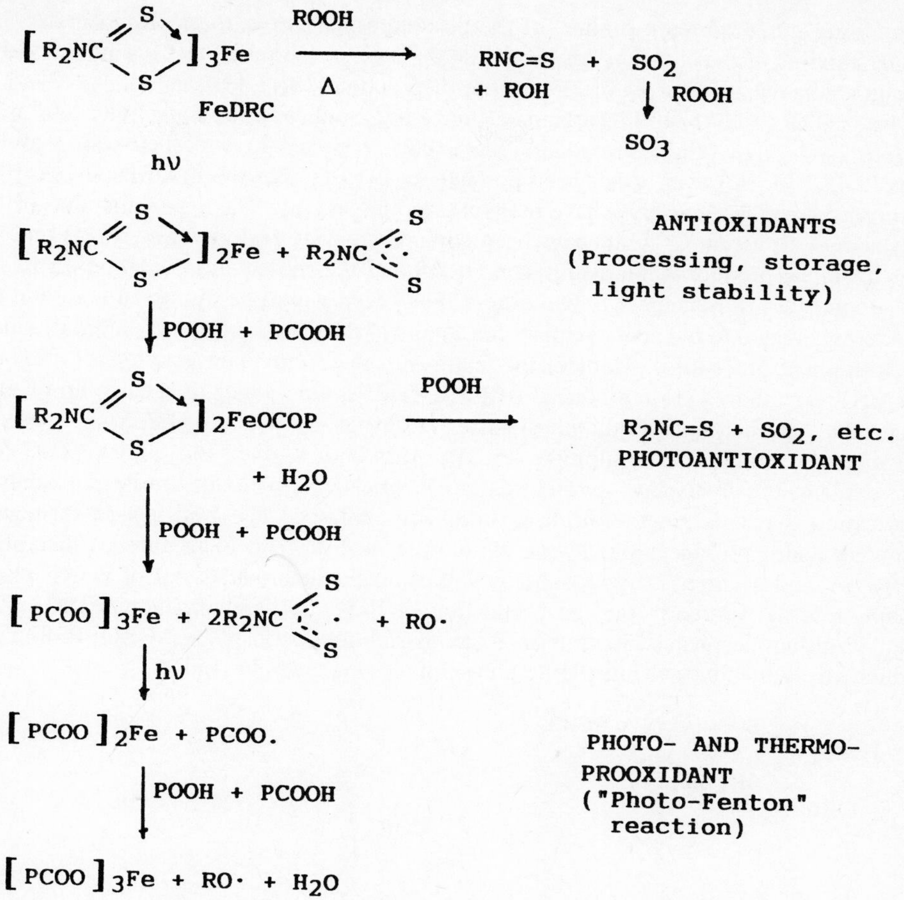

Scheme 4.7 Iron dithiocarbamate (FeDRC) photoantioxidant inversion [56]

$[R_2NC(=S)S]_3Fe$ (FeDRC) $\xrightarrow{ROOH, \Delta}$ $RNC=S + SO_2 + ROH$ $\xrightarrow{ROOH}$ SO_3

$\downarrow h\nu$

$[R_2NC(=S)S]_2Fe + R_2NC(=S)S\cdot$ — ANTIOXIDANTS (Processing, storage, light stability)

$\downarrow POOH + PCOOH$

$[R_2NC(=S)S]_2FeOCOP + H_2O$ $\xrightarrow{POOH}$ $R_2NC=S + SO_2$, etc. PHOTOANTIOXIDANT

$\downarrow POOH + PCOOH$

$[PCOO]_3Fe + 2R_2NC(=S)S\cdot + RO\cdot$

$\downarrow h\nu$

$[PCOO]_2Fe + PCOO\cdot$ — PHOTO- AND THERMO-PROOXIDANT ("Photo-Fenton" reaction)

$\downarrow POOH + PCOOH$

$[PCOO]_3Fe + RO\cdot + H_2O$

Time-controlled photooxidation now finds wide application in the environmental disposal of plastics litter both from packaging and from agricultural plastics and fibres. Mulching films are being increasingly used in agriculture in arid climates where, in combination with trickle irrigation, a controlled microclimate is produced round the roots of the growing plant. Timing of disintegration is of crucial importance for crop yield and to avoid the clogging of machinery during cropping. In general, plasticulture is practised where incident UV and temperatures are high and for this purpose a combination of NiDRC and FeDRC is required in order to extend the induction time for several months before disintegration. In exceptional cases even longer induction times may be required [56,57]. A major advantage of the FeDRC system is that ionic iron is produced by the photochemical action, and this catalyses both photooxidation and thermal oxidation so that the plastic is slowly bioassimilated to biomass by soil microorganisms [58].

The most thoroughly investigated metal dithiolates, because of their importance as antioxidants for lubricating oils, are the zinc dialkyldithiophosphates (ZnDRDP) [38-43]. The application of ^{31}P NMR [36,43,59] to the analysis of the phosphorus-containing transformation product of the dithiophosphates has markedly increased the understanding of the chemistry of the oxidation process. However, this has not served to diminish the debate over the mechanism of their action [38,42,43].

The ZnDRPs, like the dithiocarbamates, have weak radical trapping activity [60] and f values in an azo initiated oxidation have been reported to vary between 2 and 3 [36,42,45, 59]. However, it has been pointed out [31,61], that the ZnDRPs are oxidised to products which themselves have peroxyl trapping activity. The detailed studies of Al-Malaika et al. using ^{31}P NMR have been comprehensively reviewed [31,36,44] and the key chemical product is the disulphide, DRDS and the subsequent oxidation chemistry as summarised in Scheme 4.8. Only the salient features will be discussed here and the reader is directed to the above reviews for details of the characterisation and antioxidant measurement procedures. Many of the features of the antioxidant activity of MBT and ZnMBT (see above) are reproduced with the ZDRPs. One major difference however is that the parent dialkyldithiophosphoric DRDPA is also an effective peroxide decomposer. Unlike the disulphide, DRDPS, it is both a PD-C [62] and a CB-D [63] antioxidant, but it displays autoretardation kinetics in a peroxide initiated oxidation, indicating that it is not the ultimate antioxidant formed. The acid is transformed by cumene hydroperoxide after a very short time at 110°C to a mixture of disulphide (DRDS), and a variety of poly-sulphides rearrangement products, all of which play a major role as reservoirs for the production of SO_2, SO_3 and probably H_2SO_4. Zinc dialkyl dithiophosphates give intermediate "basic" dithiophosphates, $[(RO)_2PSS]_6ZnO$, which are again converted further to polysulphides and sulphur acids.

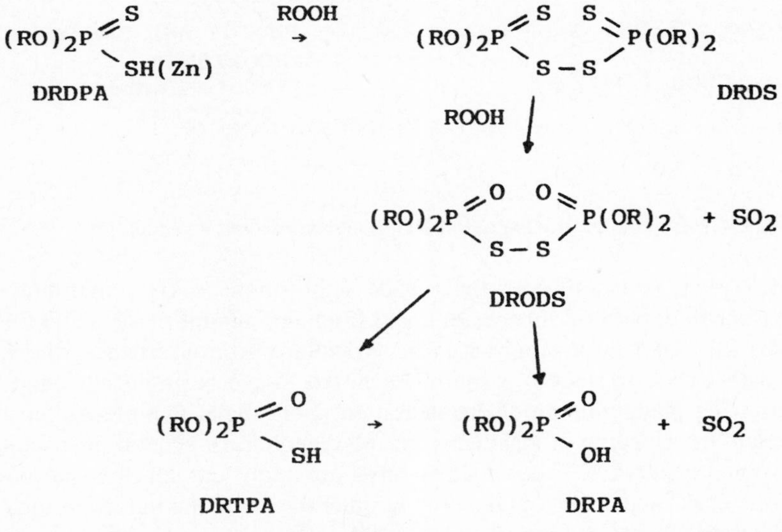

Scheme 4.8 Oxidative transformations of the dihiophosphoric acids and their zinc complexes [31,36,44]

The transition metal complexes of the dithiophosphoric acids are oxidised rapidly to DRDS at all HP/S ratios and, in the case of the nickel complex, the latter was destroyed completely before the beginning of the catalytic PD-C process. The sulphonic acid (DRTSA) is formed first and undergoes SO_2 elimination to give the thionophosphoric acid (DRTnPA) (see Scheme 4.9). The phosphoric ester, DRPA, is a stable end product which does not catalyse the decomposition of hydroperoxides [64,65].

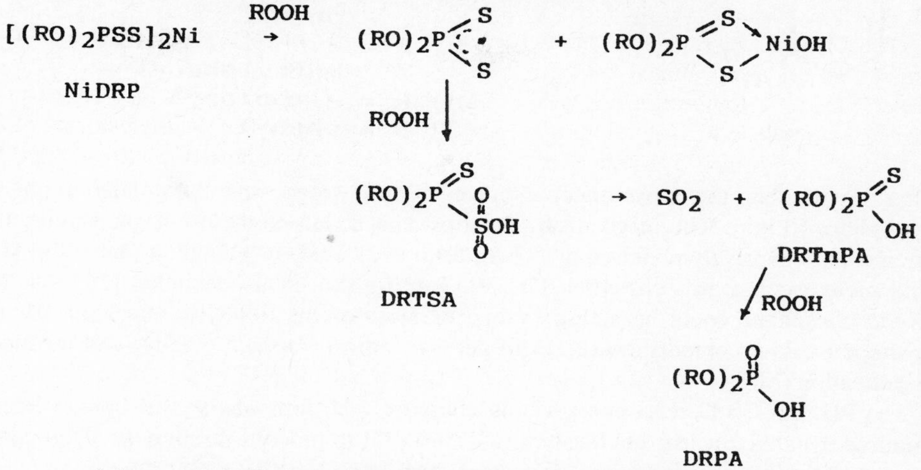

Scheme 4.9 Oxidative transformation products of nickel dithiophosphates

In a commercial application of the above chemistry to the stabilisation of polymers, Scott and Al-Malaika have found [66] that severe oxidation of DRDS during the processing of polyolefins results in a much more effective antioxidant system than DRDS itself, and the oxidation products formed during processing synergise effectively with the NiDRP and conventional UV absorbers [43,66].

4.1.4 *Phosphite and phosphate esters*
Phenyl esters of phosphorus acid (e.g. tris-nonylphenyl phosphite, TNPP, (Table 4.1) have been used for many years as antioxidants in synthetic rubbers [67] and more recently a variety of alkyl/aryl phosphites [68,69] have been used as synergists with hindered phenols and light stabilisers in thermoplastic polymers. They have relatively little thermal or photoantioxidant activity when used alone and their main function seems to be to reduce the transient oxidised products formed from the CB antioxidants, thus reducing the development of colour. Phosphites are known to act primarily by stoichiometric decomposition of hydroperoxides to alcohols (reaction 3) [70] and in the case of the simple tris-aryl phosphites, only the alcohol and the derived phosphate are formed and the latter has no antioxidant activity [71].

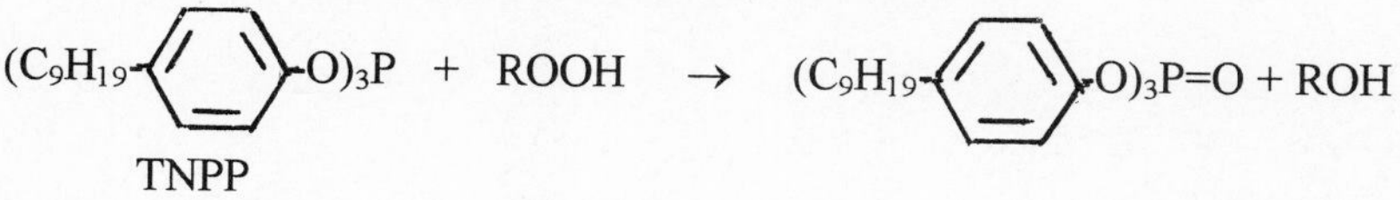

$$(C_9H_{19}\text{-}C_6H_4\text{-}O)_3P + ROOH \rightarrow (C_9H_{19}\text{-}C_6H_4\text{-}O)_3P=O + ROH \quad (3)$$

TNPP

A much more potent and mechanistically more interesting sub-group within the above class, which has been extensively studied by Kirpichnikov and co-workers since the 1960s in polyolefins [72,73] and rubbers [74], contains a cyclicphenyl phosphate function (CPP).

```
                    O
       ╱═╲     ╱                 (a)  R = 2,4,6-tri-tert-butyl
      │   │  P-OR                       phenyl
       ╲═╱     ╲                 (b)  R = 2,6-di-tert-butyl-4-
                O                       methylphenyl
                                 (c)  R = iso-propyl
         CPP                     (d)  R = phenyl
```

They found that the most effect member of this group was the hindered phenyl phosphite; CPP(a). The mechanism of antioxidant action of the cyclic phosphites has been comprehensively reviewed by Pobedimskii et al. [75], by Humphris and Scott [71] and more recently by Schwetlick [78,79]. Pobedimskii et al concluded [75] that the cyclic phosphites could be acting by one or more of the following mechanisms: (a) hydroperoxide decomposition (PD), (b) peroxyl radical trapping (CB-D) and (c) metal deactivation (MD).

The PD and CB-D reactions are both complex. The first was shown by chemically induced dynamic nuclear polarisation (CIDNP) [75] to proceed through a radical caged intermediate:

$$(RO)_3P + R'OOH \rightarrow \left[(RO)_3\overset{OH}{P} \cdot \cdot OR' \right] \rightarrow R'OH + (RO)_3P=O \qquad (4)$$

$$\downarrow$$

$$R'O\cdot + (RO)_3POH$$

Up to 10% of the caged radicals were found to escape from the cage at 75°C as measured by the polymerisation of styrene [71,75]. The radical generation process does not appear to be strongly dependant on the nature of R.

Humphris and Scott investigated in detail the transformation products formed by the oxidation of the catechol phosphites, and compared their antioxidant activity with that of the CPPs. They found that CPP(b) was a weak antioxidant in AIBN initiated cumene oxidation [71] and it is clear from reactions (5) and (6) that the reaction of alkylperoxyl with phosphite esters can only be inhibiting when RO· is a stable aryloxyl radical. However, CPP(b) was found to be converted to a much more powerful antioxidant after oxidation [76]. CPP(c) and CPP(d) were initially prooxidants but were again rapidly converted to antioxidants during oxidation. This evidence suggested that the phosphite esters themselves are not the effective antioxidants in this system.

$$R'OO\cdot + (RO)_3P \rightarrow R'O\cdot + (RO)_3P=O \qquad (5)$$
$$R'O\cdot + (RO)_3P \rightarrow R'OP(RO)_2 + RO\cdot \qquad (6)$$

Humphris and Scott isolated the oxidation products from CPP(b) [81] and examined their antioxidant activity in the oxidation of cumene initiated by both AIBN and CHP [79]. The derived phosphate, CPPO(a) caused autoretardation of AIBN-initiated cumene peroxidation without an initial prooxidant effect but a further product of this reaction, CPPO(b), was an even more powerful antioxidant. CCPO(a), unlike acyclic phosphate esters was found to be a very powerful catalyst for peroxide decomposition and gave the same pseudo first order decomposition rate as CPP(b) from which it is derived (Table 4.4).

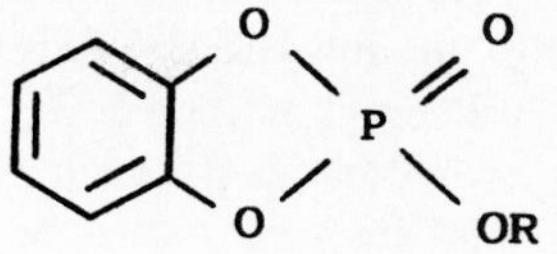

(a) R = 2,4,6-tri-*tert*-butyl phenyl
(b) R = H

CPPO

In the CHP initiated peroxidation of cumene inhibited by CPP(b), an even more marked prooxidant autoinhibition was observed [71] and this was found to be associated with the formation of CPPO(b) which proved to be a powerful PD-C antioxidant. The extent of the initial prooxidant effect observed with the catechol phosphite CPP(b) was found to be strongly dependent on the molar ratio of hydroperoxide to phosphite, HP/P. Using the diagnostic cumene decomposition procedure described earlier, it was found [81] that at HP/P<4 radical products are formed from CHP, and above this ratio the reaction is predominantly ionic.

The cyclic phosphate, CCPO(b), was formed as a major product of the reaction of hydroperoxides with CCP(b) (with liberation of 2,6-di-*t*-butyl phenol) [81]. This compound was a much more powerful ionic catalyst for cumene hydroperoxide decomposition than CPPB, CPPO(a) or the ring-opened hydrolysis product of CPPO(b), HPPO (Table 4.4).

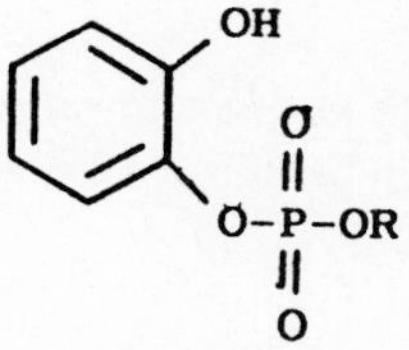

(a) R = 2,4,6-tri-*tert*-butyl phenyl
(b) R = H

HPPO

Table 4.4 Pseudo first-order rate constants for the reaction of CHP with phosphite and phosphate antioxidants in chlorobenzene at 75°C [CHP]$_o$ = 0.2M, [P]$_o$ = 0.02M [81]

Antioxidant	$10^4 k_1 s^{-1}$
CPP(b)	1.1
CPPO(a)	1.2
CPPO(b)	84
HPPO(b)	2.5

In practice there is an equilibrium between CCPO(b) and HPPO. Both were shown to be effective antioxidants in a peroxide initiated peroxidation and were considered to be the main candidates for the Lewis acid species responsible for the peroxidolytic activity of the catechol phosphites [78,79]. The mechanism proposed by Humphris and Scott to account for the antioxidant activity of the cyclic phosphite oxidation products is shown in Scheme 4.7, but it will be appreciated that in the presence of water the less powerful acyclic phosphate ester HPPO will also be involved.

In a subsequent study, Rüger et al. [82] confirmed the earlier observation that the phosphite esters, including CPP(b), have essentially no antioxidant activity in an alkylperoxyl (AIBN) initiated peroxidation but that HPPO was a very powerful CB-D antioxidant and indeed, is much more effective than its hydrolysis product, BHT. This suggests that in a "real" system, where both peroxyl and hydroperoxides are present, HPPO(a) and HPPO(b) will both be very effective autosynergists due to the presence of an "ortho activated" phenolic function which can also act as a PD-C antioxidant through the cyclic phosphate system. CCPO and HPPO were found to be effective thermo- and photoantioxidants for polyolefins [83]. Table 4.5 compares CPPO(a) with a commercial synergistic mixture of 1010 and DLTP at 150°C in polypropylene. The superiority of the phosphate ester is consistent with the suggestion that it may be acting autosynergistically.

Table 4.5 Comparison of CPPO(a) (A) with a commercial synergist (B) in polypropylene at 150°C [83]

	0.20*		0.25*		0.30*		0.35*	
	A	B	A	B	A	B	A	B
Embrittlement time, hours	170	43	263	100	355	155	448	213

```
* Antioxidant concentration, g/100g
Sample B contains DLTP and 1010 at w/w ratio 75:25
```

Schwetlick [79] has found that at 50°C HPPO(a) appears to be the end product of the hydroperoxide reaction and has suggested that, at this temperature, HPPO(a) rather than HPPO(b) is the catalytic agent in Scheme 4.10.

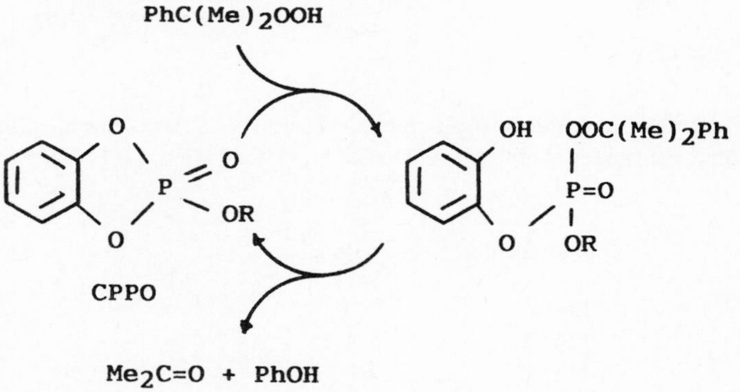

Scheme 4.10 Catalytic (Lewis acid) decomposition of cumene hydroperoxide (CHP) by a cyclic phenyl phosphate (CPPO) [71,77]

The five-membered alicyclic phosphates (e.g. CEP) and derived phosphates were also found to be efficient photoantioxidants in polypropylene [83]. Again they were catalytic peroxide decomposers at 75°C [80,81] but not at 50°C [79].

$$\text{CH}_2\text{-O} \diagdown$$
$$\quad\quad\quad\text{PO}i\text{Pr}$$
$$\text{CH}_2\text{-O} \diagup$$

CEP

The peroxidolytic activity of the cyclic phosphates has been compared with the very rapid hydrolysis of the esters of phosphoric acid derived from 1,2-diols [79,84] in biological systems and this leads to the possibility that 5-membered ring cyclic phosphates occurring *in vivo* may have peroxidolytic antioxidant activity. However, in the case of the alicyclic phosphates, this chemistry may be limited by the tendency of the β-hydroxyethylphosphate esters to polymerise to poly-phosphate esters [85].

4.2 Metal Deactivators

Metal deactivators play two distinct roles in the prevention of peroxidation of organic compounds. The first is the removal of prooxidant transition metal ions from the site where hydroperoxides are located (sequestration) so that they no longer play a significant role in radical generating redox reactions. The second is the chemical deactivation by saturation coordination of the transition metal ion so that hydroperoxides can no longer enter the outer coordination shells of the metal ions, which precedes their decomposition [86]. Transition metal ion removal by sequestration can generally be accomplished in aqueous media with an excess of a relatively weak coordinating ligand, provided the partition function of the metal complex is designed appropriately. This kind of procedure is increasingly being used *in vivo* for the therapeutic treatment of iron overload. This will be discussed further in Chapter 5.

Water soluble metal complexing agents (e.g. EDTA) are ineffective in metal deactivators in polymers when used as additives, although they have some utility in removing metal ions from aqueous polymer suspensions before coagulation, by competitive sequestration. Moreover, many transition metal compounds have little or no prooxidant activity in polymers because they are essentially insoluble [87,88]. Indeed, some of them are effective light stabilisers because they harmlessly absorb damaging incident UV light. Thus iron oxides are widely used as light stabilisers for polyolefins [638] and cause no problem with thermal oxidation unless the iron is solubilised by other ingredients within the polymer.

Pederson in a classical study of the effects of salicaldehyde-based chelating agents on the transition metal catalysed peroxidation of petroleum [89] found that di- and tetrafunctional derivatives were able to deactivate cupric oleate completely (see Table 4.6), but not all were capable of deactivating Mn, Fe, Co and Ni oleates and that some of the metal complexes (e.g. cobalt saliclylidineethylenediamine) were powerful activators for peroxidation, possibly due to increasing the solubility of the metal in the system. However, the octadentate N,N',N'',N'''-tetrasalicylidinetetra-(aminomethyl)-methane (TSTM, Table 4.6) neutralised the effect of all transition metal ions studied.

Table 4.6 Metal deactivating activities of salicylideneimines in petroleum [89]

Chelating agent (0.002%)	E_D^*				
	Mn^+	Fe^+	Co^+	Ni^+	Cu^+
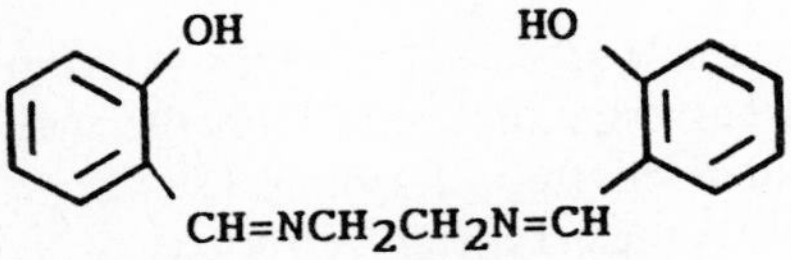 (salicylidene)-CH=NCH$_2$CH$_2$N=CH-(salicylidene)	−103	−43	−833	−	100
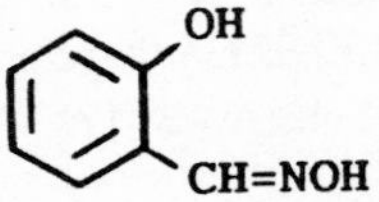 salicylaldoxime CH=NOH	−	0	−96	100	100
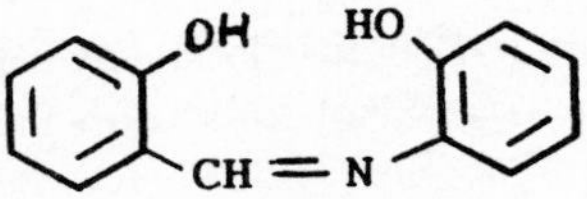 bis-salicylidene CH=N−N=CH	−84	100	96	−55	100
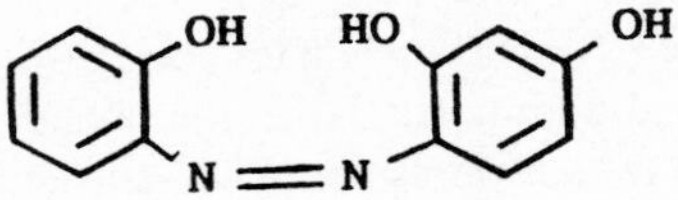 bis-salicylidene azo N=N (with extra OH)	−73	100	−	−124	100
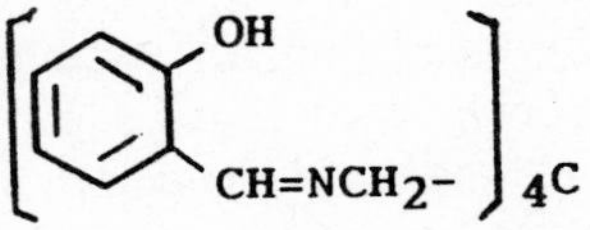 [salicylidene-CH=NCH$_2$−]$_4$C	100	100	100	100	100

*E_D = Percentage restoration of the induction time.
+ Concentration in all cases 1.6×10^{-8} M

Similar result were observed by Chalk and Smith [90] with the hexadentate 1,8-bis(salicylidineamino)-3,6-dithiaoctane (BSDTA) although this compound showed an initial prooxidant effect characteristic of the peroxidolytic sulphur antioxidants. It has since been recognised (see Section 4.1.1) that some of the most effective metal deactivators for polymers contain sulphur in the ligand and that many sulphur complexes of transition metal ions are themselves peroxidolytic antioxidants. In addition, the phenolic group has weak CB-D activity.

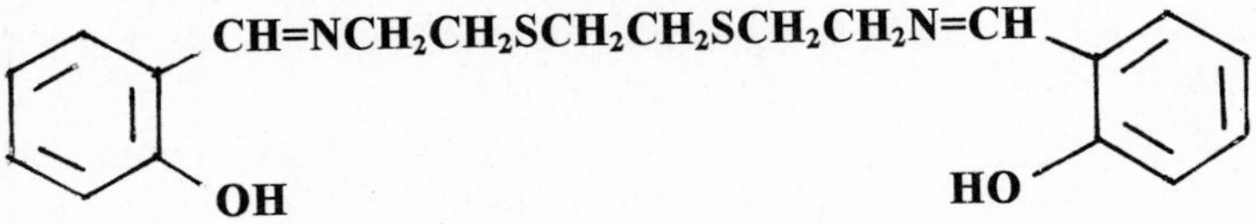

(salicylidene)−CH=NCH$_2$CH$_2$SCH$_2$CH$_2$SCH$_2$CH$_2$N=CH−(salicylidene) BSDTA

Sulphur ligands have been used for many years as copper deactivators in rubber. In particular, "efficiently" vulcanised rubbers which give rise to substantial amounts of zinc dithiocarbamates and mercaptobenzolthiazolates are very oxidatively stable even in the presence of copper ions which is normally a powerful prooxidant in rubbers. Indeed,

Pederson reported [91] that added copper ions actually increased the oxidative stability of a "sulphurless" vulcanisate in which zinc dimethyl dithiocarbamate was the main anti-oxidant species present. It was evident from these early studies that it is often difficult to distinguish between the "neutralisation" of prooxidant metal ions and the production of a more effective peroxidolytic antioxidant since both mechanisms are mutually reinforcing.

Colclough has reported similar results in lubricating oils [2,42,92]. Copper ions are effective prooxidants at low concentrations but when added to an oil containing catalytic amounts of iron and a ZnDRP (Section 4.1.2(b)), copper naphthenate (CuNaph) is much more effective antioxidant than a conventional CB-D antioxidant, OD, or than ZnDRP without copper (Table 4.7). Copper carboxylates undergo rapid metathesis with zinc dithiophosphates under these conditions [91,93] and the oil soluble copper dithiophosphates are highly effective antioxidants. It was seen earlier that copper ions can also act as catalytic CB antioxidants under conditions of low hydroperoxide concentration and it seems likely that copper dithiophosphates are also autosynergistic antioxidants in hydrocarbon media; the ligand destroys hydroperoxides by the catalytic PD-C mechanism and the Cu^+/Cu^{2+} redox system removes alkyl and alkylperoxyl radicals.

Table 4.7 Antioxidant activity of copper compounds in lubricating oils at 165°C [2]

Antioxidant	Conc^n, g/100g	Time, h	Viscosity, cP
None	-	30	500+
Cu(I)DHP	0.1	64	310
Cu(I)DOP	0.13	64	130
Cu(II)Naph	0.25	64	330
Cu(II)DRC	0.12	64	410
ZnDRP	1.2	48	500
OD (IX,Table 3.1)	0.5	30	500+

Test oil contained 1.2 g/100g ZnDRP + 40ppm Fe catalyst

It is not clear what role metal deactivation plays, if any, in this system. However, copper sequestration at the surface of copper bearings, clutch plates, etc., does involve chelation of the copper to give a protective (passivating) layer on the surface of the metal which does not "migrate" into the oil with consequent copper catalysed oxidation. In this context, the salicylidine-polyamines (Table 4.6) are particularly effective and act as both deactivators for soluble copper and as passivating agents for metallic copper [617a]. Mercaptobenzothiazole (MBT) and 2,5-dimercapto-1,3,4-thiadiazole (DMTD) and its bis-disulphide (DMTDS) are particularly effective in a copper corrosion test and almost certainly act as both sequestering agents and PD-C antioxidants [42].

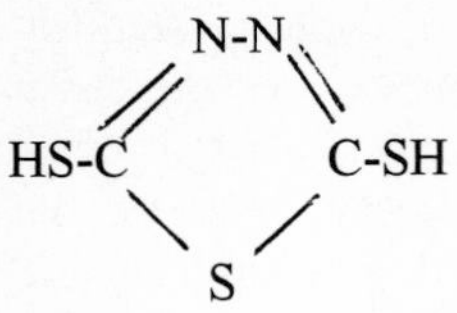

```
     N-N                      N-N            N-N
    ∥   \\                   ∥   \\         ∥   \\
HS-C     C-SH           HS-C     C-S-S-C     C-SH
    \   /                   \   /         \   /
     S                       S             S

     DMTD                         DMTDS
```

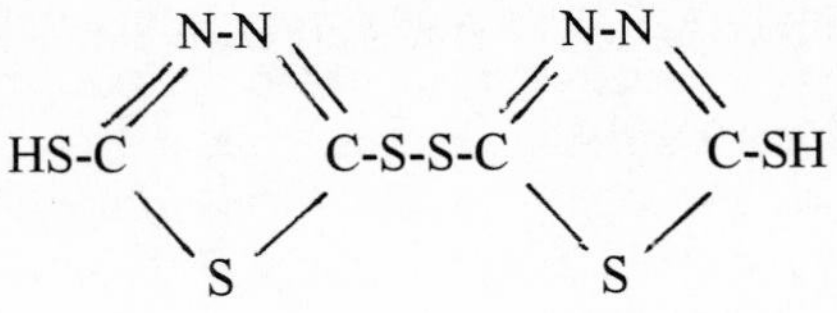

Studies in polypropylene by Scott and co-workers [94,95] have shown that zinc dimethyl dithiocarbamate completely inhibits the prooxidant activity of a 12-fold molar excess of the highly prooxidant ferric acetylacetonate (FeAcAc) during processing and in addition

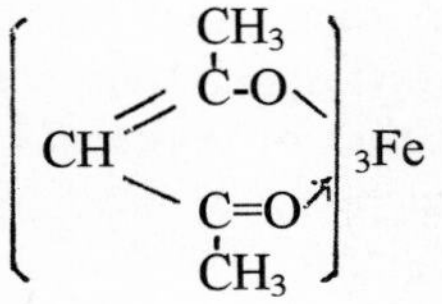

$$\left[CH \underset{C=O}{\overset{C-O}{\diagup\diagdown}} \underset{CH_3}{\overset{CH_3}{}} \right]_3 Fe \qquad FeAcAc$$

introduces an induction time to the photoprooxidant activity of the latter during photooxidation. The complete inhibition of both thermal and photooxidation again indicates that metal complexation cannot be the major mechanism involved in solution but rather the elimination of hydroperoxides.

It must be concluded from the above studies that for metal deactivation in organic solution, the peroxidolytic mechanism is much more important than metal complexing of even the most highly prooxidant transition metal ions when sulphur is a ligand. For sequestration, the reverse is probably the case.

Other autosynergistic metal deactivators have similarly been developed empirically, particularly for polyolefins, involving a combination of CB-D and metal complexing functions. Osawa and Matsuzuki [96,97] were unable to find a strong relationship between chelate stability and copper inhibiting activity in the thermal oxidation of polypropylene. 8-hydroxyquinoline, a moderately effective CB-D antioxidant, was one of the most effective compounds examined. Other phenolic chelating agents have been found to be effective copper deactivators. Hartless and Trozollo [98] found that the phenolic group was equally if not more effective in the 4-position as in the 2- position. Table 4.8 shows that complexing agents (oxamides) without a phenolic group were relatively less effective than complexing agents with a phenolic group. Hindered phenols without a complexing group were even less effective.

Table 4.8 Effectiveness of dual-action copper deactivators in polyethylene [98]

Antioxidant	T, h at 140°C
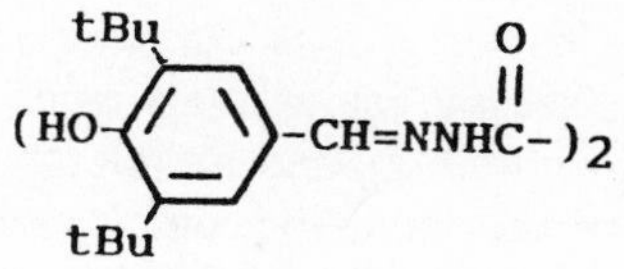 (HO-C₆H₂(tBu)₂-CH=NNHC(O)-)₂	280
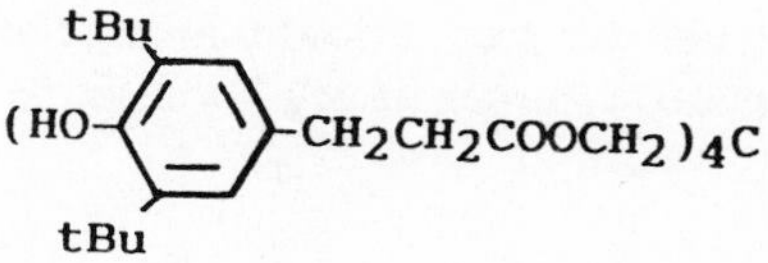 (HO-C₆H₂(tBu)₂-CH₂CH₂COOCH₂)₄C	196
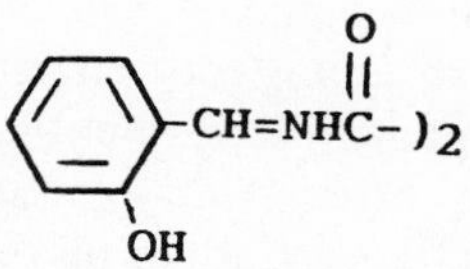 (2-HO-C₆H₄-CH=NHC(O)-)₂	74
No additive	4.5

One of the main applications of metal deactivators is in the electrical cable industry. Solublisation of copper at the metal/polymer interface can, as in the case of lubricating oils discussed above, lead to degradation of the polymer well before the end of the useful life of the cable. For this purpose, a physical requirement of a metal deactivator is that it diffuses from the polymer to concentrate at the metal surface where it immobilises the copper ions in a polymer insoluble phase [97,98]. The design of a sequestering agent for this purpose is rather different from the antioxidant function described above [99] and large and relatively insoluble molecules without CB activity appear to be particularly effective under these conditions [100].

Autosynergism (Section 4.4.3) appears to be as important in polymers as it is in model systems and lubricating oils. One of the most effective copper deactivators in polypropylene is the hydrazide derivative of the peroxidolytic antioxidant thiodipropionic acid (BTPH) and its methylene-bis derivative [99].

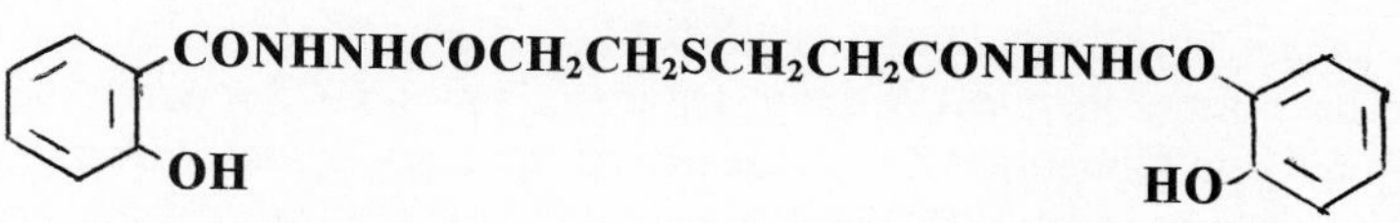

2-HO-C₆H₄-CONHNHCOCH₂CH₂SCH₂CH₂CONHNHCO-C₆H₄-OH-2

BTPH

It seems probable that as in the case of the disalicylidine derivatives, discussed above, three complementary mechanisms are involved in the activity of BTPH and its derivatives, namely, metal chelation (through the hydrazide function), peroxide decomposition (through the sulphide) and chain-breaking (through the phenol).

4.3 UV Absorbers and Screens

Any chemical compound that can absorb or reflect UV light is a potential UV stabiliser for organic substrates. Even a layer of soot and grime in the surface of plastic can lead to a lengthening of the outdoor lifetime (weathering life) in industrial environments by simply screening the polymer from UV light. Weathering involves a complex interplay of environmental factors and, as was noted in Chapter 1, it is not always possible to decide which influence is most important. It has become increasingly evident in recent years that very few light stabilisers act simply by absorbing and harmlessly dissipating light and that most "UV stabilisers" are autosynergists due to the combined effect of UV screening and one or more additional chemical mechanisms. Furthermore, all UV absorbers are not UV stabilisers and it will be seen below that a relatively small structural change in a UV absorbing compound can transform a UV stabiliser into a UV sensitiser.

4.3.1 *Pigments*

Many inorganic and organic pigments when dispersed in liquids or polymers have the ability to screen them from UV light, either by absorbing or reflecting incident radiation. By far the most effective light absorbing pigment is carbon black which is used in durable engineering rubbers and plastics. Also widely used are TiO_2, particularly in packaging and Fe_3O_4, in agricultural fibres. Carbon black, as well as being an effective absorber of light at all incident wavelengths, also contains chemical groupings, notably phenols and quinones and stable radicals with antioxidant properties [101,102]:

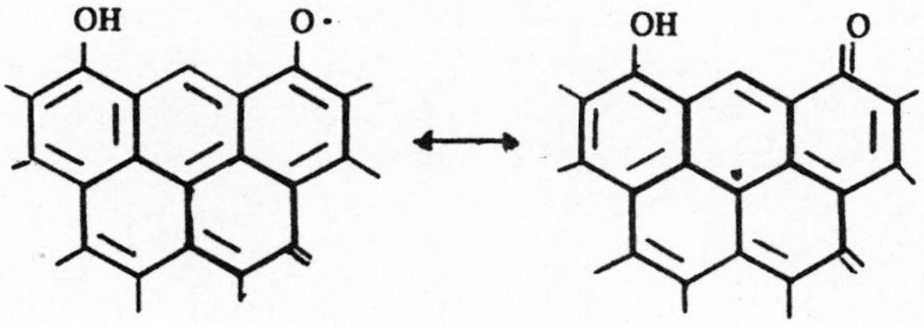

The functional groups in carbon black not only contribute to their photoantioxidant properties but also synergise with other added antioxidants, particularly sulphur compounds [101].

The limitations to carbon black are aesthetic rather than technical, and for many purposes, white or coloured pigments are used in spite of their relative technical inferiority. The effectiveness of pigments as light stabilisers for polymers has been reviewed by Rabek [103] and by Allen [88]. Many light screening pigments, unlike carbon black have little chemical activity. However, problems frequently occur due to the presence of a small amounts of UV sensitising transition metal ions in the metallic pigments and great care is taken during their manufacture to minimise this effect.

Some pigments exist in different morphological forms which also differ in photostabilising ability. For example, TiO_2 can exist as either rutile or anatase and the latter is a poor light stabiliser for polyethylene [104]. Rutile is normally used in commercial polymers, particularly in packaging which does not require a long lifetime. In long-lived products even pure rutile is not very effective due to photoexcitation of oxygen in the surface on exposure to light [103]. Hydroxyl radicals [105,106], peroxyl radicals [107], singlet oxygen [108] and even atomic oxygen [109,110] have all been reported as products of this interaction. Which of these are responsible for the sensitisation of the polymeric substrate to oxidation is far from clear, but there can be little doubt that a photoexcited state of TiO_2 is responsible for the photoprooxidant effect [111-113]. Consequently, titanium dioxide used as a whitening pigment in plastics and paint technologies is surface coated with a manganese compound [114] (normally the phosphate) to quench excited states in the titania surface, and this process can be regarded as a very specialised preventive mechanism because it precedes the generation of any reactive oxygen species.

Cadmium sulphide which is also frequently used as a pigment in engineering plastics causes photodegradation in paint film on long-term exposure to light and hydrogen peroxide has been observed as a product when it is irradiated in aqueous suspension [115].

4.3.2 *Organo-soluble nickel complexes*
A number of soluble nickel complex inhibitors of polymer photooxidation have been described by Rabek [116]. Relatively few of these have been used commercially since most of them are green. However, the long-established branched-chain nickel dialkyldithiocarbamates (NiDRC) and the nickel phenolate, 2,2'-thiobis-(4-*tert*-octyl-phenolato)-*n*-butylamine nickel(II), NiBOP, are still widely used in agricultural products.

Their main uses are in greenhouse films and horticultural twines where colour but not opacity is tolerable and green is even considered an advantage. In general nickel complexes of this type have a high extinction coefficient in the region of 330 nm. However, it has been demonstrated [117] that NiDRC have additional photoantioxidant functions other than light screening. This is illustrated in Fig. 4.7 [117] which compares the photooxidative development of carbonyl compounds in a polyethylene sample containing NiDEC (curve 6) with an unstabilised film screened from the UV source by a NiDEC-containing film at the same concentration (curve 2). Stabilisation by UV screening is clearly playing a minor role, since incorporation of NiDEC at the processing stage protects the polymer from oxidation both during processing as evidenced by the elimination of carbonyl formation, and during photooxidation it results in an induction period to carbonyl formation.

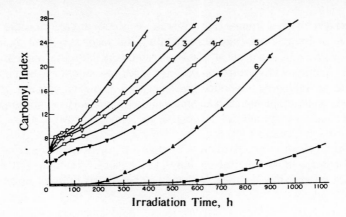

Fig. 4.7 Comparison of UV stabilisers as screens and as additives in low density polyethylene; 1, no additive; 2, NiDEC (screen); 3. HOBP (screen); 4, NiDEC + HOBP (screen); 5. HOBP (additive); 6. NiDEC (additive); 7. HOBP + NiDEC (additive). (Reproduced with permission from K.B. Chakraborty and G. Scott, *Europ. Polym. J.*, **13**, 1010 (1977)).

A combination of NiDEC with HOBP, a "UV absorber" (see below) also shows a strikingly different level of photoantioxidant activity for a film formulated with a combination of the additives compared with a PE film screened by the same combination. Similar though rather less spectacular results were observed by Ranaweera and Scott [118].

Nickel complexes are known to be effective quenchers of photochemically formed excited states, particularly of triplet carbonyl, >C=O* and singlet oxygen ($^1O_2$) [103] which are potential initiators of peroxidation (Chapter 1). However, saturated polymers, such as the polyolefins, in which these additives are mainly used, are not very reactive toward either of these oxidation sensitisers [119]. Wiles [120] was unable to find any clear correlation between photoantioxidant activity and singlet oxygen quenching ability of metal complexes and their effectiveness as UV stabilisers (Table 4.9) and some of the common commercial photoantioxidants are relatively poor quenchers.

Table 4.9 Comparison of the effectiveness of UV stabilisers in polypropylene and their $^1O_2$ quenching rate constants (k_q) [120]

Additive	Embrittlement time, h	k_q ($10^{-8} M^{-1} s^{-1}$)
None	110	-
NiDiPrP*	430	54
NiDiPrC*	1310	34
ZnDiPrC*	186	<0.1
NiBOP	250	1.3
HOBP*	350	<0.01

* See Table 4.1 for chemical structures

Some of the most effective $^1O_2$ quenchers (e.g. β-carotene and FeDRC) are not effective photostabilisers. The behaviour of the nickel dithiocarbamates is, however, consistent with the known peroxidolytic antioxidant activity of the dithiocarbamates (see Section 4.1.2(b)), but an important pre-requisite for high photoantioxidant activity of metal complexes appears to be their stability under photooxidative conditions [102]. Scott has concluded [102,119-123] that all the nickel complexes are light stable antioxidants acting by the CB-D and PD-C mechanisms discussed earlier.

4.3.3 *Phenols*
In general alkylated phenols are not effective light stabilisers. This is to a large extent due to their sensitivity to UV light. Pospišil has shown [124,125] that photo-unstable peroxidienones (e.g. BPDO, Scheme 3.3), formed during their normal peroxyl scavenging function, are effective sensitisers for photooxidation since these are themselves chromophores [102]. 2- or 4-Carbonyl phenoxyls are much more stable than those containing alkyl in the 2 or 4 positions [126] and they do not appear readily to form peroxydienones. This in part accounts for the UV stabilising effectiveness of the phenyl ester of salicylic acid (PSE), one of the earliest photoantioxidants to be used commercially in polymers [127,128].

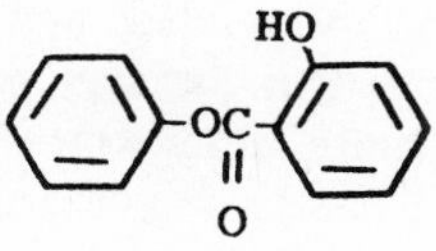

PSE

It was shown however, that the extent of intramolecular hydrogen bonding between the phenolic hydroxyl and the aromatic carbonyl group is correlated with UV stabilising effectiveness [129,130] and this was a common feature in the empirical development of the 2-hydroxy benzophenones (e.g. HRBP, Table 4.1). Benzophenones without a 2-hydroxy group are actually be photosensitisers for peroxidation and Lamola and Sharp showed [130] that the lifetime of the triplet state from a 2-hydroxy benzophenone is three orders of magnitude lower than that from an unsubstituted benzophenone, and the rapid abstraction of a hydrogen from the phenol to the photoexcited ketone (see Scheme 4.11) with subsequent thermal energy loss has been advanced to account for the harmless dissipation of the UV energy.

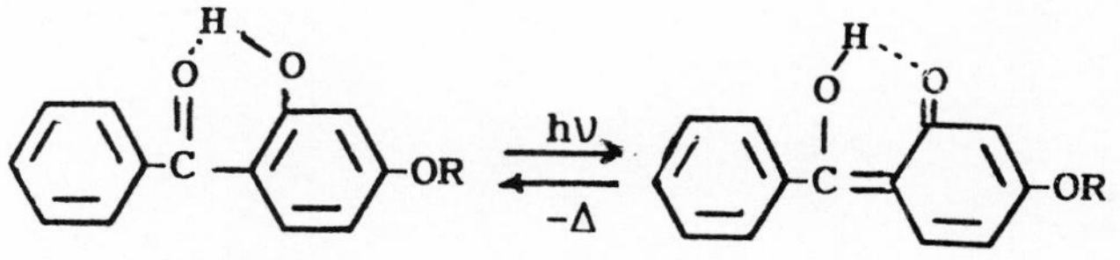

HRBP

Scheme 4.11 Energy dissipation in 2-hydroxy-benzophenones

This mechanism is supported by the fact that methylation or ionisation of the 2-hydroxy group removes the strong absorption in the vicinity of 330 nm [669], and essentially destroys photoantioxidant activity [102]. The photochemistry of the internal quenching mechanism and of this and the analogous process occurring with the 2-hydroxyphenyl-benzotriazoles (HRBT, Table 4.1) has been reviewed by Rabek [103]. The reader is referred to this for further detail.

There is a good deal of evidence to suggest that, as in the case of the nickel complexes discussed in the previous section, dissipation of UV energy is not the only mechanism involved in the photoantioxidant activity of the 2-hydroxybenzophenones and 2-hydroxybenzotriazoles [102]. 2-Hydroxy-4-octoxy-benzophenone, HOBP, is significantly more effective when incorporated into a polyethylene film than when used as a screen for an unstabilised PE film. The same was found to be true for polypropylene [131].

2-Hydroxybenzophenones have weak CB-D activity [132] and Chakraborty and Scott [133] observed that HOBP is rapidly destroyed by "oxyl" radicals in a hydroperoxide initiated peroxidation of hexane. A photosensitising ketone, benzophenone, had a similar effect. It was concluded that 2-hydroxybenzophenones are autosynergistic due to the combined effects of UV screening and CB-D antioxidant activity. The factor limiting the photoantioxidant activity of this class of UV stabiliser is not its intrinsic photostability in the absence of oxygen but the rate of sacrificial destruction by "oxyl" radicals under photooxidative conditions. This is consistent with the relative ineffectiveness of this class of compound in unsaturated and hence relatively peroxidisable substrates [133]. It is also in accord with the very powerful synergism observed between this class of photo-antioxidant and the peroxide decomposers which essentially eliminate oxyl radicals (Section 4.5).

4.4 Synergism and Antagonism

The occurrence of synergism between different antioxidant molecules is depicted generally in Fig. 4.8 and synergism is normally expressed quantitatively as the percentage improvement given by a combination of antioxidants on an additive basis and, conversely, antagonism is the percentage decrease in effectiveness compared with that expected:

$$\text{Synergism (\%)} = \frac{E - (a + b)}{a + b} \times 100 \tag{i}$$

where E is the effect of a synergistic combination, a and b are the effects of the individual antioxidants separately.

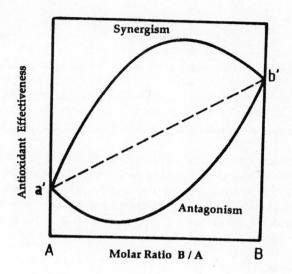

Fig. 4.8 Synergism and antagonism expressed as a function of the molar ratio of antioxidants A and B.

Two different types of synergism between antioxidants can be distinguished [134,135].

a) Homosynergism in which two antioxidants acting by the same mechanism interact, generally in a single electron or hydrogen transfer process.
b) Heterosynergism where the antioxidants act by a different mechanism and hence complement one another.

4.4.1 *Homosynergism*
The earliest and best known example of homosynergism was the seminal discovery by Olcott and Mattill that ascorbic acid synergises effectively with the tocopherols in natural oils [136,137]. This will be discussed in detail in Chapter 5, where it will be seen that regeneration of the lipid soluble α-tocopherol (α-Toc-OH) occurs at the membrane surface by reaction of α-tocopheroxyl (α-Toc-O·) with ascorbic acid $(Asc(OH)_2)$. This process constitutes one of the most powerful and all-pervasive antioxidant mechanisms in the biological cell. Both α-Toc-OH and $Asc(OH)_2$ are CB-D antioxidants and separately both form relatively stable radicals on oxidation [138-140]. *In vitro,* the addition of ascorbic acid to lard fatty acids stabilised with α-tocopherol substantially extends the oxidation induction time [144]. When a mixture of the two antioxidants is reacted with alkylperoxyl radicals, only the dehydroascorbate radical, $(Asc(OH)O·$, is observed by ESR [140]. A similar rapid hydrogen transfer from ascorbate to a hindered phenoxyl has also been noted [140] (see Scheme 4.12).

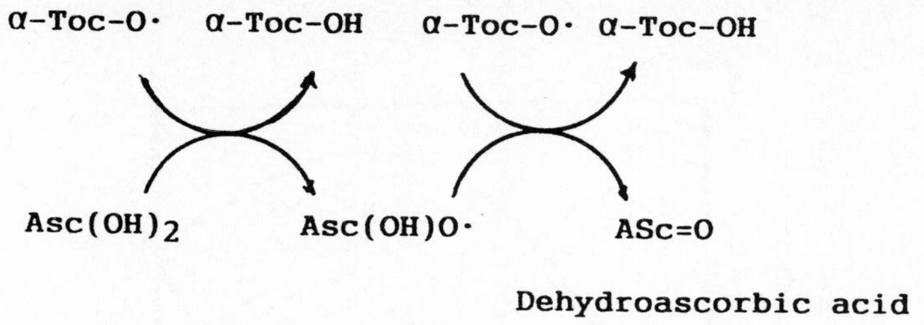

α-Toc-O· α-Toc-OH α-Toc-O· α-Toc-OH

Asc(OH)$_2$ Asc(OH)O·   ASc=O

Dehydroascorbic acid

Scheme 4.12 Homosynergism between α-tocopherol (α-Toc-OH) and ascorbic acid (Asc(OH)$_2$)

Effective homosynergism (120%) has been observed between a 2,6,-dimethyl phenol and a 2,6-di-*tert*-butyl phenol in petroleum (see ref. 134) but this is a relatively rare example of CB-D/CB-D synergism in technological systems. By contrast naturally occurring quinols and semiquinones (e.g. ubiquinols) have also been shown to act as regenerative synergists for the tocopherols (see Chapter 5) and this type of synergism appears to be much neglected in technology.

4.4.2 *Heterosynergism*
In principle synergism is possible between any two or more of the different mechanisms described earlier. In practice three types of combination have been very thoroughly investigated: in thermal oxidation, chain-breaking (CB-D) and peroxidolyic (PD-C) antioxidants and in photooxidation, UV absorbers in combination with PD or CB-D antioxidants. Some combinations; notably CB and sulphur (PD-C) in photooxidation have been found to be antagonistic.

(a) Synergism between CB-D and PD-C antioxidants
In a seminal study of antioxidants in white mineral oils at 155°C, Kennerly and Patterson [142] demonstrated (see Table 4.10) very effective synergism (180%) between a bis-phenol (DMPB) and zinc dihexyldithiophosphate (ZnDHP). The same phenol also synergised with a dialkyl sulphide (DDMS), but a combination of a dialkyl-p-phenylenediamine (DBPD) and ZnDHP was ineffective (36%). A combination of two CB-D antioxidants DMPB and DBPD was strongly antagonistic.

Table 4.10 Synergism between CB-D and PD-C antioxidants in mineral oil at 155°C

Antioxidant combination		τ,h	Synergism, %
DMBP g/100g	ZnDHP g/100g		
–	–	0.5	–
0.0033	–	6.0	
–	0.025	12.7	180
0.0033	0.025	47.5	
0.0067	–	12.7	
–	0.025	12.7	124
0.0067	–	55.0	
DMBP g/100g	DDMS g/100g		
0.0067	–	12.7	
–	0.025	12.2	128
0.0067	0.025	54.7	
DBPD g/100g	ZnDHP g/100g		
0.00.67	–	16.1	
–	0.025	12.7	36
0.0067	0.025	38.2	

DMBP = [2,2'-methylenebis(4-methyl-6-tert-butylphenol) structure: two phenol rings with tBu, OH, CH₃ substituents linked by CH₂]

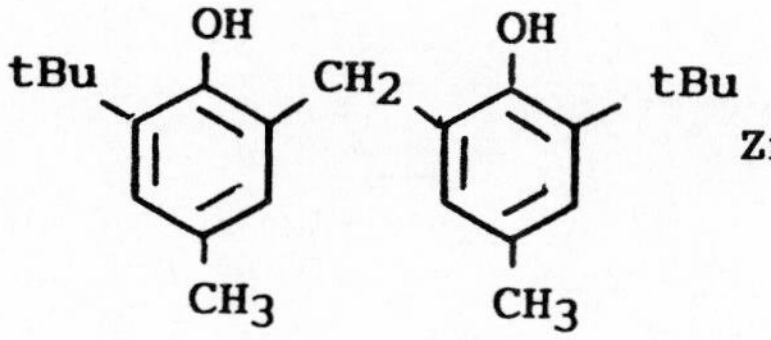

ZnDHP = $[(C_6H_{13}O)_2P(=S)S]_2Zn$

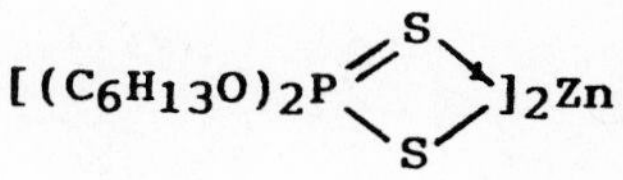

DDMS = $(nC_{10}H_{21})_2S$

DBPD = sBuNH–C₆H₄–NHsBu

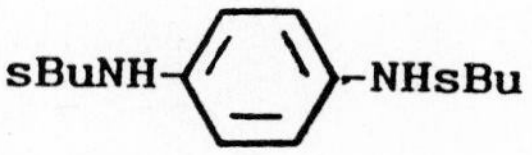

The empirical development of synergistic antioxidants in polypropylene in the 1960s [143] made possible the stabilisation of this polymer during processing and in service. Until this time, it had been doubtful whether oxidatively sensitive branched-chain polyolefins (e.g. polypropylene and poly-4-methylpentene) could ever achieve the resistance to heat and light necessary to warrant their development as commercial polymers. Detailed studies of synergistic combinations of semi-hindered phenols with dilaurylthiodiproprionate (DLTP) showed that the optimal ratio of heterosynergists was not 1:1 (see Fig. 4.8) [135,144]. As little as 10% of a CB-D antioxidant in a CB-D/PD-C mixture increased antioxidant activity five-fold.

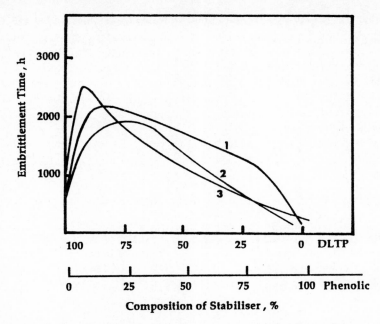

Fig. 4.9 Synergism between DLTP and phenolic antioxidants in polypropylene at constant total concentration in an air oven at 150°C. (Reproduced with permission from G. Scott in *Atmospheric Oxidation and Antioxidants*, Vol. II, Ed. G. Scott, Elsevier, Amsterdam, 1993, p.436).

In Section 4.1.1 it was seen that products formed from DRTP are very effective antioxidants but that this is preceded by a radical generating step which is particularly evident in a peroxide initiated peroxidation either in model hydrocarbons [17,23,144] or in polypropylene during processing [145]. The effect of a small proportion of a phenolic antioxidant appears to be to suppress the radical initiating processes without inhibiting the formation of the peroxidolytic antioxidant [135,144,145]. By the same token, the heterolytic destruction of hydroperoxides by the sulphur acids also prolongs the lifetime of the peroxide decomposer.

Zinc diethyldithiocarbamate (ZnDEC) which is itself a very powerful antioxidant, is also effectively synergised by hindered phenols in the thermal oxidation of both PE and PP in an air oven (Table 4.11) [117,135].

Table 4.11 Synergism between ZnDEC and Irganox 1076 in polyolefins in an air oven (concentrations 3×10^{-4} mol/100g) [117]

Antioxidant	PP Embrittlement time, h at 140°C	Syn. %	LDPE Induction time, h at 110°C	Syn. %
None	0.5		10	
ZnDEC	32		110	
1076	58		110	
ZnDEC + 1076*	165	183	300	150

* I(b), Table 3.1)

(b) Synergism between UV absorbers and PD-S or CB-D antioxidants

It was seen in Section 4.2.3 that under photooxidative conditions, the 2-hydroxy benzophenone light stabilisers (UV absorbers) are rapidly destroyed by radical generators and particularly by hydroperoxides. This process begins during high temperature processing and Table 4.12 shows that, although a typical peroxide decomposer such as ZnDEC is not a particularly active photoantioxidant in polypropylene (the same is true of polyethylene) [117], it shows powerful synergism with antioxidants.

Table 4.12 Photoantioxidant synergism between antioxidants and HOBP in propylene (concentrations 3×10^{-4} mol/110g) [117]

Antioxidant	Embrittlement time, h	Synergism, %
No additive	90	-
ZnDEC	175	-
NiDEC	500	-
1076	325	-
HOBP	245	-
ZnDEC + HOBP	700	154
NiDEC + HOBP	850	95
1076 + HOBP	650	44

Table 4.13 shows that the peroxide decomposers (Zn DEC and NiDEC) also synergise thermally with the UVA and the evidence which has been summarised by Scott et al.

[146,147] suggests that an important function of antioxidants is to protect the polymer and the UV absorbers during processing as well as during photooxidation.

Table 4.13 Thermal synergism between antioxidants (MDRCs, hindered phenol) and UV absorber (HOBP) in LDPE at 110°C [147]

Antioxidant	τ, h	Synergism, %
No additive	10	-
HOBP	10	-
1076	110	-
ZnDEC	110	-
NiDEC	105	-
1076 + ZnDEC	330	60
HOBP + ZnDEC	130	20
1076 + NiDEC	300	222

The reason for the relative ineffectiveness of ZnDEC as a photoantioxidant compared with NiDEC is its relatively poor photostability. The induction time to carbonyl formation is closely correlated with the photochemical destruction of the complexes in the polymer [148], and the reason for the synergism shown in Table 4.9 is the protection provided by the UVA for the metal complex. Hindered phenol, 1076, which is also photo-unstable [147,148], similarly synergises with HOBP but the effect is not as great. Optimal synergism in the ZnDEC/HOBP system occurs at a molar ratio [HOBP]/[ZnDEC] = 2 [135,147]. It can also be seen from Table 4.12 that the synergism observed between NiDEC and HOBP is rather less than that for ZnDEC [117,135]. This is almost certainly because NiDEC is more light stable than ZnDEC and does not benefit from the additional UV absorbing ability of HOBP.

Light stabilisers are photochemically destroyed but, if the UV energy is internally quenched as in the nickel or cobalt dialkyldithiocarbamates, then photoantioxidant activity is markedly increased relative to the zinc complexes which do not have an internal energy dissipation mechanism. Scott et al [31,122] showed that ZnDEC-amine co-complexes (e.g. ZnDEC complexed with DABCO) are considerably more photo-stable than ZnDEC itself. This was attributed in part to internal quenching of excited states of the metal complex by the amine. However, increased solubility of the co-complex in the polymer and increased rate of hydroperoxide decomposition are also involved in this synergism [135]. Ivanov and Shlyapintokh [149] have also proposed that quenching agents may improve the light stability of phenolic antioxidants but this process is strongly dependent upon the distance between the donor and acceptor [135], and where complexation is absent, relatively high concentrations of both antioxidant and quencher are necessary to achieve appreciable effects. In the above ZnDRC-DABCO complexes are well-characterised chemical compounds [31] are very strongly associated in solution. This appears to be an essential requirement for this kind of synergism.

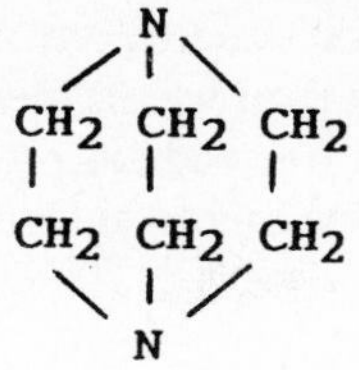

DABCO [structure: bicyclic diamine with N at top and bottom connected by three CH₂-CH₂ bridges]

UVAs have been reported to synergise with TMPs, although most effectively in relatively thick samples and with oligomeric TMPs, (see Table 4.14) [150]. By contrast, TMPs are antagonistic with sulphur antioxidants and this is believed to be due to the reduction back of the active agent, the nitroxyl, by intermediate thiyl radicals to the inactive parent amine [135]:

$$>NO\cdot + R_2NC(=S)S\cdot \rightarrow >NOSCNR_2(=S) \xrightarrow{H_2O} >NH + HOS(=O)CNR_2(=S) \quad (7)$$

Table 4.14 Synergism between HOBP and oligomeric HALS in LDPE (2mm plaques) [150]

Antioxidant	Concentration, g/100g	E_{50} (kLy)*
No additive	-	95
HOBP	0.1	265
HOBP	0.2	335
Tinuvin 622[+]	0.1	700
Tinuvin 622[+] + HOBP	0.1 + 0.1	>900

* E_{50} = Energy (kLy) to 50% retained elongation in Florida

[+] Tinuvin 622 = $-(-O-\underset{CH_3\ CH_3}{\underset{|}{\overset{|}{C}}}\underset{CH_3\ CH_3}{\underset{|}{\overset{|}{C}}}\text{piperidinyl}-NCH_2CH_2OCOCH_2CH_2CO-)_n$

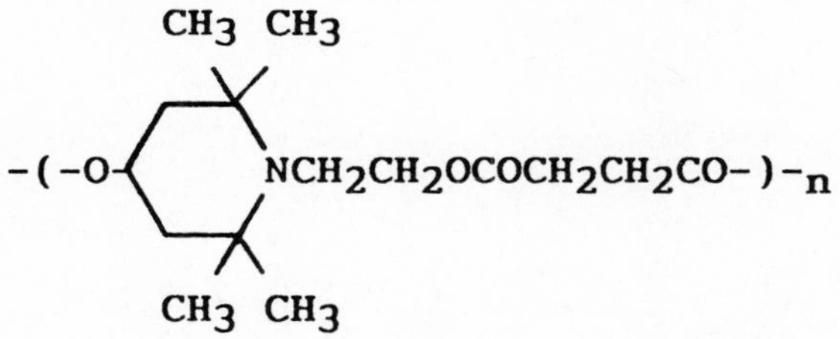

Photoantioxidant interactions can be generalised as follows [135]:

i) Chain-breaking (CB) and peroxidolytic (PD) antioxidants protect UV absorbers from destruction by hydroperoxides and their thermolysis products during processing.
ii) UV-stable UV absorbers protect CB and PD antioxidants from photolysis during environmental exposure of polymers. In some cases UV stability and antioxidant activity may be built into the same molecule.

iii) Quenching of intermediate photo-excited states of photoantioxidants may improve their effectiveness by extending their useful lifetimes.
iv) Chemical interactions between free radical products of antioxidants and stabilisers may reduce their effectiveness leading to antagonism.

4.4.3 *Autosynergism*

A number of examples of synergism between known antioxidant functions in the same molecule have been noted in the preceding sections. Typical examples are UVA/CB-D synergism in the 2-hydroxybenzophenones and 2-hydroxybenzotriazoles and UVA/PD-C synergism in the nickel dithiocarbamates and dithiophosphates and three component PD, CB-D, MD synergism in metal chelating agents (see Section 4.2). In general autosynergism is normally deduced from the fact that higher antioxidant activity results when the two functions are present in the same molecule than in the case of unifunctional molecules. For example, ZnDRC has no UV absorbing function whereas NiDRC, which has similar peroxidolytic activity, also absorbs UV and is much more UV stable [147]. The former is a unifunctional UV stabiliser, whereas the latter is difunctional and autosynergistic (Table 4.9).

CD-D antioxidants containing sulphur are generally more effective antioxidants than analogous antioxidants without sulphur. However, relatively little systematic work has been done to quantitatively measure the extent of this autosynergism. Scott et al. showed [18,151] that the autosynergist BHBM-R is approximately five times more intrinsically effective than BHT and ten time more effective than 1076 in the uninitiated peroxidation of decalin by oxygen absorption (see Table 4.15).

Table 4.15 Antioxidant effectiveness of a phenolic sulphide (all concentrations 6×10^{-4} mol/100g) [151]

Antioxidant BHBM-R (2,6-di-tBu-4-CH$_2$SR phenol)	τ, h			V g/h	S g/100g
R	D_c	PP_c	PP_o		
H	51.0	38.5	21.0	110	96
C$_2$H$_5$	47.5	33.0	5.0	100	65
C$_5$H$_{11}$	47.0	38.0	6.0	77.5	∞
C$_8$H$_{17}$	46.5	39.5	7.5	43.1	∞
C$_{12}$H$_{25}$	45.0	44.5	11.0	15.0	94
C$_{18}$H$_{37}$	44.5	27.0	15.0	2.0	57
BHT	10.0	6.0	0.4	157.5	98
1076	5.0	18.0	92.0	0.7	64

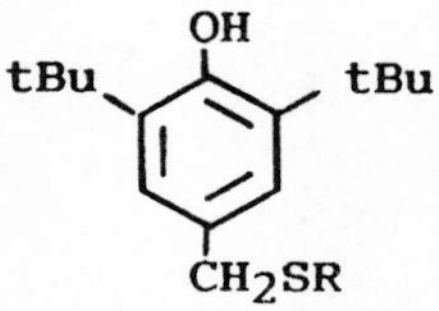

D_c, τ in decalin by oxygen absorption, PP_c, τ in polypropylene by oxygen absorption, PP_o, τ in polypropylene, in air oven by embrittlement time.

The "closed system" test (D_c in Table 4.15) provides a measure of intrinsic molar antioxidant activity without the complication of possible loss of antioxidant by volatilisation from the surface of polymers (PP_o in Table 4.15) or from partial solubility of the antioxidant in the medium (PP_c in Table 4.15). This is discussed in more detail in Section 4.5.1.

Farzaliev et al. have shown [152] that the autosynergistic antioxidants, BHBM-R (Table 4.15), catalytically destroy hydroperoxides by an ionic process at approximately the same rate (see Table 4.16) and that this rate is essentially the same for dibenzyl monosulphide (DBMS) which contains no CB-D antioxidant function. The commercial bis-phenol sulphide (BBPM), an antioxidant for polyolefins, also destroys hydroperoxides catalytically but at a rate two orders of magnitude lower. The number of moles of hydroperoxide destroyed for each mole of BHBM-R (R = $C_{12}H_{25}$), ν, was found to be extremely high (> 3 x 10^5) and the *ortho* isomers gave a similar rate. Zweifel has recently reported [153] that 2-hydroxybenzyl monosulphides are more effective in polyolefins than the 4 isomers, but in the light of the above, it seems likely that this is due to differences in physical behaviour (e.g. volatility or solubility) rather than in intrinsic antioxidant activity as has been seen in Table 4.15 for the BHBM-R analogues.

Table 4.16 Induction times (τ) and pseudo first order (catalytic) rate constants (k) for CHP decomposition by sulphide antioxidants in cumene at 70°C. (CHP concentration, 0.01M, antioxidant concentration, 0.001M) [152]

Antioxidant	τ, min	$10^2 k_s^{-1}$
R in BHBM-R		
H	25	3.4
$C_{10}H_{21}$	33	3.8
CH$_2$-(tBu)-OH(tBu)	35	3.8
DBMS	28	3.7
BBPM	–	0.01

DBMS = Ph-CH$_2$SCH$_2$-Ph, BBPM = HO-(3,5-di-tBu-phenyl)-S-(3,5-di-tBu-phenyl)-OH

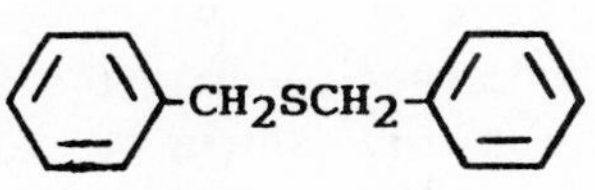

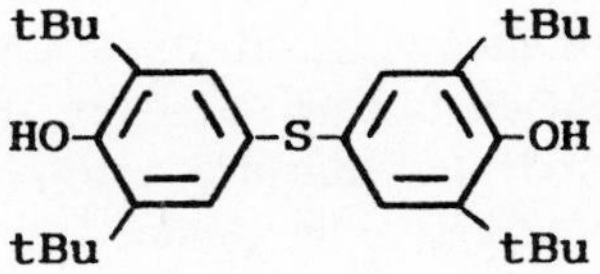

The major oxidation products identified in the oxidation of BHBM-R were the corresponding benzyl alcohol and its further oxidation products [153], indicating that the sulphur is completely eliminated from the antioxidant. In a more detailed study of BHBM-H, Scott and Suharto [154] showed that the primary product, the thiyl radical can be trapped by olefins and this reaction has been utilised in the addition of BHBM-H and related thiols under anaerobic conditions to polyunsaturated polymers (see Section

4.5.3). However, the effective catalytic agents, the peroxidolytic sulphur acids, are formed in the presence of oxygen from further oxidation products, partly by elimination of SO_2 from the unstable intermediate sulphinic acid, BHBS, see Scheme 4.13 and partly by further oxidation of the sulphinic acid to sulphonic acid, BHBSO which also eliminates SO_2 to give the benzyl alcohol.

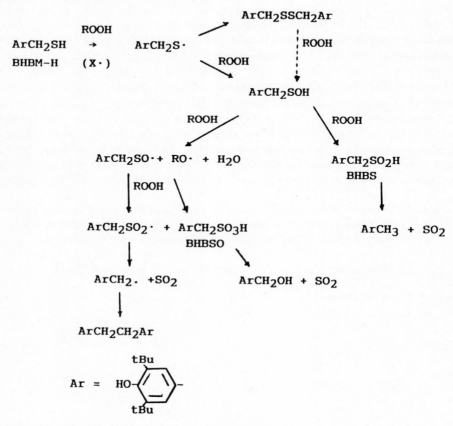

Scheme 4.13 Oxidative transformations of BHBM-H in the presence of hydroperoxides

A direct analogy was noted with the oxidation of mercaptobenzothiazole (MBT) (Section 4.1.2(a)), in which similar prooxidant and antioxidant reactions occur in parallel. A significant feature of the above work was the persistence of the hindered phenol (CB-D) structure through the intermediates and into the final sulphur-free products, indicating a prolonged synergistic interaction through scavenging of radicals produced as by-products in the redox reactions between hydroperoxides and sulphur intermediates. This, then, accounts for the very powerful autosynergism observed.

Jiráčkova and Pospíšil [155-156] have studied the oxidative reactions of the bis-phenol mono and disulphides (e.g. BBPM in Table 4.16) and their related sulphoxides and sulphones. The chemistry is very similar to that described above and in general disulphides are more effective antioxidants than monosulphides due to the lability of the intermediate thiolsulphinates (e.g. DPTS). This recalls the earlier work of Hawkins and Sautter in the non-phenolic diaryl sulphides (see Section 4.1.1).

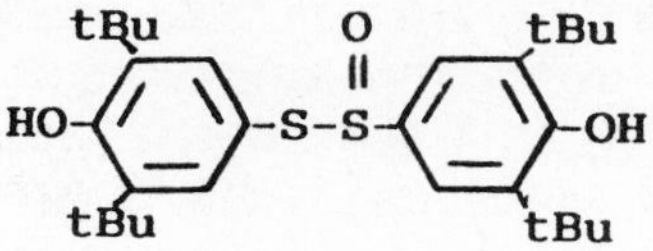

HO–⟨tBu, tBu⟩–S–S(=O)–⟨tBu, tBu⟩–OH   DPTS

A number of highly effective autosynergistic antioxidants have been developed based on mercapto esters (PMRA) containing antioxidant functions [158-162].

$AXCO(CH_2)_nSH$ PMRA

X = O or NH, A = chain-breaking antioxidant (phenol or aromatic amine) or a UV absorber (e.g 2-hydroxybenzophenone, n = 1-3

It will be seen in Section 4.5.3 that the above thiols are the basis of polymer-bound antioxidants in which the antioxidant function is directly linked to the polymer though sulphur. When the antioxidant group, A, is a UV absorber (e.g. HRBP), the latter synergises effectively with the monosulphide, both thermally and photochemically. This group of autosynergistic antioxidants also decompose hydroperoxides at the same molar rate as BHBM [163,164] and are much more effective than single function antioxidants. This is illustrated in Table 4.17 [164] which compares the autosynergistic photoantioxidant activity of EBHPT-B with a conventional polymer-soluble UV stabiliser HOBP without the sulphur (peroxidolytic) function at the same concentration. This shows that the autosynergist, even in a solvent extracted polymer which removes unbound EBHPT, is significantly more effective than HOBP. EBHPT is also a more effective thermal antioxidant than HOBP [17].

Table 4.17 Comparison of an autosynergistic UV absorber, EBHPT-ABS, with HOBP in ABS (concentration 3×10^{-3} mol/100g) [164]

UV Absorber	Induction time to carbonyl formation h	Embrittlement time, h
HOBP (unextracted)	11	35
HOBP (extracted)	–	25
EBHPT-B (unextracted)	20	52
EBHPT-B (extracted)	17	48
None	2	23

EBHPT- = ⟨⟩–CO–⟨OH⟩–OCOCH$_2$CH$_2$OCOCH$_2$S–

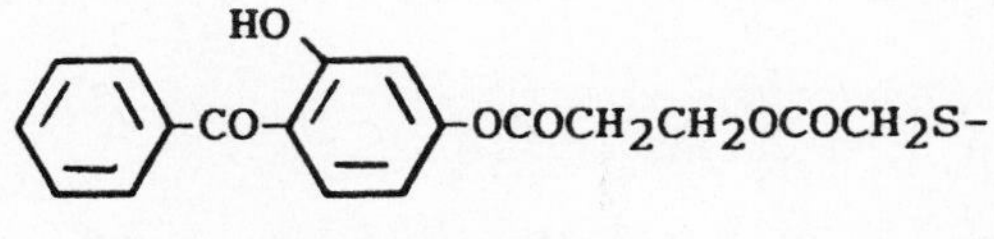

ABS = acrylonitrile-butadiene copolymer

Autosynergists may also synergise with antioxidants and stabilisers acting by different mechanisms. As discussed previously both CB-D and PD-C antioxidants synergise with UV absorbers. Thus BHBM-R is twice as effective as a synergist with HOBP in polypropylene than monofunctional phenols, BHT and 1076 [135,162,165]. However, optimal synergism between BHBM-R homologues and HOBP occurs at a 1:3 molar ratio of these antioxidants [162]. This is of some practical importance since although appreciable synergism that can be achieved at a 1:1 molar ratio this is considerably less than at a 1:3 ratio. However, the latter cannot be achieved in a single molecule and this indicates the principle limitation to the autosynergistic concept [135].

It was seen in Chapter 1 that the formation of hydroperoxides and hydrogen chloride together during the oxidative degradation of PVC leads to a radical generating redox reaction between HCl and hdroperoxides. It is difficult to exclude either of these completely during the mechanooxidation of PVC during processing, and polymer soluble metal carboxylates (e.g. zinc, calcium, cadmium and barium stearates) are moderately effective scavengers of HCl. Much more effective are alkyltin maleate esters (e.g. dibutyltinmaleate, DBTM) and the tin thioglycollates (e.g. dioctyltin dioctylthioglycollate, DOTG). Both compounds are efficient scavengers of HCl.

$$Bu_2Sn\begin{smallmatrix}O-CO\\ \\O-CO\end{smallmatrix}\begin{smallmatrix}CH\\||\\CH\end{smallmatrix} + 2HCl \rightarrow Bu_2SnCl_2 + O\begin{smallmatrix}CO\\ \\CO\end{smallmatrix}\begin{smallmatrix}CH\\||\\CH\end{smallmatrix} + H_2O \quad (7)$$

DBTM   MA

Both DBTM and derived maleic anhydride (MA) are dienophiles which react with conjugated unsaturation as it is formed in the polymer [67,166,167]. Removal of polyenes improves the colour of the polymer. In the case of DOTG, the liberated thiols are themselves PD-C antioxidants but in addition they add to monoenic unsaturation, thus removing allylic groups and reducing the rate of peroxidation (see Scheme 4.14) [168]. The monosulphides so produced are peroxidolytic antioxidants analogous to the monosulphide antioxidants discussed in Section 4.1.1, but Cooray and Scott [169] have shown that in addition the parent DOTG is directly oxidised by hydroperoxides to sulphur acids through the intermediate sulphoxides [631a5]. As in the case of the thiodipropionates, the early stages of this transformation involve the formation of radicals in the bimolecular reaction of hydroperoxide with DOTG, producing a prooxidant effect in hydrocarbon substrates: but the later stages involving sulphur acids are essentially ionic giving rise to phenol and acetone at temperatures as low as 60°C. The oxidative transformations proposed by Cooray and Scott to explain the powerful antioxidant activity of the thiotin stabilisers during processing and in service [168] are shown in Scheme 4.14.

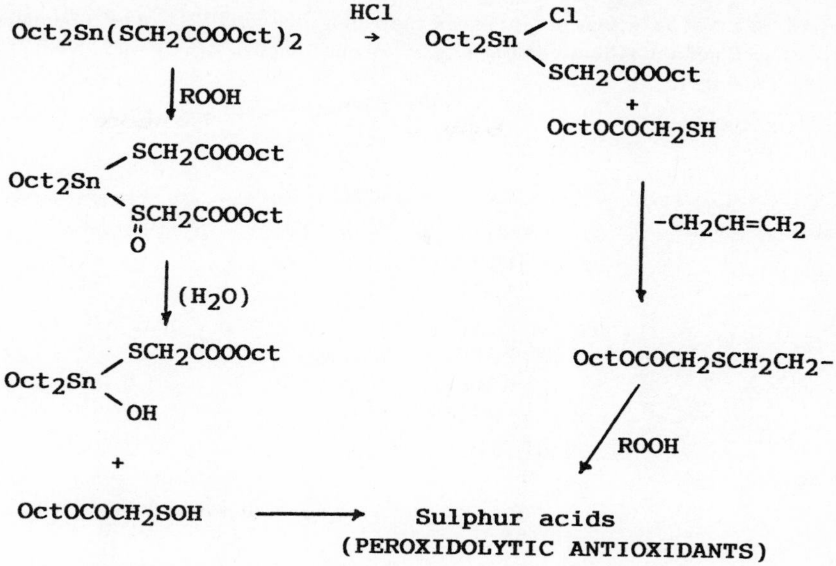

$$Oct_2Sn(SCH_2COOOct)_2 \xrightarrow{HCl} Oct_2Sn\begin{matrix}Cl\\SCH_2COOOct\end{matrix} + OctOCOCH_2SH$$

$$\downarrow ROOH \qquad\qquad\qquad \downarrow -CH_2CH=CH_2$$

$$Oct_2Sn\begin{matrix}SCH_2COOOct\\\underset{O}{\overset{\|}{S}}CH_2COOOct\end{matrix} \qquad\qquad OctOCOCH_2SCH_2CH_2-$$

$$\downarrow (H_2O) \qquad\qquad\qquad \swarrow ROOH$$

$$Oct_2Sn\begin{matrix}SCH_2COOOct\\OH\end{matrix}$$

$$+$$

$$OctOCOCH_2SOH \longrightarrow \text{Sulphur acids}$$
$$\text{(PEROXIDOLYTIC ANTIOXIDANTS)}$$

Scheme 4.14 Autosynergistic mechanism of DOTG

4.5 Physical Aspects of Antioxidant Performance

It was observed in earlier sections that molecular weight variation in a homologous series of antioxidants has a major influence on effectiveness (see for example Table 3.3). This was seen to be of particular importance under high temperature conditions and/or in a moving air stream where antioxidant volatility rather than intrinsic activity frequently dominates technological performance. However, loss of low molecular weight chemicals by leaching from packaging polymers in contact with foodstuffs is becoming increasingly important with the awareness that a substantial proportion of added antioxidants and stabilisers and of their oxidative transformation products may migrate into food during storage [170-172]. Equally important is the long-term stabilisation of biomedical polymers, particularly of polymer components of prostheses, without leaching of additives from polymer components into the body.

An appreciation of the importance of physical factors in determining antioxidant and stabiliser substantivity in polymers has developed over the past thirty years in distinct stages.

4.5.1 *Effect of molecular size on antioxidant activity*

A number of early studies, reviewed by Scott [76,144], led to the recognition that under certain conditions physical loss of antioxidants from polymers is much more important than loss by chemical reaction and that increase in molecular size may result in more effective antioxidants. Table 4.18 illustrates the effect of molecular weight for two homologous series of iso-functional antioxidants [173]: the first is the homologous series of PD antioxidants, of which the commercial antioxidant DRTP is a member (see Table 4.1) and the second is the propionate ester series of which Irganox 1076 (Table 3.1, I(b)) is a commercially important member.

Table 4.18 Effect of molecular weight on the induction time for CB-D and PD homologous series of antioxidants at 100°C (2×10^{-3} mol/100g) [173]

Antioxidant	R	Mol. wt	Induction time, h
(ROCOCH$_2$CH$_2$)$_2$S	CH$_3$	206	42
DRDP, XXIV	C$_6$H$_{13}$	346	42
	C$_{12}$H$_{25}$	514	80
	C$_{18}$H$_{37}$	702	430
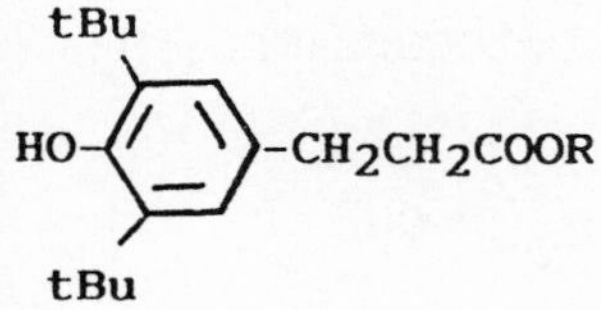 tBu-C$_6$H$_2$(OH)(tBu)-CH$_2$CH$_2$COOR	CH$_3$	292	45
	C$_6$H$_{13}$	362	80
	C$_{12}$H$_{25}$	446	400
	C$_{18}$H$_{37}$	530	>10^4

It is clear that on air-oven exposure, the lower molecular weight additives do not persist long enough in the polymer to exert their antioxidants effect. In a closed system, however, the same antioxidants may be much more effective (see Chapter 3, Table 3.3) and substrate solubility tends to dominate antioxidant effectiveness. A very similar effect of molecular size was noted in the autosynergists series, BHBM-R in the last section (see Table 4.15).

A series of structurally related but not iso-functional oligomeric antioxidants based on the condensation of *p*-cresol with formaldehyde (CPF) was also examined in the same series of complementary exposure conditions; in decalin by oxygen absorption, in polypropylene by oxygen absorption and in polypropylene in an oven ageing test (see Fig. 4.10) [76]. As outlined in Sections 3.3.2 and 4.4.3, these three tests emphasise intrinsic antioxidant activity (D_c), solubility (PP_c) and volatility (PP_o) respectively.

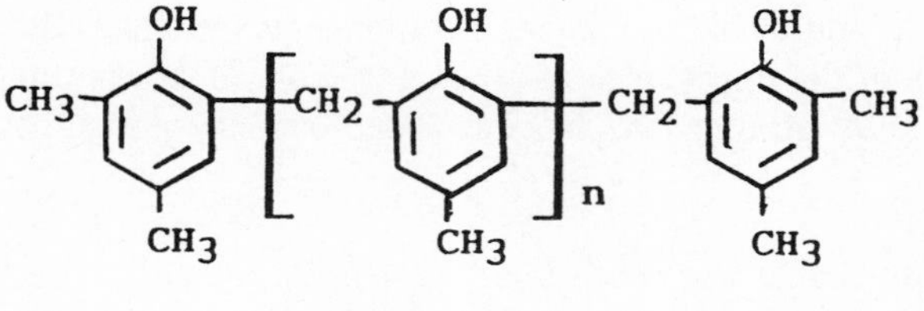

CPF

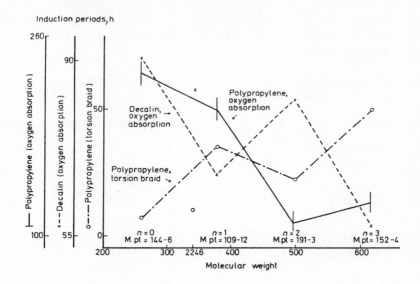

Fig. 4.10 Antioxidant activity at 140°C of homologous oligomeric phenols (CPF) in decalin (closed system, D_c), in polypropylene (closed system, PP_c) and in polypropylene (open system, PP_o). (Reproduced with permission from G. Scott, Pure and App. Chem., **30**, 275 (1972)).

This study showed the expected increase in effectiveness of an oligomeric series with increase in molecular weight in an air stream (PP_o at 140°C). However, a sharp alternation in intrinsic antioxidant activity (D_c), suggests that the in-chain and terminal phenolic groups had different activities (probably due to different hydrogen-bonding interactions). The polypropylene oxygen absorption test (PP_c) showed a very sharp decrease in antioxidant activity which can be interpreted as a composite function of alternating intrinsic activity and solubility, both of which alternate in the series.

Billingham [174] has developed a useful model to describe the behaviour of the commercial range of antioxidants and stabilisers in polymers. This incorporates the concept of diffusion rate of the antioxidant within the polymer which is particularly important under solvent leaching conditions, and together with volatility provides a complete explanation of the way in which relatively low molecular weight antioxidants behave under a wide range of conditions. Calvert and Billingham [175] recognised boundary conditions which distinguish antioxidants for which diffusion is rate-controlling (thick samples, rapid evaporation and low diffusion rate) and those for which evaporation from the surface is rate-controlling (thin samples, slow evaporation and high diffusion rate). However a number of studies have shown [31,102,34,35,174] that antioxidant solubility in the substrate dominates the activity of polymeric or oligomeric antioxidants for which loss by diffusion/evaporation is not a determining factor.

Evans and Scott [176] compared the effectiveness of polymeric and grafted 3,5-di-*tert*-butyl-4-hydroxybenzyl esters (DBBR) in polypropylene and found that the former (PDBBA, M_n = 9500) was virtually inactive, whereas the UV grafted monomer was highly effective at a similar concentration and was significantly more effective than a long-chain alkyl ester (DBBR, R = $C_{18}H_{37}$). The latter which was in turn much more effective than DBBA monomer (see Table 4.19).

Table 4.19 Effectiveness of macromolecular hindered phenols compared with low molecular weight additives in polypropylene at 120°C (air oven) [176]

Antioxidant	Conc^n (10^4 mol/100g)	Embrittlement time, h
DBBD	2	244
DBBA	2	20
Poly-DBBA	2	18
Grafted DBBA	3.18	555

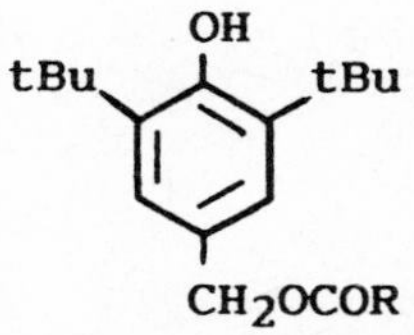

DBBD, R = $C_{18}H_{37}$
DBBA, R = $CH=CH_2$

Furthermore the activity of the polymer-bound antioxidant depended strongly on the length of the vinyl grafted segment; short grafts being more effective than long grafts.

The main conclusions to emerge from these studies were that antioxidant effectiveness under aggressive conditions is determined by the following factors:

a) Intrinsic antioxidant activity, which is primarily influenced by the structure of the molecule, including intramolecular interactions.
b) Solubility/mobility of the antioxidant in the substrate which is again determined by intra- and inter-molecular interactions of the molecule, but generally in the opposite direction to the above.
c) Volatility of the antioxidant which is determined by the molecular weight and molecular interactions with the polymer.

4.5.2 Evaluation of oligomeric antioxidants and stabilisers
Subsequent studies established that macromolecular antioxidant activity of polyvinyl antioxidants was optimal at a relatively low molecular weight. Gugumus reported that poly-(1,2,2',6,6'-pentamethyl-4-piperidyl)acrylate, DMPA, had optimal activity at $M_n \approx$ 2700 [177] and similar effects were noted in co-polymers [178]. Minagawa [179] similarly reported an optimal molecular weight $\approx$ 3000 for 3,5-di-*tert*-butyl-4-hydroxystyrene, DBHS, in polypropylene.

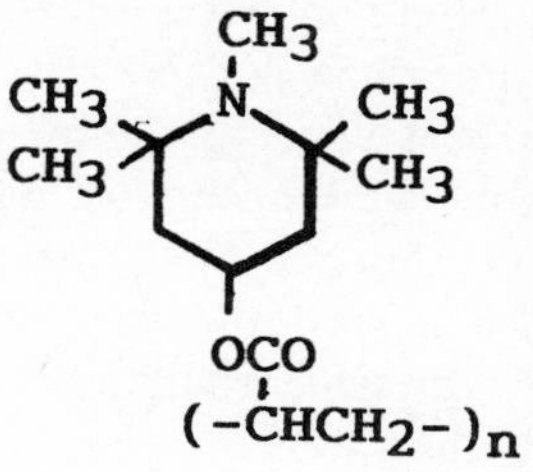

PMPA (structure shown with piperidine ring bearing N-CH$_3$, 2,2,6,6-tetramethyl, and OCO-(-CHCH$_2$-)$_n$ substituent)

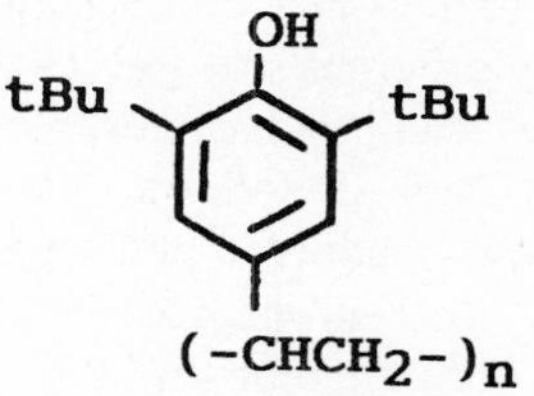

DBHS (2,6-di-tert-butylphenol with (-CHCH$_2$-)$_n$ para substituent)

Oligomeric TMPs, developed empirically, are much more substantive in polypropylene than the commercial bis-TMP, Tinuvin 770. Chimassorb 944 has been reported to have a molecular weight > 2500 and a broad melting range (100-130°C), indicating a spectrum of molecular weights [179]:

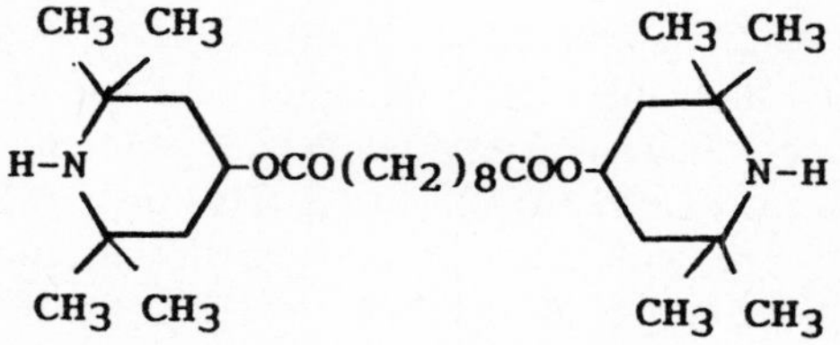

Tinuvin 770

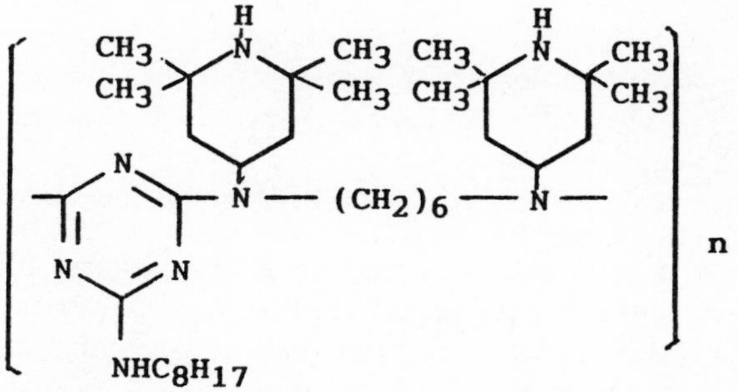

H-N(2,2,6,6-tetramethylpiperidine)-OCO(CH$_2$)$_8$COO-(2,2,6,6-tetramethylpiperidine)N-H

Chimassorb 944

[triazine ring with NHC$_8$H$_{17}$ substituent, linked via N to (CH$_2$)$_6$ spacer and to 2,2,6,6-tetramethylpiperidine units with N-H]$_n$

However, Minagawa found the optimal average molecular weight for this type of oligomeric TMP was ≈ 800 [179] and it must be concluded then that most commercial oligomeric TMPs are present as a second incompatible phase within the polymer. Thus, while the primary conditions of low volatility and low diffusion rate can be readily achieved by oligomerisation, the other essential feature required for high antioxidant activity; namely solubility in the amorphous phase of the polymer [170], steadily decreases as the molecular weight of the oligomer increases. Furthermore, it is difficult to see how this requirement can be achieved with a second polymeric substrate which has a completely different structure from the main hydrocarbon polymer. Even closely related polymers such as polyethylene and polypropylene have very limited mutual solubility [180] and it follows that the two main requirements of oligomeric antioxidants, low volatility and high solubility in the polymer, are mutually incompatible at least in hydrocarbon polymers. This pointed the way to a better stratagem to achieve maximum effectiveness under aggressive conditions involving covalent attachment of the antioxidant to the polymeric substrate.

4.5.3 *Polymer-bound antioxidants*

The recognition that antioxidant substantivity is the key to the development of stable polymers under aggressive conditions came not from the plastics industry but from the rubber industry within which the first polymer-bound antioxidants were developed. The techniques and procedures that have been used to covalently attach antioxidants and stabilisers to polymers have been reviewed by Scott [158,181-187], by Munteanu [188] and by Rabek [189]. They include the reaction of functionalised polymers with antioxidants [181,199], the copolymerisation of vinyl compounds containing antioxidant and stabiliser groups during the manufacture of vinyl polymers [181-199] and the grafting of vinyl monomers to commodity polymers [158,181-188]. The first technically successful antioxidant-modified polymer, Chemigum HR665, was designed and developed by workers at Goodyear [190-193]. This involved the co-polymerisation of the acrylamide, MDPA, with diene monomers during the synthesis of nitrile-butadiene rubber. The same principle was later applied to the UV stabilisation of vinyl polymers, notably polymethylmethacrylate and polystyrene by Vogl and co-workers, and the synthesis and copolymerisation of a range of 2-hydroxy-benzotriazoles and 2-hydroxy-benzophenones containing polymerisable vinyl groups has been described by these authors [194-198] and by Rabek [189].

$$\text{Ph-NH-C}_6\text{H}_4\text{-NCOC(CH}_3\text{)=CH}_2 \qquad \text{MDPA}$$

Although technically a very successful solution to the problem of antioxidant loss, attachment of antioxidants and stabilisers to polymers by copolymerisation is costly and can be justified only for speciality polymers where performance is of paramount importance (e.g. in engineering rubbers or polymers in space vehicles). Different techniques have been developed for commodity polymers where it is economically not feasible to develop a new polymer for every application.

At first sight, grafting of antioxidants and stabilisers into polymers has considerable attraction. Early studies showed that reactive antioxidants such as MPDA could be grafted to rubber lattices by means of redox initiators [199]. Similarly, photo-initiated grafting of vinyl antioxidants and UV stabilisers leads to covalent bonding in the surface of polymer artefacts [76,200]. However, these procedures are not very convenient for commercial application and considerable attention has been paid recently to the possibility of grafting reactive monomers to polyolefins by chemical reaction during polymer conversion [187,188]. This procedure, which has become known as "reactive processing" utilises high shear mixers or twin screw extruders to bring the reagents into intimate contact with the polymer. In the case of some chemical reactions, for example the addition of thiols across olefinic double bonds (see below) the radicals produced by shearing of the polymer chain may be sufficient to initiate adduct formation [158,187] but in most cases peroxides are added to augment radical formation [187-189]. Many patents have been published claiming the advantage of this technique but very few commercially acceptable products have emerged from this work [188]. The reason is almost certainly the low level of grafting normally achieved. Where graft yields have been reported using vinyl antioxidants, they are rarely in excess of 20%. This is due to the competition that always exists in the case of vinyl monomers between homo-polymerisation and grafting:

$$ROOR \xrightarrow{\Delta} 2RO \cdot \begin{cases} \xrightarrow{nCH_2=CHA} ROCH_2CH(-CH_2CH-)_n \text{ Ologomer} \\ A A \\ \xrightarrow{PH} P \cdot \xrightarrow{nCH_2=CHA} PCH_2CH(-CH_2CH-)n \\ A A \\ \text{Graft} \end{cases} \quad (8)$$

where PH = polyolefin

(a) Polymer adducts
Esters of maleic or fumaric acid do not readily homopolymerise but readily form adducts with polymers under reactive processing conditions. This procedure, which involves a radical chain addition of polymer (PH) across the double bond of the monomer, Scheme 4.15, was claimed by Scott and co-workers [183,201,202] to be a potential solution to the facile loss of low molecular weight TMPs from polyolefins under conditions where volatilisation (air-oven) and leaching (in contact with aqueous detergents or dry cleaning solvents) is important.

Maleic anhydride (MA) is known to react with saturated polyolefins in the presence of free radical generators. However, the yields are normally very low due to the insolubility of MA in polymers even at processing temperatures. By contrast the maleate and fumarate esters are much more soluble and react with polyolefins to a much higher degree. In the case of the piperidinyl esters, BPM and BPF, yields of adducts over 95% can be readily achieved in the presence of peroxide initiators [183,202] by the mechanism shown in Scheme 4.15.

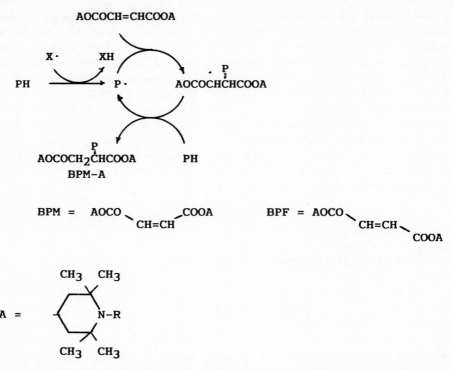

Scheme 4.15 Covalent attachment of symmetrical esters of maleic and fumaric acids containing antioxidant groups to polymers [183]

An interesting observation which is of both theoretical and practical importance is that the adducts obtained in Scheme 4.15 (e.g. BPM-A) are not only more effective than conventional TMPs (Tinuvin 770 and Chimassorb 944) after solvent extraction but are also more effective *before* extraction [183,202] (see Table 4.20).

Table 4.20 Comparison of BPM-A with conventional hals in polypropylene [183]

Antioxidant	Failure time in Xenotest, h		Failure time in air oven at 150°C, h
	Unext.	Ext.*	Unext.
BPM-A (94% bound)	1850	1250	240
Chimassorb 944	1250	635	110
Tinuvin 770	765	100	60
Control	250	100	50

* acetone/25 h

All samples contain 0.1% calcium stearate + hindered phenol (0.1% Goodrite 3114)

This is not in accord with conventional wisdom suggesting that polymer-bound antioxidants and stabilisers cannot be as efficient as low molecular weight additives due to lack of translational mobility within the polymer [178,203]. The equally plausible counter-argument is that molecular dispersion along a polymer chain results in the elimination of the physical association (agglomeration) of antioxidant molecules, allowing each molecule of the antioxidant independently to "sweep" a volume of the amorphous polymer by the "crank-shaft" movement of the polymer chains. This also occurs in grafted vinyl antioxidants but, as we have seen above, antioxidant efficiency decreases as the graft length increases. An additional but so far untested argument is that in partially crystalline polymers, where the antioxidant-modified polymer is located in the "liquid" phase, peroxyl radicals "diffuse" rapidly enough both intra- and intermolecularly by the free radical chain reaction, to permit capture by the regio-specific antioxidant.

An essential feature of the above adduct technology is that the polymer-bound antioxidant is formed as a concentrate in an appropriate polymer and is then added as a compounding ingredient to the commodity polymer during processing. This has two very important advantages. Firstly it is a more convenient and hence more economic procedure than chemically modifying the whole of the polymer feed and secondly, it has no significant effect on the crystallinity of the final product; which would be the case if the total polymer were modified.

There has been some question as to why BPM and BPF adducts are so effective as antioxidants. They behave as though the polymer-bound adducts are completely soluble in the main substrate. The most likely reason for the effectiveness of the adducts is that they form a molecularly entangled supersaturated solution during processing which persists in the final product due to lack of ability of the polymer adduct to phase separate [158].

BPF-A is currently undergoing commercial trials as VP Sanduvor PB-41 by Clariant Huningue SA as a more substantive and more effective product under service conditions than existing commercial HALS [204]. Table 4.21 compares the effectiveness in a Xenotest 450 dry/wet cycle of monomeric BPF, Tinuvin 770 and BPF adduct (BPF-A) in polypropylene. It was found that the adduct was twice as effective as Chimassorb 944 over the whole range of evaluation procedures. Although the products of reactive processing were essentially insoluble in hydrophylic solvents such as ethyl alcohol, some oligomeric material was removed by chloroform. In spite of this, however, the extracted polymer was more photostable than current commercial products, confirming the efficacy of the polymer-bound adduct.

Table 4.21 UV stability of polypropylene films containing TMPs in a Xenotest 450 (0.1% TMP + 0.1% calcium stearate + 0.05% Irganox 1010) [204]

TMP	Time to carbonyl index 0.3, h
Tinuvin 770	1300
BPF monomer	1400
BPF-A (Sandovur PB-41)	2800

A rather unexpected result from the above studies was the observation that polymer-bound TMP based on the bis-maleate structure is considerably more effective as an oven ageing antioxidant even than oligomeric Chimassorb 944 (see Table 4.20). The fact that Tinuvin 770 has essentially no thermal antioxidant activity at all suggests that the parent amine, which is not itself a photoantioxidant, is volatilised before it can be oxidised to the nitroxyl which is the effective agent.

Aromatic amines (e.g. AMI) can also be covalently attached to poyolefins by reactive processing [183].

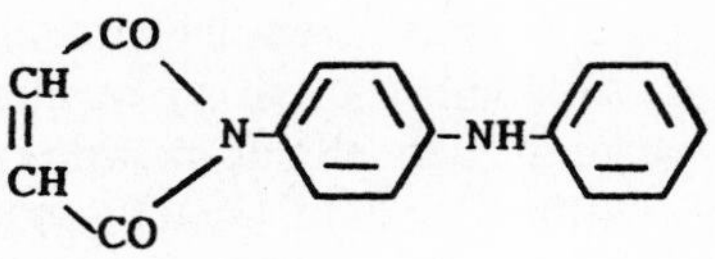

$$\begin{array}{c}\text{CH}\\ \parallel \\ \text{CH}\end{array}\begin{array}{c}\diagup\text{CO}\\ \diagdown\\ \diagup\\ \diagdown\text{CO}\end{array}\text{N}-\!\!\bigcirc\!\!-\text{NH}-\!\!\bigcirc\qquad\text{AMI}$$

AMI-A is exceptionally effective as a thermal antioxidant in polypropylene particularly under solvent extraction conditions (see Table 4.22).

Table 4.22 Comparison of AMI-A with conventional antioxidants in polypropylene in an air oven at 150°C [183]

Antioxidant	Induction period to carbonyl formation	
	Unextracted	Extracted
AMI-A	2250	2400
Irganox 1010	1350	5
Irganox 1076	1200	5
Control (no antiox)	1	1

Polymer adducts are also readily formed by the radical-initiated addition of thiols and disulphides to double bonds in polymers (see Scheme 4.16). This reaction has been extensively studied by Scott and co-workers as means of attaching antioxidants to unsaturated rubbers in rubber lattices and rubber-modified plastics using peroxide and azo initiators [181,182,205-209,211,212,222], in polyolefins initiated by UV [210] and, during reactive processing of rubbers [214,215,218], rubber modified plastics [216,217] and PVC [214,215,218].

Sec. 4.5] Physical Aspects of Antioxidant Performance

$$-CH=CHCH_2CH_2CH=CH- \xrightarrow{Shear} -CH=CHCH_2\cdot \ (M\cdot)$$
$$M\cdot + ASH \rightarrow MH + AS\cdot$$
$$M\cdot + ASSA \rightarrow MSA + AS\cdot$$

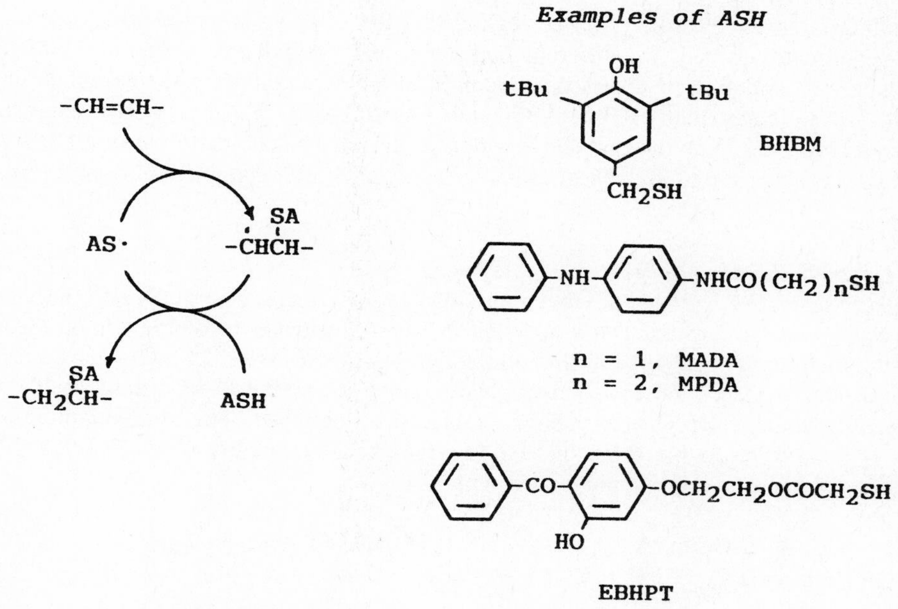

Examples of ASH

BHBM: 2,6-di-tert-butyl-4-(mercaptomethyl)phenol

MADA (n=1), MPDA (n=2): aryl-NH-aryl-NHCO(CH$_2$)$_n$SH

EBHPT

Scheme 4.16 Mechano-initiated sulphur adduct formation in unsaturated polymers

An interesting aspect of this chemistry is that both the starting thiols and disulphides and derived macromolecular sulphides (PSA) are autosynergistic (see Section 4.5.3). However, antioxidants bound through sulphur are much more effective in rubber modified polymers than synergistic combinations of low molecular weight antioxidants (see Table 4.23), and it was concluded that this results from preferential adduct formation in the rubber segment of the polymer blend which is also the more oxidatively sensitive phase.

Table 4.23 Comparison of synergistic thiol adducts, BHBM-A and EBHPT-A with conventional synergists in the photooxidation of ABS [207]

Antioxidant/Stabiliser	Induction time, h	Embrittlement time h
BHBM-A (1)+EBHPT-A (1) (U)	80	380
BHBM-A (1)+EBHPT-A (1) (E)	50	220
Conventional additives		
BHT (1)	9	34
HOBP (1)	10	40
DLTP (1)	5	25
BHT (1)+HOBP (1)+DLTP (1)	25	85

Figure in parenthesis is concentration, g/100g; U = unextracted, E = extracted.

This is an important conclusion since conventional antioxidants will normally partition between the polymer phases and, in a polymer with a minor rubber component, a considerable proportion may be in the phase where it is not required. Regio-specificity is thus a significant practical advantage of polymer-bound antioxidant in two phase polymers [158,187].

Another interesting corollary of the regio-specificity of polymer-bound antioxidants is that synergism between two different polymer-bound antioxidants acting by different mechanisms cannot occur unless they are both in the same phase. Thus Fernando and Scott [207] found that BHBM-ABS and EBHPT-ABS (Scheme 4.16) prepared separately in ABS latex did not show synergism during photooxidation until a blend of the two was subjected to high shear either in the latex or after coagulation in an internal mixer.

(b) Co-grafted antioxidants

As discussed above, grafting of vinyl antioxidants to polymers is normally an inefficient process due to the formation of a major proportion of polymer insoluble homopolymer which is relatively ineffective as an antioxidant. Scott et al. [183, 223-226] have found that in the presence of co-reactive polyfunctional monomers very high yields of covalently attached vinyl antioxidants (VA) can be obtained. The antioxidant group may optionally be in the same molecule as the polyfunctionality (e.g. nVA) or in a separate co-agent, nVR, in combination with VA:

$(CH_2=CHCOO)nRA$ $(CH_2=CHCOO)nR' + VA$
         nVA   nVR

During reactive processing of a typical nVA, AATP, in polypropylene, transient cross-linking of the polymer occurs [223,226], and the applied torque passes though a peak. If processing is discontinued at this point, the product contains cross-linked gel and cannot be processed with normal polypropylene. However, if the mechanochemical procedure is continued the oligomeric material is redistributed within the polymer by breaking and reforming chemical bonds, and the reaction product can be processed normally. AATP which forms a polymer-bound TPA when used alone is also an interlinking agent in combination with AOTP to give concentrates of highly polymer-bound TMPs which are highly effective both as a photoantioxidant and as a thermal antioxidant in polypropylene (see Table 4.24).

$CH_2=CHCOO-\underset{\underset{CH_3\ CH_3}{}}{\overset{\overset{CH_3\ CH_3}{}}{\bigcirc}}-NCOCH=CH_2$

AATP

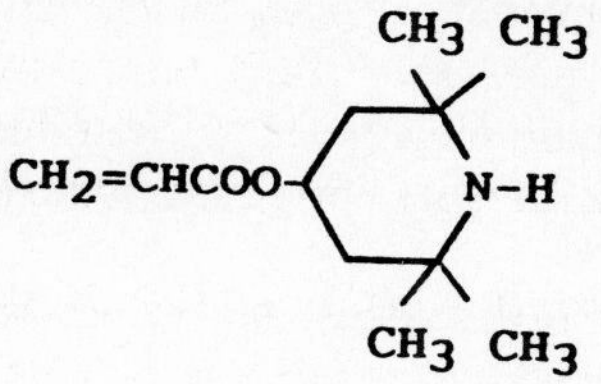

$CH_2=CHCOO-\underset{CH_3\ CH_3}{\overset{CH_3\ CH_3}{\bigg\langle}}N-H$

AOTP

Table 4.24 Co-reacted AATP and AOTP as photo- and thermal antioxidants for polypropylene (TMP = 0.4g/100g + 0.4g/100g Irganox 1076) [223]

Antioxidant	Embrittlement time, h	
	UV cabinet	Air oven at 140°C
Control (no TMP)	180	54
Co-reacted AATP + AOTP	1870	480
Tinuvin 770	1400	60

Other examples of polymer-bound light stabilisers based on nVA by reactive processing are tris-acryloyloxynitromethane, TNM [225] and nVRs include tris-acryloyloxyethane, TAE, and divinylbenzene, DVB [224].

$(CH_2=CHCOO)_3CNO_2$ $(CH_2=CHCOO)_3CCH_3$

TNM   **TAE**

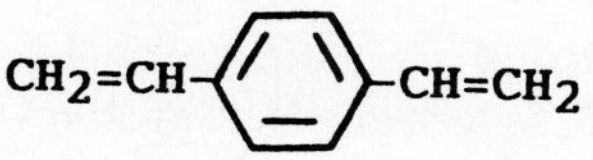

$CH_2=CH-\bigcirc-CH=CH_2$

DVB

REFERENCES

1. G. Scott, *Atmospheric Oxidation and Antioxidants*, Elsevier, 1965, p.251 et seq.
2. T. Colclough in *Atmospheric Oxidation and Antioxidants*, Vol. II, Ed. G. Scott, Elsevier, Amsterdam, 1993, Chapter 1.
3. *Oxidation Inhibition of Organic Materials*, Vols. I and II, Eds. P.P. Kremchuk and J. Pospíšil, CRC Press, Boca Raton, 1990.
4. *Aging and Stabilisation of Polymers*, Ed. M.B. Neiman, Consultants Bureau, N.Y., 1965.
5. A. Davis and D. Sims, *Weathering of Polymers*, App. Sci. Pub., London, 1983.
6. *Polymer Stabilisation*, Ed. W.L. Hawkins, Wiley Interscience, New York, 1972.
7. *Plastics Additives Handbook*, Eds. R. Gaechter and H. Muller, Hanser, Munich, 1993.
8. *Developments in Polymer Stabilisation*, Vols 1-8, Ed. G. Scott, Elsevier Sci. Pub., London, 1979-1987.
9. *Photostabilisation of Polymers*, J.F. Rabek, Elsevier App. Sci., London, 1990.
10. D. Barnard, L. Bateman, M.E. Cain, T. Colclough and J.I. Cunneen, *J. Chem. Soc.*, 5339 (1961).
11. L. Bateman, M.E. Cain, T. Colclough and J.I. Cunneen, *J. Chem. Soc.*, 3570 (1962).
12. L. Bateman and K.R. Hargrave, *Proc. Roy. Soc.*, **A224**, 389, 399 (1954).
13. D. Barnard, K.R. Hargrave and G.M. Higgins, *J. Chem. Soc*, 2845 (1956).
14. J.R. Shelton and K.E. Davies, *Int. J. Sulphur Chem.*, **8**, 197 (1973).
15. G. Scott, *Mechanisms of Reactions of Sulphur Compounds*, **4**, 99-110 (1969).
16. J.R. Shelton in *Developments in Polymer Stabilisation-4*, Ed. G. Scott, App. Sci. Pub., London, 1981.
17. J.D. Holdsworth, G. Scott and D. Williams, *J. Chem. Soc.*, 4692-99 (1964).
18. G. Scott in *Developments in Polymer Stabilisation-6*, Ed. G. Scott, Elsevier Sci. Pub., London, 1983, Chapter 2.
19. G. Scott, *Atmospheric Oxidation and Antioxidants*, Elsevier, 1965, p.363, 370.
20. C. Armstrong, M.A. Plant and G. Scott, *Europ. Polym. J.*, **11**, 161 (1975).
21. C. Armstrong and G. Scott, *J. Chem. Soc.*, 1747-52 (1971).
22. C. Armstrong, M.J. Husband and G. Scott, *Europ. Polym. J.*, **15**, 241-8 (1979).
23. M.J. Husbands and G. Scott, *Europ. Polym. J.*, **15**, 249-53 (1979).
24. E.A. Oberright, S.J. Leonardi and A.P. Kozacik, *Additives in Lubricants Symposium*, ACS, Div. Petrol Chem., Atlantic City, 1958, p. 115.
25. B.D. Flockhart, K.J. Ivin, C.R. Pink and B.D. Sharman, *Chem. Comm.*, 339 (1971).
26. A.A. Katbab, A. Ogunbanjo and G. Scott, *Polym. Deg. Stab.*, **12**, 333-47 (1989).
27. T.J. Henman in *Developments in Polymer Stabilisation-1*, Ed. G. Scott, App. Sci. Pub., London, 1979, Chapter 2.
28. W.L. Hawkins and H. Sautter, *J. Polym. Sci.*, **A1**, 4499 (1969).
29. F.A.A. Ingham, G. Scott and J.E. Stuckey, *Europ. Polym. J.*, **11**, 783-8 (1985).
30. J. Pimblott, G. Scott and J.E. Stuckey, *J. App. Polym. Sci.*, **19**, 865-77 (1975).
31. S. Al-Malaika, K.B. Chakraborty and G. Scott in *Developments in Polymer Stabilisation-6*, Ed. G. Scott, App. Sci. Pub., London, 1983, Chapter 3.
32. M.J. Husbands and G. Scott, *Europ. Polym. J.*, **15**, 879-87, (1979).

33. G. Scott, *Atmospheric Oxidation and Antioxidants*, Elsevier, Amsterdam, 1965, p. 391 et seq.
34. S. Al-Malaika, K.B. Chakraborty, G. Scott and Z.B. Tao, *Polym. Deg. Stab.*, **10**, 55 (1985).
35. S. Al-MAlaika, K.B. Chakraborty, G. Scott and Z.B. Tao, *Polym. Deg. Stab.*, **13**, 261 (1985).
36. S. Al-Malaika in *Atmospheric Oxidation and Antioxidants*, Vol. I, Ed. G. Scott, Elsevier, Amsterdam, 1993, Chapter 5.
37. S. Al-Malaika and G. Scott, *Europ. Polym. J.*, **16**, 503-9 (1980).
38. A.J. Burn, *Tetahedron*, **22**, 2153 (1966).
39. A.J. Burn, R. Cecil and V.U. Young, *J. Inst. Pet.*, **57**, 319 (1971).
40. J.A. Howard, J.H.B. Chernier and K.U. Ingold, *Canad. J. Chem.*, **51**, 1543 (1973).
41. J.A. Howard and J.H.B. Chernier, *Canad. J. Chem.*, **54**, 382,290 (1976).
42. T. Colclough, F.A. Gibson and T.F. Marsh, *UK Patent*, 2,056,148 (1981).
43. J.A. Howard in *Frontiers of Free Radical Chemistry*, Ed. W.A. Pryor, Acad. Press, 1980, p.237-282.
44. S. Al-Malaika in *Mechanisms of Polymer Degradation and Stabilisation*, Ed. G. Scott, Elsevier App. Sci., London, 1990, Chapter 3.
45. S.K. Ivanov in *Develpoments in Polymer Stabilisation-3*, Ed. G. Scott, App. Sci. Pub., London, 1980, Chapter 3.
46. D.J. Carlsson and D.M. Wiles, *Macromol. Chem.*, **C14**, 65, 155, (1976).
47. S. Al.Malaika, A. Marogi and G. Scott, *Polym. Deg. Stab.*, **30**, 789 (1985).
48. S. Al-Malaika, A. Marogi and G. Scott, *J. App. Polym. Sci.*, **33**, 1455 (1987).
49. R.P.R. Ranaweera and G. Scott, *Europ. Polym. J.*, **12**, 825-30 (1976).
50. G. Scott, *Atmospheric Oxidation and Antioxidants*, Elsevier, Amsterdam, 1965, pp. 404, 444.
51. M.U. Amin and G. Scott, *Europ. Polym. J.*, **10**, 1019-28 (1974).
52. D. Gilead and G. Scott in *Developments in Polymer Stabilisation-5*, Ed. G. Scott, App. Sci. Pub., London, 1982, p.71.
53. S. Al-Malaika, A. Marogi and G. Scott., *J. App. Polym. Sci.*, **31**, 685 (1986).
54. G. Scott, *Polym. Deg. Stab.*, **29**, 135-54 (1990).
55. G. Scott, *J. App. Polym. Sci. Symp.*, **55**, 3-14 (1994).
56. G. Scott in *Biodegradable Plastic and Polymers*, Eds. Y. Doi and K. Fukuda, Elsevier, Amsterdam, 1993 pp.79-91.
57. *Degradable Polymers: Principles and Applications*, Eds. G. Scott and D. Gilead, Chapman & Hall, London, 1995, Chapters 9,10,11 and 13.
58. R. Arnaud, P. Dabin, J. Lemaire, S. Al-Malaika, S. Chohan, M. Coker, G. Scott, A. Fauve and A. Maaroufi, *Polym. Deg. Stab.*, **46**,211-24 (1994).
59. P.A. Willermet and S.K. Kandar, *ASLE Trans.*, **27**, 67-72 (1984).
60. T. Colclough and J.I. Cunneen, *J. Chem. Soc.*, 4790-93 (1964).
61. L.R. Mahony, S. Korcek, S. Hoffman and P.A. Willermet, *Ind. Eng. Chem.* Prod. Res. Dev., **17**, 250-55 (1978).
62. S. Ivanov and I. Kateva, *Neftekchimiya*, **18**, 417 (1978).
63. S. Korcek, L.R. Mahony, M.D. Johnson and W.O. Siegl, *Soc. Automot. Eng.*, Technical Paper 810014 (1981).
64. S. Al-Malaika, M. Coker, P.J. Smith and G. Scott, *J. App. Polym. Sci.*, **44**, 1297-1305 (1992).

65. P. Sanin, I. Blagovidov, A. Vipper, A. Kuliev, S. Krein, A.K. Ramaya, G. Schor, V. Sher and Y. Zasalavsky, *Proc. Eighth World Pet. Congr.* 1971, p.91.
66. G. Scott and S. Al-Malaika, *Brit. Pat.*, 2,117,779A (1983).
67. B. Peters, *Rev. Gen. Caout.*, **34**, 1233 (1957).
68. F. Gugumus in *Developments in Polymer Stabilisation-1*, Ed. G. Scott, App. Sci. Pub., London, 1979, Chapter 8.
69. S.L. Fitton, R.N. Haward and G.R. Williamson, *Brit. Polym. J.*, **2**, 217 (1970).
70. G. Scott, *Atmospheric Oxidation and Antioxidants*, Elsevier, Amsterdam, 1965, p. 406.
71. K.J. Humphris and G. Scott, *Pure & App. Chem.*, **36**, 163-176 (1973).
72. P.I. Levin, P.A. Kirpichnikov, A.F. Lukovnikov and M.S. Khloplyankina, *Polym. Sci. USSR*, **5**, 214 (1964).
73. P.A. Kirpichnikov, L.M. Popova and P.I. Levin, *Trans. Kazansk. Khim- Tekhnol. Inst.*, **33**, 269 (1964).
74. P.A. Kirpichnikov, L.V. Verizhnikov and L.C. Angert, *Trans. Kazansk. Khim-Tekhnol. Inst.*, **33**, 287 (1964).
75. D.G. Pobedimslii, N.A. Mukmeneva and P.A. Kirpichnikov in *Developments in Polymer Stabilisation-2*, Ed. G. Scott, App. Sci. Pub., London, 1980, Chapter 4.
76. G. Scott, *Pure & App. Chem.*, **30**, 267-334 (1972).
77. K.J. Humphris and G. Scott, *J. Chem. Soc.*, Perkin Trans. II, 617-20 (1974).
78. K. Schwetlick, *Pure & App. Chem.*, **55**, 1629 (1983).
79. K. Schwetlick in *Mechanisms of Polymer Degradation and Stabilisation*, Ed. G. Scott, Elsevier App. Sci., London, 1990, Chapter 2.
80. K.J. Humphris and G. Scott, *J. Chem. Soc.*, Perkin Trans. II, 826-30 (1973).
81. K.J. Humphris and G. Scott, *J. Chem. Soc.*, Perkin Trans. II, 831-35 (1973).
82. C. Rüger, T.König and K. Schwetlick, *Acta. Polym.*, **37**, 435 (1986).
83. G. Scott and K.J. Humphris, *US Patent*, 3,920,607 (1975).
84. D.M. Brown and H.M. Higson, *J. Chem. Soc.*, 2034 (1957).
85. C. Rüger, T. König and K. Schwetlick, *J. Prkt. Chem.*, **326**, 622 (1984).
86. G. Scott, *Atmospheric Oxidation and Antioxidants*, Elsevier, 1965, p.172-80.
87. Z. Osawa in *Atmospheric Oxidation and Antioxidants*, Vol.II, Ed. G. Scott, Elsevier, Amsterdam, 1993, Chapter 6.
88. N.S. Allen in *Degradation and Stabilisation of Polymers*, Ed. N.S. Allen, App. Sci. Pub., London, 1983, Chapter 8.
89. C.J. Pederson, *Ind. Eng. Chem.*, **41**, 924 (1949).
90. A.J. Chalk and J.F. Smith, *Trans. Farad. Soc.*, **53**, 1215, 1235 (1957).
91. H.L. Pederson, *Ageing Properties of Low Sulphur Vulcanisates*, A-S Norske Kabel og Traadfabriken, Copenhagen, 1954.
92. T. Colclough, *Ind. Eng. Chem. Res.*, **26**, 1888-95 (1987).
93. H. Diamond, H.C. Kennedy and R.G. Larsen, *Ind. Eng. Chem.*, **44**, 1834-43 (1952).
94. G. Scott and D. Gilead, *Europ. Pat.*, 0236055.
95. S. Al-Malaika, A.M. Marogi and G. Scott, *J. App. Polym. Sci.*, **31**, 685-698 (1986).
96. Z. Osawa and K. Matsuzuki, *Kogyo Kagaku Zasshi* (Japan), **71**, 1536 (1968).
97. Z. Osawa in *Atmospheric Oxidation and Antioxidants*, Vol.II, Ed. G. Scott, Elsevier, Amsterdam, 1993, Chapter 7.
98. R.L. Hartless and A.M. Trozollo, *Coatings Plast. Reprints*, ACS, **34**, 177 (1974).

99. M. Minagawa, M. Akutsu and N. Kubota, *Ann. Tech Conf.*, SPE, **20**, 328 (1976).
100. C.A. Pryde and N.G. Chan, *38th Ann. Tech. Conf.*, SPE, 180 (1980).
101. G. Scott, *Atmospheric Oxidation and Antioxidants*, Elsevier, Amsterdam, 1965, p. 294
102. G. Scott in *Atmospheric Oxidation and Antioxidants*, Vol. II, Ed. G. Scott, Elsevier, Amsterdam 1993, Chapter 8.
103. J.F. Rabek, *Mechanisms of Photophysical Processes and Photochemical Reactions in Polymers*, Wiley, New York, 1987, Chapter 15.
104. R.J. Martinovich, *Plast. Tech.*, **9**, 45 (1963).
105. S.P. Pappas and W. Kühhirt, *J.Paint Technol.*, **47**, 42 (1975).
106. H.G. Völz, G. Kämpf, H.G. Fitzky and A. Klaeren in *Photodegradation and Stabilisation of Coatings*, Eds. S.P. Pappas and F.H. Winslow, ACS Symp. Ser. **163**, 163 (1981).
107. H.G. Völz, G. Kämpf, H.G. Fitzky and A. Klaeren in *Photodegradation and Photostabilsation of Coatings*, Eds. S.P. Pappas and F.H. Winslow, ACS Symp. Series, **163,** ACS, Washington D.C., 1981, p.163 et seq.
108. U.V. Curing, *Science and Technology*, Vol. II, Ed. S.P. Pappas, Stamford Tech. Marketing, Norwalk, 1984.
109. G.R. De Maré, P. Goldfinger, P. Hutbrechts, E. Jonas and M. Toth, *Ber. Bunenges, Physik. Chem.*, **73**, 867 (1969).
110. J.M. Hermann, J. Disdier and P. Pichat, *J. Catal.*, **60**, 369 (1979).
111. N.S. Allen and J.F. McKellar, *Brit. Polym. J.*, **9**, 986 (1977).
112. H.G. Völz, G. Kämpf and H.G. Fitzky, *Prog. Org. Coatings*, **1**, 1 (1973).
113. Y.S. Chow, N.S. Allen, F. Thompson, T.S. Jewitt and M.R. Hornby, *Polym. Deg. Stab.*, **34**, 243-62 (1991).
114. N.S. Allen, J.F. McKellar and D. Wilson, *J. Photochem.*, **7**, 319 (1977).
115. F. Rasti and G. Scott, unpublished work.
116. *Photostabilization of Polymers*, J.F. Rabek, Elsevier App. Sci., London, 1990, pp. 241 et seq.
117. K.B. Chakraborty and G. Scott, *Europ. Polym. J.*, **13**, 1007-13 (1977).
118. R.P.R. Ranaweera and G. Scott, *Europ. Polym. J.*, **12**, 591-7 (1976).
119. G. Scott in *Singlet Oxygen*, Eds. B. Rånby and J.F. Råbek, Wiley, Chichester, 1978, p.230.
120. D.M. Wiles in *Singlet Oxygen*, Eds. B. Rånby and J.F. Rabek, Wiley, Chichester, 1978, Chapter 34.
121. G. Scott, *S. Afr. J. Chem.*, **32**, 138-46 (1979).
122. G. Scott, *Pure & Appl. Chem.*, **52**, 365-87 (1980).
123. G. Scott in *Advances in Polyolefins*, Ed. R.S. Seymour and T. Cheng, Plenum Press, New York, 1987, p.381.
124. J. Pospíšil in *Developments in Polymer Stabilisation-1*, Ed. G. Scott, App. Sci. Pub., London, 1979, Chapter 1.
125. J. Pospíšil in *Polymer Stabilisation, Mechanisms and Applications*, Eds. N.C. Billingham and D.M. Wiles, Elsevier, London, 1991, p.91.
126. J.K. Becconsall, S. Clough and G. Scott, *Trans. Farad. Soc.*, **56**, 459-72 (1960).
127. J. Fortner, *J. Polym. Sci.*, **37**, 199 (1959).
128. G.C. Newland and J.W. Tamblyn, *J. App. Polym. Sci.*, **8**, 1949 (1964).

129. G. Scott, *Atmospheric Oxidation and Antioxidants*, 1965, Elsevier, Amsterdam, p. 183.
130. A.A. Lamola and L.J. Sharp, *Phys. Chem.*, **70**, 2634 (1966).
131. D.J. Carlsson, T. Suprunchuk and D.M. Wiles, *J. App. Polym. Sci.*, **16**, 615 (1972).
132. G.C. Newland, H.W. Patton and J.W. Tamblyn, *SPE Trans.*, **1**, 26 (1960).
133. K.B. Chakraborty and G. Scott, *Europ. Polym. J.*, **15**, 35 (1979).
134. G. Scott, *Atmospheric Oxidation and Antioxidants*, Elsevier, 1965, p.203 et seq.
135. G. Scott in *Atmospheric Oxidation and Antioxidants*, Vol. II, Ed. G. Scott, Elsevier, Amsterdam, 1993, Chapter 9.
136. H.S. Olcott and H.A. Mattill, *J. Am. Chem. Soc.*, **58**, 1627, 2204 (1936).
137. H.S. Olcott and H.A. Mattill, *Chem. Rev.*, **29**, 257 (1941).
138. P. Lambelet and J. Loliger, *Chem. Phys. Lipids*, **35**, 185-98 (1984).
139. W. Lohmann and D. Holz, *Biophys. Struct. Mech.*, **10**, 187-204 (1984).
140. A. Tkáč in *Developments in Polymer Stabilisation-8*, Ed. G. Scott, Elsevier App. Sci., London, 1987, Chapter 3.
141. C. Golumbic, *Oil and Soap*, **19**, 105 (1942).
142. G.W. Kennerly and W.L. Patterson, *Ind. Eng. Chem.*, **48**, 1917 (1956).
143. Hercules Powder Co., *Brit. Pat.*, 851670 (1960).
144. G. Scott, *Europ. Polym. J. Suppl.*, 189-213 (1969).
145. G. Scott and P.A. Shearn, *J. App. Polym. Sci.*, **13**, 1329-35, (1969).
146. G. Scott in *Developments in Polymer Degradation-1*, Ed. N. Grassie, App. Sci. Pub., London., 1977, Chapter 7.
147. K.B. Chakraborty and G. Scott, *Polym. Deg. Stab.*, **1**, 37-46 (1979).
148. G. Scott, *Pure Appl. Chem.*, **52**, 365-87 (1980).
149. V.B. Ivanov and V.Yu. Shlyapintokh in *Developments in Polymer Stabilisation-8*, Ed. G. Scott, Elsevier Sci. Pub., London, 1987, p.29.
150. F. Gugumus in *Mechanisms of Polymer Degradation and Stabilisation*, Ed. G. Scott, Elsevier App. Sci., London, 1990, Chapter 6.
151. G. Scott and M.F. Yusoff, *Europ. Polym. J.*, **16**, 497-501 (1980).
152. V.M. Farzaliev, W.S.E. Fernando and G. Scott, *Europ. Polym. J.*, **14**, 39-43 (1978).
153. H. Zweifel at "Stabilisation beyond the Year 2000", *11th Bratislava IUPAC/PECS Int. Symp., Thermal and Photoinduced Oxidation of Polymers and its Inhibition in the 21st Century*, Stará Lezná, Slovakia, June 24-28, 1996.
154. G. Scott and R. Suharto, *Europ. Polym. J.*, **20**, 139-47 (1984).
155. L. Jiráčkova and J. Pospísil, *Angew. Macromol. Chem.*, **66**, 95 (1978).
156. L. Jiráčkova, T. Jelinkova, J. Totschova and J. Pospísil, *Chem. Ind.*, 384 (1979).
157. G. Scott in *Developments in Polymer Stabilisation-8*, Ed. G. Scott, Elsevier App. Sci., London, 1987, Chapter 5.
158. G. Scott in *Atmospheric Oxidation and Antioxidants*, Vol. II, Ed. G. Scott, Elsevier, Amsterdam, 1993, Chapter 5.
159. O. Ajiboye and G. Scott, *Polym. Deg. Stab.*, **4**, 397- (1982).
160. O. Ajiboye and G. Scott, *Polym. Deg. Stab.*, **4**, 415- (1982).
161. W.S.E. Fernando and G. Scott, *Europ. Polym. J.*, **16**, 971- (1980).
162. G. Scott and M.F. Yusoff, *Polym. Deg. Stab.*, **2**, 309-19 (1980).

163. G. Scott in *Developments in Polymer Stabilisation-1*, Ed. G. Scott, App. Sci. Pub., London, 1979, Chapter 9.
164. W.S.E. Fernando and G. Scott, *Europ. Polym. J.*, **16**, 971-8 (1980).
165. M. Ghaemy and G. Scott, *Polym. Deg. Stab.*, **3**, 405 (1980-1).
166. B.B. Cooray and G. Scott, *Europ. Polym. J.*, **17**, 233-8 (1981).
167. E.D. Owen in *Degradation nd Stabilisation of PVC*, Ed. E.D. Owen, Elsevier App. Sci., London, 1984, Chapter 5.
168. B.B. Cooray and G. Scott in *Developments in Polymer Stabilisation-2*, Ed. G. Scott, App. Sci. Pub., London, 1980, Chapter 2.
169. B.B. Cooray and G. Scott, *Polym. Deg. Stab.*, **2**, 35 (1980).
170. G. Scott in *Food Additives and Contaminants*, **5**, 421-32 (1988).
171. G. Scott, *Free Rad. Res. Comm.*, **5**, 141-7 (1988).
172. G. Scott in *Free Radicals and Oxidative Stress, Biochem. Soc. Symp.*, Eds. C. Rice-Evans, B. Halliwell and G.G. Lunt, **61**, 235-46 (1995).
173. M.A. Plant and G. Scott, *Europ. Polym. J.*, 7, 1173-83.
174. N.C. Billingham in *Atmospheric Oxidation and Antioxidants*, Vol. II, Ed. G. Scott, Elsevier, Amsterdam, 1993, Chapter 4.
175. P.D. Calvert and N.C. Billingham, *J. App. Polym. Sci.*, **24**, 357 (1979).
176. B.W. Evans and G. Scott, *Europ. Polym. J.*, **10**, 453-58 (1974).
177. F. Gugumus, *Res. Disclos.* **209**, 357 (1981).
178. Š. Chmela, P. Hrdlovic and Z. Manásek, *Polym. Deg. Stab.*, **11**, 233-41 (1985).
179. M. Minagawa, *Polym. Deg. Stab.*, **25**, 121- (1989).
180. C. Sadrmohaghegh, G. Scott and E. Setudeh, *Polym. Plast. Technol. Eng.*, **24**, 149-88 (1985).
181. G. Scott in *Developments in Polymer Stabilisation-4*, Ed. G. Scott App. Sci. Pub., London, 1981, Chapter 6.
182. G. Scott in *Developments in Polymer Stabilisation-1*, Ed. G. Scott, App. Sci. Pub., London, 1979, Chapter 9.
183. G. Scott, *Makromol. Chem, Macromolecular Symposia*, **28**, 59-71 (1989).
184. G. Scott, *Chem. & Ind.*, 841-45 (1987).
185. K.B. Chakraborty, G. Scott and S.M. Tavakoli in *Advances in Elastomers and Rubber Elasticity*, Eds. J. Lal and J.F. Mark, Plenum Press, New York, 1986, pp. 189-96.
186. G. Scott , *ACS Symposium Series*, **280**, 173-96 (1985).
187. G. Scott in *Developments in Polymer Stabilisation-8*, Ed. G. Scott, Elsevier App. Sci., London, 1987, Chapter 5.
188. D. Munteanu in *Developments in Polymer Stabilisation-8*, Ed. G. Scott, Elsevier App. Sci., London, 1987, Chapter 4.
189. J.F. Rabek, *Photostabilisation of Polymers*, Elsevier App. Sci., London, 1990, Chapter 7.
190. E.H. Kleiner, *German Pat.* 1,931,452 (1970).
191. G.E. Meyer, R.W. Kavchok and J.F. Naples, *Rubb. Chem. Tech.*, **46**, 106 (1973).
192. J.W. Horvath, C.C. Grimm and J.A. Stevick, *Rubb. Chem. Tech.*, **48**, 337 (1975).
193. J.W. Horvath, J.R. Burdon, G.E. Meyer and F.J. Naples, Paper presented at ACS Meeting, Chicago, August, 1973.
194. S. Yoshida and O. Vogl, *Makromol. Chem.*, **183**, 259 (1982).
195. Z. Nir, O. Vogl and A. Gupta, *J. Polym. Sci., Polym. Chem. Ed.*, **20**, 2735 (1982).

196. F. Xi, W. Basset and O. Vogl, *Polym. Bull.*, **11**, 829 (1984).
197. S. Li, A. Gupta and O. Vogl, *J. Macromol. Sci., Chem.*, **A20**,309 (1983).
198. F. Shoukuan, A. Gupta, A-C. Ibertsson and O. Vogl in *New Trends in the Photochemistry of Polymers*, Eds. N.S. Allen and J.F. Rabek, Elsevier App. Sci., London, 1985, p.247.
199. G. Scott, *Plastics and Rubber: Processing*, 41-48 (1977).
200. H. Mingbo and H. Xingzhou, *Polym. Deg. Stab.*, **18**, 321 (1987).
201. J. Rekers and G. Scott, U.S. Pat., 4,743,657 (1988).
202. S. Al-Malaika, A.Q. Ibrahim and G. Scott, *Polym. Deg. Stab.*, **22**, 233 (1988).
203. N.C. Billingham and P. Gracia-Trabajo, *Polym. Deg. Stab.*, **48**, 419- (1995).
204. J. Malík, G. Ligner and L. Ávár at 18th International Conference on *Advances in the Stabilisation and Degradation of Polymers*, Luzerne, 1996, Adcon '96, Brussels, 1996.
205. G. Scott, *U.S.Pat.*, 4,213,892 (1980).
206. K.W.S. Kularatne and G. Scott, *Europ. Polym. J.*, **15**, 827-32 (1979).
207. W.S.E. Fernando and G. Scott, *Europ. Polym. J.*, **16**, 971-78 (1980).
208. O. Ajiboye and G. Scott, *Polym. Deg. Stab.*, **4**, 397-413 (1982).
209. O. Ajiboye and G. Scott, *Polym. Deg. Stab.*, **4**, 415-25 (1982).
210. G. Scott and M.F. Yusoff, *Polym. Deg. Stab.*, **3**, 53-9 (1980).
211. G. Scott and S.M. Tavakoli, *Polym. Deg. Stab.*, **4**, 267-78 (1982).
212. G. Scott and S.M. Tavakoli, *Polym. Deg. Stab.*, **4**, 279-85 (1982).
213. G. Scott and S.M. Tavakoli, *Polym. Deg. Stab.*, **4**, 343-51 (1982).
214. G. Scott and S.M. Tavakoli, *Polym. Deg. Stab.*, **19**, 29-41 (1987).
215. G. Scott and S.M. Tavakoli, *Polym. Deg. Stab.*, **19**, 43-50 (1987).
216. M. Ghaemy and G. Scott, *Polym. Deg. Stab.*, **3**, 405-22 (1980-81).
217. G. Scott and E. Setudeh, *Polym. Deg. Stab.*, **5**, 11-22 (1983).
218. G. Scott and E. Setudeh, *Polym. Deg. Stab.*, **5**, 81-88 (1983).
219. B.B. Cooray and G. Scott, *Europ. Polym. J.*, **16**, 1145-51 (1980).
220. B.B. Cooray and G. Scott, *Europ. Polym. J.*, **17**, 229-32 (1981).
221. B.B. Cooray and G. Scott, *Europ. Polym. J.*, **17**, 379-84 (1981).
222. E.G. Kolawole and G. Scott, *J. App. Polym. Sci.*, **26**, 2581-92 (1981).
223. G. Scott, S. Al-Malaika and A.Q. Ibrahim, U.S. Pat., 4,956,419 (1990).
224. G. Scott and S. Al-Malaika, PCT/GB89/00909 (1989)
225. G. Scott and S. Al-Malaika, U.S. Pat., 5,098,957 (1992).
226. S. Al-Malaika, A.Q. Ibrahim, M.J. Rao and G. Scott, *J. App. Polym. Sci.*, **44**, 1287-96 (1992).

5

Antioxidants in Biology

5.1 Antioxidant Mechanisms *in vivo*

The earliest studies of natural antioxidants were primarily concerned with biological substrates *in vitro* [1]. The tocopherols featured prominently, and as early as 1957, H. Dam reviewed the antioxidant role of the tocopherols *in vivo* [2]. About the same time Draper et al. [3] demonstrated that the antioxidant requirement of the lipids can be met in rats by the rubber antioxidant, diphenyl-p-phenylene diamine (DPPD, Chapter 3, Table 3.1). In spite of this fundamental observation, it was a further 20 years before antioxidants began to be mentioned in standard biochemistry textbooks in the context of oxidation and antioxidants *in vivo* and then only briefly. In 1973, Bronk stated [4] "The exact function of vitamin E is unknown, although lack of it appears to interfere with the reproduction of some animals." Two years later, Lehninger was more forthcoming [5]. Whilst still expressing ignorance of the true function of Vitamin E, he states; "Tocopherols have been found to have *antioxidant* activity; i.e. they prevent the autooxidation of highly unsaturated fatty acids when they are exposed to molecular oxygen. Such autooxidation results in the polymerisation of unsaturated fatty acids, a process similar to that occurring in the "drying" of the linseed oil in paint to produce hard tough and insoluble products." It was doubtless statements like this that led the next generation of biochemists to investigate systematically the involvement of biological antioxidants in the prevention of "plaque" formation in atherosclerosis.

An understanding of the role of peroxidation in disease has developed rapidly since 1975 and some of the evidence for this was discussed in Chapter 2, where it was seen that one of the most important criteria used to demonstrate the deleterious involvement of peroxidation was the effect of antioxidants and particularly the "essential" biological antioxidants notably vitamins C and E whose antioxidant role was a matter of considerable dispute only a few years earlier [6].

In Chapter 3, the basic mechanisms by which antioxidants act, were summarised (Chapter 3, Scheme 3.1). However, a number of additional mechanisms of free radical initiation exist *in vivo* which have no abiotic counterpart (Scheme 5.1). Most of these are concerned with the activation of ground state oxygen to reactive oxygen species (ROS) and are countered by preventive antioxidants, primarily in the aqueous phase. Of these, the most important is catalase (CAT) which catalyses the conversion of hydrogen peroxide to water without the liberation of free radicals, a peroxidolytic process which has no direct abiotic chemical analogy (Scheme 5.1) [8]. Superoxide dismutase (SOD) which catalyses the bimolecular conversion of superoxide to ground state oxygen and hydrogen peroxide is often classified as an antioxidant, but can only function effectively as an antioxidant in combination with peroxidolytic antioxidants that destroy hydrogen peroxide without the formation of free radicals, notably catalase and glutathione peroxidase [8].

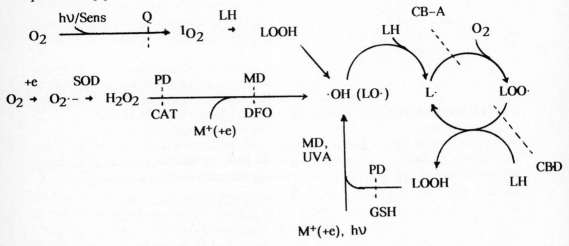

REACTIVE OXYGEN SPECIES (ROS) $O_2^{\cdot-}$, $^1O_2$, H_2O_2, $\cdot OH$, $LO\cdot$, $LOOH$

ANTIOXIDANT MECHANISMS

CB-A CHAIN-BREAKING ACCEPTOR $L\cdot + A\cdot \rightarrow [LA] \rightarrow LH\ (C=C) + AH$
  $A\cdot$ = phenoxyl, nitroxyl, etc.

CB-D CHAIN-BREAKING DONOR $LOO\cdot + AH \rightarrow LOOH + A\cdot$
  AH = phenols, aromatic amines, etc.

SOD SUPEROXIDE DISMUTASE $2\ \cdot OO^- + 2H^+ \rightarrow H_2O_2 + O_2$
  (only effective in combination with CAT)

PD PEROXIDE DECOMPOSER $LOOH \rightarrow LOH$ (non-radical)
  Catalase (CAT) for hydrogen peroxide, Glutathione peroxidase (GSH), thiols, sulphides, dithiocarbamates, etc. for alkyl hydroperoxides

MD METAL DEACTIVATOR $M^+ + Lig \rightarrow M^+Lig$ (redox inactive)
  Desferrioxamine (DFO), etc.

UVA UV ABSORBER   2-hydroxybenzophenones, etc.

Q EXCITED STATE QUENCHER   β-carotene, etc.

Scheme 5.1 Initiation and inhibition of lipid oxidation

Hydrogen peroxide is not highly reactive in the absence of transition metal ions, but in the presence of water soluble-metal ions, notably, Fe^{2+} and Cu^+, it is the major source of hydroxyl radical (·OH) and peroxidation initiation in aqueous media (Chapter 2, reaction 2). It was seen that hydroxyl radicals are highly site specific due to their great reactivity toward organic media, but the alkylperoxyl radical which is formed by attack of ·OH in a lipid medium, is much more selective and "magnifies" the initial radical attack by multiple cycling in the chain reaction (Scheme 5.1). The sequestration of transition metal ions by antioxidant proteins is therefore, after peroxidolysis, one of the most powerful preventive mechanisms *in vivo* [9].

The chemistry of the oxidation chain reaction in the lipid substrate is entirely analogous to similar peroxidation reactions in technological systems discussed in Chapter 1. Inhibition by the CB-D, PD and MD mechanisms discussed in Chapter 3 occur in parallel with the above preventive chemistry in the aqueous phase. The CB-A process has not so far been unambiguously demonstrated to occur *in vivo* with biological antioxidants, but it will be seen in Section 5.5 that spin-traps and aminoxyl radicals effectively retard peroxidation in a variety of diseases and could form the basis of antioxidant therapies complementary to those provided by the naturally occurring antioxidants in the future.

5.2 Naturally Occurring Chain-breaking Donor (CB-D) Antioxidants

The most widely distributed mammalian CB-D antioxidants are the lipid-soluble vitamin E and ubiquinol (Ubi-hq), which is formed by reduction of ubiqinone, Ubi-q, (Co-enzyme Q) and the water-soluble vitamin C. Vitamins C and E are dietary requirements in humans and their absence leads to many of the diseases associated with peroxidation discussed in Chapter 2. The ubiquinones are endogenous antioxidants involved in electron transport and their reduced forms (Ubi-hq) are primarily involved, like ascorbic acid $(Asc(OH)_2)$, in the regeneration of α-Toc-OH by reduction of the tocopheryloxyl radical (α-Toc-O·), reaction 1 [10].

$$\alpha\text{-Toc-O·} \xrightarrow{Asc(OH)_2,\ Ubi\text{-}hq} \alpha\text{-Toc-OH} \qquad (1)$$

5.2.1 Vitamin E

The Group of E vitamins fall into two classes depending on the nature of the side-chain, R. The dα-, dβ-, dγ- and dδ-tocopherols (Toc-OH) differ from the corresponding tocotrienols (Tocen-OH) only in the presence of three side-chain double bonds in the latter:

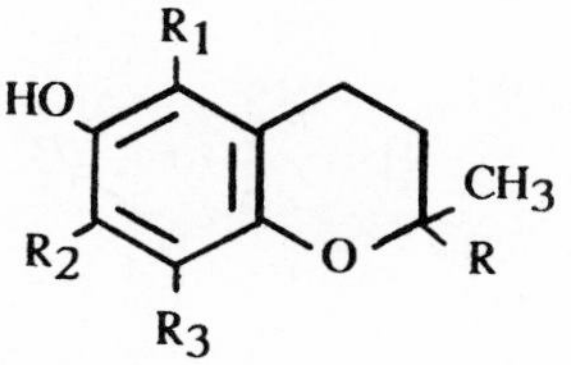

α, $R_1 = R_2 = R_3 = CH_3$
β, $R_1 = R_3 = CH_3$, $R_2 = H$
γ, $R_1 = H$, $R_2 = R_3 = CH_3$
δ, $R_1 = R_2 = H$, $R_3 = CH_3$

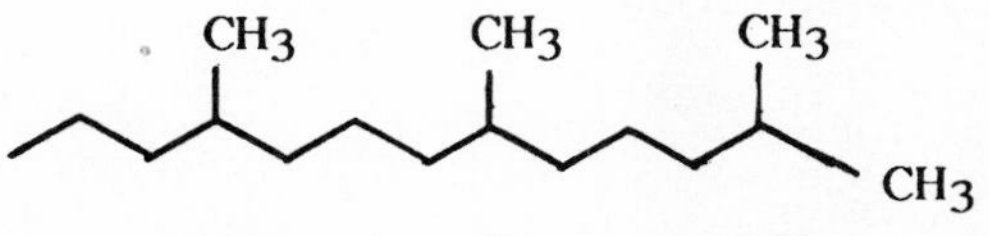

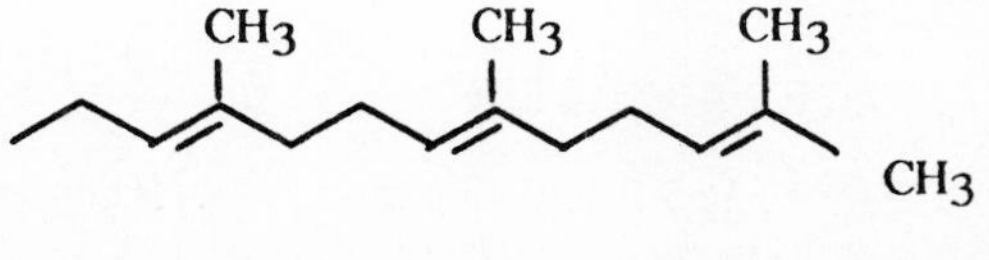

Vitamin E is the major lipid-soluble antioxidant found in the cell membrane, particularly in mitochondria [11]. The physical dimensions and conformation of the d-tocopherols particularly adapts them to be an integral part of the cell membrane lipid bilayer. The chroman structure with its phytyl "tail" is co-dimensional and soluble in the hydrophobic fatty acid residue which form the interior of the membrane, and the phenolic hydroxyl group associates with the hydrophilic glycerophosphates in the surface [12]. However, the cell membrane is thin and mobile so that the tocopherols can diffuse in and out of the membrane. They are thus able to function effectively as deactivators for peroxyl radicals as they are formed from the polyunsaturated lipids. The synthetic dl isomers of the d-tocopherols are not found in nature and although they behave very similarly to the natural isomers in model experiments *in vitro*. There is some evidence to suggest that they do not "fit" into the cell membrane as effectively as the natural isomers and are consequently excreted more rapidly [13].

All the tocopherols and tocotrienols are effective CB-D antioxidants in polyunsaturated fatty acids below a critical concentration. Thus, α-tocopherol has been found to be an antioxidant in linoleic acid at concentrations below 5×10^{-3}M but above this concentration it inverts to give prooxidant activity, almost certainly by direct interaction with oxygen and hydroperoxides [14,15]. Burton et al. [13,16,17] have measured rate constants for the reaction of ROO· with the tocopherols and structurally related synthetic antioxidants. Values of k_7 for this reaction are listed in Table 5.1. The order of CB-D activity is $\alpha > \beta \geq \gamma > \delta$, and the reported order of activity *in vivo* is the same [18]. The synthetic pentamethylhydroxychroman (PMHC) has similar activity:

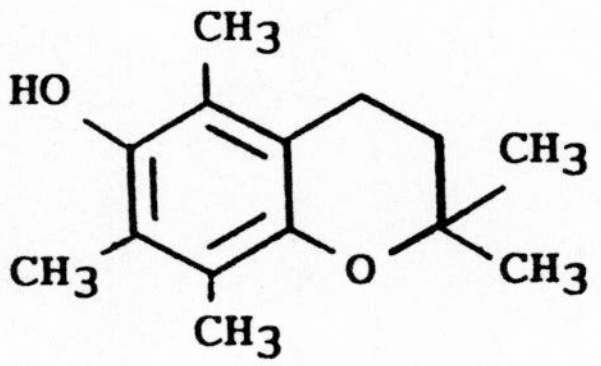

PMHC

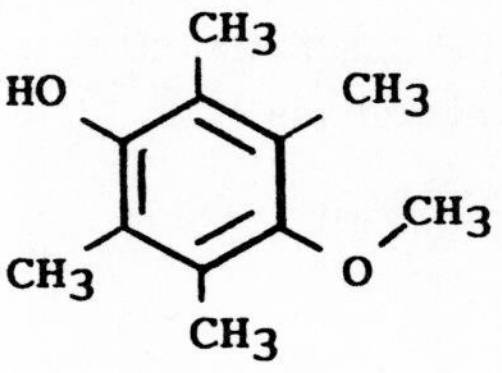

TMMP

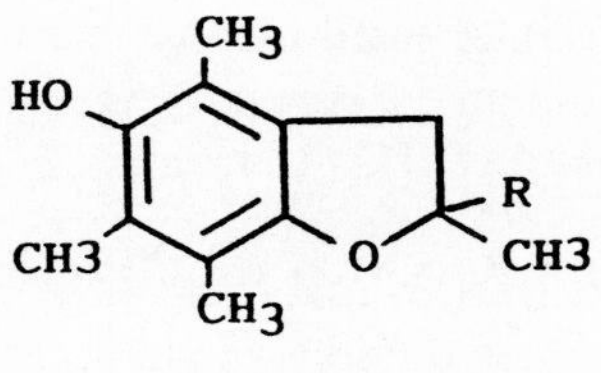

PMBF

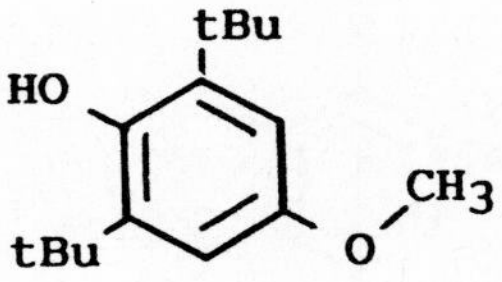

2,6-BHA

Table 5.1 Values of k_7 for the tocopherols and related phenols

Antioxidant	$10^{-4}k_7$, $M^{-1}s^{-1}$	
	[13]	[16]
α-Toc-OH	320	235
β-Toc-OH	170	166
γ-Toc-OH	130	159
δ-Toc-OH	65	65
PMHC	380	214
TMMP	39	21
PMBF	570	-
2,6-BHA	11	7.8

The acyclic analogue of α-Toc-OH, TMMP is considerably less active than α-Toc-OH but the synthetic benzofuran analogue of α-Toc-OH (PMBF) is more reactive toward ROO·. Burton et al. explain the decreased activity of TMMP and the increased activity of PMBF relative to α-Toc-OH as being due to an increase in the overlap between the 2p lone pair on oxygen with the π-bond system of the phenol from TMMP to PMBF. This is caused by increasing constriction of the oxygen in the six and five membered rings. Willson has reported a similar high activity of the hydroxy carbazole antioxidant (HDC) which is related to the antioxidant hormone, melatonin (see Section 5.2.7) [19], and has suggested [20] that the reason may be the same:

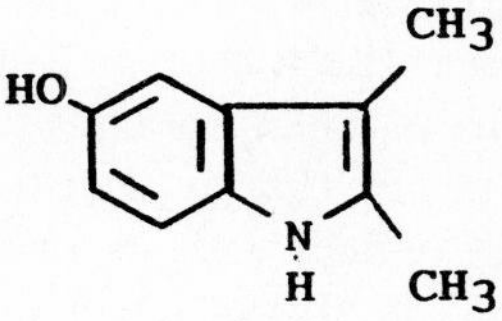

HDC

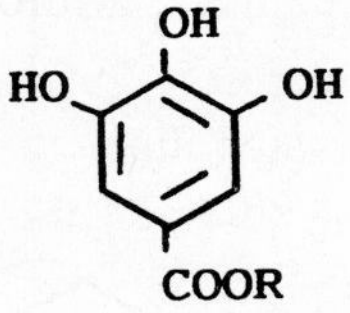

PG, R = C_3H_7
OG, R = C_8H_{17}

The activity of the tocopherols is in the reverse order in fats and oils *in vitro* [21-23]. In the Rancimat test which measures the formation of volatile oxidation products with a stream of air passing over or through the oil at elevated temperatures [23], γ-Toc-OH was found to be most effective and α-Toc-OH least effective, Table 5.2. However, all were more effective than BHT and similar to octyl gallate (OG). A small incremental increase in antioxidant effectiveness with increase in concentration is generally indicative of physical loss from the substrate due to insolubility or volatility. This is a particular feature of α-tocopherol, BHT and BHA (see $PF_{0.1}/PF_{1.0}$ in Table 5.2) and similar characteristics of these antioxidants have been seen in polymers. By contrast, γ- and δ-tocopherols show a much better response to increased concentration and these differences almost certainly have more to do with the rate of loss of the antioxidants and their oxidative transformation to higher molecular weight products than with their intrinsic antioxidant activity [21]. This recurring theme in technological testing of antioxidants was addressed in detail in Chapter 4, but the above observation illustrates the general principle that the measurement of a single kinetic parameter in solution rarely if ever provides a complete description of the activity of an antioxidant under practical conditions, whether it be in technology or in biology.

Table 5.2 Comparison of the antioxidant activities of the tocopherols with some synthetic antioxidants in lard (PF* in Rancimat test at 120°C)

Antioxidant:	Tocopherols				BHT	BHA	TBHQ	OG
Conc^n, %	α	β	γ	δ				
0.1	7.6	7.7	10.1	7.9	3.1	7.0	8.5	7.7
0.2	9.4	11.4	19.3	11.5	4.1	8.6	11.1	12.4
0.3	10.4	13.7	25.0	16.1	4.4	9.8	14.3	16.9
0.5	12.1	16.9	29.6	20.5	4.9	11.6	16.9	24.7
1.0	11.3	18.2	34.1	29.5	5.4	12.1	22.6	36.1
$PF_{0.1}/PF_{1.0}$	1.49	2.36	3.38	3.73	1.74	1.72	2.66	4.69

* PF = Protection Factor = $\tau_{antiox}/\tau_{control}$

By measuring the total radical trapping (CB-D) activities of the four tocopherols in human blood plasma and in ghost membrane extracts, Burton et al. [13,17] have concluded that "total antioxidant activity" in cells can be accounted for by this antioxidant mechanism at ambient oxygen pressures [16]. However, it is known that the concentration of α-Toc-OH remains sensibly constant in biological media in the presence of ascorbic acid Asc(OH)$_2$ until the latter is depleted [10,24-28]. This is due to regenerative synergism between α-Toc-OH and Asc(OH)$_2$ (Section 5.2.2). It is also recognised that other CB-D antioxidants, notably ubiqinol (Section 5.2.5), thiols and their oxidation products [29] and possibly even the hydroxyflavones (Section 5.2.8) may perform the same function. The antioxidant activity of α-Toc-OH in biological cells then probably has more to do with its facile regeneration from its aryloxyl by sacrificial reducing agents either present in the cell or more accessible to it than with its radical trapping efficiency. A combination of vitamin E and vitamin C is always much more effective than either antioxidant alone due to the regeneration of α-Toc-OH from α-Toc-O·, α-Toc-hq from α-Toc-q and possibly even the dehydrodimer, α-Toc-dhd from its cognate quinone, α-Toc-sd (Chapter 3) by ascorbate. This chemistry is summarised in Scheme 5.2:

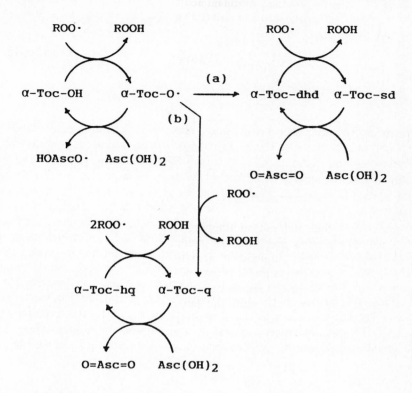

(For the structures of α-Toc-q, α-Toc-hq, α-Toc-sd and α-Toc-dhd see Chapter 3, Scheme 3.8)

Scheme 5.2 Role of ascorbic acid in the regeneration of α-tocopherol from its oxidation products

It was seen in Chapter 3 that α-tocopherol is not a very effective antioxidant in technological substrates. It is very rapidly converted to dimers and trimers which at ambient oxygen pressures have no further antioxidant activity. At low oxygen concentrations, however, the quinonoid oxidation products, α-Toc-q and α-Toc-sd are powerful catalytic antioxidants due to their facile reduction to phenols by macroalkyl radicals. It seems probable, although not yet experimentally demonstrated, that under the relatively low oxygen pressures found in the biological cell [30], reaction of that allylic radicals formed from the polyunsaturated fatty acids may act as reducing substrates for derived "stable" aryloxyl radicals in a manner analogous to galvinoxyl in polyunsaturated rubbers (Chapter 3) [31]. Whether α-Toc-O· itself will be reduced in this way will be governed by its lifetime before oxidative dimerisation and disproportionation, (Chapter 3, Scheme 3.9). However the quinonoid transformation products of α-Toc-OH, α-Toc-q and α-Toc-sd are much more stable in the presence of oxygen and are known to be readily reduced back to the parent phenols (Chapter 3).

$$-CH_2CH=CHCH=CHCH- \underset{(OO)\cdot}{|} \begin{matrix} \text{(a) } \alpha\text{-Toc-q, etc.} \\ \nearrow \\ \searrow \\ \text{(b) RH} \end{matrix} \begin{matrix} -CH=CHCH=CHCH=CH- + \alpha\text{-Toc-hq } (+ O_2) \\ \\ \underset{OOH}{|} \\ -CH_2CH=CHCH=CHCH- \\ ROOH \end{matrix} \quad (2)$$

Unexpectedly, α-tocopherol is much less effective in inhibiting peroxidation in liposomes [32] and in LDL particles [10,33-35] than it is in homogeneous solution and unlike some synthetic commercial antioxidants, notably BHA, it did not give a true induction time to autooxidation. A number of attempts have been made to explain this difference. Bowry and Stocker [10], in a detailed study of the peroxidation of LDL particles containing α-tocopherol in aqueous suspension observed substantial peroxidation before the depletion of the antioxidant from the particles. This behaviour is quite different from the peroxidation of homogenised and uniformly dispersed LDL which shows the expected induction time before hydroperoxide is formed. In attempting to explain this unexpected behaviour, Bowry and Stocker drew an analogy between the environment of an antioxidant-derived radical in an LDL particle and that of a macroalkyl radical in a growing latex particle during polymerisation. They postulated that, because there are very few α-Toc-OH molecules per LDL particle (see Fig. 5.1), the tocopheroxyl radical (α-Toc-O·) like the macroalky radical cannot terminate by dimerisation and undergoes preferential chain-transfer with the polyunsaturated components of the LDL:

$$\alpha\text{-Toc-O}\cdot + LH \quad \rightarrow \quad \alpha\text{-Toc-OH} + L\cdot \overset{O_2}{\rightarrow} \quad LOO\cdot \quad (3)$$

However, it is not obvious on the basis of this explanation why peroxyl radicals (which are clearly present in the particles since hydroperoxides are formed) should not terminate the aryloxyl in the usual way to give a peroxydienone, α-Toc-q and dehydrodimers (see Chapter 3, Scheme 3.3). Moreover, the chain transfer activity of 2,6-dimethylphenols is insignificantly small and reaction 3 is most unlikely with the relatively stable α-tocopheroxyl. It seems likely that the apparent ineffectiveness of α-Toc-OH in LDL is associated with the unusual structure of the LDL particle which is composed of a neutral lipid "core" containing a substantial proportion of cholesterol linoleate and arachidonate (see Fig. 5.1). This is surrounded by a polar lipoprotein coating intercalated with Apoprotein B. It seems likely that the hydrophilic hydroxyl of α-Toc-OH will be strongly associated with and possibly immobilised in the polar lipid coating of the particle.

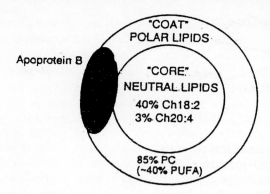

Fig. 5.1 Structure of the LDL particle. Ch18:2 = cholesteryl linoleate, Ch20:4 = cholesteryl archidonate, PC = phosphatidyl choline. Antioxidants, number per particle, α-Toc-OH: 6-12, γ-Toc-OH: 0.5, Ubi-hq: 0.5-1.2, Carotenoids: 0.4. Dimensions: diameter = 21 nm, volume = 4×10^{-24}. (Reproduced with permission from V.W. Bowry and R. Stocker, *J. Am. Chem. Soc.*, **115**, 6029 (1993)).

Barclay has suggested [32] that the lack of mobility of α-Toc-OH in this relatively viscous bi-layer may limit the ability of this molecule to terminate the peroxyl radical. Such a mobility limitation would particularly retard the dimerisation of α-Toc-O· (Chapter 3, Scheme 3.8). Antioxidants without the "phytyl tail" can distribute more rapidly both within the particle and between particles, and synthetic hindered phenols such as BHT, BHA and 2,6-BHA which are less effective CB-D antioxidants than α-Toc-OH in homogeneous solution [35], are more effective than α-Toc-OH in LDL and protect the latter from rapid depletion in the system. Greater mobility of the lower molecular weight antioxidants in the heterogeneous substrate is almost certainly part of the explanation, but it is known that the bulky ortho alkyl groups in these antioxidants also protect them from direct attack by oxygen and hydroperoxides which is the "Achilles heel" of the less hindered hydroxychromans (see Chapter 3). It is significant that probucol, which is an effective treatment for atherosclerosis, satisfies both requirements (mobility and steric hindrance).

The addition of ascorbic acid or ubiqinol to LDL [10,36] strongly protected α-Toc-OH, almost certainly by reducing back α-Toc-O· to the parent phenol by reaction 1. Since $Asc(OH)_2$ is insoluble in the LDL particle, this confirms that α-Toc-OH is substantially present in the polar "skin". Ubiquinones and quinols by contrast are located primarily in the lipid phase and the distribution of the ubiquinols changes with the length of the hydrophobic chain with associated effect on antioxidant performance. Water soluble peroxyl radicals (from 2,2'-azobis-(amidinopropane hydrochloride, AAPH) will also tend to concentrate in the polar "skin" leading to oxidative dimerisation of the immobilised antioxidant.

The conflicting evidence discussed above again illustrates the complexity of antioxidant behaviour in "real" and generally heterogeneous biological systems. Maiorino et al. have demonstrated [37] that even a variation of lipid organisation from bilayer to micellar dispersion can alter the protective affect of vitamin E and this is particularly influenced by the presence of transition metal ions (see below). It must be concluded then the measurement of the rate of radical trapping in model systems has little relevance to antioxidant effectiveness *in vivo*. Whether or not α-Toc-O· is a prooxidant as suggested by Bowry and Stocker [10] or has limited intraparticle mobility or whether it is destroyed by oxygen, localised peroxyls or hydroperoxides under these conditions, it is clearly imperative that the initial aryloxyl should be rapidly reduced back to α-Toc-OH in order to maintain its antioxidant potential. This is why homosynergism involving ascorbic acid and ubiquinol is of crucial importance to oxidative resistance *in vivo*.

α-Toc-q is readily reduced to the hydroquinone, α-Toc-hq *in vivo* [38] and there is increasing evidence that this redox couple plays a part in the antioxidant defences of the cell [39-41]. The antioxidant activity of the tocopheroquinones is however concentration sensitive in oxidatively challenged normal cells [42]. At high concentrations they are all prooxidants, but α-Toc-q less so than γ-Toc-q. However at lower oxygen concentrations, the quinonoid oxidation products (α-Toc-q and α-Toc-sd) are effective antioxidants in technological substrates and at high initiation rates (Chapter 3) and there is good evidence to suggest that they have the ability to redox cycle with alkyl and alkylperoxyl radicals under these conditions. α-Toc-q bears a close relationship to the lipid-soluble ubiquinone-10 (Ubi-q-10) which is found in cells mainly as Ubi-hq-10. The latter is an effective regenerative synergist with α-Toc-OH (Section 5.2.5) and Kohar et al. [43] have shown that α-Toc-q, like Ubi-q, is indeed reduced to α-Toc-hq in the body after ingestion.

Ubi-q-10   α-Toc-q

They also showed that more water-soluble quinone formed by oxidation of the α-Toc-OH analogue, PMHC, is reduced in blood plasma at 37°C. This could not be demonstrated for α-Toc-q presumably due to its water insolubility, but an analogous reduction must occur in the cells since it is ultimately excreted as the hydroquinone glucuronic acid conjugate [38]. The CB antioxidant activity of the redox couple dehydrodimer, α-Toc-dhd and spirodimer, α-Toc-sd (see Chapter 3, Scheme 3.8) has not been reported *in vivo* but Cillard et al. [15] noted that the peroxidation of linoleic acid was autoretarding in the presence of a-tocopherol, almost certainly due to the formation of a more effective oxidative transformation products. However, unlike α-Toc-hq, tocopherol dimers and trimers would be expected to be less "compatible" with bilipid membrane and would almost certainly be excreted more rapidly.

The presence of transition metal ions may eliminate and even invert the antioxidant activity of vitamin E. Thus, α-tocopherol actually increases the rate of hydroperoxide formation in LDL particles containing Cu^{2+} [37,44]. The evidence suggests that under these conditions, α-tocopherol is not involved in the reduction of peroxyl to hydroperoxide but is instead oxidised by Cu^{2+}, providing at the same time the highly active reduced form of copper which undergoes rapid redox reaction with hydroperoxides to give oxyl radicals. This is an important conclusion in view of the fact that antioxidant activity is frequently measured in copper-initiated peroxidations.

α-Tocopherol quenches singlet oxygen two orders of magnitude more rapidly in lipid microsomes than BHT (k_q = 1.2 x 10^8 and 0.01 x 10^8 respectively) [45] but not as effectively as β-carotene or lycopine (k_q = 40 x 10^8 and 90 x 10^8 respectively). It is tempting to suggest that this may contribute to the antioxidant effectiveness of the tocopherols, particularly in the presence of UV light, but it should be remembered that "quenching" of excited states does not always give non-radical products and in the case of α-Toc-OH, 1.5% of the energy absorbed gives radical products which are potentially capable of initiating oxidation [46].

Although the less methyl-substituted chromans are less effective as CB-D antioxidants than α-Toc-OH, γ-Toc-OH has been reported to be an effective scavenger for ·NO_2 [47]. Two products have been identified; γ-tocored and γ-tocoyellow. The mechanism proposed for their formation is shown in Scheme 5.3 and it is suggested that γ-Toc-OH may play a part in protecting cells from the effects of NO_x by reducing ·NO_2 to the less toxic ·NO [47].

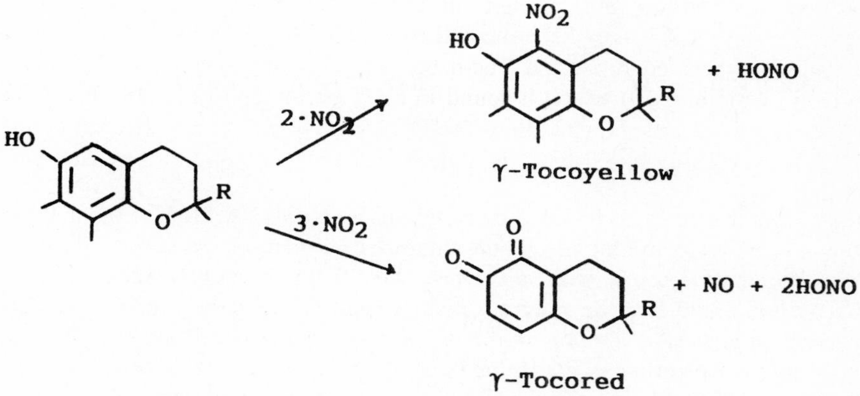

Scheme 5.3 Reduction of ·NO_2 by γ-tocopherol

5.2.2 *Vitamin C (Ascorbic acid)*

Ascorbic acid ($Asc(OH)_2$), one of the most powerful reducing agents to be found in biological systems, is water soluble and is located mainly in blood plasma [48]. It readily reduces reactive oxygen species, notably superoxide [45,49], hypochlorous acid [45,50], and water-soluble hydroxyl [45] and peroxyl [51-53] radicals to less damaging products in blood plasma and thus comes under the general classification as a CB-D antioxidant.

$$Asc(OH)_2 \xrightarrow{ROO·, ROOH} Asc(OH)O· \xrightarrow{ROO·, ROOH} O=Asc=O \quad (5)$$

Thus ascorbic acid completely inhibits peroxidation in plasma lipids due to activated neutrophils [53] and cigarette smoke [54,55]. However, a major disadvantage of $Asc(OH)_2$ as an antioxidant is its ability to act as a prooxidant in the presence of iron by reducing the relatively unreactive trivalent state of iron to Fe^{2+}. As was seen in Chapter 1, Fe^{2+} is a highly active catalyst for the formation of hydroxyl radicals from hydrogen peroxide (Fenton reaction) in iron overloaded systems. The effects of "free" ionic iron in biological systems will be discussed in Chapter 6 but it is particularly dangerous in the presence of excess vitamin C.

The major role of ascorbic acid in lipids and in LDL was seen in the previous Section (Scheme 5.2) to be its ability to regenerate the tocopherols from their phenoxyl radicals [56-59] (see also Chapter 4, Section 4.4.1). The main evidence for this in biological systems comes from the observation that a-tocopherol is not depleted from peroxidising liposomes in synergistic combination with ascorbic acid [56,60]. The rate constant for the reaction of α-Toc-O· with Asc(OH)$_2$ is high (1.6 x 10^6 M^{-1}s^{-1}) and this homosynergism is supported by epidemiological evidence (see Chapter 6). However, the reality of the regenerative phenomenon *in vivo* is still questioned by some. [61,62].

The end product of ascorbic acid oxidation is dehydroascorbic acid AscO$_2$, which

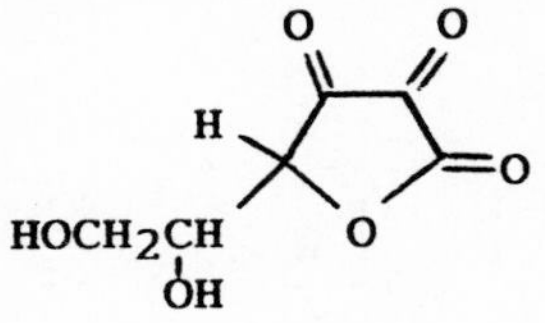

HOCH$_2$CH(OH)—[AscO$_2$ ring structure]

HOOCCH$_2$NHCOCHNHCOCH$_2$CH$_2$CHCOOH
      | |
      CH$_2$SH   NH$_2$

AscO$_2$ 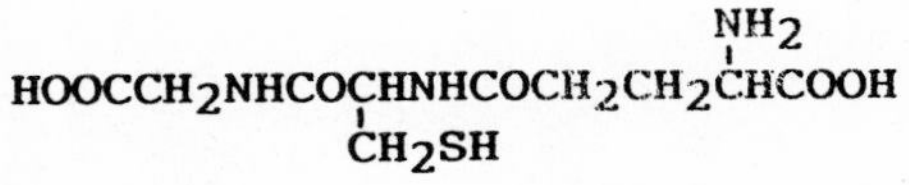   GSH

appears to have some activity in its own right as an antioxidant [63]. This may be in part due to its ability to inhibit metal ion-induced peroxidation, probably by complexation which frequently occurs with diketones, but there is also evidence [64,65] that dehyroascorbic acid may be reduced back to ascorbic acid by glutathione (GSH) and related compounds [66] which is in turn cyclically regenerated from the disulphide by the reduced form of nicotinamide adeninedinucleotide phosphate (NADPH) [65]. It appears then that ascorbic acid is not the ultimate source of electrons by single electron transfer to peroxyl radicals across the cell membrane:

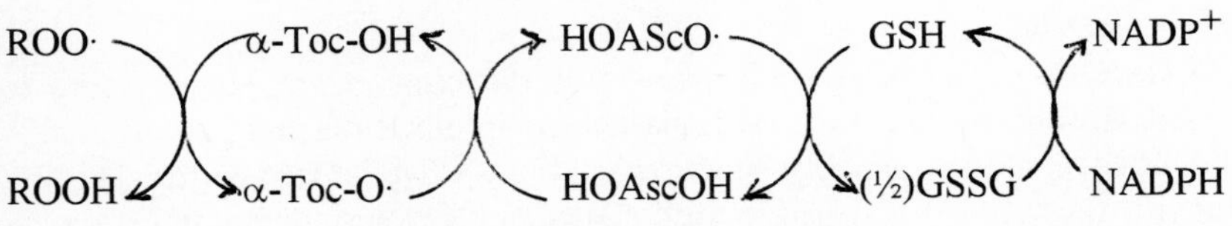

ROO· α-Toc-OH HOAscO· GSH NADP$^+$

ROOH α-Toc-O· HOAscOH (½)GSSG NADPH   (6)

Ascorbic acid and its esters have long been known to synergise effectively with the tocopherols in the preservation of fats and oils [67,68], and 2-O-octadecyl ascorbate has recently been shown to be an effective antioxidant and therapy for reperfusion injury [69-71].

5.2.3 *Tetrahydropterins and dihydropterins*

Tetrahydrobiopterin (BPH4), a co-enzyme which catalyses the hydroxylation of phenylalanine to tyrosine in liver and kidney and tyrosine to L-DOPA in the brain, has an important role in the normal functioning of the mammalian brain [72]. BPH4 has recently been shown to be a almost as effective as ascorbic acid as an antioxidant in rat brain homogenate and more effective than ascorbic acid as a scavenger of superoxide [73]. The dipyridyl herbicide, paraquat, is a potent inducer of Parkinson-like syndrome and BPH4 was found to be an effective inhibitor of its activity in cultured hepatocytes

[73]. BPH4 is oxidised to the dihydropterin, BPH2 under these conditions and the antioxidant mechanism is shown in Scheme 5.4.

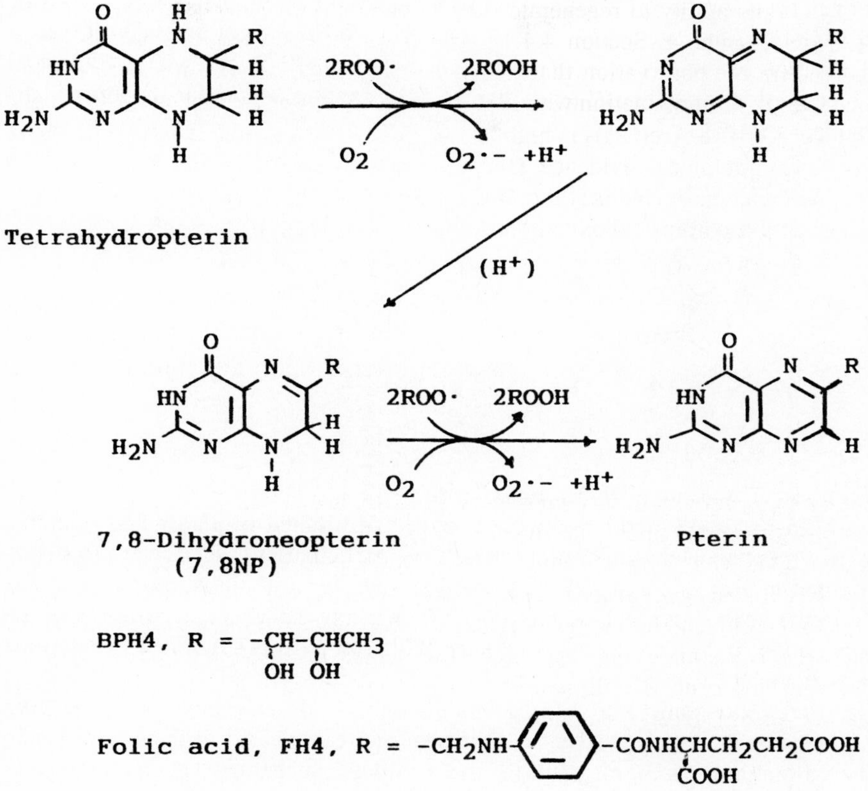

Scheme 5.4 Antiooxidant mechanism of the hydrobiopterins

However, BPH4 like folic acid (FH4) which differs only in the nature of the group in the 6 position, can also reduce molecular oxygen to superoxide and this prooxidant reaction is in competition with the antioxidant function of tetrahydropterins and indeed it is autooxidation of folic acid in foodstuffs which leads to the reduction of its vitamin activity.

7,8-Dihydroneopterin (7,8NP) is released during immune cell activation has been shown to inhibit LDL oxidation as efficiently as α-Toc-OH [74]. In combination with α-Toc-OH, 7,8NP acts synergistically and preserves α-Toc-OH during the induction period. 7,8NP is particularly effective in Cu^{2+} catalysed peroxidations and one of its functions may be to chelate transition metal ions through the pterin structure. However, pterin itself has virtually no antioxidant activity [74], showing that CB-D activity is essential to the function of the hydropterins.

5.2.4 Uric acid
The other major water soluble antioxidant found abundantly in blood plasma (160-450 μM) [48] is uric acid. Although uric acid has a similar order of radical trapping activity

to ascorbic acid in aqueous solution [75,76], unlike the latter, it appears to play only a minor role in the regeneration of α-tocopherol from its radical but it does protect ascorbic acid from oxidation in plasma [10]. Uric acid is an important antioxidant in plasma and in urine and has been shown to have peroxyl trapping activity in the latter [76]. It is also an effective iron chelating agent [77,78] and iron chelation is almost certainly its main antioxidant function in urine. However the presence of both CB-D and MD activities in the same molecule is autosynergistic (see Chapter 4) and this accounts for its antioxidant potency in aqueous media.

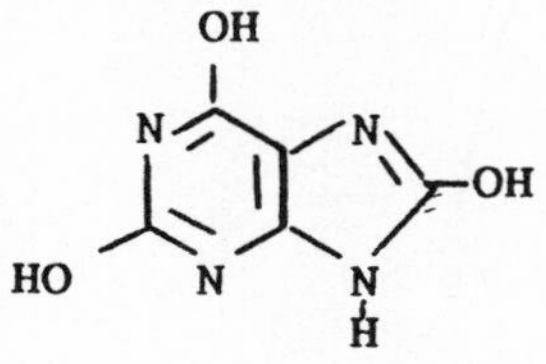

Uric acid

5.2.5 Ubiquinones (Co-enzyme Q) and ubiquinols

Ubiquinones (Ubi-q), as the name implies are widely distributed in plants and animals, particularly in mitochondrial membranes where they participate in electron transport by reversible oxidation and reduction [79]. The predominant form of ubiquinone in humans is Ubi-q-10. In the heart, liver and kidney, 70-100% is in the reduced state whereas in the brain and lung about 80% is in the oxidised state [80]. Ubi-q-10 is contained in all kinds of food but particularly meat, fish and beans.

Although the antioxidant role of the ubiquinones and ubiquinols in protecting mitochondria against oxidation has been recognised for 30 years [81], it is only relatively recently that attention has been paid to the detailed mechanism of its action in combination with the tocopherols in the cell membrane [36,82]. The ubiquinols have reactivities with peroxyl radicals comparable to α-tocopherol [36], but the direct CB-D reaction in not now thought to be the main mechanism of their action in biological membranes. The redox potential of Ubi-hq at physiological pH (-0.24V) [83] is very much lower than that of vitamin E (+0.48) [84] and the rate constant for the reaction of Ubi-hq with α-Toc-O· has been shown to be higher than that for its direct reaction with peroxyl radicals [36,85], suggesting that Ubi-hq should be able to regenerate α-Toc-OH from its aryloxyl. Convincing evidence for this homosynergistic interaction with α-tocopherol has come from a study by Kagan et al [36] of the oxidation of liposomes initiated by a water soluble azo initiator (AAPH) in the presence of both α-Toc-OH and Ubiq-hq-10 both separately and in combination. The rate of oxidation of both antioxidants was very similar when examined individually but in combination, Ubi-hq-10 was oxidised at the same rate as in the absence of α-Toc-OH, whereas α-Toc-OH was not oxidised until the concentration of Ubi-hq-10 was reduced essentially to zero in the system (see Fig. 5.2).

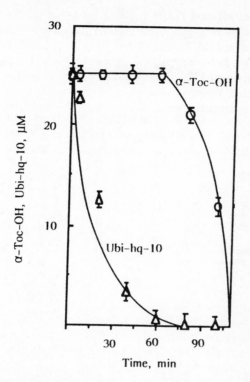

Fig. 5.2 AAPH-induced oxidation of a combination of α-Toc-OH and Ubi-hq-10 in dioleylphosphaditylcholine liposomes at 37°C. (Reproduced with permission from V.E. Kagan, D.A. Stoyanovsky and P.J. Quinn in *Free Radicals in the Environment, Medicine and Toxicology*, Eds. H. Nohl, H. Esterbauer and C. Rice-Evans, Richelieu Press, London, 1994, p.238).

This is the most convincing evidence for sacrificial homosynergism (see Chapter 4, Section 4.4.1) so far reported:

$$\alpha\text{-Toc-OH} \xrightarrow[\text{ROOH}]{\text{ROO·}} \alpha\text{-Toc-O·} \xrightarrow[\text{Ubi-sq}]{\text{Ubi-hq}} \alpha\text{-Toc-OH} \tag{7}$$

Ubi-sq = Ubisemiquinone

Kagan et al. [36] have also shown that the reduced forms of both antioxidants are present together in the lipid bilayer whereas the quinonoid oxidation products are more closely associated with the neutral lipids.

Ubiquinones are reduced to hydroquinones in the respiratory chain by NADPH but Kagan and Packer [86] found that reduction of α-Toc-O· to α-Toc-OH only occurred in liposomes in the presence of Ubi-q. A similar dependence on Ubi-q was also observed in the succinate reduction of α-Toc-O·, suggesting that Ubi-sq can oxidise these substrates but not α-Toc-O·.

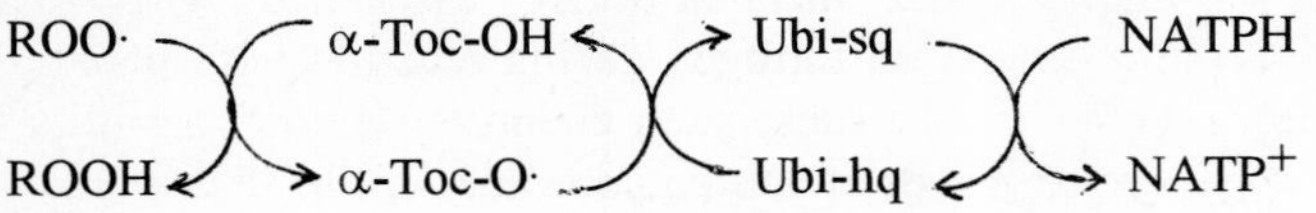

$$\text{ROO·} \quad \alpha\text{-Toc-OH} \quad \text{Ubi-sq} \quad \text{NATPH}$$
$$\text{ROOH} \quad \alpha\text{-Toc-O·} \quad \text{Ubi-hq} \quad \text{NATP}^+ \quad (8)$$

5.2.6 Bilirubin

Bilirubin is an effective CB-D antioxidant in plasma [53,87] where it is associated with albumin. It is an effective inhibitor for linoleic acid oxidation [87] and is converted to the highly conjugated biliverdin by reaction with peroxyl radicals, reaction 9:

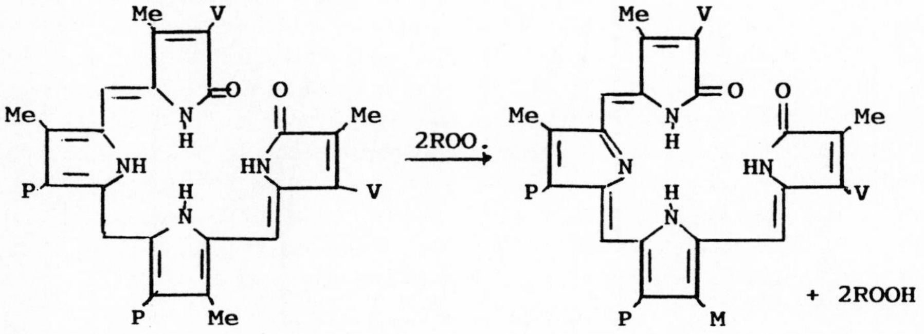

$$\text{Bilirubin} \xrightarrow{2\text{ROO·}} \text{Biliverdin} + 2\text{ROOH} \quad (9)$$

Me = methyl, V = vinyl, P = γ-propionyl

Bilirubin is also an effective metal binding agent.

5.2.7 Melatonin and Serotonin

Melatonin, N-acetyl-5-methoxy tryptamine, a hormone produced by the pineal gland from serotonin [88-90], has been reported to be a more effective CB-D antioxidant than vitamin E [91].

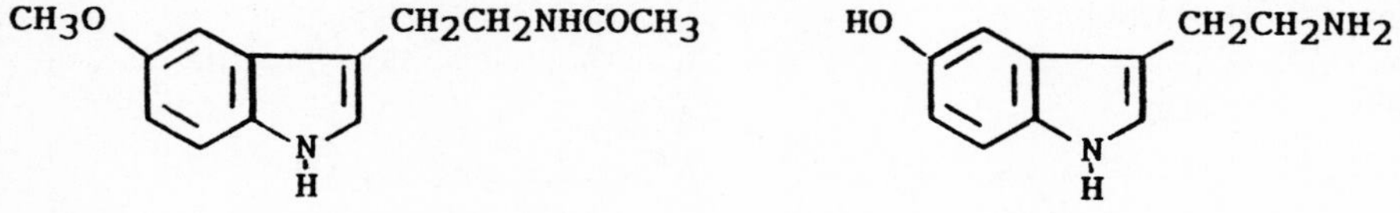

Melatonin — CH_3O-...-$CH_2CH_2NHCOCH_3$

Serotonin — HO-...-$CH_2CH_2NH_2$

Melatonin is produced in the dark and is involved in the regulation of circadian rhythm and has been reported to have anticancer [88,93] and antiaging [94] activity (see Chapter 6). Melatonon also reduces DNA damage in the presence of the carcinogen saffrole [96,97], reduces paraquat damage in the rat lung [98], and decreases lipid peroxidation in CCl_4-exposed rats and its biological effects are in general consistent with its ability to suppress oxidative damage in animals. Both melatonin and serotonin are effective CB-D antioxidants [91,92]. The former is the more efficient peroxyl radical deactivator and is structurally related to the rubber antioxidant, ethoxyquin (see Chapter 3, Table 3.1, XV). This dihydroquinoline antioxidant is one of the few that has gained acceptance for use in foodstuffs and is widely used as an antioxidant in pharmaceuticals, fish meat, edible oils, etc. [100] and in the preservation of fruit during storage [101]. The powerful CB-D antioxidant activity of both ethoxyquin and melatonin are associated with the delocalisation of the unpaired electron formed on nitrogen during reaction with peroxyl by the *para*-methoxy group (see Chapter 3). The subsequent chemistry follows a similar pattern in the indole and dihydroquinoline groups of antioxidants and will be discussed in more detail in Section 5.5.1.

Pieri et al [92] and Marshall et al [91] have shown that melatonin reacts with peroxyl ($Cl_3COO \cdot$) at a similar rate to a-tocopherol and serotonin is almost as effective as a CB-D antioxidant. However, melatonin is not as effective in Fe^{3+} catalysed peroxidation of phospholipid liposomes as serotonin and Marshall et al. have questioned whether melatonin can act as a classical CB-D antioxidant because it "does not possess a phenolic -OH group". Whilst it is certainly true that 4-hydroxy arylamines are generally more effective than unsubstituted arylamines, the latter are equally effective when activated by 4-alkoxyl and the above conclusion is difficult to understand on the basis of published work on the antioxidant activity of aromatic amines (see Chapter 3). The above experimental results are also consistent with this and It seems likely that differences in antioxidant activity of serotonin and melatonin depends upon the subsequent chemical reactivity and physical behaviour of their transformation products. Serotonin forms quinones on oxidation by peroxyl, whereas melatonin can trap alkylperoxyl, and by analogy with tetrahydroquinolines may also form aminoxyl radicals with CB-A activity (see Section 5.5.1(b)).

Melatonin is both lipid soluble and sparingly water soluble and in view of its powerful antioxidant activity and ubiquitous presence in many organs in the body, a detailed study of its antioxidant mechanisms *in vivo* would be well worthwhile because of the potential of this structure as the possible basis of new antioxidant drugs.

5.2.8 *Oestradiol*

Oestrogen, a hormone produced by the pituitary gland, is sometimes stated in the literature to have antioxidant properties. In fact only one of the group of chemicals which make up the oestrogens, namely 17β-oestradiol, has a recognised antioxidant function, and although this compound has been shown to retard the peroxidation of isolated human LDL [102-104], it is a relatively weak CB-D antioxidant since, like other 4-alkyl phenols, it lacks both an electron delocalising group and steric protection in the ortho positions to hydroxy:

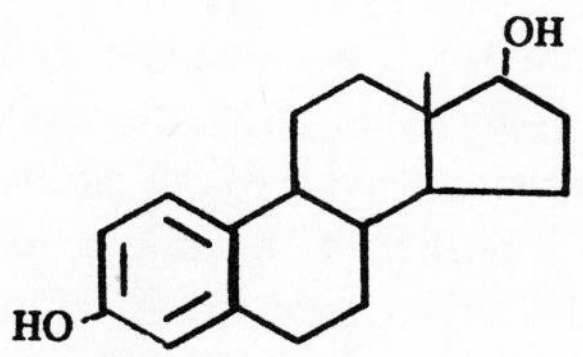

17β-oestradiol

The physiological significance of this antioxidant will be considered in more detail in Section 5.5.1 where the chain-breaking donor activity of drugs related to oestradiol will be discussed.

5.2.9 Oxides of nitrogen

The relatively stable free radical, nitric oxide ($\cdot$NO) is widely distributed in mammalian tissues and plays an important role in regulating vascular homeostasis. There has been a good deal of discussion as to whether $\cdot$NO, which is a relatively stable radical, causes or inhibits free radical tissue injury [105]. This is primarily because $\cdot$NO can act either as an electron donor, for example by reacting with superoxide, or as an electron acceptor by oxidising alkyl (or alkylperoxyl) radicals:

$$\begin{array}{c}
\cdot NO \quad \begin{array}{l} \nearrow \text{O}_2\cdot^-/\text{H}^+ \\ \rightarrow (\text{ROO}\cdot) \\ \searrow \text{R(O)}_2\cdot \end{array} \quad \begin{array}{l} \text{O=NOOH(R)} \rightarrow [\cdot\text{NO}_2 + \cdot\text{OH(R)}] \rightarrow \cdot\text{NO} + \cdot\text{OH} \\ \qquad\qquad\qquad\qquad\quad \downarrow \\ \qquad\qquad\qquad\qquad\quad \text{HNO}_3 \\ \\ \qquad\qquad\quad\;\; \text{R(O}_2)\cdot \\ \text{RN=O} + \text{O}_2 \rightarrow \text{R}_2\text{NO}\cdot + \text{O}_2 \end{array}
\end{array} \qquad (10)$$

Pernitrous acid decomposes rapidly under physiological conditions to give a hydroxyl radical which is much more reactive than either $\cdot$NO or $O_2\cdot^-$. However, this reaction like many other reactions of hydroperoxides (Chapter 1, reaction 24) almost certainly occurs in a molecular "cage" and nitric acid will be the main product. At relatively low concentrations, $\cdot$NO and its oxidation product $\cdot$NO$_2$ can react with alkyl (or alkylperoxyl) radicals to give antioxidant species (see Chapter 3, Section 3.3). Nitrosoalkanes formed in reaction 10 are of course themselves alkyl radical trapping agents giving nitroxyls which can in turn scavenge alkyl radicals. How effectively the resulting nitroxyl radicals scavenge carbon-centred radicals depends on their stabilities and subsequent disproportionation products, some of which (e.g. nitrones) are themselves alkyl radical trapping agents (CB-A antioxidants) and others (e.g. hydroxylamines) are "oxyl" radical reducing agents (CB-D antioxidants). Aminoxyls are particularly reactive toward alkenyl radicals, and the resulting hydroxylamines are again effective CB-D antioxidants, leading to catalytic antioxidant action even at ambient temperatures. The relevance of this antioxidant mechanism to polymer stabilisation was discussed in Chapter 3 (see Schemes 3.10 and 3.11).

$\cdot$NO$_2$ in spite of its damaging effects in the lungs of animals [106] may also be an antioxidant in oxidations at ambient oxygen pressures, particularly in the presence of light where it is reduced to nitrous acid which has been shown to be an effective CB-D antioxidant (see Chapter 3, Scheme 3.18). One of the critical factors determining whether NO and NO$_2$ behave as prooxidants or antioxidants is without doubt the concentration of oxygen in the tissues concerned and the availability of enzymic reducing agents. Ways of utilising the antioxidant activities of oxides of nitrogen could emerge as the means of delivering this radical trap at the site where it is required are better understood.

5.2.10 Polyhydroxyphenols

A large number of polyhydroxyphenols with antioxidant activity are widely distributed in fruit, tea and red wines. Because of their non-toxicity, many of these have been evaluated in the food industry as antioxidants for oils and fats. Early studies of the hydroxyflavones suggested that they are highly effective autosynergists because of their CB-D activity coupled with the ability of the polyphenol structure to complex transition metal ions [107].

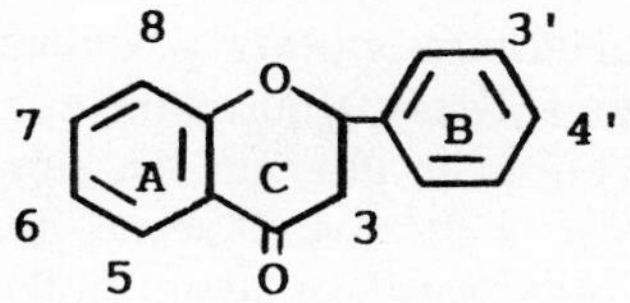

Flavones

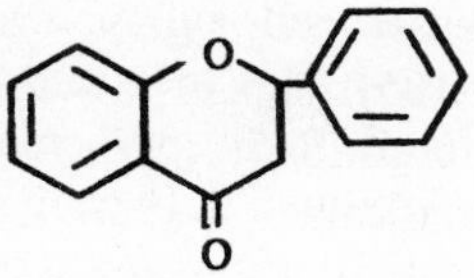

Flavanones

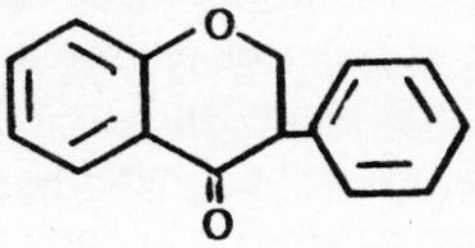

Isoflavones

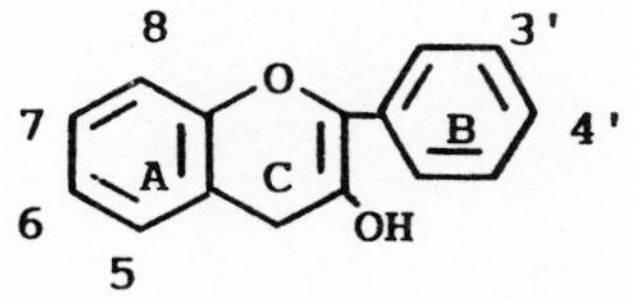

Flavanes

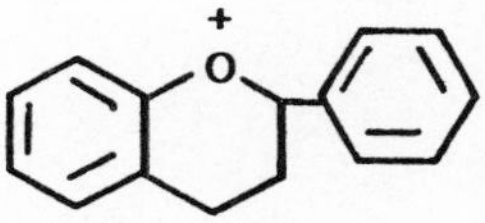

Anthocyanidines

Iron [108-110] and copper [111] chelating ability has been reported for a variety of flavonoids during lipid peroxidation in cells. However, it has also been reported [112] that quercetin, myricetin and gossypol, a polyphenol isolated from the cotton plant, may increase hydroxyl radical formation in the presence of Fe^{3+}-EDTA.

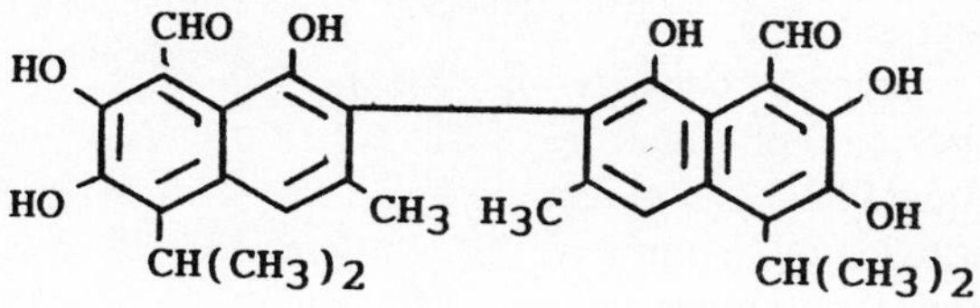

Gossypol

Bleomycin, an iron chelator used as an anticancer agent, is also activated to hydroxyl radical generation by quercetin and myricetin [113]. Its is believed to bind to DNA with iron in the reduced state where it induces DNA strand breakage by producing hydroxyl radicals.

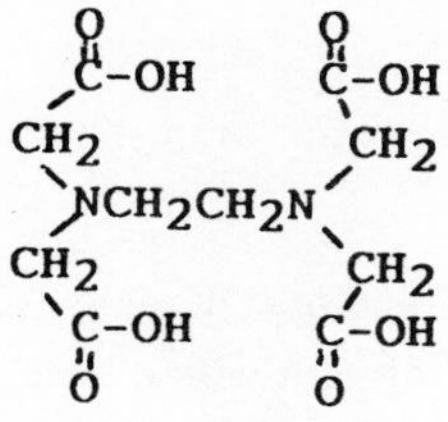

EDTA

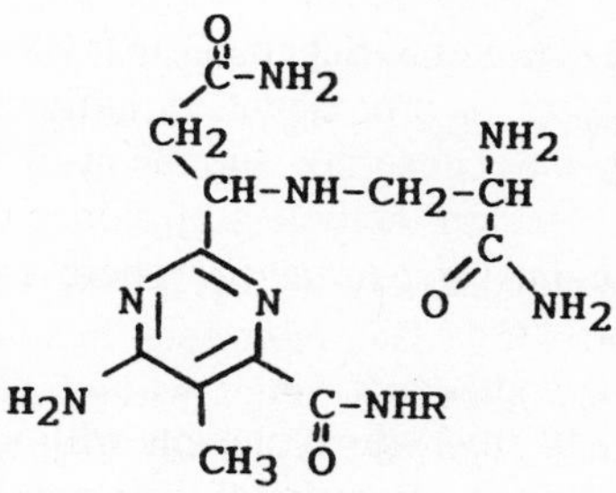

Bleomycin

The radical promotion effect of the flavonoids in the presence of transition metal ions appears therefore to be a very specific reaction with nitrogen chelating agents in which the iron is preferentially reduced to the Fenton-reactive Fe^{2+} by the catechol system [112].

The CB-D antioxidant activity of the hydroxyflavones in foodstuffs appears to be related to the number of ortho-dihydroxy groupings in A and B rings. Gossypetin, robinetin and myricetin (see Table 5.3) were found to be particularly effective although the simple gallate esters (e.g. propyl gallate, Section 5.2.1) were almost as effective. It seems likely that the solubility of the antioxidants in the medium may also contribute significantly to antioxidant activity.

Table 5.3 Antioxidant activities of flavonoids in natural oils and fats at 60°C

Antioxidant type and hydroxyl substitution									Common name	τ^*,h	PF^+
3	5	6	7	8	2'	3'	4'	5'			
Flavone											
OH	-	-	-	-	-	OH	OH	-		145	0.6
OH	-	-	OH	OH	-	-	-	-		438	3.9
OH	-	OH	OH	-	-	-	-	-		145	0.6
OH	-	-	OH	-	-	OH	OH	-		320	2.6
OH	-	-	OH	OH	-	OH	OH	-		555	5.2
OH	OH	-	OH	-	-	OH	OH	-	Quercetin	410	3.6
OH	-	-	OH	-	-	OH	OH	OH	Robinetin	855	8.5
OH	OH	-	OH	-	-	OH	OH	OH	Myricetin	685	6.6
OH	OH	OH	OH	-	-	OH	OH	-	Quercitagetin	698	6.7
OH	OH	-	OH	OH	-	OH	OH	-	Gossypetin	1050	10.2
OH	-	-	OH	OH	-	OH	OH	OH		595	5.6
Flavanone											
OH	OH	-	OH	-	-	OH	OH	-	Taxifolin	350	2.7
Flavane											
OH	OH	-	OH	-	-	OH	OH	-	D-Catechin	275	2.1
OH	OH	-	OH	-	-	OH	OH	-	L-Epicatechin	220	1.5
									Propyl gallate	680	8.5
									No antioxidant	90	-

$^*\tau$ = time in hours to reach peroxide value of 25

$^+$PF = protection factor = $\tau_a - \tau_o / \tau_o$ where τ_a = induction time with antioxidant and τ_o = induction time without antioxidant.

The hydrogen donor activity of the flavonoid antioxidants has also been examined in aqueous solution [75]. The reasoning behind this approach was that, due to their hydrophylicity, the phenolic groups will be associated with the surface of the phospholipid layer and will therefore be ideally located to scavenge ROS in the aqueous phase. Using a water soluble radical cation formed by oxidation of 2,2'-azino-bis(3-ethylbenzothiazoline-6-sulphonate (ABTS$^+$), the reducing capability of a number of polyphenols was compared with the water soluble analogue of α-tocopherol, Trolox (6-

hydroxy-2,5,8-tetramethyl chroman-2-carboxylic acid) to obtain their Trolox equivalent antioxidant activities (TEAC). These are listed in Table 5.4. The flavanone structure (e.g. taxifolin) was found to be significantly less effective than the correspondingly substituted flavone (e.g. quercetin), and this was attributed to the lack of conjugation between the B and C rings in the former [75,114]. However, taxofolin is not significantly inferior to quercetin in fatty ester (see Table 5.4) and care must be taken in extrapolating from the hydrophylic to hydrophobic media and from electophilic (ROO·) to nucleophilic (ABTS$^+$) radicals.

Table 5.4 Hydrogen transfer activity to flavonoid antioxidants relative to trolox (TEAC) [75]

3	5	6	7	8	2'	3'	4'	5'	Common name	TEAC
\multicolumn{11}{l}{Antioxidant type and hydroxyl substitution}		
\multicolumn{11}{l}{Flavone}		
-	OH	-	OH	-	-	-	OH	-	Apigenin	1.45
-	OH	-	OH	-	-	-	-	-	Chrysin	1.43
OH	OH	-	OH	-	-	OH	OH	-	Quercetin	4.70
OH	OH	-	OH	-	-	OH	OH	OH	Myricetin	3.10
OH	OH	-	OH	-	-	-	OH	-	Kaempferol	1.34
Rut	OH	-	OH	-	-	OH	OH	-	Rutin	2.42
\multicolumn{11}{l}{Flavanone}		
-	OH	-	OH	-	-	-	OH	-	Naringenin	1.53
OH	OH	-	OH	-	-	OH	OH	-	Taxifolin	1.90
-	OH	-	Gly	-	-	-	OH	-	Naringin	0.97
\multicolumn{11}{l}{Isoflavone}		
-	OH	-	OH	-	-	-	OH	-	Genistein	1.0
-	OH	-	-	-	-	-	OH	-	Genistin	0.79
\multicolumn{11}{l}{Flavane}		
OH	OH	-	OH	-	-	OH	OH	-	Catechin	2.2
OH	OH	-	OH	-	-	OH	OH	-	Epicatechin	2.5
\multicolumn{11}{l}{Anthocyanidin}		
OH	OH	-	OH	-	-	OH	OH	-	Cyanidin	4.2
-	OH	-	-	-	-	-	OH	-	Apigenidin	2.35
OH	OH	-	OH	-	-	-	OH	-	Perlagonidin	1.3
OH	OH	-	OH	-	-	OMe	OH	-	Peonidin	2.22
OH	OH	-	OH	-	-	OMe	OH	OMe	Malvidin	2.06

Other donor antioxidants	TEAC
α-Tocopherol	0.97
Ascorbic acid	0.99
Uric acid	1.02
Glutathione	0.9
Bilrubin	1.5
Albumin	0.69

Gly = glycoside, Rut = rutinoside

Quercitin and myricetin which contain the 3',4'-dihydroxy structures in the B ring have the highest CB-D activity of the flavonoids studied due to the activating (electron delocalisation) of an ortho or para hydroxyl group. Quercitin, kaempherol and rutin have been reported to scavenge superoxide [115,116] and it has been suggested [117] that they may quench singlet oxygen in the skins of fruits where they are found in relatively high concentrations. However, the singlet oxygen quenching ability of the flavonoids is similar to synthetic phenolic antioxidants, two orders of magnitude lower than α-tocopherol and three orders of magnitude lower than β-carotene [80]. Furthermore, as was noted earlier (Section 5.2.1), quenching of singlet oxygen by phenolic antioxidants, unlike quenching by β-carotene, occurs by a combination of physical and chemical interactions and it seems probable then that in general, excited state quenching is less important than peroxyl radical scavenging [118-122].

The chalcones and dihydrochalcones have been found to be more effective antioxidants in the rancimat test in lard at 120°C than the flavanones to which they are structurally related (see Table 5.5) [123]. Nevertheless, the profound effect on antioxidant activity of the 3,4-dihydroxy structure in the B ring was observed in the chalcones and hydrochalcones as well as the flavanones. Interestingly, unlike the flavones [75], the effect of conjugated ketone in the 1 position of the B ring in the chalcones was found to reduce CB-D activity. This is consistent with the effects of electron attracting carbonyl groups on CB-D activity in synthetic antioxidants (see Chapter 3) which is blocked in the dihydrochacones. It seems likely that this reflects the difference between the attack of an electrophilic peroxyl radical in the fatty esters and the nucleophilic ABTS+ cation in aqueous solution.

Table 5.5 Comparison of antioxidant activity of hydroxychalcones and hydroxydihydrochalcones with the hydroxyflavones in lard at 120°C

Antioxidant type and hydroxyl substitution								Common name	τ, h at concn		
									0.25%	0.05%	0.1%
Flavanone											
3	5	6	7	8	2'	3'	4'	5'			
-	OH	-	OH	-	-	-	OH	-	Naringenin	0.35	0.4	0.4
-	OH	-	OH	-	-	OH	OMe	-	Hesperatin	0.5	0.6	0.7
-	-	-	OH	OH	-	OH	OH	-	-	7.8	15.4	19.7
-	OH	-	OH	-	-	OH	OH	-	Eriodictyol	3.9	7.0	9.5
Chalcone											
3	5	6	7	8	2'	3'	4'	5'			
OH	OH	OH	-	-	-	OH	OH	-	Okanin	9.9	18.2	21.8
OH	-	OH	-	OH	-	OH	OH	-	-	4.5	11.0	15.6
-	-	-	-	-	-	OH	OH	-	-	10.3	17.9	26.6
-	-	OH	-	-	-	OH	OH	-	-	12.3	22.8	27.2
Dihydrochalcone											
2'	3'	4'	5'	6'	2	3	4	5			
OH	-	OH	-	OH	-	-	OH	-	-	1.0	1.3	1.7
OH	-	OH	-	OH	-	OH	OMe	-	Dihydro-hesperatin	1.5	2.2	4.4
OH	OH	OH	-	-	-	OH	OH	-	Dihydrookanin	12.5	22.3	31.6
OH	-	OH	-	OH	-	OH	OH	-	-	12.8	24.1	44.2
-	-	-	-	-	-	OH	OH	-	-	12.0	20.1	29.1
-	-	OH	-	-	-	OH	OH	-	-	15.1	24.0	29.9
None											0.35	

The flavonoids have been reported to protect α-tocopherol in LDL [80]. Whether this is through competitive scavenging of peroxyl radicals or by regenerative synergism as in the case of ascorbic acid and the ubiquinols is not yet clear. However, in a solution study, Duthie et al. [124] have found that like other hydroxybenzenes, catechin and quercetin rapidly reduce galvinoxyl to hydrogalvinoxyl. The rate of this reaction with the most reactive flavonoids (e.g. morin and myricetin) is over an order of magnitude higher than that of α-tocopherol with galvinoxyl [125], suggesting that α-tocopheroxyl itself should be readily reduced by these antioxidants, leading to homosynergism in the surface of the cell membrane.

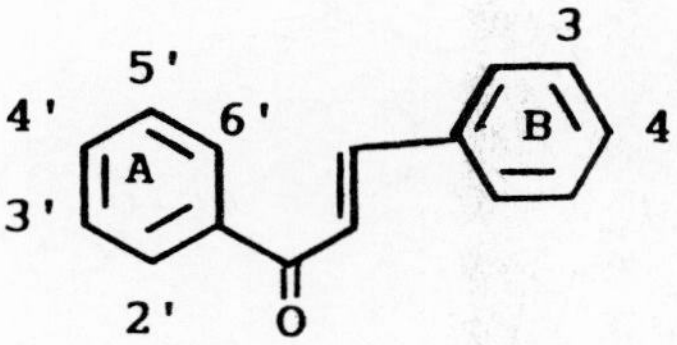

Chalcones

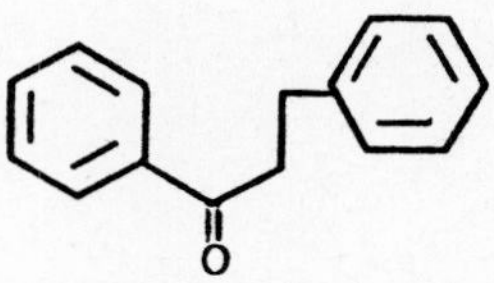

Dihydrochalcones

In summary, the flavonoids are particularly good examples of autosynergistic antioxidants operating by the combined CB-D and MD mechanisms. The two functions cannot be separated in practice since they both operate together in a mutually protective manner in transition metal ion catalysed systems.

5.2.11 *Herbiforous antioxidants*

A variety of naturally occurring phenols are widely distributed in plants [127,128]. These include gallic acid and its derivatives which are important components of tannin and which are isolated from Chinese galls (swellings on the stems of *Caesalpina*).

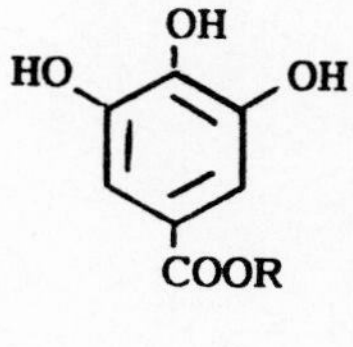

Gallic acid esters

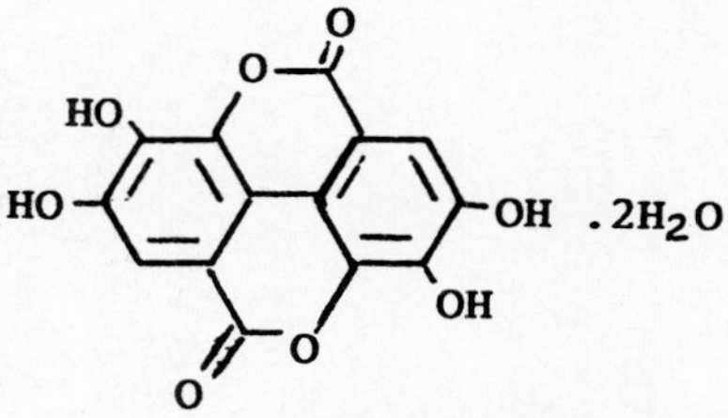

Ellagic acid

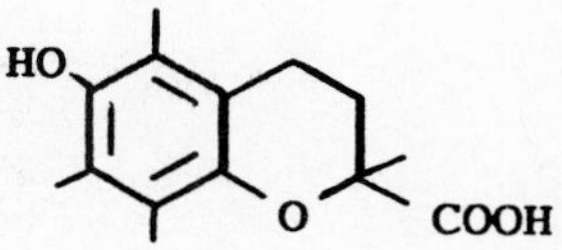

Trolox C

Tannin is also found in tea and is responsible for its astringent taste. It is also present in the barks, leaves and fruits of many other plants. Tannic acid is a highly effective CB-D antioxidant in aqueous media with activity similar to Trolox C, the water soluble α-tocopherol analogue [129], but the oil soluble esters which are highly effective as antioxidants for fatty foods (see above) are an order of magnitude less effective in aqueous media. The dimerised derivative of gallic acid, ellagic acid found in soft fruits and vegetables also has antioxidant activity similar to α-tocopherol and has been shown to have anticancer activity [130].

The flavonoids and cetechins (Section 5.2.10), which contain similar di- or trihydroxybenzene structures are also isolated from seeds, barks and leaves (e.g. green tea). Sesame seed oils are a rich source of active phenolic antioxidants, notably sesamol and sesamolinol:

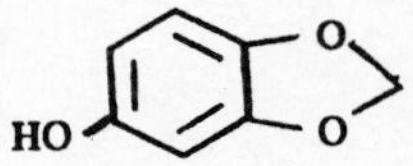

Sesamol

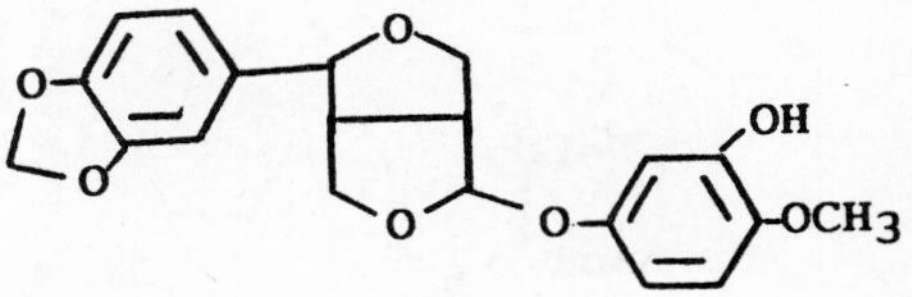

Sesamolinol

Rosmary contains a number of structurally related catechol derivatives of which carnasoic acid, rosmarinic acid and rosmanol are typical [127]:

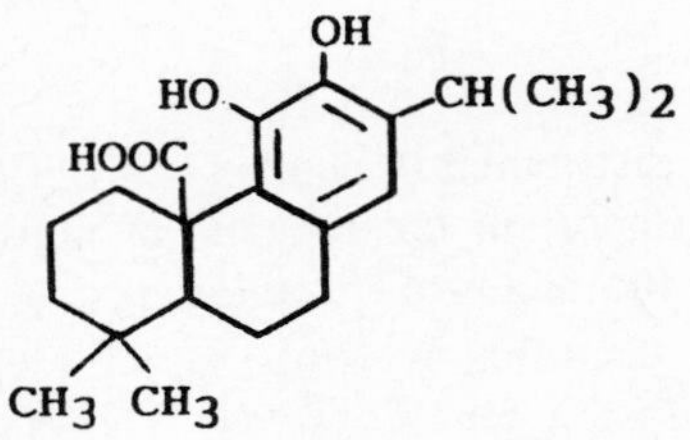

Carnosoic acid

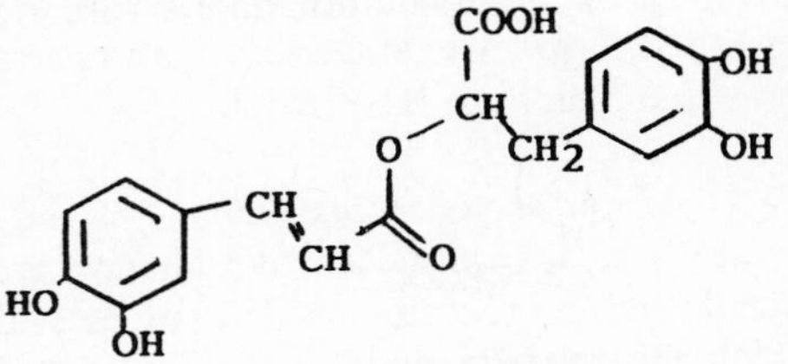

Rosmarinic acid

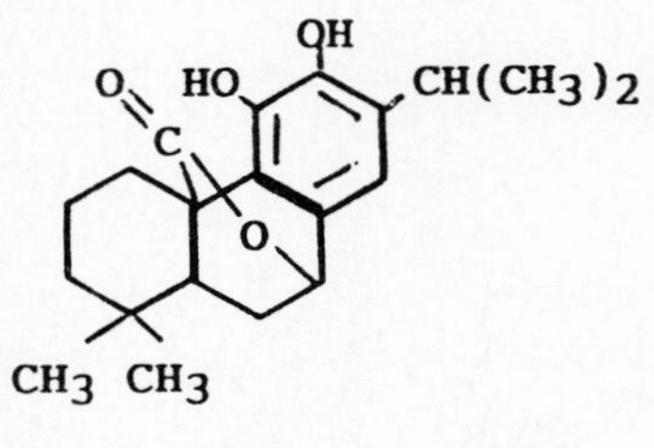

Rosmanol

However, carnasoic acid is only 5% as effective as a-tocopherol and 25% as effective as ascorbic acid as a CB-D antioxidant [128], whereas the flavonoids are often as effective or even more effective than α-tocopherol under similar conditions [75,128].

Monohydroxy phenols such as eugenol from cloves, thymol from thyme, vanillin from vanilla and 6-gingerol from ginger are also widely used therapeutically or as flavourings but they are only moderately effective as antioxidants (vanillin has about 20% of the activity of α-tocopherol and gingerol only 1% [129]). Since spices are generally imbibed in very small quantities and are not highly effective as antioxidants, they are only a minor source of antioxidants in the normal diet [128].

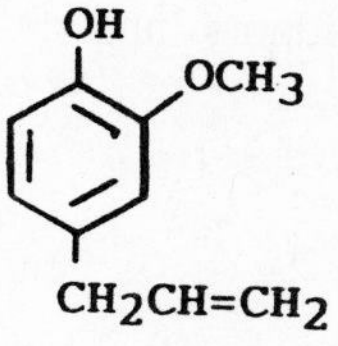

Eugenol 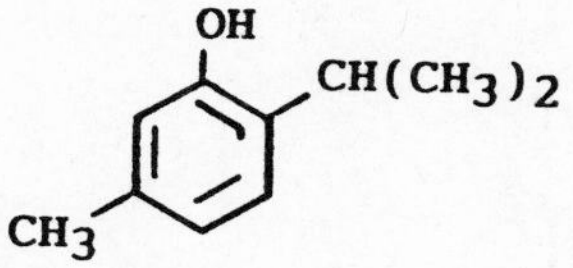   **Thymol** 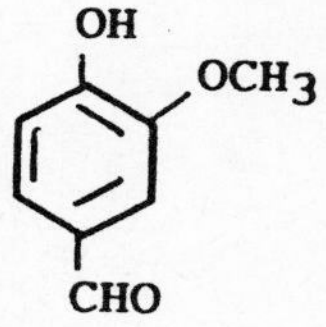   **Vanillin**

Phenols derived from cinnamic acid, which contain a more extended conjugated system than the monohydroxy phenyl shown above, have been reported to be more potent antioxidants [131,132]. Coumaric acid, caffeic acid and ferulic acid are plentifully available as constituents of plant cell walls [133] and are potentially available by hydrolysis as "natural" antioxidants for foods [132]:

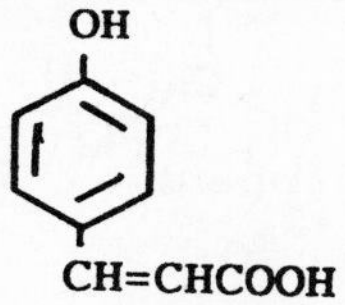

Coumaric acid 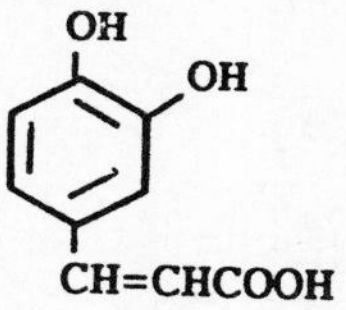   **Caffeic acid** 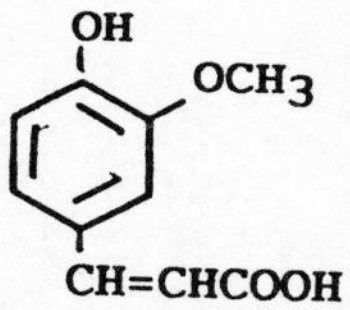   **Ferulic acid**

Caffeic, coumaric and ferulic acids have also been identified as components of olive oil [134-138] and are believed to be at least partly responsible for the oxidation resistance of this component of the "Mediterranean diet" [134,135,138] (Chapter 6). Nardini et al. [138] found that caffeic acid was a very effective inhibitor of Cu^{2+} catalysed LDL peroxidation and compared favourably with vitamin E. It was somewhat less effective than vitamin E in an azo (ROO·) initiated peroxidation and it is clear that a major function of caffeic acid is metal chelation through the catechol group. Cinnamic acid, coumaric acid and ferulic acid which do not contain this function were relatively ineffective.

5.2.12 *Carotenoids and retinoids*
The carotenoids are a group of highly conjugated compounds which have been shown on the basis of epidemiological evidence to be protective against diseases of oxidation, notably cancer and atherosclerosis (Chapter 6):

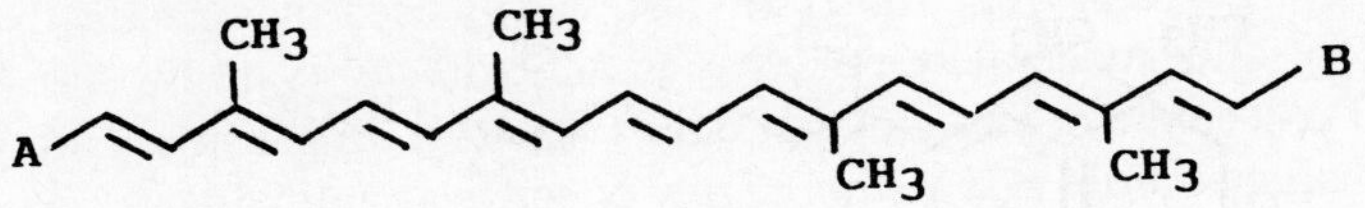

[Structural diagrams of carotenoid backbone with end-groups A and B, showing:

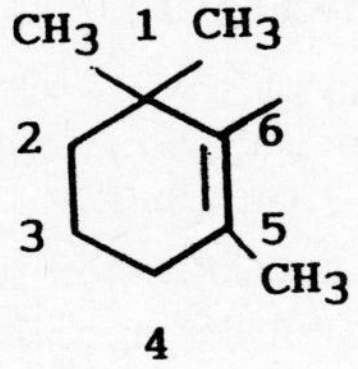

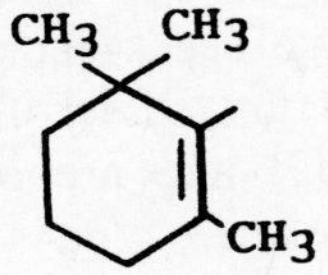

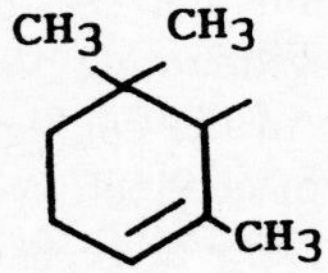

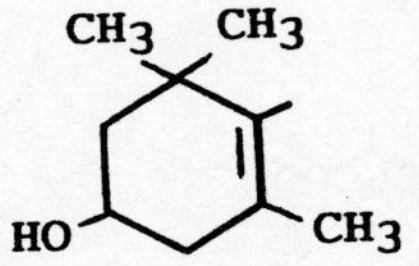

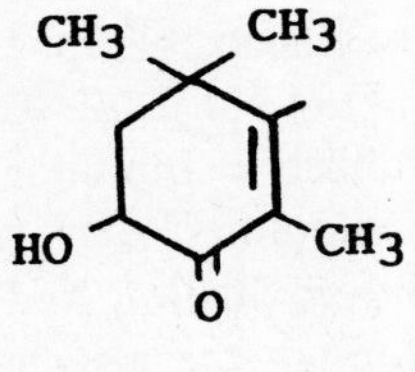

- **β-Carotene**: A,B = β-ionone ring (CH3 positions at C1, C1, C5; numbered 1–6)
- **α-Carotene**: A = β-ionone ring; B = α-ionone ring (double bond shifted)
- **Lycopene**: A,B = open-chain isoprenoid end
- **Astaxanthin**: A,B = 3-hydroxy-4-keto-β-ionone ring (HO– and =O substituents)
- **Canthaxanthin**: A,B = 4-keto-β-ionone ring (=O substituent)]

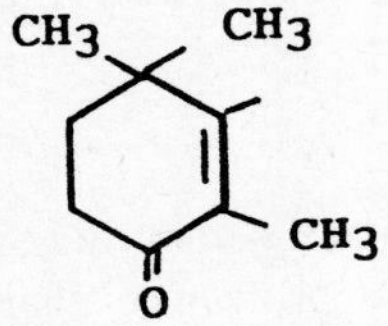

The most important member of this class is β-carotene which is commonly referred to as an "antioxidant". As will be discussed below, the more correct designation is "retarder" but in order not to confuse the reader, the common designation will continue to be used. Retinol (Vitamin A) is formed by enzymic cleavage of several (but not all) carotinoids in the body and shows similar behaviour in a peroxidative environment to the carotinoids [139].

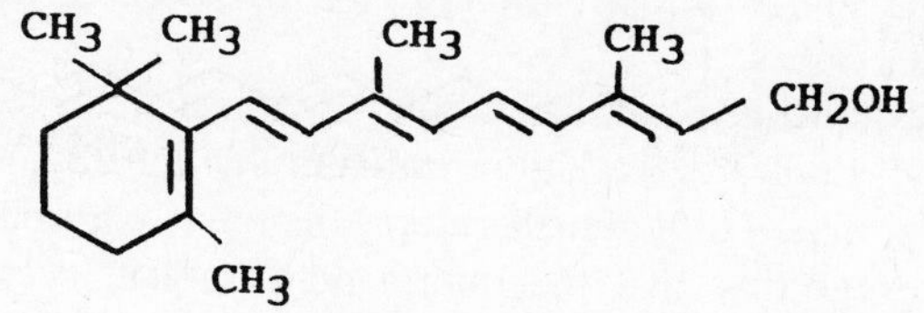

Retinol (Vitamin A)

β-carotene is very rapidly peroxidised in the presence of AMVN with appreciable absorption of oxygen during the retardation period [140-142]. In linoleate micelles, the amount of oxygen absorbed increases with β-carotene concentration [143], but the rate of linoleate peroxidation decreases, suggesting that extensive peroxidation of the β-carotene molecule occurs during retardation and the involvement of β-carotene oxidation products in the overall propagation and retardation process [142]. 5,6-Epoxy-β-carotene is a product of β-carotene peroxidation [140], suggesting that initial attack of peroxyl occurs at the end double bonds in the conjugated system giving a highly resonance stabilised allylic radical. Addition of alkylperoxyl or molecular oxygen to conjugated unsaturation (Scheme 5.5) can in principle be repeated many times until the conjugation is destroyed by oxidation.

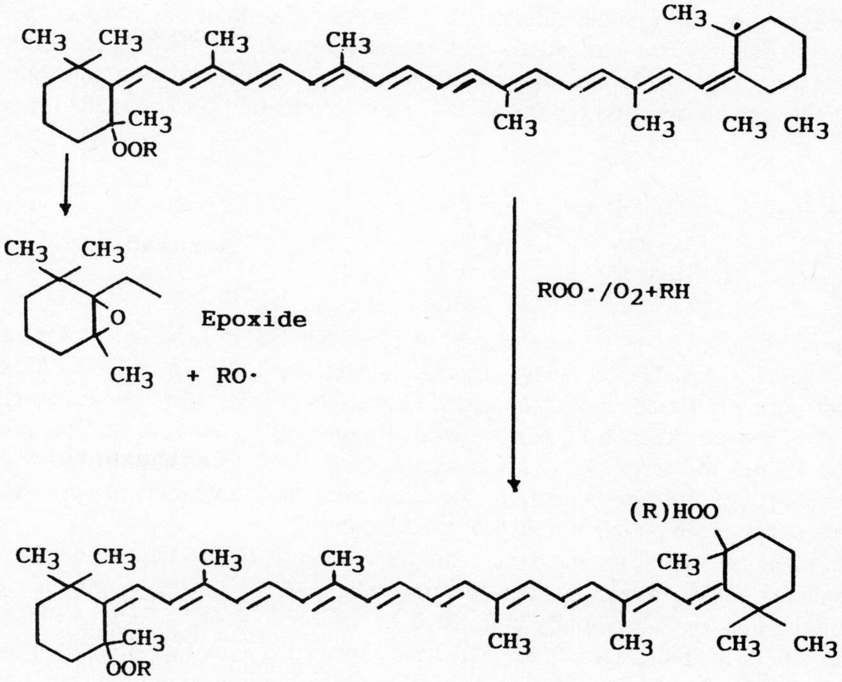

Scheme 5.5 Peroxidation of β-carotene

The peroxides and particularly the hydroperoxides formed in this process are unstable and will break down during the course of retardation to give the observed polar oxidation products, notably alcohols and ketones.

The keto carotenoids, astaxanthin and canthaxanthin have been reported to be more effective inhibitors of methyl linoleate oxidation initiated by an organosoluble azo compound, AMVN, than β-carotene [139]. Furthermore, they are oxidised more slowly under the same conditions, suggesting that the intermediate radical species are more stable. The structure of the initially formed canthaxanthin polyenoxyl (CXPE) provides an explanation for this since this radical is a "stable" polyenyloxyl similar to the aryloxyls formed in the reactions of alkylperoxyls with phenols.

Canthaxanthin polyenoxyl (CXPE)

It seems possible, although it has not yet been established experimentally, that the polyenols and polyeneones, formed in the oxidation of the carotenoids may play a major part in the subsequent activity of this class of antioxidant.

The carotenoids and retinoids are strictly retarders rather than inhibitors of oxidation since they are co-oxidised with polyunsaturated fatty acids. Their main function then is to introduce relatively stable carbon-centred radicals into a peroxidising substrate, thus slowing the reaction with oxygen (reaction 12) and terminating peroxyl radicals (reaction 13):

$$ROO\cdot + RH \rightarrow ROOH + R\cdot \qquad (11)$$
$$R\cdot + O_2 \rightarrow ROO\cdot \qquad (12)$$
$$R\cdot + R'OO\cdot \rightarrow ROOR' \qquad (13)$$

Reactions 12 and 13 are in competition and reaction 13 is favoured at low oxygen concentrations [147]. For the above reasons, ß-carotene, unlike the CB-D antioxidants does not normally introduce a characteristic induction period before peroxidation [139] but it does reduce the rate of AIBN initiated oxidation by 80% at oxygen pressures between 15 and 150 torr [143]. This compares with 40-93 torr found in the veins and arteries [144]. At ambient oxygen pressures however, the carotenoids are prooxidant in linoleate esters due to preferential hydroperoxidation [67].

β-Carotene shows a strong synergistic interaction with CB-D antioxidants, notably α-tocopherol [145], suggesting that the prooxidant effects of the carotinoids due to reaction 11 may be substantially eliminated by hydrogen donors. Retinol (vitamin A) behaves similarly, and Livrea et al [146] have shown that 1 mol of all-*trans* retinol per mol of LDL in combination with α-tocopherol produces an induction time over 250 times longer than that expected for α-tocopherol alone at 37°C. Moreover, under conditions of oxygen saturation, α-tocopherol sacrificially protected retinol during the inhibition time resulting from this combination. However, α-tocopherol was also protected by retinol during peroxidation under the above conditions, suggesting another possible mechanism for the synergistic interaction. It is known that allyl and allylperoxyl radicals can act as reducing agents for phenoxyl and nitroxyl radicals (Chapter 3). This suggests the possibility that α-Toc-O· may compete with oxygen for carbon-centred radicals leading to the regeneration of α-Toc-OH by reaction with the highly labile methylene groups in the conjugated β-carotenyl radical (Scheme 5.6). This could lead to a stationary concentration of α-tocopherol until the β-carotene is ultimately destroyed by oxidation.

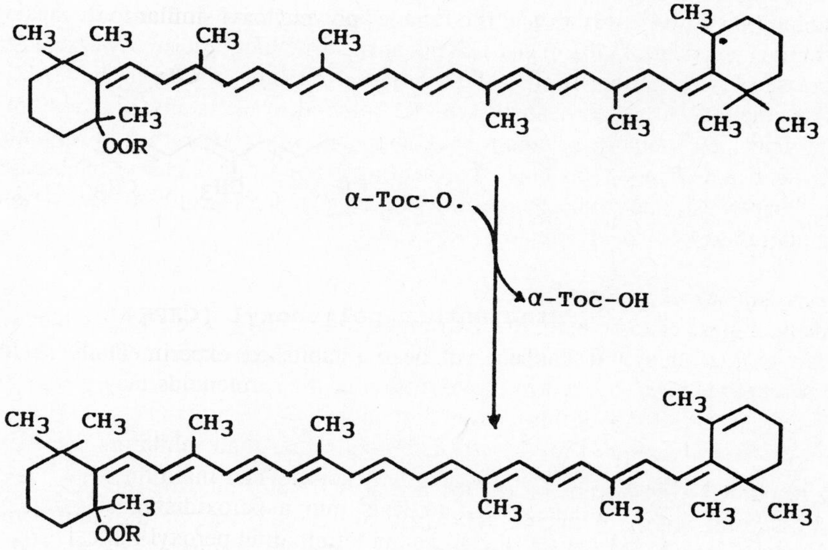

Scheme 5.6 Possible regeneration of α-tocopherol from its radical by β-carotene

Further studies in model compounds are required to elucidate the nature of the synergism between the tocopherols and carotenoids in view of epidemiological evidence that combinations of these antioxidants are much more important than any individual antioxidant nutrient in the maintenance of human health.

5.3 Naturally Occurring Preventive Antioxidants and Synergists

In antioxidant technology, the most important preventive antioxidants are the hydroperoxide decomposers (PDs), the transition metal deactivators (MDs) and the UV absorbers (UVAs) (Chapter 4). These are also protective mechanisms *in vivo*, but other cooperative mechanisms operate in parallel and may be synergistic with them. For example catalase (Cat) and glutathione peroxidase ($GSHP_x$) are the main enzymic defence against hydrogen peroxide and alkyl hydroperoxides (Scheme 5.1). The former which acts primarily against hydrogen peroxide has no analogy in technology whereas $GSHP_x$ which can also efficiently destroy alkyl hydroperoxides shows similarities to peroxidolytic antioxidants in lubricating oils and polymers. Moreover, Cat and $GSHP_x$ are frequently found together in the same organs (e.g. the liver) but in different parts of the cell. Thus hydrogen peroxide produced in the cytosol is disposed of by catalase, whereas hydrogen peroxide produced in the mitochondria is destroyed by GSH [148]. These two enzymes therefore act in tandem to fulfil the same function in a complementary way.

In some cases, perhaps more frequently than with technological antioxidants, biological antioxidants may operate in parallel by different mechanisms. Thus, glutathione is a reducing agent which serves to regenerate ascorbic acid from its oxidation products by a homosynergistic CB-D mechanism, see reaction (6), but it is also an essential component of the preventive antioxidant $GSHP_x$. Some CB-D antioxidants such as albumin, bilirubin and the flavonoids also have the ability to complex with and remove ferrous ions thus preventing the activation of peroxides to free radicals [149].

Unlike the vitamin antioxidants discussed above, the building blocks for the enzyme antioxidants are plentifully available in the food supply and none of them are considered to be "essential" nutrients [150]. Thus $GSHP_x$ is readily "induced" by oxidative stress, giving increased concentrations in the cell than under "resting" conditions [151,152]. An exception to the above generalisation are the mineral constituents of some of the antioxidant enzymes, notably selenium, a component of some forms of glutathione peroxidase, whose deficiency has been shown to give rise to an increase in diseases of oxidation (Chapter 6), and zinc, magnesium and copper are essential to the actions of some mammalian superoxide dismutases (see below).

5.3.1 Superoxide dismutase (SOD)

Superoxide dismutase or to be more precise dimutases are highly specific enzymes whose sole role in biological systems appears to be the rapid conversion of superoxide to hydrogen peroxide [153].

$$2O_2^{\cdot -} + 2H^+ \rightarrow H_2O_2 + O_2 \qquad (14)$$

Although reaction 14 does occur in the absence of SOD, this enzyme increases the rate constant by several orders of magnitude. There are three distinctly different forms of SOD. The first, which is widely distributed in animals, contains copper and zinc (CuZnSOD) and the copper appears to be intimately involved in the dismutation reaction [148]. Scheme 5.7 illustrates in simplified form the cyclical oxidation and reduction of copper in superoxide leading to the catalytic transformation of the latter to hydrogen peroxide.

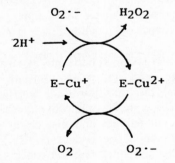

Overall: $2O_2^{\cdot -} + 2H^+ \rightarrow H_2O_2 + O_2$

Scheme 5.7 Catalytic mechanism of CuZnSOD

Two other types of SOD contain manganese and iron respectively in place of Cu/Zn. FeSOD has not been found in mammals but is the main SOD in bacteria, sometimes in combination with Mn. MnSOD also found in bacteria but is also present in animals and vegetables [148].They are not affected by inhibitors for CuZnSOD (e.g. CN^- or diethyl dithiocarbamate salts which are very powerful complexing agents for Cu^{2+}). Although most diets provide an adequate daily amount of zinc, copper and manganese there is an increasing tendency to supplement zinc and manganese but not copper due to its powerful redox interaction with hydroperoxides. Daily intakes of 15 mg of Zn and 5 mg of Mn are generally recommended [128].

As has been seen earlier, hydrogen peroxide, the primary product formed by SOD is potentially very toxic due to its ability to undergo redox reactions with uncomplexed transition metal ions, notably iron and copper. Unless it is efficiently removed it is a potential source of initiating hydroxyl radicals by the Fenton reaction (Chapter 2).

5.3.2 *Catalase*

Catalase (Cat) and glutathione peroxidase (GSH-P_x) are both known to destroy hydrogen peroxide and hydroperoxides at approximately the same rate ($\cong 10^7 M^{-1} s^{-1}$) without the external liberation of hydroxyl radicals. Catalase contains four haeme groups (see below) buried in the hydrophobic interior of protein with part of the haeme periphery accessible to peroxides and these reach the metal centre through narrow channels permeable to water [154,155].

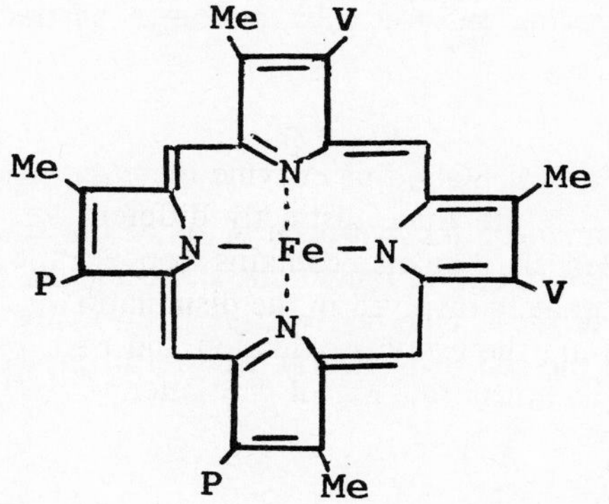

```
Me = methyl
V  = vinyl
P  = γ-propionyl
```

Ferriphotoporphorin

The iron in haeme therefore has limited accessibility to large molecules and its main function then is to deactivate hydrogen peroxide itself, and this can be represented in simplified form by the cyclical mechanism shown in Scheme 5.8.

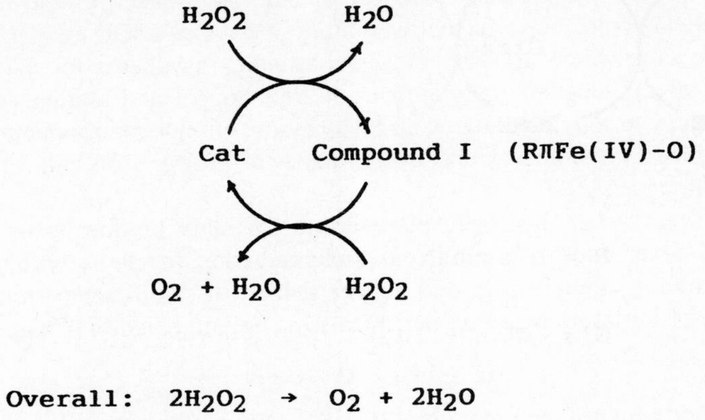

$$H_2O_2 \quad H_2O$$

$$Cat \quad Compound\ I \quad (R\pi Fe(IV)\text{-}O)$$

$$O_2 + H_2O \quad H_2O_2$$

Overall: $2H_2O_2 \rightarrow O_2 + 2H_2O$

Scheme 5.8 Simplified mechanism of hydrogen peroxide decomposition by catalase

The oxidised catalase structure Cat-I has a nominal valency Fe(V), but the ferriprotoporphorin structure is strongly electron delocalising and Cat-I is often depicted as Rπ-Fe(IV)-O. However, catalase is not only a catalyst for peroxide decomposition; it will also, like peroxidase, oxidise some reducing substrates by hydrogen abstraction [155]:

$$\text{Cat-I} + \text{RH} \rightarrow \text{Cat-II} + \text{R}\cdot \qquad (15)$$

In principle catalase is in competition with ionic iron for hydrogen peroxide but since it catalyses hydrogen peroxide decomposition five orders of magnitude more rapidly than does Fe^{2+} [156] and since the concentration of the haeme peroxidases is normally very much higher than that of free ferrous ions in the cell, it is clearly a powerful protectant against the *in vivo* Fenton reaction.

5.3.3 *Glutathione peroxidase*
The chemistry of glutathione peroxidase action can be formally represented as in Scheme 5.9 which summarises the reduction of hydrogen peroxide to water at the expense of sugar phosphates. This is an oversimplification in a number of respects [157].

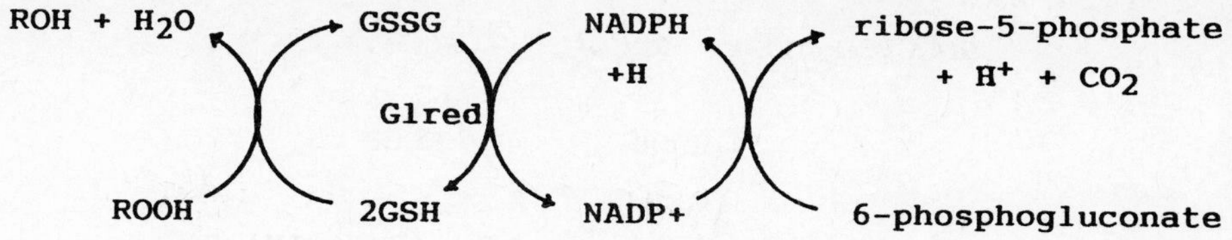

Scheme 5.9 Peroxidolytic mechanism of glutathione

ROH + H$_2$O → GSSG → NADPH+H → ribose-5-phosphate + H$^+$ + CO$_2$

ROOH → 2GSH (Glred) → NADP+ → 6-phosphogluconate

R = H or alkyl (normally lipid)

G = $^-$OCOCH(CH$_2$CH$_2$CNHCHCNHCH$_2$COOH)
    with $^+$NH$_3$, O, and CH$_2$- side groups

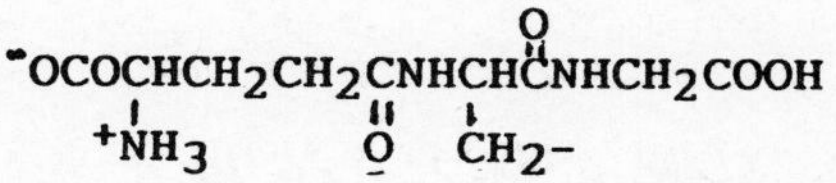

Glred = glutathione reductase

Scheme 5.9 Peroxidolytic mechanism of glutathione

a) In the absence of the enzyme, glutathione, like all thiols, does react with hydroperoxides but very much more slowly [158]. Furthermore the product formed in the presence of the enzyme appeared to be exclusively disulphide, whereas in its absence a wide variety of further oxidation products (e.g. thiolsulphinates, thiolsulphonates, sulphur acids, (RS(O)$_x$H), SO$_3$ and H$_2$SO$_4$) are formed (Scheme 5.10). These are the basis of the peroxidolytic activity of thiols, monosulphides and disulphides in technological substrates (Chapter 4). However, the sulphenic, sulphinic and even sulphonic acids also have CB-D activity and a catalytic cycle involving R(SO)$_x$. has also been proposed to account for the mechanoantioxidant behaviour of sulphur compounds in rubbers (Chapter 4, reaction (2)).

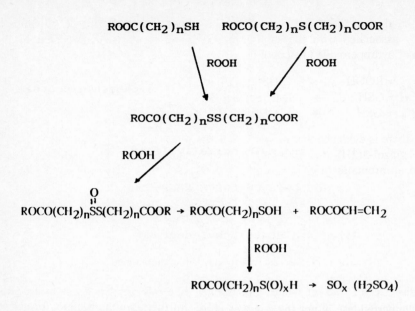

Scheme 5.10 Abiotic oxidation of thiols, monosulphides and derived disulphides

b) Other thiols can replace glutathione in the enzyme [157], but none is as effective as GSH (see Table 5.6). Surprisingly, methylmercaptoacetate (MMA), which is closely related to the abiotic sulphur antioxidants described in Scheme 5.10 is more effective than cisteinyl esters. So far no facile abiotic reduction of the disulphide back to thiol has been reported in the absence of the enzyme. In practice, up to one third of the disulphide may be derived from other thiol-containing amino acids which nevertheless undergo reduction back to the respective thiols in the presence of glutathione.

Table 5.6 Rates of reaction of some thiols with hydrogen peroxide relative to GSH (100) in the presence of **bovine GSH peroxidase** [157]

Thiol	% of GSH turnover
Mercaptoacetic acid	2.1
3-Mercaptopropionic acid	1.1
Methylmercaptoacetate (MMA)	28.0
Methyl-3-mercaptopropionate	2.5
L-Cyteine methyl ester	3.6
L-Cysteinylglycine	6.8
γ-L-Glutamyl-L-cyteine methyl ester	26.0

c) Although non-selenium $GSHP_x$ is widely distributed in the organs of animals and is even the major peroxidolytic enzyme in some (e.g. rat testis), in other organs (e.g. rat lung, spleen and heart) selenium is an essential co-agent. It is believed that selenium acts as a catalyst in the above cycle. Flohé [157] has proposed that the selenium is chemically associated with the sulphur of GSH and is directly implicated in the reduction of hydroperoxide to alcohol.

$$E\text{-}GSSeH + ROOH \rightarrow E\text{-}GSSeOH + ROH \tag{16}$$
$$E\text{-}GSSeOH + GSH \rightarrow E\text{-}GSSeSG + H_2O \tag{17}$$
$$E\text{-}GSSeSG + GSH \rightarrow E\text{-}GSseH + GSSG \tag{18}$$

However, there is evidence that *in vivo* some selenium compounds (e.g. 2-phenyl-1,2-benzisoselenazol-3(2H)-one, Ebselen) in which the selenium is firmly covalently bonded to an aromatic ring, in the presence of GSH, mimic the effects of $GSeHP_x$ [148].

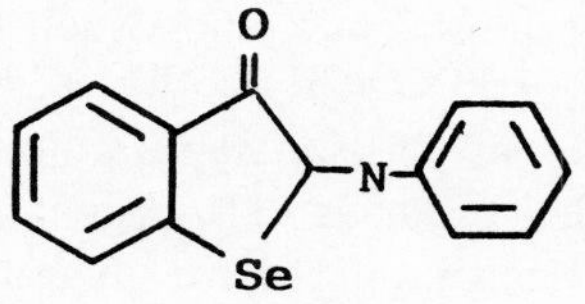

Ebselen

Glutathione peroxidase is not the only selenoprotein in higher animals [159] and it seems likely that selenium, like sulphur has a broader antioxidant function than that outlined above. The effect of selenium compounds in biological systems is reminiscent of the catalytic abiotic action of selenium compounds (e.g. zinc diselenocarbamates) as peroxidolytic antioxidants [160] and may again involve the formation of selenic acids analogous to the sulphenic acids and their action may not be limited to GSH as a thiol co-agent [161].

d) Although $GSHP_x$, like catalase will decompose hydrogen peroxide with equal facility, $GSHP_x$ is approximately ten time more effective in removing hydroperoxides than catalase and this is believed to occur predominantly in the surface of the cell membrane [157].

5.3.4 α-Lipoic acid

α-Lipoic (α-LA) is an endogenous a co-factor for the oxidative decarboxylation of α-keto acids [162] with vitamin C-like activity [163]. The antioxidant behaviour of α-lipoic acid also shows certain similarities to that of glutathione in that it is readily reduced to its dithiol form, dihydrolipoic acid (DHLA) under physiological conditions [164] and it is an effective CB-D antioxidant in this form (Scheme 5.11) [165,166]. Administration of α-LA to vitamin E deficient mice has been known for many years to prevent symptoms of vitamin E deficiency [163] even though α-LA is not a vitamin and is biosynthesised from unsaturated fatty acids in the body [167].

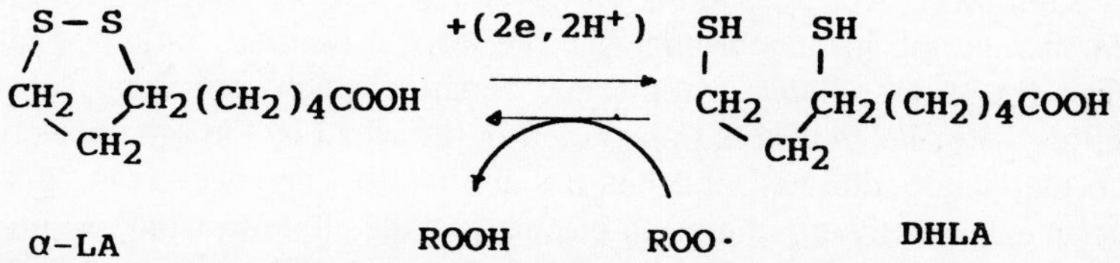

Scheme 5.11 Redox antioxidant activity of α-lipoic acid (α-LA) and its reduced form, dihydrolipoic acid (DHLA)

DHLA is a more powerful reducing agent than GSH (redox potential DHLA/α-LA = -0.32V compared with GSH/GSSG = -0,24V) and readily reduces GSSG to GSH [168] and superoxide to hydrogen peroxide [166,169]. α-LA by contrast does not appear to reduce either peroxyl radicals [165] or superoxide [169].

α-LA and DHLA have both been reported to react with other ROS (for recent review see Packer et al. [164]), for example hydroxyl radicals, HOCl, $^1O_2$, etc. However, the rates of these reactions is unremarkable compared with other antioxidant. For example, singlet oxygen reacts with α-LA at about the same rate as α-tocopherol but two orders of magnitude more slowly than β-carotene [170]. DHLA also chelates transition metal ions and, like the thiol compounds discussed in Chapter 4, it appears to be an effective copper deactivator. In this context there is an obvious structural and mechanistic analogy between DHLA and 2,3-dimercaptosuccinic acid which is a therapeutic deactivator for lead (Section 5.5.2)

Another autosynergistic role of DHLA appears to be the regeneration of other antioxidants from their oxidation products. The regeneration of GSH from GSSG was referred to above and Bast and Haenen [171] have suggested that this is important *in vivo* since a combination of DHLA and GSSG but not DHLA alone prevented Fe^{2+}/Asc(OH)$_2$ induced lipid peroxidation. There is also evidence that α-LA can regenerate Ubi-hq from Ubi-q in mice under oxidative stress [172] and DHLA regenerates ascorbate in oxidatively stressed LDL. It seems then that α-LA by reduction by NADPH can replace GSSG in the regeneration of α-Toc-OH discussed in Section 5.2.2 (reaction (6)) [173].

Although sulphur acids have not been reported as oxidation products of α-LA, it seems highly likely in the light of the chemistry discussed in Chapter 4 (Section 4.4.1) that these are formed (Chapter 4) and probably contribute to the activity of this antioxidant. There is little doubt that both α-LA and DHLA are multifunctional autosynergists with CB-D, PD and MD activity. In addition DHLA can homosynergise with electron donors.

5.3.5 *Metal chelating agents*
It was noted in earlier chapters that ionic iron is a powerful catalyst for radical formation, particularly in the presence of hydrogen peroxide and alkyl hydroperoxides. Consequently, the deactivation or removal of "Fenton reactive" iron from extra-cellular fluids is a major mechanism of antioxidant action. About two thirds of the 3.5-4.5 g of iron in the body is present as haemoglobin, and a smaller amount as myoglobin [174] in which the iron is complexed in such a way that it does not normally participate in the Fenton reaction. Iron is also an important component of catalase (Section 5.3.2) and is present in many oxidase enzymes, notably the peroxidases which in association with peroxides play a major part in controlled radical formation as part of the body's defence mechanisms. Much of the remaining iron is stored in the protein ferritin which contains up to 4,500 atoms of iron as Fe^{3+} [175] and is again inert toward hydrogen peroxide up to the "saturation" concentration. However, in diseases due to iron overload, notably thalassaemia and haemochromatosis, the ferritin becomes saturated and the excess iron appears as water-soluble cytoplasmic ferritin which can catalyse radical generation reactions. Non-haem iron is transported in the blood by the glycoprotein transferrin. This is normally not saturated and has the ability to "scavenge" iron salts from the serum. Other metal deactivators found in plasma include albumin which is a weak binder of iron but a strong deactivator of copper [176]. However, albumin is destroyed sacrificially by hydroxyl radicals and is also frequently associated with the chain-breaking antioxidant,

bilirubin which it transports in the blood and bilirubin in addition to being a CB-D antioxidant (see Section 5.2.6) has much of the iron-binding capacity of haemin from which it is derived. The CB-D antioxidant, uric acid (Section 5.2.4), which is found in appreciable concentrations in extra-cellular fluids is also autosynergistic due to its ability to bind both iron and copper [176], and caeruloplasmin which is structurally related to SOD both binds and oxidises Fe^{2+} while at the same time scavenging $O_2\text{-}$.

5.3.6 *Photoantioxidants*
Since UV light is absorbed by the skin and apart from the eyes, light has little direct effect on biological cells beneath the surface of the skin, it is the epidermis and the lens of the eye that primarily require protection from the effects of UV light. By far the most important UV screening chemical in the body is the intensely coloured polymer, eumelanin which is also used as an artist's pigment (Natural Brown 9). Since the lens of the eye has to remain essentially colourless, the protection of the eye against photooxidation cannot be achieved simply by screening UV and is performed biologically by a combination of antioxidants which includes melanin precursors.

Melanin is an oxidation product of tyrosine. The chemistry of its formation via intermediate phenoxyl radicals has analogies to the oxidation of the antioxidant alkyl phenols discussed in Chapter 3. Tyrosine itself is a relatively weak antioxidant since its rate of reaction with alkylperoxyl radicals is slow compared with the alkylated and hydroxylated phenols. However, it is readily enzymically oxidised to the neurotransmitter 3,4-dihydroxyphenylalanine, DOPA. This compound and some of its further oxidation products are much more powerful sacrificial antioxidants than tyrosine due to delocalisation of the phenoxyl unpaired electrons in the highly conjugated oligomeric products (see Scheme 5.12).

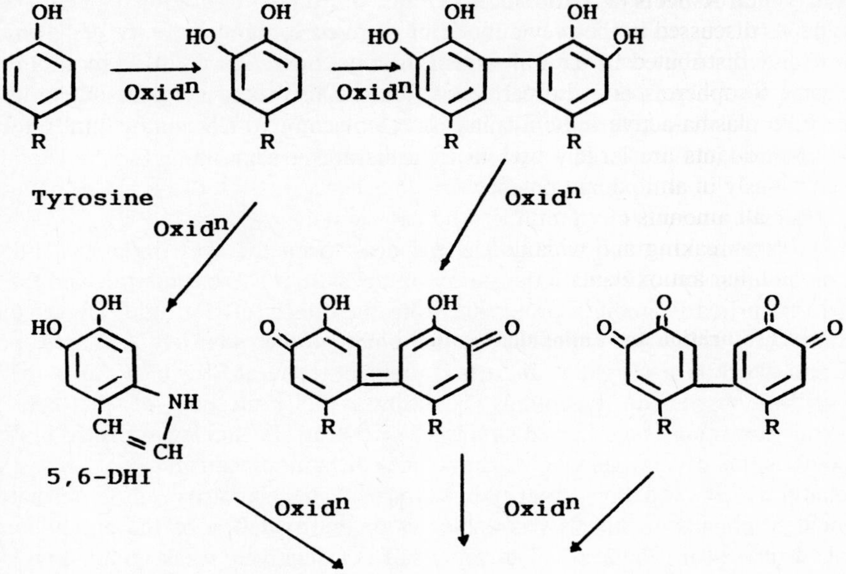

MELANIN

OLIGOMERIC PHENOLS, INDOLES,
QUINONES, QUINONIMIDES, SEMIQUINONES

L-DOPA, R = $CH_2CH(NH_3^+)COO^-$

Dopamine, R = $CH_2CH_2NH_2$

Scheme 5.12 Oxidation of tyrosine and dopamine to melanin

Melanin is both an oxidising agent and a reducing agent [177] and has been shown to contain relatively stable free radicals due to their delocalisation in the conjugated system. In this respect it shows basic similarities to carbon black which is the most effective light stabiliser for polymers (Chapter 4) and is again known to have photoantioxidant and sun-screening properties similar to melanin. In a study of the redox properties of melanin by ESR, Sarna et al. found [177,178] that it could both reduce stable nitroxyls and oxidise the cognate hydroxylamines. UV and visible light accelerated these redox processes and also modified the redox equilibrium. Thus although eumelanin produces superoxide and hydrogen peroxide on irradiation [65a3], it also scavenges it [177,179] and these two opposed effects are probably associated with different redox centres in the macro-molecule. Overall, however, melanin is an effective chain-breaking antioxidant, and the very powerful UV protective effect found in the skins of dark coloured people is due to autosynergism between its UV screening effect and chain-breaking antioxidant activity. This is again analogous to the autosynergism observed in hydroxybenzophenone UV absorbers (Chapter 4) and the latter are used as sun-screens in cosmetic preparations for light skinned people who wish to avoid the risk of UV skin damage (Chapter 6).

5.4 Nutritional Aspects of Antioxidants

Antioxidants discussed in the previous Sections are either synthesised in the human body or are widely distributed in natural foodstuffs. In general, the lipid-active antioxidants, notably the tocopherols and the carotenoids, are taken into the body in fats and oils, whereas the plasma-active antioxidants, ascorbic acid, the flavonoids and other water soluble antioxidants are largely present in fruits and vegetables. However oils and fats vary enormously in antioxidant potency (Table 5.7). In general, the saturated fats contain relatively small amounts of vitamin E, whereas the polyunsaturated fats are a rich source of these chain-breaking antioxidants [181] and as was seen above (Section 5.2.9), some also contain other antioxidants.

Table 5.7 Unsaturation and antioxidant contents of natural oils and fats [180]

Oil/fat	S g/100g	MU g/100g	PU g/100g	VE g/100g	VA g/100g	Ratio VE/PU
Butter	54.0	19.8	2.6	2.0	750	0.77
Coconut oil	85.2	6.6	1.7	0.7	0	0.41
Cod liver oil	N	N	N	20.0	18000	-
Corn oil	12.7	24.7	57.8	17.24	0	0.30
Cottonseed oil	25.6	21.3	48.1	42.77	0	0.89
Olive oil	14.0	69.7	11.2	5.1	0	0.46
Palm oil	45.3	41.6	8.3	33.12	0	4.00
Peanut oil	18.8	47.8	28.5	15.16	0	0.53
Rapeseed oil (HEA)	5.3	64.3	24.8	22.1	0	0.89
Rapeseed oil (LEA)	6.6	57.2	31.5	22.1	0	0.70
Safflower oil	10.2	12.7	72.1	40.68	0	0.56
Sesame oil	14.2	37.3	43.9	N	0	-
Soya oil	14.5	23.2	56.5	16.29	0	0.29
Sunflowerseed oil	11.9	20.2	63.0	49.22	0	0.78
Wheatgerm oil	18.8	15.9	60.7	136.65	0	2.25

S = saturated, MU = monounsaturated, PU = polyunsaturated, N = not measured,
VE = vitamin E, VA = vitamin A

It was seen in Chapter 1 that methylene groups activated by two double bonds, as for example in linoleic, linolenic and arachidonic acids, are ten times more reactive toward peroxidation than the isolated double bond found in oleic acid. Consequently, arachidonic acid oxidises forty times more rapidly than oleic and the 1,4-polyunsaturates act as "initiators" for the peroxidation of the more saturated fats. Fig. 2.1 (Chapter 2) shows that there is a general relationship between the polyunsaturated oil content of fats and the concentrations of vitamin E. To a first approximation then, the rate of oxidation of a mixture of saturated (S), monounsaturated (MU) and polyunsaturated (PU) fats depends on the VE/PU ratio in the mixture. Evolution has ensured that the most oxidatively unstable oils, notably safflower and sunflower oils, are also more effectively protected by antioxidants. Homo sapiens evolved as a fruit and vegetable eater and presumably this diet was substantially protective against the relatively unpolluted environment in which primitive man lived. In modern life, a combination of environmental oxidative stresses have increased the demand on man's biological antioxidant defences. A parallel development is the emergence of the large scale food processing industry with its requirement for cheapness and food "purity" and with it the inevitable consequence that many antioxidant nutrients were removed. The simple operation of "steam refining" at 280°C leads to almost 50% depletion of vitamin E in 30 minutes (see Fig. 5.3) and treatment at 300°C for two hours essentially removes all the antioxidants [181]. The tocopherols are highly effective in combating atherosclerosis by inhibiting the peroxidation of the polyunsaturated esters of cholesterol and in retrospect, it must be judged that the emphasis on polyunsaturated fats without maintaining their Vitamin E protective complement was an unfortunate diversion from a fully nutritious diet. As will be seen in Chapter 6, this is supported by epidemiological evidence.

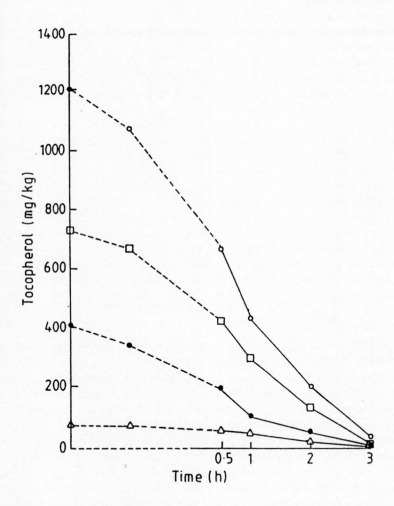

Fig. 5.3 Reduction in tocopherol concentrations in soyabean oil with time of steam refining at 280°C. -o-, total tocopherol, -□-, γ-tocopherol, -●-, δ-tocopherol, -△-, α-tocopherol, ---- indicates the normal range of refining times. (Reproduced with permission from Leatherhead Food Research Association Report No. 563, courtesy R. Swift).

Some food manufacturers rely on the small amount of vitamin E that survives the refining process. However, the antioxidant content of edible oils depends not only on the processing operation [181] but also, in the case of spreads based on the sunflower oil, on where the oil plant is grown and on the time of year when it was harvested [182]. Most "block" and "tub" margarines which are widely sold in shops and supermarkets as non-proprietary products, contain no supplemental vitamin E although they may be highly polyunsaturated. Thus non-branded Sunflower Tub Margarine contains only 10 mg/100 g of vitamin E [183] which is probably just about sufficient to protect them before use. Other oils used in spread manufacture, notably, palm, rapeseed and soya oils also contain varying amounts of vitamin E. Rapeseed and soya oils are highly polyunsaturated and are chosen because of their "liquid" characteristic at refrigerator temperatures. Even before processing these oils have a net negative antioxidant protection (that is the level required to protect the polyunsaturation in the oil from peroxidation, Chapter 6, Section 6.3.1). Some branded products based on these oils are not at present supplemented with vitamin E or vitamin A. Palm oil contains an exceptionally high VE/PU ratio (Chapter 2, Fig. 2.2) and as it is a relatively non-oxidisable oil, it contains a large excess antioxidant potential.

There has in the past ten years been an overemphasis on the importance of polyunsaturates in nutrition. However, the impact of recent research is now beginning to be reflected in formulation of fats in the food industry, and the emphasis is changing from the role of the polyunsaturates to the need to supplement with vitamin E at least to the level which they contain in the natural state. Table 5.8 shows a selection of brand margarines and spreads some of which are supplemented with vitamins E as well as vitamin A. In some cases, β-carotene is added, primarily as a colourant [183]. It is interesting to note that there is now a good relationship between polyunsaturated content and vitamin E supplementation in these basic foodstuffs. Some, but by no means all, branded margarines and spreads are thus making a significant contribution to human antioxidant defences. Equally important is the increasing use of the more saturated oils, such as virgin olive oil, which in nature contains an excess of α-tocopherol over that required for its own protection and which in addition contains other phenolic antioxidants. Olive oil has been shown to have a beneficial effect in the diet by displacing the polyunsaturates in membranes, where it could provide substantially increased protection against LDL peroxidation. This will be discussed further in Chapter 6.

Table 5.8 Unsaturation and antioxidant contents of fats and spreads at January 1st 1996

	TF	S	MU	PU	VE	VA	VE/PU
		g/100g			mg/100g	µg/100g	mg/g
Margarines							
Blue Band[1]	80.0	16.0	40.0	19.0	20*	800*	1.05
Stork[1], Echo[1]	80.0	34.0	4.0	5.0	10*	800*	2.00
Fat Spreads							
Mono Rapeseed oil[2]	75.0	11.5	35.0	14.7	15.0	830	1.02
Flora Sunflower[2]	70.0	15.0	19.5	35.0	38.0	800*	1.09
Utterly Butterly[2]	69.0	10.6	32.2	13.5	15.0	830	1.11
Krona[1]	70.0	28.0	31.0	7.0	10*	800*	1.42
I can't believe it's not butter[1]	70.0	16.5	19.5	33.5	25*	800*	0.75
Vitalite[3]	70.0	16.0	19.0	35.0	25*	-	0.71
Reduced Fat Spreads							
Summer County, Stork light blend[1]	60.0	11.0	31.0	14.0	10*	800*	0.83
Olivio[1]	60.0	14.0	34.5	11.0	10*	800*	0.91
Pact[4]	60.0	15.0	30.0	15.0+	33.6	-	2.24
Low Fat Spreads							
Gold[2]	39.0	10.6	15.5	6.4	7.5	830	1.17
Gold Sunflower[2]	39.0	9.2	8.3	17.0	25.0	830	1.47
Flora extra light[1]	39.0	9.0	9.5	20.0	38.0	800*	1.90
I can't believe it's not butter light[1]	40.0	9.5	16.0	14.0	12*	800*	0.63
Delight[1]	39.0	14.5	18.0	6.0	10*	800*	1.67

TF = total fats, S = saturated fats, MU = monounsaturated fats, >95% cis,
PU = polyunsaturated fats
1. Courtesy Van den Bergh Foods Ltd   2. Courtesy St. Ivel Ltd
3. Courtesy Kraft Jacobs Suchard Ltd   3. Courtesy MD Foods, Plc

Nuts are also a rich source of unsaturated oils but unlike the processed vegetable oils they contain several other antioxidants in addition to vitamin E (see Table 5.9). Selenium is present in relatively high concentrations in brazil, cashew pecan and walnuts. β-Carotene is present in pecan, pine and pistachio [181]. Dietary fibre (DF) is present in all nuts where it has been measured. Dietary or soluble fibre consists largely of non-starch polysaccharides (NSP) and although, unlike starch, this is not digested as a primary source of energy and passes through the body essentially unchanged, it reduces the risk of coronary heart disease [185] and cholesterol levels [186].

Table 5.9 Fatty acid, dietary fibre and antioxidant contents of nuts [180]

Nut	S g/100g	MU g/100g	PU g/100g	VE mg/100g	Se µg/100g	DF g/100g	β-C µg/100g
Almond	4.7	34.4	14.2	23.98	4.0	(7.4)	0
Brazil	16.4	25.8	23.0	7.18	230-5300	4.3	0
Cashew	10.1	29.4	9.1	1.3	34.0	3.2	0
Chestnut	0.5	1.0	1.1	1.2	Tr	4.1	0
Coconut	59.3	3.9	1.6	1.4	(2.0)	N	0
Hazelnut	4.7	50.0	5.9	24.98	Tr	6.5	0
Macadamia	11.2	60.8	1.6	1.49	7.0	5.3	0
Peanuts	8.2	21.1	14.3	10.09	3.0	6.2	0
Pecan	5.7	42.5	18.7	4.34	12.0	4.7	50
Pine	4.6	19.9	41.1	13.65	N	1.9	10
Pistachio*	4.1	15.2	9.8	2.28	(3.0)	3.3	71
Walnuts	5.6	12.4	47.5	3.83	19.0	1.5	0

S = saturated fats, MU = monounsaturated fats, PU = polyunsaturated fats
Se = selenium, DF = dietary fibre
* in shells

The mechanism of the antioxidant activity of DF has not yet been demonstrated since polysaccharides have no recognised chain-breaking or peroxidolytic antioxidant activity. It seems likely, however, that they chelate damaging transition metal ions since typically dextran, a glucose homopolysaccharide, is used to introduce iron in the case of iron deficiency and it seems likely that it will similarly assist in the elimination of iron when in excess.

Vitamin C is highly concentrated in many fruits, notably blackcurrants, strawberries, oranges, lemons, guava, paw-paw and kiwi fruit (Table 5.10). These fruits do not normally contain nutritionally significant concentrations of vitamin E, but some do, notably avocados and olives which unlike most fruits contain a substantial amount of unsaturated fatty acids [181]. Olives is one of the few fruits that contain no vitamin C but, as noted previously, they contain high levels of oil soluble phenolics (Section 5.2.11).

Table 5.10 Vitamin C (VC), Vitamin E (VE), Selenium (Se), β-carotene (β-C) and Dietary fibre (DF) content of fruits [180]

Fruit	VC mg/100g	VE mg/100g	SE μg/100g	β-C μg/100g	DF[e] g/100g
Apples, cooking	14	0.27	Tr	(17)	1.6
eating	10	0.59	Tr	18	1.8
Apricots	6	N	(1.0)	200-3370	1.2
Avocado	6	3.2	Tr	16	3.4
Bananas	11	0.27	(1.0)	21	1.1
Blackberries	15	2.37	Tr	80	3.1
Blackcurrants	150-320	1.0	N	100	3.6
Cherries	11	0.13	(1.0)	25	0.9
Clementines	54	N	N	75	1.3
Damsons	5	0.6	Tr	265	1.6
Dates	12	N	(3.0)	15	1.5
Figs	1	N	Tr	(64)	7.5
Gooseberries, cooking	14	0	Tr	110	2.4
dessert	27	0	Tr	(18)	1.7
Grapefruit	36	0.19	(1.0	17	1.3
Grapes	3	(0.19)	(1.0)	17	0.7
Guava	9-410	N	N	435	3.7
Kiwi fruit	59	N	N	37	1.9
Lemons	58	N	(1.0)	18	N
Lychees	45	N	N	0	0.7
Mangoes	37	1.05	N	1800	2.6
Melon	9-17	0.1	Tr	230-1000	1.0
Nectarines	37	N	(1.0)	58	1.1
Olives	0	2.0	N	180	2.9
Oranges	44-79	0.24	(1.0)	28-155	1.7
Passion fruit	23	N	N	750	3.3
Paw-paw	60	N	N	810	2.2
Peaches	31	N	(1.0)	58	1.5
Pears	6	0.5	Tr	18	2.2
Pineapple	12	0.1	Tr	18	1.2
Plums	4	0.61	Tr	295	1.6
Prunes	Tr	N	≅3.0	140	2.4
Raspberries	32	0.48	N	6	2.5
Rhubarb	6	0.2	Tr	60	1.4
Strawberries	77	0.2	Tr	8	1.1

N = no reliable information,   ( ) = estimated,   Tr = trace,   e = Englyst method

Many vegetables also contain substantial concentrations of vitamin C (Table 5.11) and in particular the deep green vegetables such as spring cabbage, Brussels sprouts and water cress. Both green and red peppers are also an important source of vitamin C and it is noteworthy that high ascorbate fruits and vegetable also frequently contain relatively high

Table 5.11 Vitamin C (VC), Vitamin E (VE), Selenium (Se), β-carotene (β-C) and Dietary fibre (DF) in vegetables [180]

Vegetable	VC mg/100g	VE mg/100g	SE µg/100g	β-C µg/100g	DF[e] g/100g
Asparagus	12	1.16	(1.0)	315	1.7
Aubergine	4	0.03	(1.0)	70	2.0
Beans, broad	8	0.61	N	225	6.5
green	12	0.2	N	330	2.2
soya	Tr	2.9	14.0	12	15.7
Beetroot	5	Tr	Tr	20	1.9
Broccoli	87	(1.3)	Tr	575	2.6
Brussels sprouts	115	1.0	N	215	4.1
Cabbage, average	50	0.2	(1.0)	385	2.4
spring	180	N	N	2630	3.4
Carrots, old	6	0.56	1.0	8115	2.4
Cauliflower	43	0.22	Tr	50	1.8
Celery	8	0.2	(3.0)	50	1.1
Cucumber	2	0.07	Tr	60-260	0.6
Garlic	17	0.01	2.0	Tr	4.1
Leeks	7	0.92	(1.0)	735	2.2
Lentils	Tr	N	105.0	N	8.9
Lettuce	5	0.57	(1.0)	50-910*	1.2
Marrow	11	0	N	110	0.5
Mushrooms	1	0.12	9.0	0	1.1
Mustard and cress	33	0.70	N	1280	1.1
Okra	16	N	(1.0)	465	4.0
Onions, bulb	5	0.31	(1.0)	10	1.4
spring	26	N	N	620	1.5
Parsnip	17	1.0	2.0	30	4.6
Peas	24	0.21	(1.0)	300	4.7
Peppers, green	120	N	N	175	N
red	140	0.8	Tr	3840	1.6
Plantain	15	0.2	2.0	360	1.3
Pumpkin	14	1.06	N	450	1.0
Potatoes, new	16	0.06	1.0	Tr	1.0
main crop	11	0.06	1.0	Tr	1.3
fried in corn oil	9	4.90	2.0	Tr	2.2
fried in dripping	19	0.06	(2.0)	N	2.2
Radish	17	0	(2.0)	Tr	0.9
Spinach	26	1.71	(1.0)	3535	2.1
Swede	31	N	(1.0)	350	1.9
Sweet potato	23	4.56	(1.0)	1820-16,000	2.4
Sweetcorn	14	N	N	140	1.5
Tomato	17	1.22	Tr	640	1.0
Turnip	12	Tr	(1.0)	20	2.4
Watercress	62	Tr	N	2520	1.5
Yam	4	N	N	Tr	1.3

N = no reliable information, ( ) = estimated, Tr = trace, e = Englyst method
* = most carotene found in outer green leaves of lettuce.

levels of β-carotene (e.g. oranges, black currants, paw-paw, spring cabbage, water cress and red peppers and green but not white lettuce). However, the presence of β-carotene can often be discerned by the intense yellow-red colour (e.g. carrots, apricots, melons, mangoes, oranges, etc.).

Most fruits and vegetables are also an important source of dietary fibre. Noteworthy among the vegetables (Table 5.11) are the pulses (beans and lentils) and one of the more important food additives, pectin, an important NSP derived from gallactose is present in many fruits and vegetables and is commercially extracted from apples.

Meats are rich in fats but relatively deficient in polyunsaturated fats and for this reason contain relatively small amounts of vitamin E and of course no vitamin C. They do however contain some selenium (Table 5.12). Seafoods are relatively deficient in fats but they are relatively more unsaturated then those in meat (Table 5.13). The nutritional investigations into fish have concentrated on the highly unsaturated components of the oils. Oily fish are a rich source of eicosapentaenoic acid (EPA, C20:5 n=3) and docosahexaenoic acid (DHA, C22:6 n=3) which have recently been claimed to have

Table 5.12 Vitamin E (VE) and selenium (Se) content of meats [180]

Meat	S g/100g	MU g/100g	PU g/100g	VE mg/100g	Se μg/100g
Bacon	2.7	3.1	0.8	0.05-0.14	(1.0-4.0)
Beef (cooked)	26.9	30.4	2.6	0.55	2.0
Chicken	5.9	7.5	3.3	0.06-0.15	6.0-7.0
Duck	11.6	23.1	5.1	0.02	N
Lamb (cooked)	31.5	24.5	3.0	0.05-0.3	Tr
Pork (cooked)	23.0	25.1	9.3	0.12	N
Rabbit (stewed)	3.2	1.5	2.5	N	17.0

Table 5.13 Fatty acid, vitamin E (VE) and selenium (Se) content of seafoods [180]

Fish	S g/100g	MU g/100g	PU g/100g	VE mg/100g	Se μg/100g
Cod	0.1	0.1	0.3	0.44	28
Haddock	0.1	0.1	0.2	N	22
Herring	5.3	8.5	3.1	0.21	34
Lemon sole	0.2	0.3	0.5	N	(44)
Mackerel	3.3	8.0	3.3	N	30
Plaice	0.2	0.6	0.5	N	36
Salmon	2.2	5.1	3.4	N	20
Trout	(0.7)	(1.1)	(1.0)	N	16
Crustacea					
Crab	0.7	1.4	1.5	N	17
Lobster	N	N	N	1.5	N
Prawns	0.4	0.5	0.4	N	18
Shrimps	0.1	0.2	0.3	N	49
Molluscs					
Cockles	0.1	Tr	0.1	N	45
Mussels	0.4	0.3	0.7	N	45

anticancer activity (Chapter 6). Information is at present unavailable as to how effectively they are protected by endogenous antioxidants at source. However, most fish do contain substantial amounts of selenium (Table 5.13) suggesting that peroxidolytic antioxidants may be a major peroxidation inhibitors in fish oils.

Cereals which again contain relatively small amounts of oils which are substantially polyunsaturated but are relatively well protected by a combination of vitamin E and selenium. In human nutrition, wheat germ is a particularly rich source of vitamin E and this combined with the relatively high levels of DF make wholemeal wheat bread and rye bread particularly nutritious components of the popular diet.

As was mentioned earlier, herbs and spices are generally not taken in high enough quantity in the average diet to make a very significant nutritional contribution. Nevertheless, several spices contain relatively high concentrations of β-carotene (Table 5.14) and this in combination with the phenolic antioxidants discussed in Section 5.2.12 makes an additional synergistic contribution to the protection of foods.

Table 5.14 Antioxidant contents of herbs and spices [180]

Herb/spice	VC mg/100g	VE mg/100g	β-C μg/100g	Se μg/100g
Chilli powder	0	N	21000	N
Cinnamon	0	N	155	(15)
Curry powder	0	N	100	N
Garam masala	0	N	340	N
Mint	31.0	5.0	740	N
Nutmeg*	0	N	60	N
Paprika	0	N	36250	N
Parsley	190.0	1.7	4040	(1.0)
Pepper, black	0	N	115	3.0
white	0	N	Tr	3.0
Rosemary	0	N	1880	N
Sage	0	N	3540	N
Thyme	0	N	2280	N

* also contains 28.5g/100g fats

There is accumulating evidence that many fruits (e.g. apples and citrus fruits) and some naturally derived beverages (e.g. tea and red wines) contain substantial quantities of polyhydroxyphenols. The food tables have not yet begun to list the abundance of the flavonoids in fruits, vegetables and beverages, but 4000 have been isolated from natural sources [187]. Until recently, the bioflavonoids were considered by the medical profession to be unimportant although their antioxidant effectiveness in fats and oils has been known for many years [107], but they must now be considered to be of comparable significance to ascorbic acid as water soluble antioxidants in the human diet. The bioflavonoids have attracted popular attention because of their presence in teas [188-190] and wines [75,120,191] (Table 5.15) and because of their possible connection with the beneficial effects of the "European Diet" [120]. They are widely distributed in fruits and vegetables, particularly in coloured fruits but they are also present in dark green vegetables of the brassica family. Although their full significance in the human diet has yet to be fully evaluated, they have been shown to have anticancer activity when consumed as part of a vegetable diet [192] and like ascorbic acid, they have been shown to inhibit the oxidative modification of LDL [119] and reduce the risk of coronary heart disease in the elderly [193].

Table 5.15 Some dietary sources of polyphenolic antioxidants

Antioxidant Class*	Source
Flavone	
Apeginin	Parsley, celery
Chrysin	Fruit skins
Quercetin	Onion, lettuce, broccoli, tomato, cranberry, apple peel, olive oil, red wine, tea, berries
Kaempferol	Endive, leek, broccoli, radish, black tea, grapefruit.
Rutin	Buckwheat
Hesperatin	Lemon, sweet orange
Flavanone	
Naringenin	Eucalyptus
Taxifolin	Citrus fruits
Naringin	Citrus fruit peels
Flavane	
Catechin	Tea (Camellia)
Epicatechin	Tea (Camellia)
Anthocyanidin	
Cyanidin	Cherry, raspberry, strawberry
Apegenidin	Coloured fruits
Perlagonidin	Perlargonium, scarlet rose
Malvidin	Blue grapes

* See Tables 5.3-5.5 for antioxidant activities

Some widely publicised "health food" products are known to be important sources of flavonoids. Notable among these are royal jelly and propolis and even honey [150], all produced by bees and used in the nutrition and protection of the hive. They are so far poorly characterised and consequently very variable in performance and depend significantly on where the pollen, from which many of the constituents originate, is gathered. Poplar and fir trees appear to be the most potent sources.

5.5 The Antioxidant Potential of Drugs

Although a variety of antioxidants have been shown to have beneficial effects in animals, very few have been developed as therapeutic agents in man. The reason for this is almost certainly associated with the reports over the past ten years that the widely used industrial phenolic antioxidant, BHT, has been reported to have carcinogenic properties when fed to animals at very high dosages [194]. The dosages used in these tests are unrealistically high (1-2%) and animals most closely related to man (monkeys and dogs) did not show such effects. Consequently, BHA, a rather more effective analogue of BHT is now permitted as a food additive in the USA. There also appears to be some evidence that

BHT and to a lesser extent BHA and ethoxyquin may be implicated as co-promoters of tumours in combination with carcinogens [194]

5.5.1 *Drugs with chain-breaking antioxidant activity*
(a) Probucol

Surprisingly, in view of the above, probucol, a lipid soluble antioxidant with a chemical structure similar to antioxidants used in hydrocarbon polymers and containing the same chain-breaking function as in BHT is used as a "cholesterol-reducing" drug in the treatment of artherioslerosis (Chapter 6). However, phenolic sulphides are very effective autosynergists (Chapter 4) combining the CB-D activity of BHT with the peroxidolytic activity of alkyl sulphides. Analogous compounds are highly effective thermal antioxidant for polymers [195,196].

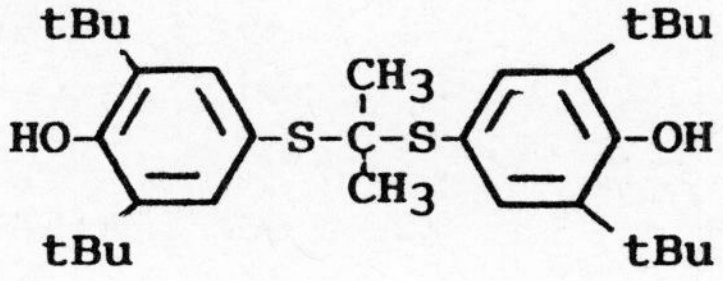

$$\text{HO} - \underset{tBu}{\overset{tBu}{\bigcirc}} - S - \underset{CH_3}{\overset{CH_3}{C}} - S - \underset{tBu}{\overset{tBu}{\bigcirc}} - \text{OH}$$

Probucol

Probucol effectively inhibits LDL peroxidation [197] and like BHT it is oxidised to a relatively stable aryloxyl [198]. It is much more effective in reducing monocyte adhesion to the aortic endothelium than conventional "cholesterol lowering agents" (Chapter 6) and its therapeutic activity almost certainly has more to do with the inhibition of LDL peroxidation than with cholesterol lowering. Furthermore, the probucol analogue, MDL 29,311 in which the sulphur atoms are linked by methylene rather than *iso*-propylidine, is as effective as probucol as an antioxidant but has no effect on cholesterol levels [199]. The peroxidolytic antioxidant mechanism of probucol, by analogy with that of related alkylaryl and diaryl sulphides (including ibuprofen, Section 5.5.2(c)) [195] almost certainly involves the formation of sulphur acids with both CB-D and PD activity, (Scheme 5.13).

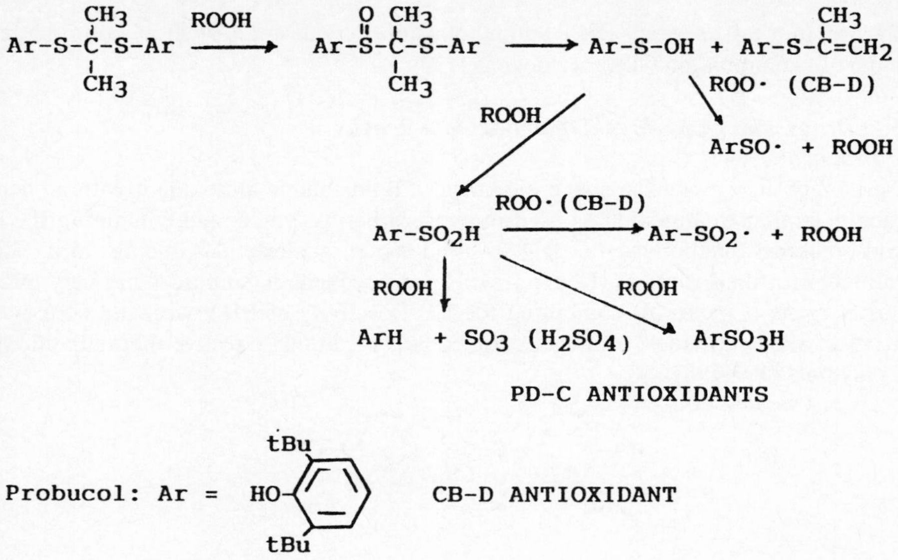

$$Ar-S-\underset{\underset{CH_3}{|}}{\overset{\overset{CH_3}{|}}{C}}-S-Ar \xrightarrow{ROOH} Ar-S-\underset{\underset{CH_3}{|}}{\overset{\overset{O}{\|}}{C}}-S-Ar \longrightarrow Ar-S-OH + Ar-S-\underset{ROO\cdot (CB-D)}{\overset{\overset{CH_3}{|}}{C}}=CH_2$$

$$\downarrow ROOH \qquad\qquad ArSO\cdot + ROOH$$

$$Ar-SO_2H \xrightarrow{ROO\cdot (CB-D)} Ar-SO_2\cdot + ROOH$$

$$ROOH \downarrow \qquad\qquad ROOH$$

$$ArH + SO_3 \ (H_2SO_4) \qquad ArSO_3H$$

PD-C ANTIOXIDANTS

Probucol: Ar = HO–⟨tBu, tBu⟩ **CB-D ANTIOXIDANT**

Scheme 5.13 Autosynergistic antioxidant mechanisms of probucol

(b) Dihydroquinolines

The dihydroquinoline (HQ) group of antioxidants, of which ethoxyquin is the most important example, have been used for many years in rubber technology (Chapter 3) and more recently it has gained wide acceptance as an antioxidant in foodstuffs because of its low toxicity [100]. Part of the effectiveness of the 2,2'-substituted dihydroquinolines is their ability to form relatively stable aminoxyls which are CB-A antioxidants in polyunsaturated systems at ambient temperatures [101] (see Section 3.5.1(b) and Table 3.10). The HQs form oligomeric arylamines and aminoxyls on oxidation by peroxyl radicals and their versatility is thus due to their ability to act as both reducing agent and oxidising agents for peroxyl radicals, Scheme 5.14. [200-202].

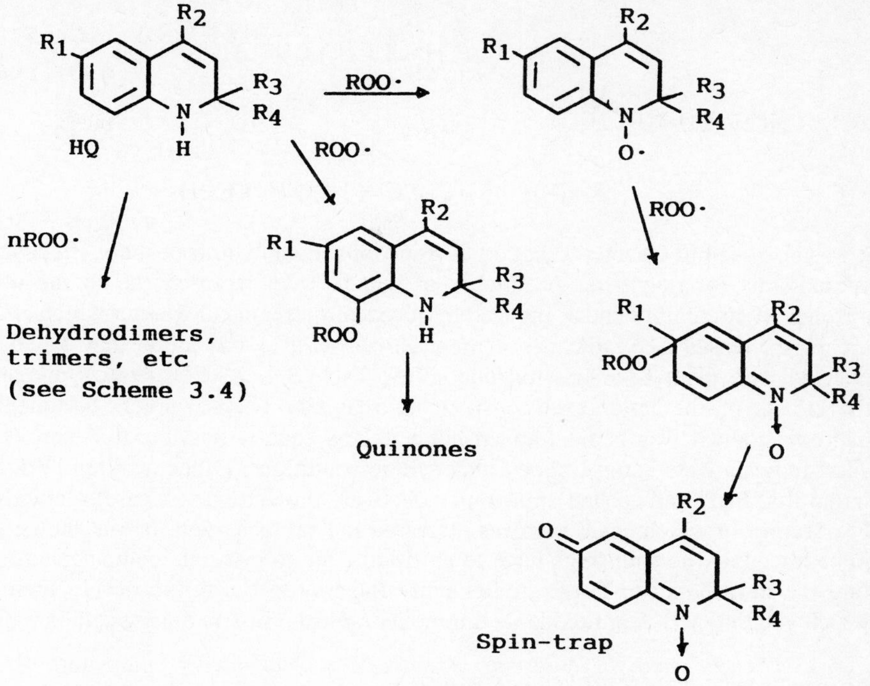

Scheme 5.14 Oxidation of dihydroquinoline antioxidants

Ethoxyquin, R_1 = OCH_2CH_3, R_2 = R_3 = R_4 = CH_3
QAO, R_1 = H, R_2 = OCH_2CH_3, R_3 = R_4 = Ph

Tanfani [203] has recently observed that the nitroxyl derived from one of the dihydroquinolines (QAO, R_1 = H, R_2 = OEt, R_3 = R_4 = Ph) is a very effective antioxidant for serum albumin by trapping ROO·. Again, the end products are quinone nitrones which themselves have the ability to trap alkyl and alkylperoxyl radicals. Aminoxyl radicals are tolerated well in the body and do not reduce cell survival and they appear to offer prospects for the treatment of a variety of diseases involving peroxidation.

(c) Spin-traps
It was seen in Chapter 2 that carbon-centred radicals can be identified by spin-trapping with 2-methyl-2-nitrosopropane (MNP) during the peroxidation of arachidonic acid. This is an important observation since it implies that carbon spin trapping by nitoso compounds or nitrones can compete with the propagation process even in the presence of excess oxygen:

$$\text{-(CH=CHCH}_2\text{)CH=CHCH-} \xrightarrow{\text{(OO)·}} \begin{matrix} \xrightarrow{\text{ST}} \text{-(CH=CHCH}_2\text{)}_3\text{CH=CHCH-} + O_2 \\ \\ \xrightarrow{\text{RH}} \text{-(CH=CHCH}_2\text{)}_3\text{CH=CHCH-} + R· \end{matrix} \quad (16)$$

(with ST· on the upper product and OOH on the lower product)

It was also seen in Chapter 3 that other spin-traps (notably nitrones) are effective catalytic antioxidants for polymers. A number of papers have reported the therapeutic use of nitrones in circulatory shock induced by physical or chemical trauma which is known to involve increased free radical formation. Novelli et al. [204] found that, when rats were injected with phenyl-*tert*-butylnitrone (PBN, Table 3.4, XXIII) before traumatic shock, the effects of the latter was considerably reduced. These workers assumed that the hydroxyl radical was being trapped but it seems equally possible that peroxyl or even alkyl radicals were being trapped under these conditions. Other workers [205,206] have found that PBN and related spin-traps effectively protect against cardio-toxicity induced by adriamycin which again involves increased radical formation. It was shown [206] that to be effective, the spin-traps have to partition from to cytosolic compartment of the cell into the mitochondria. These studies point the way to the future development of drugs which generate CB-A antioxidants (aminoxyl radicals) in-situ during cell damage.

(d) Tamoxifen and Oestrogen antagonists

Tamoxifen is widely used in the treatment of breast cancer [207-211] and has also been reported to have beneficial effects against CVD and in particular CHD [212]. It retards the peroxidation of human LDL but 4-hydroxytamoxifen (4-HT), a metabolite of tamoxifen was an order of magnitude more effective in inhibiting Cu^{2+} catalysed LDL peroxidation than was tamoxifen itself, and 4-HT was also almost twice as effective as 17β-oestradiol [211].

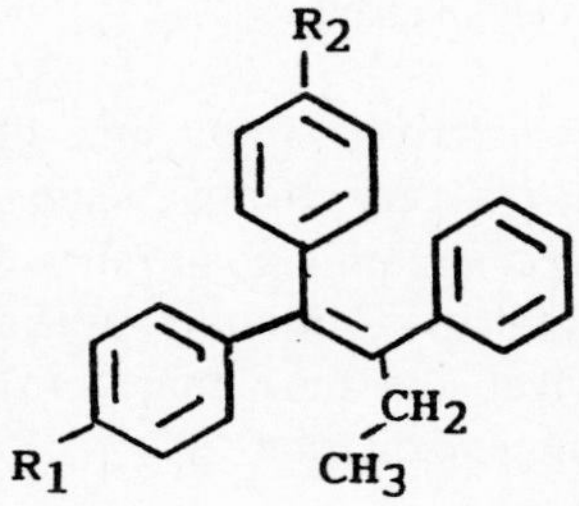

Tamoxifen R_1=H, R_2=OCH$_2$CH$_2$N(CH$_3$)$_2$

4-Hydroxytamoxifen (4HT)
   R_1=OH, R_2=OCH$_2$CH$_2$N(CH$_3$)$_2$

Nafoxidine, R_1=OCH$_3$, R=OCH$_2$CH$_2$N(CH$_3$)$_2$

Unlike oestradiol, 4-HT gives a highly delocalised phenoxyl and in consequence is a better hydrogen donor, Scheme 5.15:

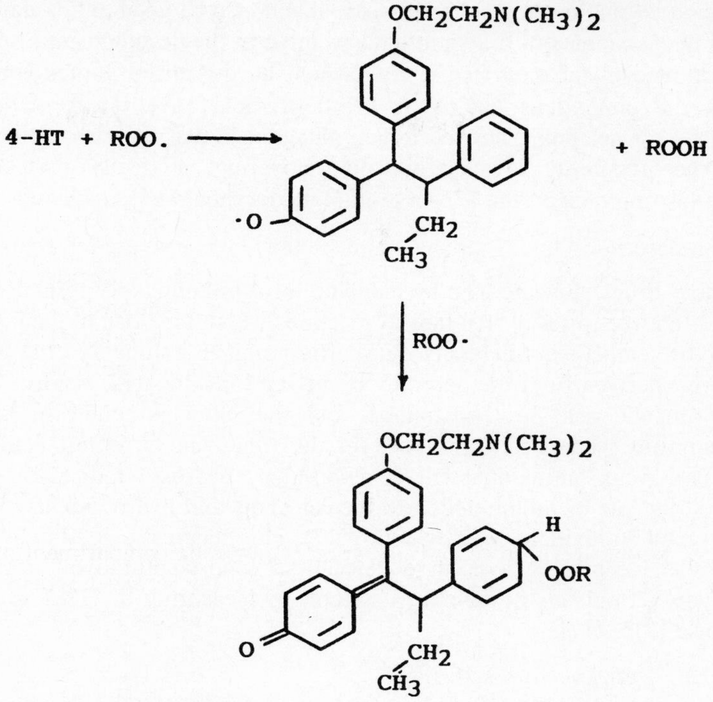

Scheme 5.15 Chain-breaking mechanism of 4-HT

However, tamoxifen and the structurally related nafoxidine do not have a phenolic hydroxyl group and it seems likely that the weak antioxidant activity of these compounds is due to ß-carotene-like radical trapping activity; that is they form a peroxypolyene adduct in which two peroxyl groups are trapped (Scheme 5.16).

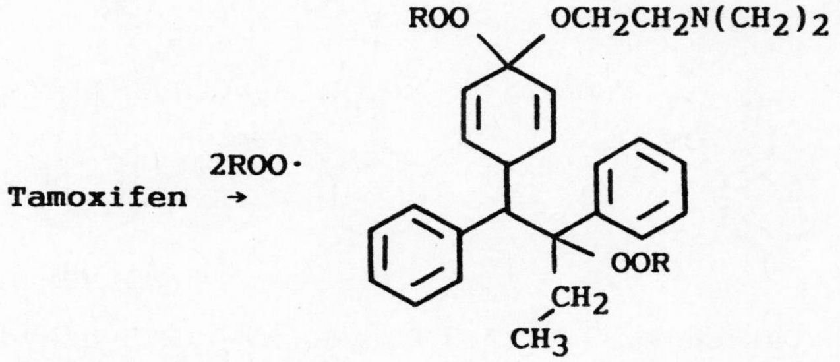

Scheme 5.16 Radical trapping potential of tamoxifen

Tamoxifen and nafoxidine are weaker antioxidants than 4-HT because, as is the case with the carotenoids, the intermediate carbon-centred radicals involved in Scheme 5.16 can also react with oxygen whereas this is less likely with the stabilised polyenoxyl from 4-HT where the unpaired spin density is primarily on oxygen (Scheme 5.15).

It has been proposed that oestrogen antagonists should also be good antioxidants [211]. However recent developments in this field have resulted in the development of oestradiol derivatives substituted in the saturated B ring which have reduced antioxidant activity compared with the tamoxifens and even 17β-oestradiol [211]. There seems to be considerable scope for designing new oestrogen antagonists that are also more powerful antioxidants. The possibility of electronically activating, sterically protecting and autosynergising (as in probucol) the 17β-oestradiol model should be considered.

(e) Aspirin

Anti-inflammatory drugs alleviate pain by reducing local inflammatory responses. It was see in Chapter 2 that the series of free radical reactions which give rise to the prostanoids can be inhibited by synthetic antioxidants. The earliest anti-inflammatory drug was acetyl salicylate (aspirin) derived from the naturally occurring salicylic acid. Aspirin is rapidly hydrolysed to salicylic acid in the stomach and the latter is found in plasma in appreciable quantities when aspirin is used in the treatment of inflammation [213]. Salicylic acid is a weak antioxidant but is oxidised by hydroxyl radicals at diffusion controlled rate at the site of inflammation to the catechols and hydroquinones which are much more powerful antioxidants (see Scheme 5.17). This classical oxidation by hydroxyl radicals [214] has recently been used to detect and semi-quantitatively measure the formation of hydroxyl radicals in biological systems by measuring by HPLC the amount of DHBAs formed [215-217].

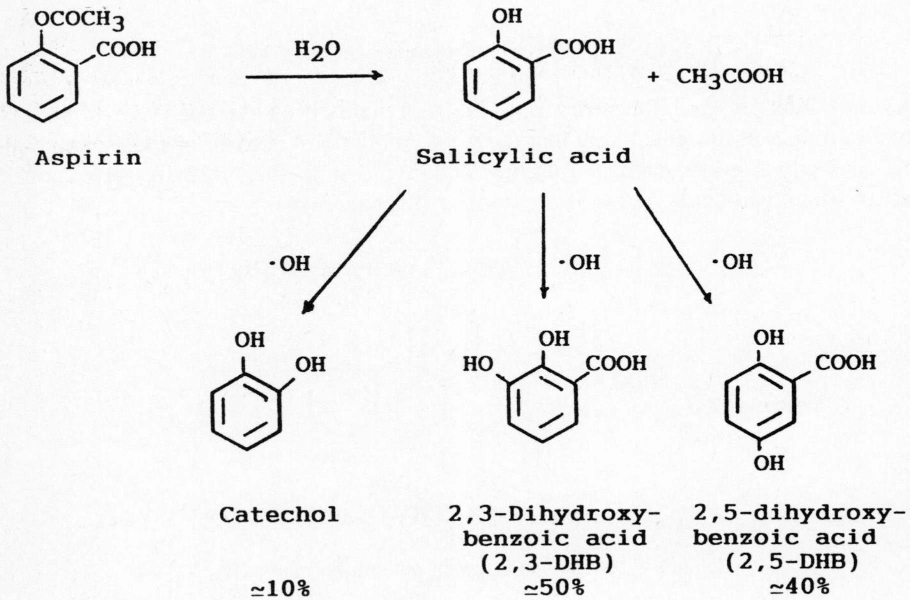

Scheme 5.17 Antioxidants derived from salicylic acid

However, this may not be a very accurate measure of hydroxyl radical formation due to its rapid transformation to further oxidation products *in vivo* [218]. Hydroxylation is not restricted to hydroxy benzenes and the hydroxyl radical will oxidise most aromatics to phenols [213]. Many non-steroidal anti-inflammatory drugs (NSAIDs) are aromatic compounds which either already contain antioxidant structures (e.g. mefenamic acid), can be hydrolysed to antioxidants (e.g. indomethacin) or can be hydroxylated to phenolic antioxidants by the above mechanism (e.g. naproxen, etc.).

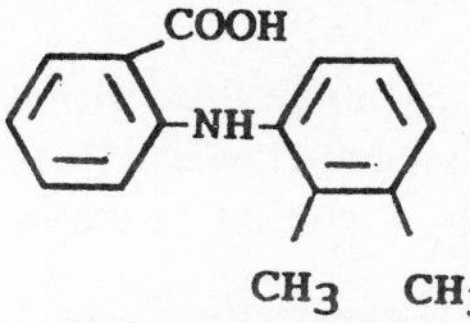

Mefenamic acid

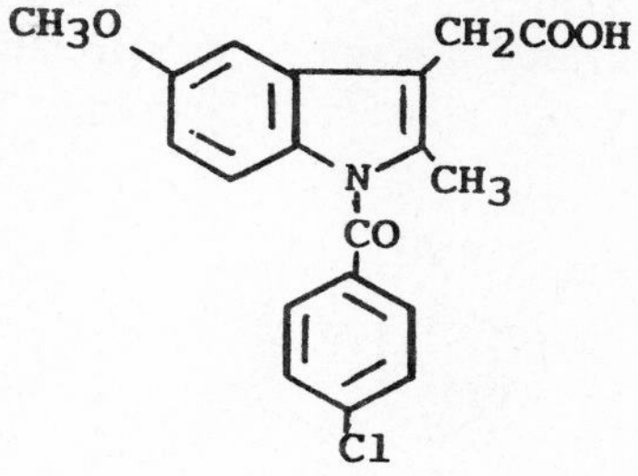

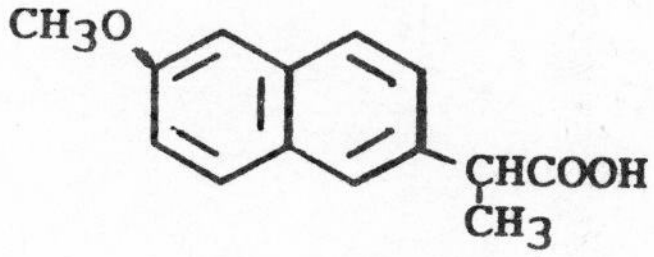

Indomethacin

Naproxen

(f) Paracetamol (Acetamidophenol)

Paracetamol (4-acetamidophenol) is closely related to 4-acylamidophenols (Suconox-4, 9, 12 and 18) used as antioxidants in rubber and lubricating oil technologies. The 4-acylamidophenols are much more powerful chain-breaking antioxidants than the pro-antioxidants discussed in the previous section and do not require activation by hydroxyl radicals to become effective. Paracetamol is known to rapidly form the transient free aryloxyl on oxidation by PHS during inhibition of prostaglandin synthesis [219]. The end products of this process, as in the case of analogous antioxidants in technological systems, are dimers and oligomers together with disproportionation products [220]. The mechanism of these processes are summarised in Scheme 5.18.

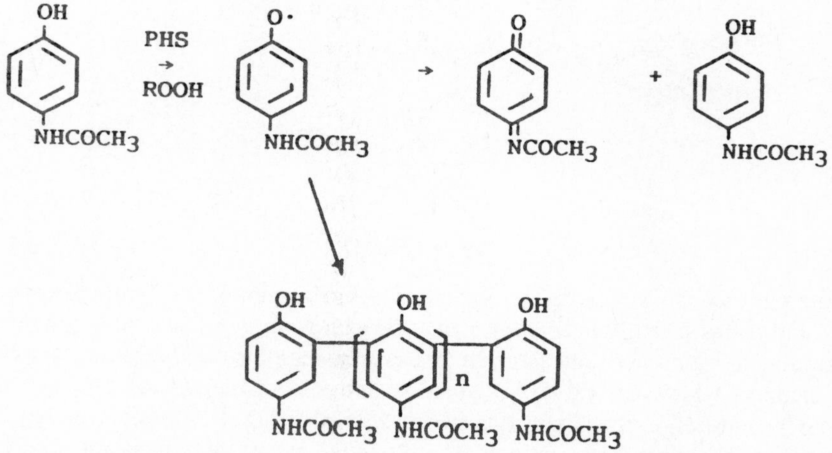

Scheme 5.18 Oxidation of paracetamol by peroxidase

Paracetamol is therefore not a very effective CB-D antioxidant for the cycloxygenase step in prostaglandin synthesis at high peroxide concentrations [221], but at lower hydroperoxide concentrations it becomes more efficient. This may be the reason for its relatively weak anti-inflammatory activity.

(g) Allopurinol
Allopurinol is a powerful inhibitor of the oxidation of hypoxanthine, a major source of superoxide and hydrogen peroxide under conditions of ischaemia-reperfusion [222]. It is converted by hydroxyl radicals to oxypurinol, which has a similar structure to uric acid (Section 5.2.4).

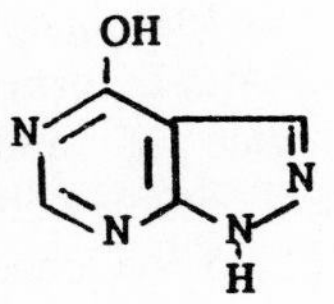

Allopurinol 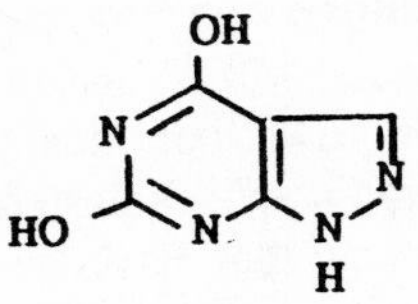   **Oxypurinol**

Allopurinol is tightly bound to the site of oxidation damage and like SOD and catalase is an effective inhibitor of reperfusion-induced increase in myeloperoxidase during ischaemia-reperfusion [222]. It seems likely that allopurinol is acting both as a multi-stage sacrificial hydrogen donor [223] and as an iron chelating agent [224].

(h) Dipyridamole
Dipyridamole is used as a vasodilatory drug with antiplatelet activity [225]:

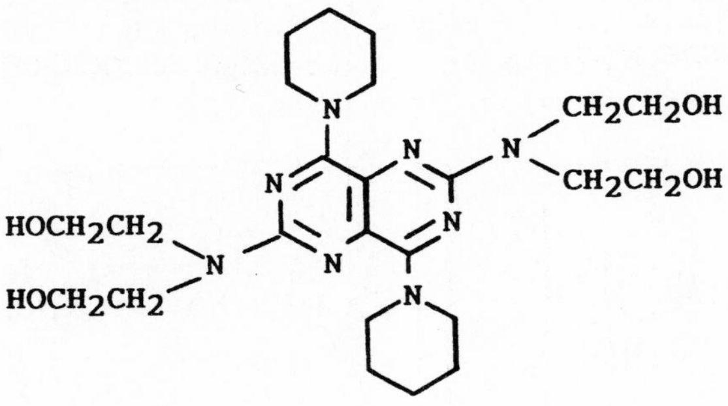

Dipyridamole

This compound has been shown to be an effective inhibitor of lipid peroxidation [226,227] and it has recently been shown that dipyridamole is a CB-D antioxidant with activity similar to vitamin E and like VE it has a stoichiometric inhibition factor of 2 [228]. It contains no recognised CB-D function but is nevertheless an effective "oxyl" radical trap by addition to the conjugated pyrimazole structure. Its shows some structural resemblance to the pyrimidine antioxidants discussed below and it seems likely that secondary hydroxylated products are formed following the initial trapping reaction, but these have not yet been identified [228].

5.5.2 Drugs with preventive antioxidant activity
(a) Thiocarbamoyl compounds

It has been known for over twenty years that bis-dialkythiuram disulphides (e.g. Disulfiram) and the chemically related bis-alkylxanthogen disulphides effectively inhibit the carcinogenic effects of procarcinogens such as 1,2-dimethyl hydrazine (DMH) [229] and it has been shown that disulfuram inhibits the oxidation of chemical carcinogens *in vivo*.

More recently, attention has centred on the activity in the aqueous phase of sodium diethyldithiocarbamate, NaDEC, as an inhibitor of foam cell formation [230] and the related inhibition of LDL peroxidation [230,231]. This PD antioxidant effectively inhibited the oxidation of cholesterol linoleate and foam cell formation [230]. Antioxidant mechanisms that have been proposed without evidence include metal ion chelation [232], scavenging of HOCl, ·OH and peroxynitrite [231], nitric oxide scavenging [233] and superoxide scavenging [234]. The decomposition of hydrogen peroxide and hydroperoxides which was seen in Chapter 4 to be the primary antioxidant mechanism of the MDRCs is mentioned by only one group [233] and for some reason the peroxidolytic mechanism appears to have been overlooked or ignored as a potential explanation for the antioxidant activity of the dithiocarbamates *in vivo*. It was seen in Chapter 4 the metal dithiocarbamates are relatively ineffective as peroxyl scavengers compared with their very high activity as peroxide decomposers [235-240]. Transition metal deactivation can be a contributory reason for their antioxidant activity and the compounds are autosynergistic for this reason.

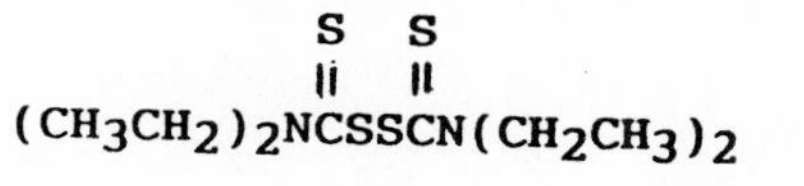

$$\underset{\text{Disulfuram}}{(CH_3CH_2)_2N\overset{\overset{S}{\|}}{C}S\overset{\overset{S}{\|}}{SC}N(CH_2CH_3)_2} \qquad \underset{\text{Diethylxanthogen}}{CH_3CH_2O\overset{\overset{S}{\|}}{C}S\overset{\overset{S}{\|}}{SC}OCH_2CH_3}$$

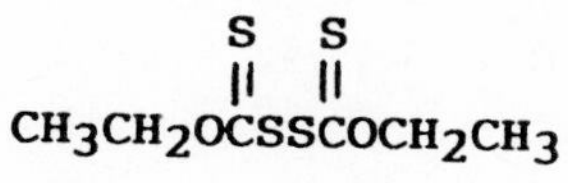

$$R_2NC\overset{S}{\underset{S}{\diagdown}}M\overset{S}{\underset{S}{\diagup}}CNR_2 \qquad M = Na, Zn, Ca, Ni, Mn, \text{etc.}$$

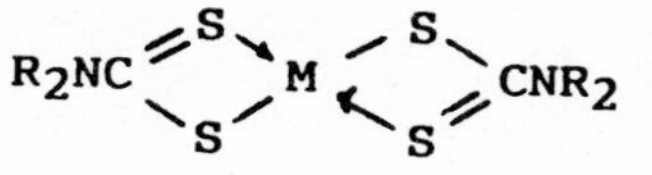

Metal dialkyldithiocarbamates, MDRC

(b) Oltipraz

Oltipraz is a thiocarbonyl disulphide which is effective against carcinogen-induced neoplasia [130]. It is closely related chemically and mechanistically to the disulphide antioxidants discussed in Chapter 4 and almost certainly acts by the same peroxidolytic mechanism. However, in its reduced form it may also be a CB-D/metal chelating antioxidant similar to a-lipoic acid (Section 5.3.4).

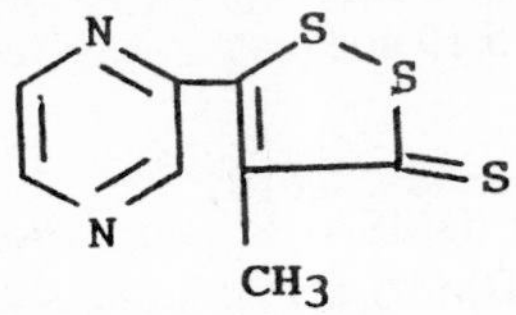

Oltipraz

(c) Ibuprofen

Ibuprofen is a widely used as a non-steroidal anti-inflammatory drug. Its structure is quite analogous to the diaryldisulphide antioxidants used in polymers (Chapter 4) and the mechanism of its antioxidant action shows similarities to probucol (Section 5.5.1). Diaryl disulphides (ArSSAr) are not themselves antioxidants but are oxidised to thiolsulphinates and sulphur acids by hydroxyl radicals, hydroperoxides and other reactive oxygen species [237]. The oxidation products can act by peroxidolytic and to a lesser extent, chain-breaking mechanisms (see Schemes 5.19 and 5.13).

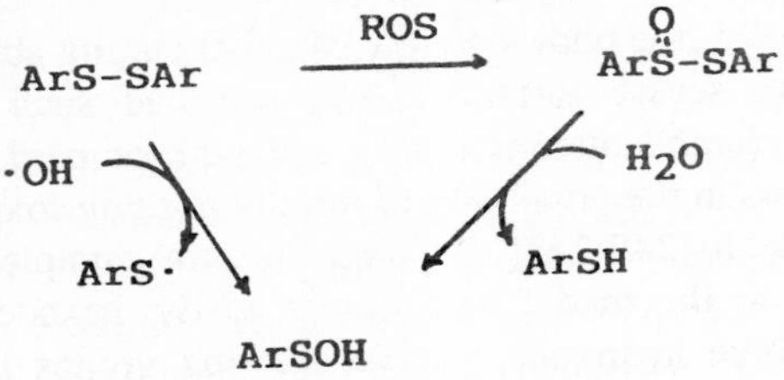

For subsequent PD chemistry of ArSOH, ArSO· etc., see Scheme 5.13.

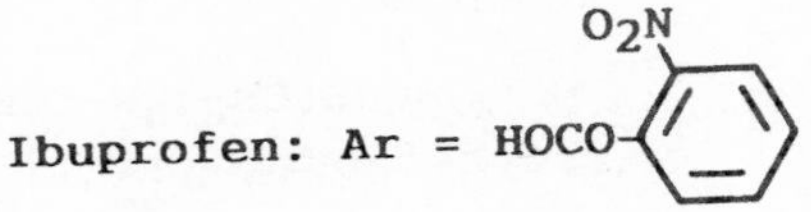

Ibuprofen: Ar = HOCO-(2-nitrophenyl with O_2N)

ROS include $O_2^{·-}$, H_2O_2, ·OH, ROO·, ROOH, etc.

Scheme 5.19 Antioxidant mechanism of ibuprofen

Hawkins and Sautter showed that diphenyldisulphide has no significant antioxidant activity until it is oxidised to the thiolsulphinate, which is a powerful peroxidolytic antioxidant which oxidises further to the ultimate peroxidolytic antioxidants, sulphonic acids and H_2SO_4 [242]. In effect this means that ibuprofen is inactive until it reaches the site of inflammation where it is triggered by ROS to give the effective peroxidolytic antioxidants. In this respect it differs from probucol which, due to the presence of the phenolic function, is a CB-D antioxidant *ab initio*.

It has been reported [243] that ibuprofen also maintains thiol levels in the liver after circulatory trauma and may also act by chelating iron, presumably through the 2-nitrocarboxy group [244]. However, its main function is to inhibit lipid peroxidation at the site of inflammation [245] and there can be little doubt that in the chemistry summarized in Schemes 5.14 and 5.19 adequately describes its therapeutic effect.

(d) Dimercaptosuccinic acid

Meso-2,3-dimercaptosuccinic acid (DMSA) is an approved oral chelating agent for lead poisoning [246]. Like ibuprofen, DMSA protects thiols in liver and brain. Treatment of lead loaded mice with DMSA significantly decreased MDA levels [246[and it seems likely that the thiol groups are again giving rise to chain-breaking and peroxidolytic sulphur acids. This is a further example of an autosynergistic (CB-A, PD-C, MD) antioxidant system.

(e) Singlet oxygen quenchers

Singlet oxygen quenchers, notably gold compounds, have been reported to be effective in the treatment of rheumatoid arthritis [247], but surprisingly there have been no systematic studies of the effects of carotenoid supplementation [248].

(f) Metal chelating agents

As was seen in Section 5.3.5, the body's natural metal chelating ability is not sufficient to protect against the more severe forms of iron overload such as thalassaemia and haemochromatosis. Considerable attention has therefore been paid, particularly by Hider, Porter and their co-workers to the possibility of developing non-toxic iron chelating drugs that can be given by mouth [249-253]. To date the only complexing agent which has proved to be acceptable to the medical profession is the hexadentate desferrioxamine (DFO) which contains three hydroxamate co-ordinating groups in the same molecule. DFO is isolated from *Streptomces pilosis* and both it and its iron complex (ferrioxamine) is rapidly excreted in urine and in bile. However, for successful treatment of iron overload, it has to be administered by intravenous transfusion.

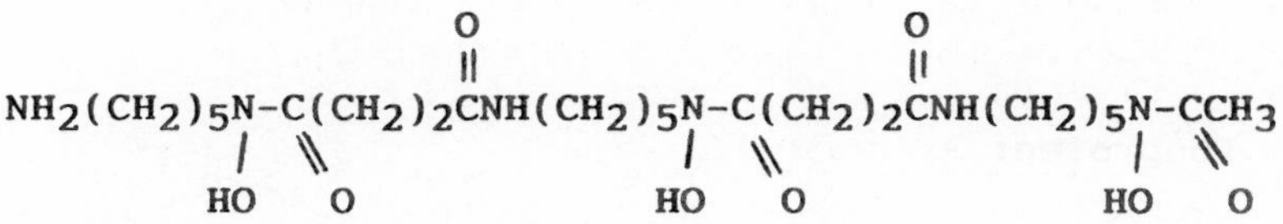

$$NH_2(CH_2)_5N-C(CH_2)_2CNH(CH_2)_5N-C(CH_2)_2CNH(CH_2)_5N-CCH_3$$

Desferrioxamine, DFO

Porter, Huehns and Hider [249] have outlined the requirements of a therapeutically acceptable iron chelating drug as follows;

1. It should be able to cross biological membranes enabling it to be absorbed from the intestinal tract and to enter the cells of many tissues (including the liver where iron is substantially deposited in iron overload) and the heart. It was concluded that neutral chelators permeate cells more rapidly than charged molecules and should therefore be more orally active.
2. The chelator should not metabolise in the body, for example by hydrolysis in the stomach or by microbial breakdown.
3. Both the ligand and the metal complex should be water soluble without containing a charged group (e.g. carboxylate).

4. The partition coefficient (K_{part}) between lipid and water (at pH7) should be 0.2-1.0 to assure ability to penetrate hepatocytes without causing acute toxicity.
5. Iron redistribution should not occur from relatively robust sites such as the liver to more vulnerable ones such as the heart. Thus ligand with intermediate lipid solubility that gives an iron complex with somewhat lower lipid solubility represent a realistic compromise.
6. In the case of chronic thalassaemia, a chelator must be able to mobilise at least 0.5 mg/kg in order to achieve a negative iron balance and the dose must be as low as possible to limit toxic side-effects.
7. An iron chelator must be specific for iron. That is, the stability constant (K = [Fe complex]/[Fe^{3+}] x [Ligand]) should be considerably greater than those of other biologically important metals (e.g. Zn, Ca, Cu, Mn, etc.).
8. Iron chelates should not catalyse radical formation from hydrogen peroxide or oxygen.

These limitations place a serious restriction on the chemical classes of chelating agent that can be considered. Some, but not all requirements are satisfied by DFO. It is a very powerful neutral chelator but it is only 10% efficient so that a daily intravenous dose of 40-60 mg is required in the case of severe iron overload. It chelates other metal ions several orders of magnitude less effectively, which is one of the reasons for its relatively low toxicity. However, the ester groups in the linear molecule tend to hydrolyse during metabolism and the need to introduce it intravenously is its major disadvantage.

DFO has other antioxidant activity beside metal chelation. It is also an effective chain-breaking (CB-D) antioxidant (Scheme 5.20) due to the presence of the oxime group which gives a relatively stable nitroxyl ($\tau_{1/2} \cong$ 10 min) [254]. Although this is an apparent advantage due to the synergism between the two antioxidant functions in the same molecule (autosynergism), it is not necessarily so since the peroxyl-trapping process is sacrificial and, by reducing the number of hydroxamate groups available for chelation, may reduce the ability of the molecule to scavenge iron (Scheme 5.20). Furthermore, it has been reported that the nitroxyl is damaging to enzymes [254].

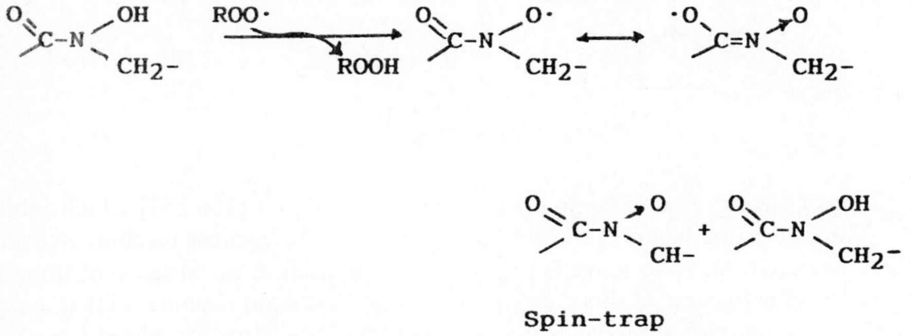

Spin-trap

Scheme 5.20 Antioxidant (CB-D) activity of DFO

Levin [255] has reported that the ability of chemicals to penetrate the rat blood-brain barrier was linearly related to K_{part} for compounds of molecular weight < 300, but there was no such relationship for molecules of molar mass > 500. However, most tridentate molecules, like DFO, have a molar mass > 500. Attempts to use water soluble analogues of naturally occurring siderophores (e.g. enterobactin) by sulphonation was not successful since these are highly polar and K_{part} is reduced [250].

Following an extensive evaluation of iron chelators, Hider and co-workers [250-253] have concluded that the bidentate 3-hydroxypiridin-4-one (HPO) and to a lesser extent the tridentate desferrithiocin (DFT) structure possess many of the necessary properties to satisfy the requirements outlined above. Their effectiveness is strongly dependent on the chelate-metal ratios and due to the time taken (typically 1 ms) to form the fully co-ordinated complex (FeL_3 in the case of HPO), some redox reactions may occur to give radicals.

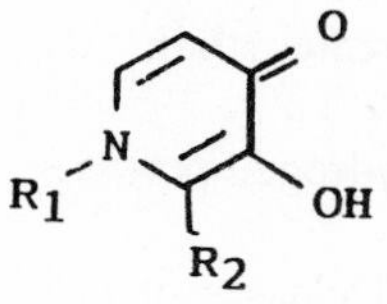

(a) $R_1=R_2=Me$; CP20
(b) $R_1=R_2=Et$; CP94

HPO has a high affinity for iron and has been shown to remove iron from the serum of overloaded animals under biological conditions [252] and is efficiently absorbed [250]. The iron complexes exist as FeL_3 at pH7 and are relatively inert toward peroxides as compared with EDTA. As in the case of the hexadentate ligands (e.g. DFO), the iron is sterically protected from attack by peroxidic species. By contrast, catechols form appreciable amounts of the 3:1 complexes only above pH 8 and are much less effective in inhibiting iron catalysed redox reactions.

Drugs developed for the treatment of acute brain and spinal cord injury contain the pyrimidine structure, shown typically for U74006F:

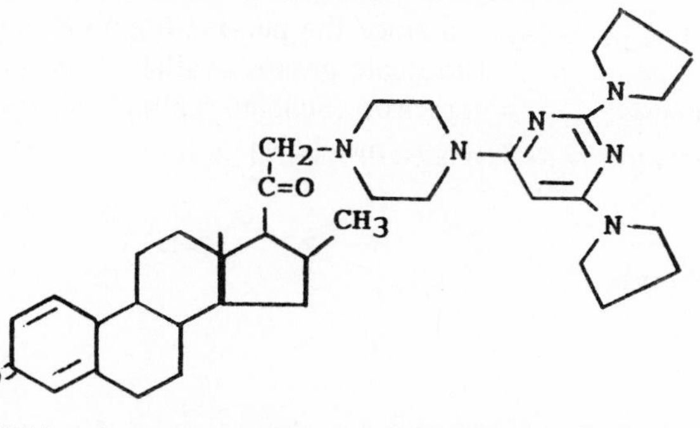

Trilazad mesylate
U74006F

U74006F has been reported to be an effective lipid antioxidant [256,257] which inhibits the iron induced peroxidation of the rat brain. It is effective against cerebral ischaemia [258] and has been shown to scavenge peroxyl radicals even in the absence of iron. The reason for the effectiveness of this compound is not at first sight obvious since it appears to contain no recognised antioxidant group. The pyrimidine structure might have weak metal chelating properties but would not be expected to have the ability to deactivate iron as effectively as DFO or HPO above. However, Hall has reported [259] that during the *in vitro* Fenton reaction, U74006F is oxidised by hydroxyl radicals to the corresponding 2,4,6-triamino-5-hydroxypyrimidine. He suggested that this derivative is formed by hydroxyl radicals attack during ischaemia and is more effective than the parent pyrimidine as an antioxidant. This is entirely analogous to the hydroxylation of salicylic acid to 2,3- and 2,5-DHB during cerebral ischaemia in the cortex of anaesthetised rats [260].

2,4,6-Triamino-5-hydroxypyrimidines should indeed be more effective both as metal chelators and as CB-D antioxidants than the parent pyrimidine (see Scheme 5.21). The heteroaryloxyl radical is extensively delocalised in the pyrimidine structure and on the 2,4,6-triamine substituents. The analogy with uric acid (Section 5.2.4) and allopurinol (above, Section (g)).

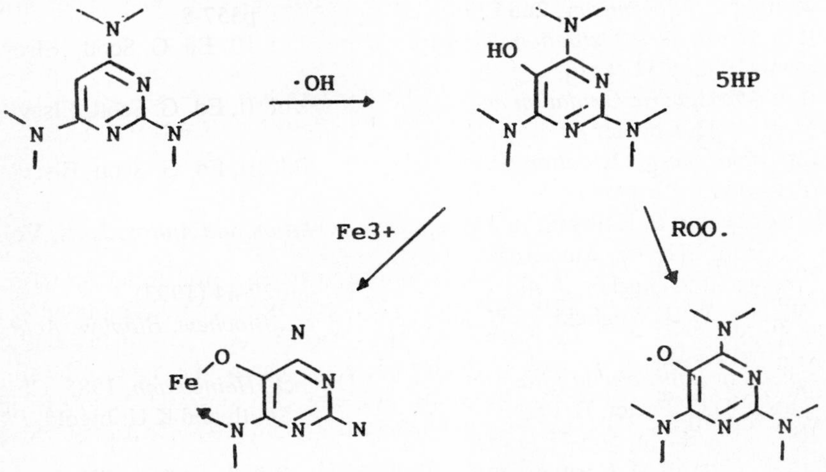

Scheme 5.21 Antioxidant activity of 5-hydroxypyrimidines

The hydroxylation of U74006F thus produces a site specific autosynergist. However, the hydroxyl group is involved in both metal deactivator (MD) and chain-breaking donor (CB-D) roles and an improved autosynergist, U78517F, which contains the essential tetramethylchroman structure of a-tocopherol has been developed by Hall et al. [257]:

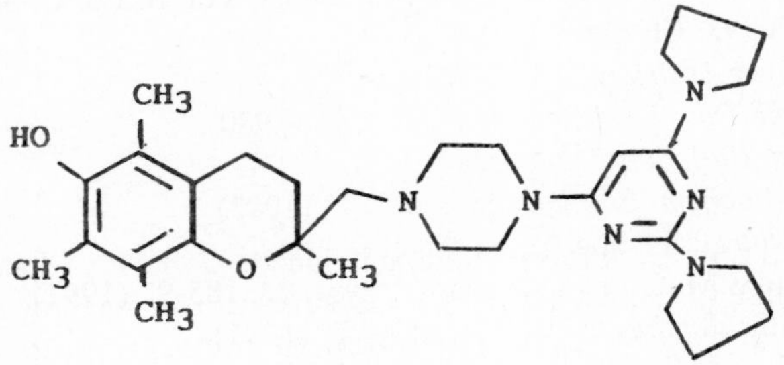

In this compound the MD and CB-D functions are not competitive and it is effective at concentrations an order of magnitude lower than U74006F [261,262]. U785171 is more effective at 10 µM than vitamin E at 50 µM in inhibiting calcium influx in peroxidising human astrocytoma cells initiated by the Fenton reaction [262]. This points an important way forward in the development of new autosynergistic antioxidant drugs by combining the known activity of non-toxic naturally occurring antioxidants in the same molecule with the correct lipophilicity.

REFERENCES

1. H.S. Olcott and H.A. Mattill, *Chem. Rev.*, **29**, 257 (1941).
2. H. Dam, *Pharmacol. Revs.*, **9**, 1 (1957).
3. H.H. Draper, S. Goodyear, K.D. Barbee and B.C. Johnson, *Brit. J. Nutr.*, **12**, 89-97 (1958).
4. J.R. Bronk, *Chemical Biology*, Macmillan, 1973, p.154.
5. A.L. Lehninger, *Biochemistry*, 2nd Edition, Worth, 1977, p357-8.
6. G. Scott in *Atmospheric Oxidation and Antioxidants*, Vol. III, Ed. G. Scott, Elsevier, Amsterdam, 1993, p.32 et seq.
7. G. Scott in *Atmospheric Oxidation and Antioxidants*, Vol. II, Ed. G. Scott, Elsevier, Amsterdam, 1993, Chapter 8.
8. E. Niki in *Atmospheric Oxidation and Antioxidants*, Vol. III, Ed. G. Scott, Elsevier, Amsterdam, 1993, Chapter 1.
9. J.M. Gutteridge and B. Halliwell in *Atmospheric Oxidation and Antioxidants*, Vol. III, Ed. G. Scott, Elsevier, Amsterdam, 1993, Chapter 3.
10. V.W. Bowry and R. Stocker, *J. Am. Chem. Soc.*, **115**, 6029-44 (1993).
11. M.M. Oliveira, W.B. Weglicki, A. Nason and P.P. Nair, *Biochem. Biophys. Acta*, **180**, 98 (1969).
12. A.T. Diplock in *Fats Soluble Vitamins*, Ed. A.T. Diplock, Heinemann, 1985, Chap 3.
13. G.W. Burton, D.O. Foster, B. Perly, T.F. Slater, I.C.P. Smith and K.U. Ingold, *Phil. Trans. Roy. Soc.*, **B 311**, 565-78 (1985).
14. J. Cillard, P. Cillard, M. Cormier and L. Girre, *J. Am. Oil Chem. Soc.*, 252-5 (1980).
15. J. Cillard, P. Cillard and M. Cormier, *J. Am. Oil Chem. Soc.*, 255-62 (1980).
16. G.W. Burton and K.U. Ingold, *J. Am. Chem. Soc.*, **103**, 6472-7 (1981).
17. G.W. Burton, L. Hughes and K.U. Ingold, *J. Am. Chem. Soc.*, **105**, 5950-1, (1983).
18. B. Century and M.K. Horwitt, *Fed Proc.* (1), **24**, 906-11 (1965).
19. R.L. Willson, *Ciba Symposium, Biology of Vitamin E*, Pitman, 1983, 19-44.
20. R.L. Willson, *Phil. Trans. Roy. Soc.*, **B311**, 576 (1985).
21. S.P. Kochhar in *Atmospheric Oxidation and Antioxidants*, Vol. II, Ed. G. Scott, Elsevier, Amsterdam, 1993, Chapter 2.
22. M.T. Juillet, *Fette Seifen Anstrich.*, **77**, 101 (1975).
23. D.B. Parrish, *CRC Crit. Rev. Food Sci. Nutr.*, **13**, 161 (1980).
24. W.D. Fitter, *Brit. Med. Bull.*, **49**, 545-55 (1993).
25. G.R. Buettner, *Arch. Biochem. Boiphys.*, **300**, 535-43 (1993).
26. B. Frei, *Am. J. Clin. Nutr.*, **54** (suppl.), 1113S-8S (1991).
27. I. Jialal, G.L. Vega and S.M. Grundy, *Atherosclerosis*, **82**. 185-91 (1991).
28. I. Jilal and S.M. Grundy, *J. Clin. Invest.*, **87**, 597-601 (1991).
29. A.L. Tappel, *Vitamins and Hormones*, **20**, 493-510 (1962).
30. L. Stryer, *Biochemistry*, Freeman, 1975, pp. 72-6.
31. G. Scott, *Phil. Trans. Roy. Soc.*, **B311**,577 (1985).
32. L.R.C. Barclay, *Can. J. Chem.*, **71**, 1-16 (1993).
33. K. Sato, E. Niki and H. Shimasaki, *Arch. Biochem., Biophys.*, **279**, 402-5 (1990).
34. E. Niki and E. Komura in *Oxygen Radicals in Biology and Medicine*, Eds. M.G. Simic, K.A. Taylor, J.F. Ward and C.V. Sonntag, Plenum, New York, 1989, pp.561-6.

35. W.A. Pryor, J.A. Cornicelli, L.J. Devall, B. Tait, B.K. Trivedi, D.T. Witiak and M. Wu, *J. Org. Chem.*, **58**, 3521-32 (1993).
36. V.E. Kagan, D.A. Stoyanovsky and P.J. Quinn in *Free Radicals in the Environment, Medicine and Toxicology*, Eds. H. Nohl, H. Esterbauer and C. Rice-Evans, Richelieu, London, 1994, pp.221-48.
37. M. Maiorino, A. Zamburlini, A. Roveri and F. Ursini, *Free Rad. Biol. Med.*, **18**, 67-74 (1995).
38. C.K. Chow, H.H. Draper, A.S. Csallany and M. Chiu, *Lipids*, **2**, 390-6 (1967).
39. J.A. Lindley, H. Zhang, H. Kaseki, N. Morisaki, T. Sato and D.G. Cornwell, *Lipids*, **20**, 151-5 (1985).
40. V.A. Liepkalns, B. Icard-Liepkalns and D.G. Cornwell, *Cancer Lett.*, **15**, 173-8 (1982).
41. D.E. Thornton, K.H. Jones, Z. Jiang, H. Zhang, G. Liu and D.G. Cornwell, *Free Rad. Biol. Med.*, **18,** 963-976 (1995).
42. V.W. Bowry, K.U. Ingold and R. Stocker, *Biochem. J.*, **288**, 341-4 (1992).
43. I. Kohar, M. Baca, C. Suarno, R. Stocker and P.T. Southwell-Keely, *Free Rad. Biol. Med.*, **19**, 197-207 (1995).
44. M. Maiorino, A. Zamburlini, A. Roveri and F. Ursini, *FEBS Lett*, **330**, 174-6 (1993).
45. K. Briviba and H. Sies in *Natural Antioxidants in Human Health and Disease*, Ed. B. Frei, Academic Press, San Diego, 1994, Chapter 4.
46. S. Kaiser, P. DiMascio, M.E. Murphey and H. Sies, *Arch. Biochem. Biophys.*, **277**, 101-8 (1990).
47. R.V. Cooney, P.J. Harwood, A.A. Franke, K. Narala, A.K. Sundström, P.O. Beggren and L.T. Mordan, *Free Rad. Biol. Med.*, **19**, 259-269 (1995).
48. J.F. Keaney and B. Frei in *Natural Antioxidants in Human Health and Disease*, Ed. B. Frei, Academic Press, San Diego, 1994, Chapter 11.
49. M. Nikishimi, *Biochem. Biophys. Res. Comm.*, **63**, 463-8 (1975).
50. B. Halliwell, M. Wasil and M. Grootveld, *FEBS Lett.*, **213**, 15-17 (1987).
51. B. Frei, L. England and B.N. Ames, *Proc. Natl. Acad. Sci.* USA **86**, 6377- 81 (1989).
52. J.E. Packer, T.F. Slater and R.L. Willson, *Nature*, **278**, 737-8 (1979).
53. B. Frei, R. Stocker and B.N. Ames, *Proc. Natl. Acad. Sci.* USA, **85**, 9748-52 (1988).
54. B. Frei, *Am. J. Clin. Nutr.* **54**, Suppl., 1113S-11183S (1991).
55. B. Frei, T.M. Forte, B.N. Ames and C.E. Cross, *Biochem. J.*, **277**, 133-8 (1991).
56. E. Niki, J. Tsuchiya and Y. Kamiya, *C. Chem. Lett.*, 789-92 (1982).
57. E. Niki, T. Saito, T. Kawakami and Y. Kamiya, *J. Biol. Chem.* 259, 4177-82 (1984).
58. J. Neužil and R. Stocker, *J. Biol. Chem.*, 269, 16712-9 (1994).
59. T. Doba, G.W. Burton and K.U. Ingold, *Biochim. Biophys. Acta*, **835**, 298-303 (1985).
60. H. Wefers and H. Sies, *Europ. Biochem. J.*, **174**, 353-7 (1988).
61. G. Burton, U. Wronska, L. Stone, D.O. Foster and K.U. Ingold, *Lipids*, **25**, 199-210 (1990).
62. H.H. Draper in *Atmospheric Oxidation and Antioxidants*, Vol. III, Ed. G. Scott, Elsevier, Amsterdam, 1993, Chapter 10.
63. J.F. Keaney and B. Frei in *Natural Antioxidants in Human Health and Disease*, Ed. B. Frei, Academic Press, San Diego, 1994, Chapter 11.
64. B.S. Winkler, *Biochem. Biophys. Acta*, **925**, 258-64 (1987).

References

65. J.M. May, Z-C. Qu, R.R. Whitesell and C.E. Cobb, *Free Rad. Biol. Med.*, **20**, 543-51 (1996).
66. J. Han, J. Martensson, A. Meister and O.W. Griffith, *FASEB J.* **6**, 5631 (1992).
67. G. Scott, *Atmospheric Oxidation and Antioxidants*, Elsevier, Amsterdam, 1965, Chapter 8.
68. S.P. Kochar in *Atmospheric Oxidation and Antioxidants*, Vol. II, Ed. G. Scott, Elsevier, Amsterdam, 1993, Chapter 2.
69. K. Kato, S. Terao, N. Shimamoto and M. Hirata, *J. Med. Chem.*, **31**, 793-8 (1988).
70. H. Tada, Y. Kutsumi, T. Misawa, N. Shimamoto, T. Nakai and S. Miyabo, *J. Cardiovasc. Pharmacol.*, **16**, 984-91 (1990).
71. M. Tanabe and G. Kito, *Japan. J. Pharmacol.*, **50**,.467-76 (1989).
72. J.A. Blair and G. Farrar in *Atmospheric Oxidation and Antioxidants*, Vol. III, Ed. G. Scott, Elsevier, Amsterdam, 1993, Chapter 6.
73. S. Kojima, I. Ona, I. IIZuka, T. Arai, H. Mori and K. Kubota, *Free Rad. Res.*, **23**, 419-30 (1995).
74. S.P. Gieseg, G. Reibnegger, H. Wachter and H. Esterbauer, *Free Rad. Res.*, **23**, 123-36.
75. C.A. Rice-Evans, N.J. Miller, P.G. Bolwell, P.M. Bramley and J.D. Pridham, *Free Rad. Res.*, **22**, 375-83 (1995).
76. C.A. Rice-Evans in *Biochem. Soc. Symp.*, **61**, 103-16 (1995).
77. A. Sevanian, K.J.A. Davies and P. Hochstein, *Free Rad. Biol. Med.*, **1**, 117-24 (1985).
78. K.J. Davies, A. Sevanian, S.F.M. Muakkassah-Kelly and P. Hochstein, *Biochem. J.* **235**, 747-54 (1986).
79. A.L. Lehninger, *Worth*, Second Edition, 1975, p.493 et seq.
80. K. Briviba and H. Sies, *Natural Antioxidants in Human Health and Disease*, Academic Press, San Diego, 1994, Chapter 4.
81. A. Mellors and A.L. Tappell, *J. Biol. Chem.*, **241**, 4353-56 (1966).
82. P. Forsmark-Andrée, G. Dallner and L. Ernster, *Free Rad. Biol. Med.*, **19**, 749-57 (1995).
83. P.R. Rich and D.S. Bendall, *Biochem. Biophys. Acta*, **592**, 506-511 (1980).
84. P. Neta and S. Steenken, *J. Phys. Chem.*, **93**, 7654- 9 (1982).
85. K. Mukai, S. Kikuchi and S. Urano, *Biochem. Biophys. Acta*, **1035**, 77-83 (1990).
86. V.E. Kagan and L. Packer in *Free Radicals and Antioxidants in Nutrition*, Eds. F. Corogiu, S. Banni, M.A. Dessi and C. Rice-Evans, Richelieu Pess, London, 1993, p.27-36.
87. R. Stocker and B.N. Ames, *Proc. Natl. Acad. Sci.*, USA, 84, 8130-4 (1987).
88. R.J. Reiter, *BioEssays*, 14, 169-75 (1992).
89. R.J. Reiter and B.A. Richardson, *FASEB J.* 6, 2283-7 (1992).
90. R.J. Reiter, *J. Pineal Res.*, 18, 1-11 (1995).
91. K-A. Marshall, R. Reiter, B. Poeggeler, O.I. Aruma and B. Halliwell, *Free Rad. Biol. Med.*, **21**, 307-15 (1996).
92. C. Pieri, M. Marra, F. Maroni, R. Decchioni and F. Marcheselli, *Life Sci.*, **55**, 271-6 (1994).
93. D.E. Blask in *Melatonin*, Eds. H.S. Yu and R.J. Reiter, CRC Press, Boca Raton, 1993, pp.447-76.
94. W. Pierpaoli and W. Regelson, *Proc. Nat. Acad. Sci.*, **91**, 787-91 (1994).

95. H.S. Dweik and G. Scott, *Rubb. Chem. Tech.*, **57**, 735-43 (1984).
96. D.X. Tan, B. Poeggeler, R.J. Reiter, L.D. Chen, L.C. Manchester and L.R. Barlow-Walden, *Cancer Lett.*, **70**, 65-71 (1993).
97. D.X. Tan, R.J. Reiter, L.D. Chen, B. Poeggeler, L.C. Manchester and L.R. Barlow-Walden, *Carcinogenesis*, **15**, 215-8 (1994).
98. D. Melchiorri, R.J. Reiter, A.M. Attia, M. Hara, A. Burgos and G. Nistico, *Life Sci.*, **56**, 83-85 (1995).
99. W.M.U. Daniels, R.J. Reiter, D. Melchiorri, E. Sewerinek, M.I. Pablos and G.G. Ortiz, *J. Pineal Res.*, **19**, 1-6 (1995).
100. S. Torisson, F. Gunstone and R. Hardy, *J. Am. Oil Chem. Soc.*, **69**, 806 (1992).
101. P.M. Chen, D.M. Varya, E.A. Mielke, T.J. Facteau and S.R. Drake, *J. Food Sci.*, **55**, 167, 171 (1990).
102. L.A. Huber, E. Scheffler, T. Poll, R. Ziegler and H.A. Dressel, *Free Rad. Biol. Med.*, **8**, 167-73 (1990).
103. V.A. Rifici and A.K. Khachadurian, *Metabolism*, **41**, 1110-14 (1990).
104. H. Wiseman, G. Paganga, C. Rice-Evans and B. Halliwell, *Biochem. J.*, **292**, 635-8 (1993).
105. H. Rubbo, M. Tarpey and B.A. Freeman, *Biochem. Soc. Symp.*, **61**, 33-45, (1995), Free Radicals and Oxidative Stress, Eds. C. Rice-Evans, B. Halliwell and G.G. Lunt, Portland Press, London, 1995.
106. C.A. O'Neill, A. van der Vliet, J.P. Eiserich, J.A. Last, B. Halliwell and C.E. Cross, Biochem. Symp., **61**, 139-152 (1995).
107. G. Scott, *Atmospheric Oxidation and Antioxidants*, Elsevier, Amsterdam, 1965, p.214-5.
108. I.B. Afanas'ev, A.I. Dorozhko, A. Kostyuk and A.I. Ptapovitch, *Biochem. Pharmacol*, **38**, 1763-9 (1989).
109. M.J. Laughton, P.J. Evans, M.A. Morony, J.R.S. Holt and B. Halliwell, *Biochemical Pharmacology*, **42**, 1673-81 (1991).
110. I. Morel, G. Lescoat, P. Cognel, O. Sergent, N. Pasdeloup, P. Brissot, P. Cillard and J. Cillard, *Biochemical Pharmacol.*, **4**, 13-19 (1993).
111. H. Mangiapane, J. Thompson, A. Salter, S. Brown, G.D. Bell and D.A. White, *Biochem. Pharmaocol.* **43**, 445-50 (1992).
112. M.J. Laughton, B. Halliwell, P.J. Evans and J.R.S. Hoult, *Biochem. Pharmacol.*, **38**, 2859-65 (1989).
113. B. Halliwell and J.M.C. Gutteridge, *Free Rad. Biol. Med.*, Clarenden Press, 1989, p.484.
114. W. Bors, W. Heller, C. Michel and M. Saran in *Methods in Enzymology*, Academic Press, **186**, 343-55 (1990).
115. U. Takahama, *Photochem. Photobiol.*, **42**, 89-91 (1985).
116. C. Yuting, Z. Rongliang and J. Yong, *Free Rad. Biol. Med.*, **9**, 19-21 (1990).
117. J.B. Harborne in *Plant Pigments*, Ed. T.W. Goodwin, Academic Press, San Diego, 1988, p.299-343.
118. J. Terao, M. Piskuli and Q. Yao, *Arch. Biochem. Biophys.*, **308**, 278-84 (1994).
119. C. DeWhalley, S.M. Rankin, J.R.S. Hoult, W. Jessup and D.S. Leake, *Biochem. Pharmacol.*, **39**, 1743-50 (1990).

120. E.N. Frankel, J. Kanner, J.B. German, E. Parks and J.E. Kinsella, *Lancet*, **341**, 454-7 (1993).
121. S.M. Rankin, C.V. DeWhalley, J.R.S. Hoult, W. Jessup, G. Wilkins, J. Collard and D.S. Leake, *Biochem. Pharmacol.*, **45**, 67-75 (1993).
122. J. Pincemail, C. Deby, Y. Lion, P. Braquet, P.Hans and R.Goutier in *Flavanoids and Bioflavanoids*, Eds. L. Parkas, M. Gabor and F. Kallay, Elsevier, 1986, p.423-36.
123. S.Z. Dziedzic, B.J.F. Hudson and G. Barnes, *J. Agric. Food Chem.*, **33**, 244-6 (1985).
124. G.G. Duthie, D.B. McPhail and P.C. Morrice, *Symposium on Antioxidants, Lipids and Disease*, SCI Lecture Paper 0018, December 1993.
125. G.G. Duthie, private communication.
126. S.A.B.E. van Acker, D-J. van den Berg, M.N.J.L. Tromp, D.H. Griffionen, W.P. van Bennekom, W.J.F. van der Vijgh and A. Bast, *Free Rad. Biol. Med.*, **20**, 331-42 (1996).
127. S.P. Kochar in *Atmospheric Oxidation and Antioxidants*, Ed. G. Scott, Elsevier, Amsterdam, 1993, Chapter 2.
128. *Antioxidant Nutrition*, R. Greer and R. Woodward, Souvenier Press, 1995, Chap. 5.
129. O.I. Aruoma, J.P.E. Spencer, J. Butler and B. Halliwell, *Free Rad. Res.*, **22**, 187-90 (1995).
130. T.W. Kensler and K.Z. Gutyon in *Atmoshperic Oxidation and Antioxidants*, Vol. III, Ed. G. Scott, Elsevier, Amsterdam, 1993, Chapter 12.
131. E. Graf, *Free Rad. Biol. Med.*, **13**, 435-48.
132. B.C. Scott, J. Butler, B. Halliwell and O.I. Aruoma, *Free Rad. Res. Comm.*, **19**, 241-53 (1993).
133. R.D. Hartley and E.C. Jones, *Phytochem.*, **16**, 1531-4 (1977).
134. G. Papadopoulos and D. Boskou, *J. Am. Oil. Chem. Soc.*, **68**, 669-71 (1991).
135. C. Nergiz and K. Unal, *Food Chem.*, **39**, 237-40 (1991).
136. G. Montedoro, M. Servili, M. Baldioli and E. Miniati, *J. Agric. Food Chem.*, **40**, 1571-6 (1992).
137. C. Scaccini, M. Nardini, M. D'Aquino, G. Ventili, M. Di Felici and G. Tomassi, *J. Lipid Res.*, **33**, 627-33 (1992).
138. M. Nardini, M. D'Aquino, G. Tomassi, V. Gentill, M. Di Fekice and C. Scaccini, *Free Rad. Biol. Med.*, **19**, 541-52 (1995).
139. J. Terao, *Lipids*, **24**, 659-61 (1989).
140. T.A. Kennedy and D.C. Liebler, *Chem. Res. Toxicol.*, **4**, 290-5 (1991).
141. G.J. Handelman, F.J.G.M. van Kuik, A. Chatterjee and N. Krinsky, *Free Rad. Biol. Med.*, **10**, 427-37 (1991).
142. O.A. Ozhogina and O.T. Kasainka, *Free Rad. Biol. Med.*, **19**, 578-81 (1995).
143. W.A. Pryor, T. Stickland and D.F. Church, *J. Am. Chem. Soc.*, **110**, 2224-9 (1988).
144. G.W. Burton and K.U. Ingold, *Science*, **224**, 568-74 (1984).
145. E.L. Gibbs, G.W. Lennox, L.F. Nims and F.A. Gibbs, *J. Biol. Chem*, **144**, 325-32 (1942).
146. P. Palozza and N. Krinsky, *Arch. Biochem. Biophys.*, **297**, 184-7 (1992).
147. M.A. Livrea, L. Tesoriere, A. Bongiorno, A.M. Pintaudi, M. Ciaccio and A. Riccio, *Free Rad. Biol. Med.*, **18**, 401-9 (1995).
148. B. Halliwell and J.M.C. Gutteridge, *Free Radicals in Biology and Medicine*, Second Edition, Clarendon Press, Oxford, 1989, Chapter 3.

149. J.M.C. Gutteridge and B. Halliwell in *Atmospheric Oxidation and Antioxidants*, Vol. III, Ed. G. Scott, Elsevier, Amsterdam, 1993, Chapter 3.
150. *Antioxidant Nutrition*, R. Greer and R.Woodward, Souvenier Press, 1995, Chapter 6.
151. J.M. Stark, S.K. Jackson and J. Parton in *Free Radicals: Chemistry, Pathology and Medicine*, Eds. C. Rice-Evans and T. Dormandy, Richelieu Press, London, 1988, pp. 187-210.
152. H. Melhorn and A.R. Wellburn in *Free Radicals: Chemistry, Pathology and Medicine*, Eds. C. Rice-Evans and T. Dormandy, Richelieu Press, London, 1988, pp.253-270.
153. I. Fridovich, *Meth. Enzymology*, **58**, 61 (1986).
154. T.J. Reid, M.R.N. Murty, A. Sicignanao, N. Tanaka, W.D.L. Musick and M.G. Rossman, *Proc., Natl. Acad. Sci. USA*, **78**, 4767-71 (1981).
155. D. Metodiewa and B. Dunford in *Atmospheric Oxidation and Antioxidants*, Vol. III, Ed. G. Scott, Elsevier, Amsterdam, 1993, Chapter 11.
156. C. Walling, *Acc. Chem. Res.*, **8**, 125-31 (1975).
157. L. Flohé in *Free Radicals in Biology*, Ed. W.A. Pryor, Academic Press, New York, Vol V, Chapter 7.
158. L. Flohé, W. Günzler, G. Jung, E. Schaich and F. Schneider, *Hoppe Seiler's Zeit. Physiol. Chem.*, **352**, 159 (1971).
159. T.C. Stadtman, *Ann. Rev. Biochem.*, **49**, 93 (1980).
160. Z. Osawa and K. Nakano, *J. Polym. Sci. Symp.*, **57**, 267 (1976).
161. F. Ursini, M. Mariorino and C. Gregolin, *Biochim. Biophy. Acta*, **839**, 62 (1985).
162. L.J. Reed, *Acc. Clin. Res.*, **7**, 40-46 (1974).
163. H.R. Rosenberg and R. Culik, *Arch. Biochem. Biophys.*, **80**, 86-93 (1959).
164. L. Packer, E.H. Witt and H.J. Tritschler, *Free Rad. Biol. Med.*, **19**, 227-50 (1995).
165. V.E. Kagan, A. Shvedova, E. Serbinova, S. Khan, C. Swanson, R. Powell and L. Packer, *Biochem. Pharmacol.*, **44**, 1637-49 (1992).
166. Y.J. Suzuki, M. Tsuchiya and L. Packer, *Free Red. Res. Comm.*, **18**, 115-22 (1993).
167. J.P. Carreau, *Meth. Enzymol.*, **62**, 152-8 (1979).
168. P.C. Jocelyn, *Europ. J. Biochem.*, **2**, 327-31 (1967).
169. Y.J. Suzuki, M. Tsuchiya and L. Packer, *Free Rad. Res. Comm.*, **15**, 255-63 (1991).
170. B. Stevens, S.R. Perez and R.D. Small, *Photochem. Photobiol.*, **19**, 315-6 (1974).
171. A. Bast and G.R.M.M. Haenen in *Antioxidants in Therapy and Preventive Medicine*, Eds. I. Emerit, L. Packer and C. Auclair, Plenum Press, New York, 1990, pp. 111-6.
172. M.E. Gotz, A. Dirr, R. Burger, B. Janetzky, M. Weinmuller, W.W. Chan, S.C. Chen, H. Reichmann, W.D. Rausch and P. Riederer, *Europ. J. Pharmacol.*, **266**, 291-300 (1994).
173. L. Packer in *Tioocseure: 2 International thioctic acid workshop*, Eds. K. Schmidt and H. Ulrich, Universimed Verlag Gmbh, 1992, 35-44.
174. J.B. Porter, E.R. Huehns and R.C. Hider in *Baillièe's Clinical Haematology*, Vol. 2, No.2, 1989, Chapter 2.
175. M.C. Linder and H.N. Munroe, *Analyt. Biochem.*, 241, 295 (1972).
176. B. Halliwell and J.M.C. Gutteridge, *Free Rad. Biol. Med.* Second Edition, Clarendon Press, Oxford, pp. 249-260.
177. T. Sarna and H.M. Swartz in *Atmospheric Oxidation and Antioxidants*, Vol. III, Ed. G. Scott, Elsevier, Amsterdam, 1993, Chapter 5.
178. T.Sarna, W.Korytowski and R.C. Sealey, *Arch. Biochem. Biophys.*, **239**, 226 (1985).

179. R. Dunford, E.J. Land, M. Rozanowska, T. Sarna and T.G. Truscott, *Free Rad. Biol. Med.*, **19**, 735-40 (1995).
180. McCance and Widdowson's *The Composition of Foods*, Fifth Edition, Eds. B. Holland, A.A. Welch, I.D. Unwin, D.H. Buss, A.A. Paul and D.A.T. Southgate, Royal Society of Chemistry and Ministry of Agriculture, Fisheries and Food, 1994.
181. Personal communication, R. Swift, Food Research Association, Leatherhead, 1995.
182. Personal communication, A. Sinclair, Safeway Stores plc (1995).
183. Personal communication, G. Burrows, Matthews Foods plc (1995).
184. D. Kromhout, E.B. Bosschieter and C. deLezenne, *Lancet*, **ii**, 518-22 (1982).
185. C.M. Ripsin, J.M. Keenan, D.R. Jacobs, P.J. Elmer, R.R. Welch et al., *JAMA*, **267**, 3317-25 (1992).
186. J.N. Morris, J.W. Marr and D.G. Clayton, *BMJ*, **2**, 1307-14 (1977).
187. E. Middleton, *Pharmaceutical News*, **1**(3), 6-8 (1994).
188. B. Zhao, X. Li, S. Cheng and X. Wenjuan, *Cell Biophys.*, **14**, 175-185 (1989).
189. F. Nanjo, M. Honda, K. Okushio, N. Matsumoto, F. Ishigaki, T. Ishigami and Y. Hara, *Biol. Pharm. Bull.*, **16**, 1156-9 (1993).
190. B.J.P. Quartley, M.N. Clifford, R. Walker and C.M. Williams in *Recent Advances in the Chemistry and Biochemistry of Tea*, SCI Lecture Papers, 0029 (1994).
191. S. Renaud and M. de Lorgeril, *Lancet*, **339**, 1523-6 (1992).
192. M.G.L. Hertog, P.C.H. Hollman, M.B. Katan and D. Kromhaut, *Nutr. Cancer*, **20**, 21-29 (1993).
193. M.G.L. Hertog, M.G.L. Fesrens, P.C.H. Holman, M.B. Katan and D. Kromhout, *Lancet*, **342**, 1007-11 (1993).
194. T.W. Kensler, N.E. Davidson and K.E. Guyton in *Atmospheric Oxidation and Antioxidants*, Vol. III, Ed. G. Scott, Elsevier, Amsterdam, 1993, Chapter 12.
195. G. Scott and R. Suharto, *Europ. Polym. J.*, **20**, 139 (1984).
196. A. Ogunbanjo and G. Scott, *Europ. Polym. J.*, **21**, 541-4 (1985).
197. S. Parthasarathy, *J. Clin. Invest.* **89**, 1618-21 (1992).
198. R.H. Bisby, S.A. Johnson and A.W. Parker, *Free Rad. Biol. Med.*, **20**, 411-420 (1996).
199. S.J.T. Mau, M.T. Yates, R.A. Parker, E.M. Chi and R.L. Jackson, *Arterioscler. Thromb.*, **11**, 1266-75 (1991).
200. L. Tamir, M. Prusiková and J. Pospíšil, *Angew. Makromol. Chem.*, **190**, 53 (1991).
201. J.A. Malkin, N.U. Piroga, J.A. Ivanov, I.E. Pokrovskaya and V.A. Kuzmin., *Izv. Akad. Nauk, ser. khim.*, 2008 (1981).
202. D.V. Fentsov, T.B. Lobanova and O.T. Kasaikina, *Neftekhymia*, **30 (1)**, 103 (1990).
203. F. Tanfani, P. Carloni, E. Damiani, L. Greci, M. Wozniak, D. Kulawiak, K. Jankowski, J. Kaczor and A. Matuszkiewics, *Free Rad. Res.*, **21**, 309-15 (1994).
204. G.P. Novelli, P. Angiolini, R. Tani, G. Consales and L. Bordi, *Free Rad. Res. Comms.*, **1**, 321-37 (1986).
205. E. Monti, G. Paracchini, G. Perletti and F. Piccinini, *Free Rad. Res. Comms.* **14**, 41-45 (1991).
206. F. Piccinini, S. Bradamante, E. Monti, Y-K. Zhang and E.G. Janzen, *Free Rad. Res.*, **23**, 81-7 (1995).
207. V.C. Jordan, *Brit. J. Pharmacol.*, **110**, 507-17 (1993).
208. R.J. Chlebowski, J. Butler, A. Nelson and L. Lillington, *Cancer*, **72**, 1032-7 (1993).

209. H. Wiseman, *Tamoxifen: Molecular Basis of Use in Cancer Treatment and Prevention*, Wiley, Chichester, 1994.
210. H. Wiseman, G. Paganga, C. Rice-Evans and B. Halliwell, *Biochem. J.*, **292**, 635-8 (1993), H.Wiseman, Biochem. Pharmacol., **47**, 493-8 (1994).
211. H. Wiseman in Biochem. Soc. Symp. **61**, *Free Radicals and Oxidative Stress*, Eds. C. Rice-Evans, B. Halliwell and G.G. Lunt, Portland Press, London, 1995, pp.209-219.
212. C.C. McDonald and H.J. Stewart, *BMJ*, **303**, 435-7 (1991).
213. B. Halliwell and J.M.C. Gutteridge, *Free Rad. Biol. Med.* (Second Edition), Clarendon Press, 1989, p.429.
214. W.A. Waters, *Chem. Soc. Ann. Rep.*, **42**, 130-55 (1945).
215. B. Halliwell, *FEBS Lett.*, **92**, 321-6 (1978).
216. R.A. Floyd, J.J. Watson and P.K. Wong, *J. Biochem. Boiphys. Meth.*, **10**, 221-35 (1984).
217. R.A. Floyd, R. Henderson, J.J. Watson and P.K. Womg, *Free Rad. Biol. Med.*, **2**, 13-18 (1986).
218. M. Ingelman-Sundberg, H. Kaur, Y. Terelius, J.O. Persson and B. Halliwell, *Biochem. J.*, **276**, 753-7 (1991).
219. R.P. Mason and V. Fischer, *Fed. Proc.*, **45**, 2493-9 (1986).
220. G. Scott, *Atmospheric Oxidation and Antioxidants*, Elsevier, 1965, pp.125 et seq.
221. W.E.M. Lands, R.J. Kulmacz and P.J. Marshall in *Free Radicals in Biology*, Vol. VI, Ed. W.A. Pryor, Academic Press, New York, 1984, Chapter 2.
222. N.R. Harris, B.J. Zimmerman and D.N. Granger in *Oxygen Free Radicals in Tissue Damage*, Eds. M. Tarr and F. Samson, Birkhäuser, 1992, Chapter 7.
223. P.C. Moorhouse, M. Grootveld, B. Halliwell, J.G. Quinlan and J.M.C. Gutteridge, *FEBS Lett.*, **213**, 23-8 (1987).
224. K.M. Ko and D.V. Godin, *Biochem. Pharmacol.*, **40**, 803-9 (1990).
225. G.A. Fitzgerald, *N. Engl. J. Med.*, **316**, 1247-57 (1987).
226. N. Morisaki, J.M. Slitts, L. Bartels-Tomei, G.E. Milio, R.V. Panganamala and D.G. Cornwell, *Artery*, **11**, 88-107 (1982).
227. L. Iuliano, F. Violi, A. Ghiselli, C. Alessandri and F. Balsano, *Lipids*, **24**, 430-3 (1989).
228. L. Iuliano, J.Z. Pedersen, G. Rotilio, D. Ferro and F. Violi, *Free Red. Biol. Med.*, **18**, 239-47 (1995).
229. L.W. Wattenberg, *J. Nat. Cancer Inst.*, **54**, 1005-6 (1975).
230. J. Liu, M.K. Shigenaga, L-J. Yan, A. Mori and B.N. Ames, *Free Rad. Res.*, **24**, 461-72 (1996).
231. D. Schulz, J.T. Skamarauskas, N. Law, M.J. Mitchinson and J.V. Hunt, *Free Rad. Res.*, **23**, 259-71 (1995).
232. Y.P. Vedernikov, P.I. Mordvinteev, I.V. Malenkova and A.F. Vanin, *Europ. J. Pharmacol.*, **212**, 125-8 (1992).
232. F.W. Sunderman, *J. New Drugs*, **4**, 151-61 (1964).

234. S. Mankhetkorn, Z. Abedinzadeh and C. Houee-Levin, *Free Rad. Biol. Med.*, **17**, 517-27 (1994).
235. *Atmospheric Oxidation and Antioxidants*, G. Scott, Elsevier, Amsterdam, 1965, 193.
236. J.D. Holdsworth, G. Scott and D. Williams, *J. Chem. Soc.*, 4692-9 (1964).

237. G. Scott in *Mechanisms of Reactions of Sulphur Conpounds*, Ed. N. Kharash, Vol. 4, 1969, pp. 99-110.
238. C. Armstrong, F.A.A. Ingham, J.G. Pimblott and J.E. Stuckey, *Proc. Int. Rubb. Conf.*, F2.1 (1972).
239. G. Scott, *S. Afr. J. Chem.*, **32**, 137-46 (1979).
240. S. Al-Malaika, K.B. Chakraborty and G. Scott, *Developments in Polymer Stabilisation-6*, Ed. G. Scott, Applied Science Pub., London, 1983, Chapter 3.
241. W.L. Hawkins and H.J. Sautter, *J. Polym. Sci.*, **A1**, 3499 (1969).
242. I. Kúdelka, J. Brodilová, M. Prusiková and J. Pospísil, *Polym. Deg. Stab.*, **17**, 287-301 (1987).
243. P.H. Ward, M. Maldonado, J. Roa, V. Manriques and E. Vivaldi, *Free Rad. Biol. Med.*, **22**, 561-69 (1995).
244. T.P. Kennedy, N.V. Rao, W. Noah, R.J. Michael, M.H. Jafri, G.H. Gurner and J.R. Hoidal, *Investigation*, **86**, 1565-73 (1990).
245. A.G. Coran, E.A. Litchen, J.J. Paik and D.G. Remick, *J. Surg. Res.*, **53**, 272-9 (1992).
246. N. Ercal, P. Treeratphan, T.C. Hammond, R.H. Matthews, N.H. Grannemann and D.R. Spitz, *Free Rad. Biol. Med.*, **21**, 157-61 (1996).
247. E.J. Corey, R.A.Mehrota and A.U. Kahn, **236**, 68-70 (1987).
248. A. Bendich in *Natural Antioxidants in Human Health and Disease*, Ed. B. Frei, Academic Press, San Diego, 1994 Chapter 15.
249. J.B. Porter, E.R. Huehns and R.C. Hider in *Ballière's Clinical Haematology, Vol.2, No. 2*, 1989, Chapter 2.
250. P.S. Dobbin and R.C. Hider, *Chem. in Brit.* 565-8 (June, 1990).
251. R.C. Hider, S. Singh and J.B. Porter, *Proc. Roy. Soc. Edinburgh*, **99B(1/2)**, 137-68 (1992).
252. R.C. Hider, J.B. Porter and S. Singh in *The Development of Iron Chelators for Clinical Use*, Eds. R.J. Bergeron and G.M. Brittenham, CRC Press, 1994, Chap 16.
253. S. Singh, H. Khodr, M.I. Taylor and R.C. Hider in *Free Radicals and Oxidative Stress, Environment, Drugs and Food Additives*, Eds. C. Rice-Evans, B. Halliwell and G.G. Lunt, Biochem. Soc. Symp., **61**, 127-37 (1995).
254. M.J. Davies, R. Donker, C.A. Dunster, C.A. Gee, S. Jones and R.L. Willson, *Biochem. J.*, **246**, 725-9 (1987).
255. V.A. Levin, *J. Med. Chem.*, **23**, 682-7 (1980).
256. E.D. Hall, P.A. Yonkers, P.K. Andrus, J.W. Cox and D.K. Anderson, *J. Neurotrauma*, **9**, 5425-41 (1992).
257. E.D. Hall, J.M. Braughler and P.A. Yonkers, *J. Pharmacol. Exp. Ther.*, **258**, 688-94 (1991).
258. E.D. Hall, J.M. Braughler and J.M. McCall, *J. Neurotrauma*, **9**, 5165-72 (1992).
259. E.D. Hall in *Oxygen Free Radicals in Tissue Damage*, Eds. M. Tarr and F. Samson, Birkhäuser, 1992, Chapter 9.
260. C-S. Yang, N-N. Lin, P-J. Tsai, L. Liu and J-S. Kuo, *Free Rad. Biol. Med.*, **20**, 245-50 (1996).
261. E.D. Hall, K.E. Pazara, J.M. Braughler, K.L. Linseman and E.J. Jacobson, *Stroke*, **21**, II-83-III-87 (1990).
262. P.L. Munns and K.L. Leach, *Free Rad. Biol. Med.*, **18**, 467-78 (1995).

6

Antioxidants in Disease and Oxidative Stress

6.1 Epidemiological Studies

It will be clear from Chapter 2 that the detrimental role of free radicals and peroxidation in biological systems has frequently been clarified by the intervention of known antioxidants *in vivo* and, in some acute diseases involving peroxidation, antioxidant therapy is already becoming well established (Chapter 5, Section 5.5). However, an important corollary of the above concept is that the prevalence and severity of diseases involving oxidation might be expected to be associated with deficiencies of antioxidants in the diet. A considerable number of such epidemiological comparisons have been carried out both within countries and between different areas of the world where major differences in diet are known to exist. Particularly interesting has been the evidence for increase in cardiovascular disease in countries like Israel whose dietary pattern has changed markedly over a thirty year period from a substantially fruit and vegetable diet to the fat and meat western diet [1]. Similarly convincing is the migration of individuals from Japan to the USA with associated increase in CVD [2].

Regional differences have attracted a good deal of attention from epidemiologists and in particular the increase in heart disease and cancer in the northern countries of Europe compared with southern European countries [3-5]. Fig. 6.1 compares the mortality from coronary heart disease for 31 countries for both men and women [5]. Scotland has the highest death rate closely followed by the USSR and Czechoslovakia. These countries contrast sharply with Japan and the southern European countries, notably France, Portugal, Spain and Italy. The mid-European countries and Australia and the USA fall in the middle band. The incidence of cancers shows a similar trend between northern and southern countries although there are marked variations depending on the type of cancer.

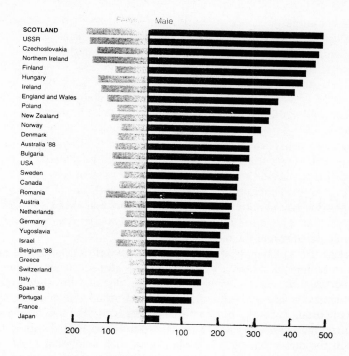

Fig. 6.1 Incidence of coronary heart disease (1989). Mortality per 100,000, age 49-69. (Reproduced with permission from The Scottish Diet, Report of a Working Party to the Chief Medical Officer for Scotland, Scottish Office Home and Health Department, 1994, p.10.)

There is a significant difference in CHD between Scotland on the one hand and England and Wales on the other, which tends to rule out genetic differences as a major cause; and in confirmation of this, there are also very significant differences favouring the remoter rural populations such as Orkney, Shetland and the Western Isles and the major conurbations (e.g. Greater Glasgow), and it is concluded that inner city social deprivation and consequently diet is a major factor in determining mortality as reflected in the diet of the respective populations [5]. This is again confirmed by the incidence of cancer in different social groups. Stomach cancer is most closely associated with social class. Thus in 1971 mortality was 50% greater among manual workers than in non-manual and this difference does not appear to change appreciably with time. However, using other types of cancer as criteria it was found that lung cancer correlates inversely with social prosperity, but malignant melanoma correlates directly. This seems to indicate the secondary correlations with smoking for lung cancer and exposure to UV for melanoma. However, the same evidence cannot exclude a contribution from pollution by the inner urban environment.

In cross-cultural comparisons, a vegetarian diet has been associated with lower mortality from ischaemic heart disease; and mortality from heart disease in England, Wales, Scotland, Norway and Israel has been reported to be inversely correlated with the **calculated** ascorbic acid intake from fresh fruits and green vegetables based on the food tables [6]. Other surveys have shown a similar correlation between fruit and vegetable intake and the incidence of cancers [7-10]. However, it is recognised that as discussed in Chapter 5, the food tables have inherent inaccuracies due to losses during storage, processing, cooking, etc. Furthermore, fruits and vegetables contain a variety of protective agents, such as flavanoids, fibre, etc. and this has led to over-prescriptive government guide-lines on the daily intake of specific items of foodstuffs by the general public [11], some of which (notably the emphasis on polyunsaturates without due consideration to antioxidant content) will almost certainly turn out to be misplaced during the next few years. Some campaigning organisations go further and even recommend taking a pocket calculator to the supermarket to minimise fat intake with no reference at all to what essential antioxidant nutrients the fats contain [12]. This kind of advice is almost irrelevant to those members of society most at risk and places a strong emphasis upon supplementation of staple foods based on a thorough scientific understanding of the role of antioxidant nutrients in the modern diet. It was seen in the previous chapter that some essential antioxidant nutrients are removed in modern food processing to the detriment of the modern diet.

The importance of vitamin E as an antioxidant component of the diet of animals was recognised by Dam [13] and others in the late 1950s (for a concise review see Tappel [6]). It was not until the early 1980s however that Gey and his co-workers began to systematise the epidemiological study of the role of antioxidants in specific diseases such as atherosclerosis, by correlating antioxidant status in diet and in plasma with the incidence and severity of the disease in human populations [14]. Thurnham et al. [15] recognised that the concentration of vitamin E by itself in plasma is not a sufficient criterion of antioxidant status and that this has to be related to the polyunsaturated lipid concentration. The importance of lipid-adjusted vitamin E as opposed to absolute vitamin E concentration in cross-cultural epidemiology can be illustrated by the fact that in China where there is a low prevalence of CHD, the absolute α-tocopherol levels in plasma are relatively low [16,17], but because of a low fat intake in the diet the α-toco-pherol/cholesterol ester ratio is relatively high (5.35 µmol/mmol). Antioxidant status must therefore be "normalised", either using the vitamin E/LDL-cholesterol ratio [14,15,18] or by adjusting to an average European level of 220 mg/dl cholesterol + 110 mg/dl triglycerides [15,19,20].

In view of the limitations of epidemiological studies noted above it must be accepted that this approach to the cause of human illness cannot alone establish an unequivocal link between cause and effect, since, unlike laboratory animals it is not possible to exclude confounding factors which result from the environment and personal habits of the human subject. Confounding factors can be reduced by including carefully matched controls or by interrogative analysis of the environment and personal life-style of the individual, but studies of human populations can never match in deductive ability the study of laboratory animals under carefully controlled conditions or the behaviour of isolated cells in the laboratory. It is likely that in the final analysis, the "proof" that human beings demand to guide their own behaviour can only come from such rigorous scientific investigations interpreted in the light of epidemiological studies.

Epidemiological investigations of the role of antioxidants in disease prevention have progressed over the past ten years from descriptive studies to randomised intervention trials where antioxidants and placebos are given to members of a population on a "double blind" basis. The progress in 1994 toward this situation has been outlined by

Gaziano et al. [21]. These investigations are categorised in order of increasing analytical accuracy as follows:

a) Descriptive studies of populations which are further divided into those based on dietary intake and those involving analysis of antioxidants in blood plasma or tissues.
b) Case-control studies, in contrast to descriptive studies of populations, data obtained from individuals with a particular disease are matched with appropriate controls in the healthy population and antioxidant levels in both groups are measured at the time of diagnosis. However, the effectiveness of this procedure still depends on the correct selection of controls which are representative of the healthy population in order to eliminate confounding factors.
c) Prospective (cohort) studies in which antioxidant status is measured prior to the development of disease. This procedure has the considerable advantage over case-control studies in that it reduces the influence of the disease itself on individual behaviour or antioxidant status in the plasma. Prospective studies are further divided into two categories, those involving only dietary intake as obtained from extended questionnaires, which are only as reliable as the individuals who complete them, and those based on measurement of plasma antioxidants. The latter are clearly more definitive but do give rise to problems if the plasma is stored before analysis since antioxidants and in particular vitamin C are slowly oxidised even when frozen.
d) Randomised intervention trials involve the use of diet supplements on "double blind" administration have the advantage that confounding variables are randomly distributed between supplemented and non-supplemented subjects.

6.2 Epidemiological Studies of Cardiovascular Disease
6.2.1 *Diet-based descriptive studies*
Five studies were carried out in the USA [22,23], UK [24,25] and Israel [1] between 1975 and 1985. Inverse associations were shown between CVD and fruit and vegetable consumption [1,23-25] and vitamin C intake [22].

6.2.2 *Plasma-based descriptive studies*
Gey and co-workers have carried out a cross-cultural comparison of vitamin C, vitamin E and selenium plasma levels between communities in northern and southern Europe which are known to be associated with high and low incidence of CVD respectively [3,4]. Fig. 6.2 correlates plasma vitamin E concentrations for a number of populations with CHD mortality [20]. This shows a reasonable correlation with the CHD mortality incidence data in Fig. 6.1. However, the deficiencies of "broad-brush" descriptive epidemiology are evident, notably the poor correlation observed with the Irish data, and this illustrate the difficulties of excluding confounding factors from this type of investigation. Also excluded from the data in Fig. 6.2 are anomalous results from Finland which show mortalities much higher than that expected on the basis of plasma vitamin measurements. This is believed but not so far proved to be due to a genetic abnormality (see below).

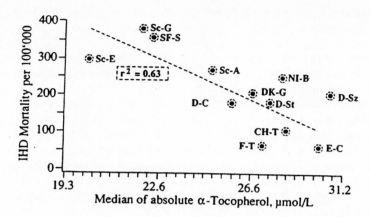

Fig. 6.2 Cross-cultural comparisons of the MONICA vitamin sub-study. Correlation between age-specific IHD mortality and the median of absolute α-tocpherol concentration. CH-T, Switzerland-Thun; D-C, Germany-Cottbus; D-St, Germany-Schwedt; D-Sz, Germany-Schleiz; DK-G, Denmark-Glostrup/Copenhagen; E, Spain, Catalonia, north of Barcelona; F, France, Toulouse/Haut Garonne; NI-B, Northern Ireland-semi-urban Belfast; Sc-A, Scotland-Aberdeen; Sc-E, Scotland-Edinburgh; Sc-G, Scotland-Glasgow. (Reproduced with permission from K.F. Gey, U.K. Moser, P. Jordan, H.B. Stählin, M. Eichholzer and E. Lüdin, *Am. J. Clin. Nutr.*, **57** (Suppl.), 789S (1993).

A much larger international study, within the WHO MONICA project (MONICA, MONItoring determinant and trends of CArdiovascular disease trends) following up Gey's hypothesis of the relationship between atherosclerosis and the antioxidant/fatty acid composition of diet [14] began in the late 1980s and is not yet completed. However, some very significant conclusions are beginning to emerge [16].

1. In European study populations with similar plasma lipoproteins (including the incidence of hypo and hypercholesterolemia) there was a strong inverse correlation between CHD mortality and vitamin E status [18,20]. Vitamin E status is a stronger predictor of CHD mortality than any of the classical risk factors, including plasma cholesterol, diastolic blood pressure, and smoking. Other essential antioxidants made relatively smaller contributions.
2. The tentative ranking order for CHD risk factors from 16 study populations is as follows:

    VE > Cholesterol > ß-C > VC > diastolic blood pressure > VA
    (vitamin antioxidants all inversely correlated)

3. The four communities with the lowest CHD mortality (Spain, France, Italy and Switzerland) had an "optimal" status of both VE and VC and three out of the four had very high β-C levels. However, Spain which had a low β-C status (comparable with Scotland), had a particularly high VC status [20], suggesting that low β-C levels become important only at sub-optimal VC levels and that VC can compensate for β-C deficiency but not necessarily the reverse.
4. Combining lipid standardised VE, VC and β-C in a multiple-regression analysis for all European communities but excluding Finland (see below), Gey showed that 88%

of the differences in ischaemic heart disease could be accounted for by differences in these antioxidants. Selenium appears to be much less significant. Although contributions from classical predictors were marginal, their inclusion explained a total of 90% of the differences.

5. Three Finnish populations did not fit the above analysis. The reason for this is not yet fully understood. The "Finland factor" is partially but not fully explained by hypercholesterolaemia. Genetic factors might be involved but there is no real evidence for this [20]. There is however evidence [26] that in a Finnish population with a very high level of CVD, selenium concentration was low. It is equally possible however that major dietary changes in recent years may have confused the outcome of the analysis. Furthermore, until the epidemiological studies are enlarged to take into account the relative importance of the flavonoids and dietary fibre as compared with the vitamin antioxidants, it will not be possible to assess their contribution in a particular population [20].

6. Compensation for deficiency of one antioxidant by an augmented level of another has been observed in both humans and animals for combinations of VC/VE [27-29], VE/β-C [30], VC/β-C [31] and for VE/GSH [32]. Although *in vitro* studies would suggest that all of these interactions could be synergistic (see Chapters 4 and 5), it is not clear from *in vivo* studies which of them are synergistic and which represent an additive effect of one antioxidant on another. Indeed, in the case of ß-carotene, it has been questioned whether the preventive effect of fruit and vegetables in cancer is due to the carotenoid itself or some so far unidentified factor in the diet.

7. Nevertheless, these preliminary data have provided information relating CHD mortality with antioxidant status [20]. Table 6.1 shows the levels of β-carotene, retinol, α-tocopherol and ascorbic acid required to reduce CHD mortality from >250/100,000 to <130/100,000.

Table 6.1 Plasma antioxidant status and CHD incidence [20]

Antioxidant	Moderately increased risk of CHD mortality (>250 deaths/100,000)	Low risk of CHD mortality (<130 deaths/100,000)
β-Carotene[*]	<0.3 μmol/l	> 0.4-0.6 μmol/l
Retinol[+]	<0.21 μmol/l	> 2.2 μmol/l
α-Tocopherol[+]	<24 μmol/l	> 27-28 μmol/l
α-Tocopherol/cholesterol ratio	<4.1 μmol/mmol	> 4.8-5.6 μmol/mmol
Ascorbic acid	<24 μmol/l	≅ 35-55 μmol/l

[*] Containing 15-25% α-carotene.
[+] Lipid standardised to 210 mg/dl cholesterol and 110 mg/dl triglycerides.

6.2.3 *Case-control studies*

Gey and co-workers [3,4] studied the relationship between plasma levels of vitamins A, C, E, β-carotene and risk of angina pectoris in Scottish males compared with healthy controls and found a significant inverse correlation. The odds ratio between the lowest and highest quartiles of lipid normalised VE was 2.98 (95% confidence interval (CI) = 1.07-6.7). VE and β-C showed similar relationships but VA did not. The antioxidant status of the lowest risk category (fifth quintile), see Table 6.2, corresponded closely with those found in the MONICA project but the greatest relative risk category was associated with lower antioxidant plasma levels than in the cross-cultural comparison.

Table 6.2 Plasma antioxidant status and relative risk of angina morbidity (Edinburgh case-control study [16])

Antioxidant	Plasma concn (µmol/l) for greatest relative risk (first quintile, RR=1.4-2.7)	Plasma concn (µmol/l) for lowest relative risk (fifth quintile, RR<1.2)
β-Carotene (with 15-25%α-)	< 0.26	> 0.50
Retinol	< 1.93	> 2.69
α-Tocopherol	<18.9	>28.2
α-Toc/cholesterol	< 3.1 (µmol/mmol)	> 4.6 (µmol/mmol)
Vitamin C	<13.0	>41.5

Smoking was found significantly to reduce the odds ratio, see Table 6.3 which includes data from another Scottish study [33], confirming the importance of smoking as an important cause of peroxidation and hence CVD risk by reducing antioxidant levels. Again sub-optimal antioxidant status is a better predictor of CHD than decrease in classical risk factors [16]. In Scottish males in the study age group, cholesterol levels in excess of 8 mmol/l occur in only in 10% of subjects and >10 mmol/l in only 1%.

Table 6.3 Relation between estimated daily dietary intake and plasma levels of antioxidant nutrients in smoking and non-smoking Scotsmen [3,306,308]

	β-Carotene		Vitamin C		Vitamin E		
	Intake mg	Plasma µmol/l	Intake mg	Plasma µmol/l	Intake mg	Plasma µmol/l	Ratio µmol/mmol
Edinburgh Angina Case-control Study							
Non-smokers	-	0.42	-	41	-	26	3.99
Smokers	-	0.34	-	24	-	24	3.63
Healthy, middle-aged Scotsmen (Glasgow/Aberdeen)							
Non-smokers	3.5	0.44	61	37	7.6	25	1.8
Smokers	2.9	0.31	49	18	7.6	25	1.6
Dietary survey, Aberdeen							
Non-smokers	2.5	-	61	45	5.7	15	2.75
Smokers	2.5	-	63	27	6.0	15	2.62

Other case-control studies were less successful in demonstrating a link between VE status and CVD in this case by the measurement of nutrients in adipose tissue [34].

6.2.4 *Prospective (cohort) studies*
Five major studies of antioxidant intake based on food questionnaires and CVD between 1983 and 1993 involving very substantial populations showed an inverse correlations

between VC, VE and β-C [35,36], VC [37], β-C [38], VC [39]. Only one showed no association (with VC intake) [40]. In three studies in which plasma was stored at low temperatures until analysis was required [41-43] only one [41] showed a significant inverse relationship between antioxidant (β-C) and CVD. In the case of the non-correlating studies [42,43], serum was stored at -20°C as compared with -70° for the positive result.

In the very large prospective Basel study (2974 middle-aged men) by Gey and his co-workers [44,45], plasma VC, VE and β-C were measured at baseline (i.e. at the beginning of the fourteen year period during which the development of IHD and stroke were monitored). Both VC and β-C showed a significant inverse correlation with both pathologies but VE did not. However, another study involving the measurement of thickening of the carotid artery showed an inverse correlation with both α-tocopherol, β-carotene and selenium [46]. An important finding from Gey's work is that in a prosperous and well nourished society as in the Basel study, the effects of vitamin antioxidant variation may be minimal compared with more socially deprived societies [4]. Cholesterol-standardised vitamin E in particular showed no concentration mediated effect above an α-tocopherol/cholesterol ratio of 5.2. Similar conclusions were reached from studies in Holland [42] and Germany [47].

6.3 Antioxidant Intervention and Supplementation
6.3.1 *Naturally occurring antioxidants*
The epidemiological studies outlined in the previous section raise important questions as to the adequacy of human diet. It might be expected that more developed societies would have a diet which is able to better resist degenerative diseases associated with peroxidation of cells. In fact the evidence seems to be in the opposite direction and, as will be discussed in more detail below, diet in the USA is much less adapted to providing the antioxidant nutrients demanded by the modern industrial environment than is the diet of southern Europe in spite of the highly prescriptive advice given by the US National Committees on Diet and Health [48].

A survey of all the epidemiological evidence referred to above by Gey [16] has indicated threshold plasma concentrations of antioxidants for minimum risk (optimal levels) defined as levels below which increase in antioxidant concentration results in a reduced the CVD or cancer risk. Gey has suggested [16,49] that the plasma concentrations of the four major risk antioxidants shown in Table 6.4 provide a basis for the formulation of a Recommended Optimum Intake (ROI). Inverse correlations have been found particularly for the carotenoids in the prevention of cancer and for vitamin E in protection against CVD [49], but a sub-optimal level of any single antioxidant significantly increases relative risk and combinations of sub-optimal levels of two antioxidants increases risk more than additively [16,20,50].

Table 6.4 Recommended optimal intake (ROI) of antioxidant vitamins [16]

Vitamin C   > 50 μmol/l
Vitamin E   > 30 μmol/l
α-Tocopherol/cholesterol ratio   > 5.2 μmol/mmol
Vitamin A   > 2.2 μmol/l
Total carotene > 0.5 μmol/l and β-carotene > 0.4 μmol/l

It might be expected that the diets of developed societies would match up to or exceed these minimum requirements. However, preliminary results from the US National Health and Nutrition Examination Survey (NHANES) published in 1992 [37] does not bear this out. A representative sample of US adults taking polyvitamin supplements was compared with subjects who were not. The relative risk of death from all causes in the polyvitamin group was reduced to 65% of the non-supplemented group and of CVD to 58%. The supplementing group had a daily intake of $\geq$ 130 mg of vitamin C but the non-supplementing group had only a mean of 22 mg/day. However, in the case of subjects with a daily intake of $\geq$ 130 mg VC from diet only, which should provide a plasma level well above 50 µmol/l, there was no significant reduction in CVD. A daily intake of 130 mg of vitamin C provides a plasma level well above the optimal 50 µmol/l "safety level" given above, whereas <50 mg does not supply the ROI even in non-smokers. In smokers the position is even worse (see Table 6.3). The position of non-supplemented subjects with a high vitamin C status (i.e. >130 mg) is at first sight difficult to explain until it is realised that the multivitamin supplement also contained 20 mg of α-tocopherol (RDA = 10 mg) and 1.5 mg vitamin A (RDA = 1.0 mg). This reinforces the importance of the synergistic (regenerative) interactions of vitamins C and E discussed in Chapter 5.

Even more telling was the Harvard Health Professionals multivitamin self-supplementation study [36]. This involved 39,910 health-oriented males (only 9% smoked) who took physical exercise and whose diet was relatively low in fat and high in fibre. Daily vitamin C intake even in non-supplementing subjects was high (mean >92 mg) so that the majority had a plasma level >50 µmol/l, which contrasts with that of the average American (see above). Not surprisingly then, there was no significant effect of vitamin C variation on CVD (infarcts, cardiac deaths) over a four year follow-up period. By contrast, vitamin E showed a significant inverse correlation. Thus a daily consumption of >60IU (>40 mg RRR-α-tocopherol, equivalent to >60 mg dl-α-tocopheryl acetate) reduced the multivariate risk of CVD to 0.64 relative to subjects consuming 1.6-6.9 IU. Even more significantly, daily supplements of >100 IU vitamin E reduced the risk of all coronary events to 0.54 compared with non-supplementing subjects.

The reasons for the imbalance between vitamins C and E in American (and probably also some European) diets is worth exploring in view of the strong recommendations that have been made toward a "healthy diet". The main reason for this is almost certainly the emphasis on fruit and vegetable (and hence vitamin C) and polyunsaturated oils and fats intake without giving due consideration to vitamin E requirements. Indeed, a major report from the British Department of Health [11] seriously undervalues the role of vitamin E in countering the adverse prooxidant effects of polyunsaturates. Consideration of Tables 5.10 and 5.11 in Chapter 5 shows that most vegetables and fruits contain relatively minor amounts of vitamin E and it would require considerably more than the recommended "five servings of fruits and vegetables daily" to provide the necessary protection [48]. Even in the case of the common "high" vitamin E fruits (blackberries, blackcurrants, mangoes, olives, etc.) and vegetables (brussel sprouts, parsnips, tomatoes or pumpkins), more than a kilogram would be required to supply the ROI of vitamin E from any or a mixture of these.

The most prolific source of vitamin E is polyunsaturated oils, but as was noted in Chapter 5, a significant proportion of the antioxidant is removed during processing and conversion to edible products such as margarines, spreads, mayonnaise, etc. Recently more health conscious sections of the food industry have taken action to replace these losses but even today the cheaper grades of these staple foodstuffs are not supplemented even to the minimum level to protect the polyunsaturated content of the fat from oxidation (see below), and the vitamin E or α-tocopherol content of the food is rarely listed on the package. Supplementation of products with α-tocopherol at least to the level of the original oil would seem to be a prudent first step. However, it has to be recognised that even this may not be sufficient in highly polyunsaturated products, since although the amount of vitamin E in the original oil was sufficient to delay rancidity in derived products under ambient conditions, it is almost certainly not sufficient to protect the same polyunsaturated esters under the more oxidatively aggressive conditions in the tissues of humans with a generally unhealthy life-style (particularly smokers see below). It has been estimated [51] that 1g of linoleic acid requires 0.6 mg α-tocopherol and 1g linolenic acid requires 0.9 mg α-tocopherol even in the healthy body to prevent peroxidation of the polyunsaturated component of the cell.

Gey [52] has pointed out that a major difference between the dietary components of corn and soybean oils which are the staple polyunsaturates in the USA and in sunflower and safflower oils consumed in southern Europe, is that the former contain γ-tocopherol as the main vitamin E antioxidant whereas α-tocopherol, which has been shown to be more effective in the cell, is the main vitamin E constituent in sunflower and safflower oils. This results in a much higher α-tocopherol/PUFA ratio in the European fats diet than in US fats (see Table 6.5). Even more significantly, unrefined olive oil (extra virgin) provides an even higher α-Toc-OH/PUFA ratio and refining reduces this ratio substantially. It was noted in Chapter 6 that olive oil also contains caffeic acid which will contribute toward its oxidative stability. However both of these antioxidants are reduced by the refining process. The consequent sub-optimal α-tocopherol intake in the USA may well account for the higher incidence of CVD in the USA and for the good response to supplementation by α-tocopherol in the US Professionals and US Nurses studies.

Table 6.5 α-tocopherol/PUFA ratios in European vegetable oils [52]

	[α-Tocopherol] mg/100g	[PUFA] g/100g	[α-Toc-OH]/[PUFA] mg/g
Sunflower	49-69	61	0.8-1.1
Safflower	35-48	77	*0.5-0.6*
Grapeseed	20-56	61	*0.3-0.9*
Cottonseed	39	49	0.8
Corn	11-31	62	*0.2-0.5*
Rapeseed	3-11	32	0.1-*0.3*
Soyabean	8-12	61	0.1-*0.2*
Olive			
refined	≤5	9	≤*0.5*
virgin (Sasso)	12	9	1.3
extra virgin (Crete)	16	9	1.8
extra virgin (France)	19	9	2.1
extra virgin (Apulia, Italy)	21	9	2.3
extra virgin (Southern France)	23	9	2.5

The preferred minimum [α-Toc-OH]/[PUFA] ratio is ≥0.6 and numbers in italics indicate a sub-optimal ratio.

The adequacy of dietary vitamin E then depends on the intake of vitamin E in foods but particularly on the *net* R,R,R-α-tocopherol; i.e. the difference between gross R,R,R-α-tocopherol equivalent (TE) and TE required for the protection of the associated PUFAs in the diet [53]. Thus all oils with a TE/PUFA ratio ≤0.5 have net negative vitamin E content and may actually withdraw vitamin E from body stores in protecting the ingested fats against peroxidation. Table 6.6 taken from Gey [52] lists the net vitamin E in some of the popular vegetable oils marketed in Europe and it can be seen that corn, peanut, rapeseed and soybean oils do not provide the net protection required in the cell. Only sunflower and extra virgin olive oils provide the necessary TE/PUFA ratios (~1.0 and 2.0 respectively) to adequately protect cell function.

Table 6.6 Net α-tocopherol provided by European vegetable oils [52]

	PUFA g/100g	α-Tocopherol mg/100g		
		Requirement	Total	Net
Sunflower	61	37	49-69	+12 to +22
Olive	9	6	12-23	+6 to +17
Cottonseed	49	30	39	+9
Safflower	75	35	35-48	0 to +13
Peanut	34	21	13-20	-1 to -8
Corn	62	38	11-31	-7 to -27
Rapeseed	32	22	3-11	-11 to -19
Soyabean	61	38	8-12	-26 to -30

Net α-tocopherol is total α-tocopherol minus that required for protection of the PUFA in the oil (i.e. 0.6 mg RRR-α-tocopherol/g linoleic acid + 0.9 mg RRR-α-tocopherol/g linolenic acid)

Most of the commodity oils which are the basis of margarines and spreads have a vitamin E/PUFA ratio between 0.3 and 0.9 before refining and even butter lies in this range (see Chapter 5, Table 5.7). Assuming an average antioxidant requirement of the same polyunsaturates in the human cell of 0.75 mg/g, it seems evident that most of the natural oils provide no net surplus antioxidant capacity [16,53] above their own requirements. Exceptions to this are palm oil and wheatgerm oil which contain a substantial excess antioxidant potential, and on this basis a VE/PU ratio of between 2 and 4 would seem to be not unreasonable in a commercial food product to allow for losses that occur during storage, cooking, etc. Thus a daily consumption of 15 g of margarine reinforced with 2.7 mg/g of RRR-α-tocopherol could supply the necessary net level of α-tocopherol in the diet to achieve the 50% reduction of CVD referred to above. It should be noted that 2-4 mg/g is the level of antioxidant required to protect a polyunsaturated rubber from oxidation under relatively mild conditions (e.g. during manufacture and storage), but would be much lower than that required in a synergistic combination in a motor car tyre subjected to physical stresses and higher than ambient temperatures during service (20-40 mg/g).

Table 5.8 in Chapter 5 shows that many branded grades of margarines and spreads do satisfy the minimum requirements outlined above. All but one have a VE/PU ratio of 0.75 mg/g or greater and thus have the antioxidant capacity to neutralise the "prooxidant" effects of their own intrinsic polyunsaturated esters. It must be re-emphasised that the above generalisation does not apply to most cheaper "own brand" margarines and spreads found on the supermarket shelves which are normally not supplemented and rely instead on the vitamin E that escapes the refining process.

It has become evident from epidemiological studies that the ability of vitamin E to protect the polyunsaturates in cells from peroxidation is compromised by smoking.. Thus in the Health Professionals study [36], the reduction of CVD in smokers by supplementation (>100 IU) was less (0.67, 95% CI 0.34-1.31) than that of non-smokers (0.52, 95% CI 0.34-0.78). It is known that smokers have an increased level of plasma conjugated dienes [54] and TBARs [55], markers of oxidative stress, compared to non-smokers, both of which are reduced by substantial vitamin E supplementation. Supplementation of both smokers and non-smokers by a daily intake of 400 IU (280 mg) of dl-α-tocopherol for 10 weeks reduced the plasma conjugated dienes of both non-smokers and smokers to the same level although there was a significant difference in non-supplemented controls. The same level of supplementation increased plasma α-tocopherol concentrations to 20 mg/l (48 μmol/l) [54] which is superior to the "optimal" level for normal subjects discussed above. The increased requirement for vitamin E in smokers is a reflection of increased radical generation in human tissues [55-58]. Whether the primary ROS or secondary peroxyl radicals resulting from an increased rate of peroxidation in the cells are the reason for antioxidant depletion is not yet clear (see discussion of β-carotene below), but it does argue the case that radical reactions induced by cigarette smoke are a major cause of CVD and that it can be countered by antioxidant supplements at higher levels than those supplied by a normal diet.

Carpentier et al. [59] have shown that the ratio of α-tocopherol to cholesterol is consistently low in atherosclerotic lesions compared with the ratio in the normal arterial wall, probably as a result of oxidation within the lesion.

Like vitamin C, β-carotene is not normally deficient in normal populations subjected to a dietary regimen of regular fruit and vegetable intake. Thus in non-smoking Scotsmen [60], the ROI (>0.4 μmol/l) for this antioxidant (Table 6.4) is achieved by a daily intake of 2.5 mg β-C and in the more diet-conscious US Health Professionals, a mean daily consumption of 6.1 mg provided a plasma concentration of 0.46 μmol/l which is well above the ROI [61]. The position is, however, quite different in the case of smokers [16]. In the Health Professionals study [36], smokers required an increased β-carotene intake (4.2-5.8 mg) to achieve a reduced risk of 0.46 and this was further reduced to 0.3 by supplementation to >8.6 mg daily. This is believed to be due to the lower plasma levels ($\cong$ 40% lower) achieved by smokers than by non-smokers [62]. It was seen in Chapter 5 that β-carotene is a retarder of peroxidation, rather than an inhibitor and that it undergoes co-oxidation with other cell components and requires the presence of α-tocopherol to realise its full antioxidant potential. This is confirmed in a study of German smokers [63] where it was found that when plasma concentration of vitamins C and E were optimal, smokers and non-smokers did not differ appreciably in their plasma β-carotene levels. It must be concluded therefore [16] that, as in similar studies elsewhere [60,64], plasma response to β-C in the diet is associated with a sub-optimal level of VE and to a lesser extent with sub-optimal VC.

In a trivitamin supplementation study, Calzada et al. [65] found that while each of the antioxidants (VE, VC and β-C) when used alone became elevated in plasma and in LDL, the antioxidant:cholesterol ratios were also increased (1.5-2 for α-tocopherol). There is some evidence that at moderate doses, one antioxidant will provide some protection to another.

6.3.2 *Synthetic antioxidants*
Although there is circumstantial evidence that a small amount of BHT, BHA and possibly EQ, absorbed from foodsfuffs [66] and foodstuffs packaging have a beneficial effect in atherosclerosis [67], only a few systematic supplementation trials of synthetic antioxidants have so far been carried out. One such study in rabbits involving dietary supplementation of BHT at 10g/kg in a diet which also increased cholesterol (10g/kg diet), significantly decreased aortic thickening compared with controls without BHT [68]. This was associated with decreased formation of cholesterol oxidation products.

Probucol (Chapter 5, Section 5.5.1), a drug originally reported to reduce cholesterol levels [69] was later shown to be an effective lipid antioxidant [70]. It is an effective autosynergistic antioxidant, closely related to the bis-phenol sulphides widely used for many years in technological media (Chapter 5) and it is surprising that analogous sulphide autosynergists have not been investigated, at least in animals for antiatherogenic activity. A related compound, MDL 29,311, (see Chapter 5, Section 5.5.1) has been reported to behave similarly to probucol in inhibiting TBARS formation in LDL in supplemented rabbits, but unlike probucol it did not lower cholesterol [71]. Furthermore, this compound provided immediate protection from aortic lesions in supplemented animals.

Sodium diethyldithiocarbamate (NaDEC), a water soluble member of a powerful group of peroxidolytic antioxidants *in vitro* (see Chapter 4, Table 4.1) has been shown to inhibit foam cell formation at very low concentrations (1-5 μM) [72]. Metal chelation was ruled out as a primary mechanism although, as was seen in Chapter 5, Section 5.5.2), this may be producing an autosynergistic co-operative interaction with hydroperoxide decomposition. In view of the exceptional potency of the dithiolates there is clearly a need to examine the effectiveness of related dithiolates (Table 4.1) as antiatherogenic drugs *in vivo*.

In summary, evidence from recent work suggests that antiatherogenic drugs do not need to be anticholesterologenic. In a study in which probucol was fed to monkeys previously fed a high cholesterol diet, probucol did not reduce cholesterol but did reduce aortic intimal lesion size, and the degree of this protection was directly related to the induction time to peroxidation as measured by the formation of conjugated dienes in LDL.

6.4 The Effects of Antioxidants on CVD in Animals
Studies of cardiovascular disease in experimental animals offers the clear advantage of controlled laboratory conditions and the possibility of direct observation of tissue damage as a result of diet, but it suffers from the obvious disadvantage that animals are biologically different from human beings, and what is observed in animals does not always apply to man. This is particularly true in the case of vitamin C where rabbits and some other small animals unlike man are able to synthesise this chemical. Animal studies are therefore frequently carried out in guinea pig or in a mutant rat strain,

neither of which can synthesise VC [73]. Nevertheless animal studies are a very valuable way of directly observing the effects of adding antioxidants to diet, and following changes at the cellular level and results up to 1994 have been reviewed by Lynch and Frei [72].

Early studies in guinea pigs [74,75] and hypercholesterolaemic rats [76] showed the importance of vitamin C in preventing atherosclerosis. More recently the effect of vitamin C deficiency and supplementation has been studied in larger mammals (e.g. pigs [77]) with similar conclusions [73]. Kimura et al. [78] found that dietary ascorbic acid reduced LDL peroxidation in genetically scorbutic rats, and Bocan et al. [79] showed that supplementation with both VC and VE inhibits the formation of atherosclerotic lesions in the hypercholesterolaemic New Zealand White rabbit. Vitamin E and selenium supple-ments independently also reduced aortic lesions in rabbits fed a high cholesterol diet [80] to 57% and 39% respectively of the controls, whereas supplementation with both antioxidants reduced the extent of atherosclerosis even further (to 28% of control). It may be concluded from these studies that supplementation with VE and/or Se decreases atherosclerotic lesions by inhibiting plasma lipid peroxidation **in spite of elevated serum cholesterol levels** [73]. Watanabe rabbits which suffer from familial hypercholesterolaemia also showed an attenuation of early atherosclerotic lesions when fed a vitamin E diet [81]. Lynch and Frei [73] have drawn the following conclusions.

a) The totality of the evidence suggests that vitamin E supplementation has an antiatherogenic effect and that vitamin C deficiency is proatherogenic.
b) For antioxidant supplementation to have a maximal inhibitory effect on atherosclerosis, it may have to be commenced during an early stage of the disease when peroxidation is the key mediator of atherosclerotic lesion formation.

To the above must be added that combinations of different antioxidants generally produce effects that are greater than those of individual antioxidants, and that future work should concentrate on the development of synergistic optima initially in animals in order to assist in the design of supplements for humans

6.5 Epidemiological Studies of Cancer
6.5.1 *Cancer and diet*
Epidemiology has so far been less successful in relating carcinogenesis to antioxidant status than it has in the case of CVD. Studies up to 1994 have been extensively reviewed by Fontham for vitamin C and β-carotene [82], by Knekt for vitamin E [83] and by Garland et al. for selenium [84]. Unlike the more comprehensive studies for CVD, the tendency has been to study the status of antioxidants individually and not to correlate them with other antioxidants. This is particularly true of ß-carotene which in the 1980s gained a "magic bullet" reputation, due to reports by Bjelke [85], Peto et al [86] that vitamin A and β-carotene might reduce the incidence of lung cancer. Subsequent case-control studies involving food questionnaires have shown that VC which is also present in the same vegetable and a high fruit diet is equally negatively correlated with lung cancer risk as β-C [82,87]. A high intake of fruits in Louisiana was associated with a significant reduction in risk of all lung cancers (relative risk compared with controls, RR = 0.67), and a similar reduction was found for VC. Adjustment for

fruit and VC intake eliminated the effects of vegetables and β-C. In similar studies in China [88], no protective effects were observed from carotene-rich vegetables, but dark green vegetables and some specific fruits (e.g. apples, pears and bananas) were particularly associated with a decreased risk of lung cancer [89]. It is interesting to note that some of these are a rich source of the flavonoid antioxidants (Chapter 5, Table 5.15), and these have not so far been independently correlated with lung cancer incidence although they frequently occur in the same foodstuffs as VC and β-C. Furthermore, they play a very similar role to the former in the protection of the cell and may even be involved in the regeneration of α-tocopherol from its aryloxyl radical (Chapter 5).

The majority of case-control epidemiological studies of vitamin E status and lung cancer reviewed by Knekt [83] have shown a considerably lower level of VE levels among lung cancer cases than among controls. In spite of this, Knekt concludes that existing data on plasma vitamin E status and lung cancer do not provide a conclusive answer to the question of possible protection by VE. A confounding factor in many of these studies is the possibility that VE may have been lost in stored plasma. This does not apply to the multifactorial Basel cohort study discussed below where antioxidants were measured at the beginning. Case-control studies have also shown an inverse correlation between serum selenium and lung cancer [90].

Of the large number of epidemiological studies reviewed by Fontham [82] on the association between fruit and vegetable consumption and cancers of the respiratory-digestive tract, stomach, pancreas, colon and rectum, cervix, bladder, breast, thyroid and epidermis, over a hundred have shown an inverse relationship, and many were statistically significant after adjustment for potential confounders. Where vitamin C plasma levels have been included in the analysis, a strong protective effect has been observed in cancers of the oral cavity/pharynx, esophagus, stomach and pancreas and available evidence also supports a protective effect of vitamin C against cancers of the larynx, lung, rectum, cervix and breast (see reference 82 for a detailed discussion). Evidence is less strong for cancers of the colon and bladder. There are some reports of a lack of protective effect or even a promotion of some cancers (e.g. stomach, colon and rectum in Japan), but in spite of this, Fontham concludes that protective effects have been demonstrated in widely divergent populations throughout the world [82] and that a dose-response relationship has been found in many of the studies.

In spite of the ability of the body to tolerate levels of vitamin C much higher than international RDAs, relatively few randomised double-blind trials have so far been reported although a number are in progress [82], generally as part of a multivitamin study. This is important, in view of the evidence already discussed and the theory of synergism that underlies it, that the use of single "magic bullet" chemopreventive agent cannot provide a solution to cancer prevention, or indeed to the reduction of any other peroxidation-associated degenerative disease.

Similar although rather less positive conclusions have been reached by Knekt [83] in reviewing the preventive effects of vitamin E in the same wide range of cancers. The strongest inverse relationship between VE status and cancer risk was found with cancers of the lung, cervix, head and neck and with melanoma. No significant association was found for gastrointestinal and breast cancers although inclusion of combined VE and selenium has shown a positive protective effect [91]. This again indicates the potential importance of synergistic combinations of antioxidants acting by different mechanisms outlined in Chapter 4.

Although not originally designed for the study of cancer, the Basel prospective CVD study discussed above has provided important evidence on the relationship between base-line age-adjusted plasma antioxidant concentrations and cancers [44,45,92] (see Table 6.7). The cumulative index of antioxidant vitamins (CIAVIT) was found to be significantly more sensitive than plasma concentrations of individual antioxidant and was low in all major types of cancer and in non-malignant lung disease. However, it does not follow that a low concentration of one antioxidant can be fully compensated for by increasing that of another [4,50], nor would this be expected on theoretical grounds since in a multicomponent synergistic combinations the optimal effect always exceeds the activity of individual antioxidants several-fold and is generally found to require at least 20% of any individual antioxidant (see Chapter 4). However, optimisation of synergistic antioxidant combinations is difficult and generally can only be done at present by empirical observation.

Table 6.7 Deaths due to cancer in the prospective basel study.
Age-adjusted plasma antioxidant mean values (µM) at baseline [4]

Cause of death	N	β-C	VA	VC	VE	CIAVIT*
All causes	268	0.30	2.82	43.1	35.0	450
Cancer						
All	102	0.28	2.77	41.5	33.7	400
Lung	37	0.25	2.85	43.5	33.5	362
Stomach	17	0.23	2.48	35.8	31.3	268
Colorectal	9	0.40	2.96	44.4	31.1	581
Survivors	2707	0.35	2.83	47.0	35.5	581

* Cumulative index of antioxidant vitamins; product of the micromolar concentrations of VC, VE and β-C.

Gey et al [4] have concluded that maximal preventive effect of a combination of antioxidants probably requires an optimal concentration of every antioxidant. They also conclude that a higher antioxidant status is required for the prevention of cancer than for the inhibition of atherogenesis. Other studies in which cancer incidence was related to serum antioxidant concentration also showed an inverse correlation between the vitamin antioxidants in cancer of the colon, lung and stomach [93] and the lung [94]. However, some early studies may have been rendered unreliable because of the decay in concentration of some of the natural antioxidants during storage before measurement [95].

In Section 6.1 the limitations of diet-based epidemiological studies was discussed with particular relevance to cardiovascular disease. The same principles apply in the consideration of the relationship between diet and cancer. However, the position is even more complex due to the multiplicity of potential sites and the different dependence of each site on different components of the diet [96]. This is exacerbated by the unreliability of diet assessment based on questionnaires due to the changes that may occur over a long period of time due to the development of the disease. This approach must therefore by complemented by the measurement of biological markers in the blood plasma, ideally both at the beginning of a prospective trial and at intervals during the follow-up. This, however, brings its own problems due to the need to store very large numbers of blood samples under conditions where no change over many years can be guaranteed. In the European Prospective Investigation into Cancer and Nutrition (EPIC) begun in 1989 in seven European countries (France, Germany, Italy, Netherlands, Spain, Greece and UK) it is proposed to measure markers of food intake (e.g. fatty acid composition and urinary nitrogen), markers of exposure to xenobiotics by DNA transformation products, levels of serum antioxidants (VE, VC, carotenoids, ceruloplasmin, transferrin, glutathione) and most importantly, markers of antioxidant status, notably MDA and total antioxidant capacity of serum [96]. These procedures require new techniques for the convenient removal of samples of plasma, serum, etc. from liquid nitrogen storage at -196°C without exposure of the remainder to air at ambient temperatures. This will be achieved by storage of multiple 0.5 ml samples in separate micro-containers. EPIC is expected to produce definitive results within the next 10-15 years.

Intervention trials, however, represent a much more direct way of arriving at scientific conclusions provided they are properly designed. A number of major diet supplementation trials are already in progress in populations with an elevated risk of cancer [97-99]. These involve, in addition to vitamin E, β-carotene, retinol, vitamin C and selenium. The results of one of these multivitamin intervention trials have already been published with encouraging results [99]. Almost 30,000 adults in Linxian, China, a population with high rates of oesophageal and stomach cancers showed a significantly reduced incidence of stomach cancer after supplementing with VE, β-C and Se for five years but no reduction in oesophageal cancer in this period.

In contrast to multi-supplement intervention trials, a selective supplement trial based on β-carotene and/or retinol in a high risk population of US smokers and asbestos workers (CARET) [100] has given negative results and has been terminated early [101] because of increased incidence of lung cancer and deaths. This seems to have laid to rest the "magic bullet" approach to cancer prevention and this is entirely consistent with the known maverick behaviour of both β-carotene and vitamin A as a retarders of oxidation at low oxygen pressures and prooxidants at normal pressures. It further suggests that associated antioxidants in foods rich in the carotenoids may be much more important than β-carotene itself (e.g. cantaxanthin and astaxanthin are more effective antioxidants than β-carotene, see Chapter 5, Section 5.2.12 and the flavonoid antioxidants, see Chapter 5, Section 5.2.10) In spite of these reservations about single supplement trials, it was seen in Chapter 5 that β-carotene and vitamin E are strongly synergistic. The same principle almost certainly holds for combinations of β-C with other CB-D antioxidants (e.g. vitamin C and the flavonoids), and the lesson to be drawn from negative trials is the importance of understanding how antioxidants act and when and why they synergise or antagonise. These principles have been learned, largely empirically, by technologists under conditions where failure may have led to financial loss but not normally to loss of life and are codified in general guidelines as outlined in Chapters 3 and 4 of this book. The epidemiological studies of Gey and his collaborators have shown (Section 4) that same principles that govern the design of antioxidant protective systems *in vitro* also apply *in vivo*.

6.5.2 *Melatonin and Cancer*

Melatonin is a hormone produced by the pineal gland which is known to have oncostatic activity [102] (see Chapter 5). It is formed at night from serotonin, and its synthesis is inhibited by light and by very weak magnetic fields. The incidence of leukaemia in children has been shown epidemiologically to be associated with increased exposure to very low power frequency magnetic fields [103], and it has been proposed [104] that this is a consequence of the known reduction of melatonin production by magnetic fields. Melatonin is a very powerful antioxidant (Chapter 5) and may prove to be an important example of an endogenous antioxidant involved in oncostasis in many organs in the body [105,106].

6.6 Effects of Antioxidants on Cancer *in vivo*

It was seen in Chapter 2 that diet has a great influence upon the incidence of many cancers, particularly if this contains chemical carcinogens. It is not surprising then that many studies of the effects of antioxidants in animals have been concerned with the co-administration of chemicals which affect the initiation stage of carcinogenesis. Typical of these are 7,12-dimethylbenz[a]anthacene (DMBA), benzpyrene (BP), 1,2-dimethylhydrazine (DMH), 2-acetylaminofluorene (2-AAF), aflatoxin B_1 (AFB_1), N-nitroso-bis-(2-oxopropyl)amine (BOP), dimethylnitrosamine (DMN), diethylnitrosamine (DEN), methylazoxymethanol (MAM), 3'-methyl-4-dimethylaminoazobenzene (DAB), etc. [107]. Thus orally administered VE resulted in fewer and smaller tumours induced by DMBA in hamsters [108,109] and at lower DMBA concentrations, VE was able completely to prevent the development of epidermoid cancer [110].

The effect of two levels of ascorbate (0.25% and 1% by weight) in the diet on colon carcinogenesis in rats induced by DMH completely eliminated tumour formation compared with 26% in the controls [111]. Other carcinogens that are blocked by VC include BP [112], heavy metals [113], pesticides [114], etc. However, an important function of VC appears to be to block the conversion of nitrates (from foods) to nitrites and hence reduce carcinogenic nitrosamine formation [115,116].

Ellagic acid (Chapter 5, Section 5.2.11), a very powerful antioxidant closely related to gallic acid, is found in fruits and vegetables and has been shown to inhibit the mutagenic activity of N-nitroso compounds, AFB_1 and BP. The current consensus from animal studies [107] seems to be that the primary function of ellagic acid is to prevent the interaction of the carcinogenic metabolite with critical sites in the DNA. However, the similarity in activity and physical behaviour of this antioxidant to that of vitamin C and other water soluble antioxidants suggest that it may also be involved in the prevention of the oxidation of the procarcinogens to their active agents.

β-Carotene has been found to be effective in retarding the appearance and number of initiated tumours in animals [117]. In cell culture experiments initiated by X-rays, canthaxanthin was found to be as effective as β-carotene at a concentration an order of magnitude lower [118]. Canthaxanthin was also found to be more effective than β-C against UV initiated skin tumours in mice [119]. These finding accord with the relative antioxidant effectiveness of the two carotenoids (Chapter 5) but it is possible that under these conditions the carotenoids are also functioning as singlet oxygen quenchers.

Antioxidants have been shown to inhibit many of the biochemical changes associated with tumour promotion [107]. 12-O-tetradecanoylphorbol-13-acetate (TPA), a widely used tumour promoter in animal studies, has been shown [120] to stimulate the reduction of [GSH]/[GSSG] in epidermal cells and this is inhibited by α-tocopherol as well as GSH, and both antioxidants inhibited the reduction in $GSHP_x$ activity induced by TPA treatment [121]. It seems likely then that GSH or α-Toc-OH scavenge destroy or scavenge reactive oxygen species by replenishing the intracellular GSH level.

Protective effects are not limited to the naturally occurring antioxidants. Tumour suppression has been observed with a range of the above carcinogens in the stomach, colon and lung of mice by the synthetic antioxidants BHA, BHT and ethoxyquin (see Chapter 3 for chemical structures) [122] and also by the sulphur compound, Oltipraz which is related to the peroxidolytic thiocarbamate antioxidants used in rubber technology (Chapter 5, Section 5.5.2).

It is clear that anticancer activity is not limited to any one class of antioxidants since BHA, BHT and EQ are chain-breaking (CB-D) antioxidants whereas Oltipraz is a preventive (PD) antioxidant. It would be of considerable theoretical and practical interest to examine the possibility of synergism between Oltipraz and the CB-D antioxidants. The major mechanism of action of the above antioxidants is the inhibition of the oxidation of the pro-carcinogens to the true initiator of carcinogenesis [122] or by inhibiting radical formation from tumour promoters. Thus both BHA and BHT were effective in inhibiting DMBA-induced mammary cancers in a post-initiation dietary administration of antioxidant; and Moore et al. [122] have found that BHA and α-tocopherol inhibited the incidence of both liver and pancreatic lesions in hamsters in the later stages of carcinogenesis, and in the absence of any discrete exposure to tumour promoters. Similarly, rats treated with BHA after exposure to DMBA developed fewer mammary carcinomas than did control animals [124]. Glutathione was found to be a highly effective inhibitors of the progression of papillomas to carcinomas [125] and appears to be the main endogenous antioxidant protecting cells during tumour promotion.

There have been reports of BHT and to a lesser extent BHA acting as carcinogens [107] but this is generally at much higher concentrations than are likely to be encountered in clinical practice. However, Chajès et al. [126] have shown that α-tocopherol inhibits the cytotoxic effects of peroxidation products of γ-linolenic acid in breast cancer cells (Chapter 2) and conclude from this that α-tocopherol may be antagonistic to cancer therapy which depends on the formation of peroxidation products.

6.7 Antioxidants and Ageing

A number of degenerative diseases increase with age. Notable among these are cancer and cardiovascular disease, discussed in the previous sections and Altzheimer's and Parkinson's diseases. It has been proposed that this reflects increased free radical formation and/or reduced activity of endogenous antioxidants with age [127-130]. Specific environmental and dietary factors were seen to be involved in the development of CVD and cancer but no analogous external factors have so far been demonstrated to be associated with the development of PD and AD. There is, however, increasing evidence that the decline in neural and immune functions in old age may involve increasing radical "leakage" in the mitochondrial electron transport system [131]. Mitochondria are required for optimal cellular metabolism and regulation and are therefore present at relatively high levels in all cells and yet 90% of the oxygen utilised by the cell is absorbed by the mitochondrial electron transport system with the associated potential for the generation of reactive oxygen species ($O_2\cdot^-$, H_2O_2, $\cdot OH$, etc.). The proliferation of the "killer" lymphocytes, B and T of the immune system is markedly inhibited by exposure to ROS. Thus fish oil polyunsaturates which are rich in the highly peroxidisable 20:5 and 22:6 fatty esters decrease T lymphocyte proliferation [132], but this was restored to the level seen in a placebo group by 200 mg/day of vitamin E. Lymphocyte depression appears to involve the oxidation of surface thiols and can be reversed by endogenous thiols such as cysteine and glutathione [133,134]. However, the same effect is achieved by the synthetic antioxidant 2-mercaptoethanol which acts by destroying peroxidic species. It seems likely then that a general peroxidolytic antioxidant effect is operating with all the thiols. Ubiquinone-10 (Ubi-q-10) whose antioxidant activity in cells was discussed in Chapter 5, like α-tocopherol has also been found to be immunoprotective [135,136]. Dietary studies have shown that supplementation with glutathione reverses the age-associated decline in immune responsiveness in mice [137] and α-tocopherol [138] and β-carotene [139] had the same effect in humans.

The age dependent accumulation of lipofuscin and TBARs formed by peroxidation of the lipids (Chapter 2), and damage to DNA as evidenced by the increased formation in the rat liver and urine of oxidation products such as 8-hydroxy-2'-deoxyguanosine (oxo^8dG) [140], provide confirmation that the processes of ageing and the development of degenerative diseases of old age have a similar chemical basis. Cutler and co-workers found that the concentrations of several antioxidants in the livers of primates and rodents are strongly correlated with life-span energy potential (LEP) [141]. This is shown typically for SOD in Fig. 2.4. A similar relationship with LEP has also been found for uric acid and a positive but non-linear relationship for β-carotene. α-Tocopherol concentrations in the plasma of mammals also showed a good correlation, but ascorbic acid in mammalian tissues did not.

The antioxidant hormone melatonin (Chapter 5) which is produced from serotonin by the pineal gland at night and circulates to all organs of the body in the plasma is a powerful antioxidant related to ethoxyquin. The amount produced and the duration of its production are both decreased in older human beings [142] (Fig. 6.3) and in older animals [143]. Mice given melatonin in their drinking water every night were found to live 20% longer than untreated mice [144].

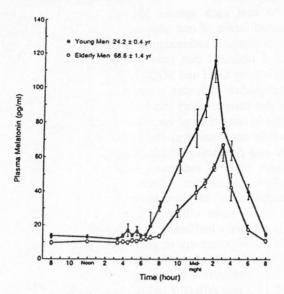

Fig. 6.3 Blood melatonin concentrations in young (~24 years) and elderly (~68.5 years) human males. (Reproduced with permission from R.J. Reiter, *BioEssays*, **14**, 171 (1992), adapted from Ref 142).

Food restriction in rats resulted in a significant preservation of melatonin levels compared to rats fed an unrestricted diet [145] and this resulted in an increase in lifespan from 30 to 44 months. *Ad libitum* fed rats exhibited numerous tumours at death, whereas the calorie restricted diet rats displayed very few tumours at the same age [146]. Mice reared on a restricted diet exhibit lower lipid peroxidation as evidenced by TBARs and increased liver catalase activity [147]. Similarly calorie restricted mice had lower levels of oxo^{8}dG in both nuclear and mitochondrial DNA isolated from liver [148] suggesting that these tissues have lower spontaneous mutation rates. Tian et al. [149] have shown that caloric restriction decreases the rate of lipid peroxidation and protein degradation. These data are entirely consistent with the information discussed above. In particular, the effects of antioxidants on cancer development is consistent with their beneficial effect on the immune system [149] and provides a rational basis for understanding why food restriction both increases maximum life-span of animals and reduces the incidence of experimentally induced cancers [146].

A consequence of the above observations should be that supplementation of the diet of animals by antioxidants should lead to an increase in LSP. One of the earliest proponents of the life extending effects of antioxidants was Harman [127], who fed a synthetic peroxide decomposer (2-mercaptoethylamine, 2-MEA) and two synthetic chain-breaking donor antioxidants (ethoxyquin, and 2-ethyl-6-methyl-3-hydroxy-pyridine) to mice and rats and found an increase in mean life-span. These results have since been questioned on the grounds that the taste or smell of the antioxidants may have reduced the animals food intake with the effects described above. Vitamin E supplementation of the diet of mice halved the rate of accumulation of lipofuscin but did not increase either the mean life-span or the life-span potential of the animals [150]. Cutler has suggested [151] that a possible reason for the ineffectiveness of antioxidant supplementation is that each species has an automatic compensatory mechanism whereby an increased intake of one antioxidant results in a homeostatically controlled decrease of other endogenous antioxidants. There is evidence to support this in that rats fed on a vitamin E deficient diet show a sharp increase in the antioxidants which control peroxides, notably GSH and SOD [150]. From this it was concluded that the net level of tissue antioxidant defences remains constant over a wide range of dietary uptakes, and that this compensatory mechanism accounts for the observed constancy of species ageing rates in spite of large variations in dietary intake of specific antioxidants. This argument is fully consistent with the genetic control theory of antioxidant defence.

On mechanistic and practical grounds, however, the homeostatic theory is open to question. The above study of melatonin whose level of formation is age dependent shows that supplementation can increase life span under favourable conditions. Furthermore, combinations of synergistic antioxidants acting by complementary mechanisms are much more effective than individual antioxidants. For example, vitamin E alone is relatively ineffective in the absence of its primary synergist, ascorbic acid and peroxide decomposers are in general less effective in the absence of a CB-D antioxidant. This reasoning leads to the argument advanced repeatedly throughout this book that heavy supplementation with a single antioxidant may be counter-productive since it may result in a *less efficient* rather than a *more efficient* control of oxidation. The theory of antioxidant synergism (Chapter 4) suggests that that supplements should normally be administered as balanced combinations of chain-breaking and preventive antioxidants.

The homeostatic theory implies a "feed-back" modulation of antioxidant synthesis or absorption. There is at present no evidence that it applies to all antioxidants and in particular to synthetic antioxidants which may function by mechanisms different from those already existing *in vivo*. For example, it was seen in Chapter 2 that the carbon radical trap, MNP, is an effective inhibitor of the inflammatory prostaglandin cascade probably by the catalytic mechanism discussed in Chapter 3 but not so far studied *in vivo*. If the catalytic effect is confirmed *in vivo*, then this opens up quite novel approaches to increasing antioxidant defences in humans. At the same time it could solve another problem which faces the antioxidant therapist, namely the difficulty of targeting the site of oxidation within the cell. It is known that vitamin E is effective in the lipid bi-layer because it is physically an integral part of it [43]. By contrast, synthetic antioxidants do not normally have the correct hydrophobicity and conformation to fit easily into the phospholipid bi-layer and are consequently less effective not because they are intrinsically less active but because they are not in the right place. Synthetic antioxidants designed to be part of the cell membrane lipids might be expected for this reason to be much more effective (see Chapter 4, Section 4.5). It is premature to conclude then that a correctly designed life-extending synergistic antioxidant supplement has yet been tested.

6.8 Parkinson's Disease

There is increasing evidence that the ageing brain undergoes enhanced oxidative stress [152] and in Parkinson's disease the oxidation of L-DOPA seems to be particularly implicated [153,154]. L-DOPA is an antioxidant and its oxidation products are thought to be involved in the subsequent pathology; consequently the control of this process is of great practical importance. SOD appears to be the most effective of the biological protective agents [152]. L-DOPA and its transformation products dopamine and melanin are themselves antioxidants (Chapter 5, Scheme 5.12) but like many catechols, they can also be a prooxidant under certain conditions [154]. The rate constant for the reaction of L-DOPA with $Cl_3COO\cdot$ was found to be 1.3×10^7 $M^{-1}s^{-1}$, about an order of magnitude lower than that of α-tocopherol and ascorbic acid. However, dopamine reacts directly with oxygen to give hydrogen peroxide and hydroxyl radicals [155]. Methylation of one hydroxyl group significantly reduces the ability of dopamine to activate Cu^{2+} to radical generation [156] and the hormone antioxidant, melatonin (see Chapter 5) reduced the prooxidant effects of both dopamine and L-DOPA. Miller et al. [156] have suggested that O-methylation of the catecholamines and production of melatonin by the pineal gland "may be important components of the brain's antioxidant defences".

The activity of glutathione peroxidase has been found to be reduced in the substantia nigra of PD patients [157] whereas the concentration of MnSOD is increased compared to controls without neurological disease. The concentrations of the chain-breaking antioxidants (VE,VC) appear to be similar to those of controls [158].

Intervention trials with vitamin E (3200 IU/day) and vitamin C (3000 mg/day) in newly diagnosed PD subjects, increased by 2.5 years the period before it became necessary to commence replacement therapy with L-DOPA compared with controls [159]. A subsequent study supplementing vitamin E alone at a lower level (2000 IU/day) did not produce a similar improvement [160] indicating once again the importance of synergistic combinations in intervention trials. Since iron concentrations in the brains of PD patients is elevated (Chapter 2) and hydrogen peroxide resulting from the reaction of superoxide with SOD is also increased while glutathione peroxidase concentrations are reduced, this must indicate an increased potential for the Fenton reaction and there is a strong argument for including in future intervention trials not only one of the new autosynergistic lazaroid drugs (e.g. U78517F, Chapter 5), but in addition selenium to activate $GSHP_x$ in combination with VE and VC.

6.9 Alzheimer's Disease
Alzheimer's disease shows some general chemical similarities to Parkinson's disease [158]. Again there appears to be an elevation of SOD and an increase in melanin formation. Plasma levels of vitamins E and A were found to be deficient in Alzheimer's patients [161,162], although no deficiency was found in the brain [163].

6.10 Antioxidants in Inflammation
6.10.1 *Rheumatoid arthritis*
Oxidative damage of synovial fluid results from free radical production from the activated neutraphils and macrophages (Chapter 2, Section 2.7.1). Consequently, in epidemiological studies of RA patients [164], the levels of antioxidant nutrients, notably VE and VC were found to be depressed, and in a cohort study of RA patients, serum Se concentrations was found to be inversely related to the severity of the disease [165]. The activities of the antioxidant enzymes, SOD, CAT and $GSHP_x$ were also found to be significantly depressed in erythrocytes from RA patients compared to controls [166]. Consistent with above observations was the demonstration by Thurnham et al. [167] that the total radical trapping potential (TRAP assay) of RA patients was significantly lower than that of healthy controls.

The vitamin antioxidants have some activity in decreasing the effects of peroxidation during rheumatoid arthritis. A number of intervention trials with RA subjects have shown the benefits if VE supplementation in pain relief at levels between 100 and 600 IU/day [168]. Supplementation with Se in deficient subjects [169] led to removal of the deficiency and repair of GSHP$_x$ activity but did not lead to clinical improvement. Injection of SOD into rheumatoid joints has been reported to reduce inflammation [170]. However, the naturally occurring iron complexing proteins, notably lactoferrin and transferrin, were found to be very effective in inhibiting iron catalysed peroxide decomposition (Chapter 2) and the suppressive action of the iron chelator, deferrioxamine on chronic inflammation was found to be sufficiently encouraging for trials in human patients to be carried out [171]. The rapid restoration of haemoglobin concentrations in human subjects after injection of DFO resulted from the suppression of inflammation. However, side-effects were observed at the treatment levels of DFO required (3 g/day for up to 3 weeks) and further progress awaits the development of the less toxic chelators discussed in Chapter 5.

The most successful treatments for rheumatoid arthritis involve interference with prostanoid production which was seen in Chapter 2 to involve a free radical mechanism. The corticosteroids (e.g. cortisol) and their synthetic analogues (e.g. prednisolone and cortisol) act by inhibiting the release of arachidonic acid by interfering with the formation of an essential enzyme (phospholipase A$_2$) may in the most general sense be regarded as "preventive" antioxidants by decreasing the concentration of a highly oxidisable intermediate. However, this is not within the definition of antioxidants used in this work and they will not be discussed further. A second group of compounds, the non-steroidal anti-inflammatory drugs (NSAIDs) do interfere directly with the peroxidation process and the CB-D mechanism of their action was discussed in Chapter 5, Section 5.5.1. Most favoured among the NSAIDs is aspirin because of its relatively low toxicity. However, the antimalerials, choroquin and hydroxychloroquine, both of which have CD-D activity due to the presence of the aromatic alkylamino groups, have also been used in the treatment of rheumatoid inflammation [170];

Chloroquin, R=CH$_2$CH$_3$
Hydroxychloroquin, R=CH$_2$CH$_2$OH

Gold compounds (e.g. gold sodiumthiomalate) have long been used in the treatment of rheumatoid arthritis [172]. Their mechanism of action has not been established. They may act as singlet oxygen quenchers (Chapter 5) [168], but the presence of sulphur is probably significant in view of its relationship to the peroxidolytic antioxidant, penicillamine and its corresponding disulphide.

$$Na_{2n} \left[\begin{array}{c} COO^- \\ Au\text{-}S\text{-}CH \\ CH_2 \\ COO^- \end{array} \right]_n \qquad \begin{array}{c} (CH_3)_2\ C\text{-}CHCOOH \\ |\ \ \ \ S\ NH_2 \\ |\ \ \ \ S\ NH_2 \\ (CH_3)_2\ C\text{-}CHCOOH \end{array}$$

Gold sodiumthiomalate   Penicillamine disulphide

It should be noted that both of the above contain the "delayed action" sulphur antioxidant system similar to that in ibuprofen discussed in Chapter 5.

6.10.2 *Hypoxia-reperfusion*

The main evidence for the formation and damaging effect of oxyl radicals during hypoxia-reperfusion (HR) came from the study of the effects of antioxidants. Thus Jolly et al. [173] found that a combination of superoxide dismutase and catalase protected the dog heart from reperfusion injury, similar effects have been observed after reperfusion of the rat kidney [174,175] and the canine pancreas [175].

Supplemented vitamin E in animals has been found to preserve microvascular perfusion after experimental gastric [176] or spinal cord [177] injury, after coronary occlusion [178] and after myocardial IR [179]. Specific evidence has been found for ubiquinone (Ubi-q or Co-q) deficiency in humans undergoing cardiac surgery [180,181]. Subsequent therapeutical studies with Co-q-10 showed considerable benefits from using this antioxidant in ischemic heart disease [182-186] and in heart failure [187-9]. Patients suffering from angina were found to recover from the effects of exercise much more rapidly when pre-treated with Co-q-10 for up to 4 weeks [186]. Similar beneficial effects in work capacity were observed with patients with previous myocardial infarction [185] and, very significantly in this case, the improvement was associated with a decrease in plasma MDA compared with a placebo group. Although there is still a general reluctance to attribute the beneficial effects of Co-q (or more correctly, its reduced form, Co-hq or Ubi-hq) to its antioxidant activity, there seems to the objective observer to be no doubt that this is a primary function in ishaemia reperfusion. It was seen in Chapter 5, Section 5.2.5, that not only is Co-hq a powerful CB-D antioxidant in its own right but it also plays an important homosynergistic role in the maintenance of vitamin E concentration by reducing tocpheroxyl radicals to the parent phenols. This process will be facilitated by the reducing conditions of ischaemia [190], and will occur in competition with the reduction of oxygen to superoxide (Chapter 2) which is the major cause of damage in reperfusion. Ubiquinol may also be involved in the direct reduction of ferrylmyoglobin to metmyoglobin [190].

A number of animal studies have shown that there is a decrease in antioxidant status in the central nervous system following injury. Ames and co-worker [191] noted reductions in ascorbic acid, α-tocopherol and ubiquinols in rats following spinal cord injury, and other workers [192] have found that glutathione concentrations are reduced in the rat brain following cerebral ischaemia in addition to vitamins C and E. Conversely, the administration of α-tocopherol either alone [193] or with selenium [194] before trauma was found to reduce peroxidation in traumatised spinal cord tissue.

α-Lipoic acid (α-LA) or its reduced form, DHLA, have been found to protect against HR in a number of animal studies [195-199]. The evidence again suggests that the primary function of DHLA is to preserve or regenerate vitamin E since the effect of DHLA is dependent upon the presence of VE [196].

Nitrone spin-traps, which act by the complementary chain-breaking (CB-A) mechanism, were seen in Chapter 5 (Section 5.5.1(c) to be protective against the effects of ischaemia reperfusion.

Desferrioxamine has been found to be effective in inhibiting the effects of reduced iron formed during ischaemic injury to the brain following cardiac arrest [200]. Intravenous injection of DFO (50 mg/kg) substantially reduced the concentration of ionic iron and, although the iron level was not reduced to that of the non-ischaemic controls, the degree of peroxidation was also substantially reduced compared with standard intensive care treatment, Table 6.8.

Table 6.8 Effect of DFO on brain iron concentrations and lipid peroxidation products after ischaemia reperfusion [52]

Conditions	Low molar mass Fe nmol/100mg	MDA nmol/100mg	Conjugated dienes nmol/100mg
Non-ischaemic controls	9.05	7.32	6.04
Standard intensive care	37.04	12.2	12.43
Standard intensive care + DFO (50mg/kg)	24.3	9.4	6.42

The orally active metal chelators based on 3-hydroxypiridin-4-one (Chapter 5, Section 5.5.2(f)) have also been observed to inhibit post-ischaemic cardiac injury in the rat [201] and are at present being investigated in the treatment of heart attack in humans. Other drugs with iron chelating activity, notably the lazaroids (Chapter 5, Section 5.1.2(f)), have also been found to inhibit ischaemic damage [202,203].

By contrast, SOD was found to be antagonistic toward reducing antioxidants (notably GSH), probably due to the reduction of copper in the enzyme to Cu^+ [204].

6.10.3 *Pancreatitis*

Newly diagnosed patients with acute pancreatitis had virtually no ascorbic acid in their sera [205] but Braganza has reported [206] that daily injection of similar subjects with 0.5 g of ascorbic acid normalised serum levels within 4 days, and has suggested that this treatment should be the first line of defence in the treatment of patients with acute pancreatitis. Sulphur compounds also have a pronounced therapeutic effect [206], notably N-acetylcysteine, an immediate precursor of glutathione peroxidase, which like ascorbic acid was reduced in concentration in the sera of acute pancreatitis. Other antioxidants deficient to a lower degree than ascorbic acid were selenium, β-carotene and vitamin E [207]. The administration of a preparation containing vitamins A (1500 IU), C (90 mg) and E (45 mg), together with selenium (100 μg) and additional methionine (up to 2 g), resulted in a significant reduction in number of attacks and in background pain [208] without serious side-effects. Replacement of vitamin A by β-carotene produced similar results. It was concluded that oral antioxidant therapy offers a preferable and viable alternative to pancreatic surgery in the treatment of chronic pancreatitis and points the way forward to the treatment of acute pancreatitis and associated cystic fibrosis and kwashiorkor (see below).

6.10.4 *Cystic fibrosis*

It was seen in Chapter 2, Section 4.4.4 that the concentrations of lipid and DNA oxidation products (hydroperoxides, TBARS, oh^{8}dG, etc.) were elevated in the plasma of CF patients in spite of the presence of near normal concentrations of α-tocopherol. Ascorbic acid, uric acid and thiol-derived antioxidants were elevated in children with CF were higher than controls [209] indicating an antioxidant imbalance rather than a deficiency. Selenium levels have been reported to be somewhat decreased [210] and it has been suggested that the "normal" plasma antioxidant concentrations are not sufficient to counter the severe oxidative environment observed in the lungs of CF patients [211]. In confirmation of this it has been shown [212] that correcting the vitamin E deficiencies in CF patients decreased the rate of peroxidation of their LDL. In view of the close association between CF and acute pancreatitis, it has been suggested [213] that the use of substantial antioxidant supplements, similar to those found to be effective in pancreatitis, may be justified both on theoretical grounds and on the basis of clinical experience [214]. However, it is probably of critical importance to carry out model studies first *in vitro* to determine which antioxidant mechanism is most compromised in CF subjects because of the danger of exacerbating the problem.

6.10.5 *Disorders of prematurity*

It was noted in Chapter 2, Section 2.7.5, that premature infants have higher than normal plasma levels of vitamin C and that they require a longer time than full-term infants to achieve the normal level of vitamin E. In the absence of ferrous iron, this would not normally lead to oxidative stress, but in preterm infants, circulating levels of the main iron binding protein, transferrin and of caeruloplasmin which has the ability to oxidise Fe^{2+} to Fe^3, thus acting as a preventive antioxidant, are frequently low [215]. Silvers et. al. [216] have shown that the survival rate of premature infants is significantly correlated with the caeruloplasmin:vitamin C ratio and this in turn correlates with the ability of the plasma to inhibit the formation of TBARS.

Kelly [217] has concluded that the only serious deficiency in the endogenous antioxidant defences in preterm infants is catalase, the activity of which was found to increase markedly throughout gestation [218]. Catalase is also a preventive antioxidant acting in a co-operation with the iron deactivators discussed above. They both ameliorate the Fenton chemistry but in complementary ways and open up prospects for the design of improved preventive synergists.

Retinopathy and intraventricular haemorrhage in premature infants involves damage to the endothelial cells in the former case and to the endothelial cells in the brain in the latter. Both occur in the perinatal period and there is evidence that both hypoxia and ischaemic reperfusion may play a part in the cell damage. Vitamin E supplementation has been shown to have some effect in reducing the severity of both retinopathy [219,220] and haemorrhage [221,222]. Kelly [217] has advocated the use of multiple antioxidant supplementation in the treatment of preterm infants and in view of the known importance of co-enzyme Q in ischaemia reperfusion (Section 6.10.2), the latter should be included.

6.10.6 *Inflammatory bowel disease*
IBD subjects show a deficiency of all the main vitamin antioxidants; vitamins A, C, E and β-carotene [223]. Vitamin C supplementation gives no benefit [224], but by contrast, vitamin E supplementation does show therapeutic benefit [225] as do ebselen, a selenium compound [226] (Chapter 5, Section 5.3.3) and SOD [227].

6.10.7 *Kwashiorkor*
It was seen in Chapter 2 (Section 2.7.8) that kwashiorkor is associated with a deficiency of most of the nutritional antioxidants leading to repeated infections which precipitate the disease. At the same time, increased levels of stored iron have been reported in children in regions where kwashiorkor is prevalent [228]. In other words there is an excess of prooxidants over antioxidants; a situation crying out for antioxidant therapy. Injection of children suffering from kwashiorkor with desferrioxamine leads to an increased urinary excretion of iron as compared with normal children and Golden et al. [229] suggest that the first line of defence against kwashiorkor should be DFO injection to allow the treatment of infections and the recovery of the iron binding proteins.

Pancreatic atrophy is almost universally associated with kwashiorkor [230] and Braganza has proposed [206] that since the cause of the pathologies is the same, a synergistic combination of chain-breaking and preventive antioxidants [231.232] might be equally as effective in the treatment of kwashiorkor as it is in the treatment of pancreatitis. Systematic supplementation of synergistic combinations of antioxidants have not yet been reported.

6.10.8 *Iron Overload*
The most common disease of iron overload in the West, haemochromatosis can be readily treated, particularly in the early stages by venesection (Chapter 2, Section 2.8.1). In this process, the endogenous porphyrin ring acts as a preventive antioxidant by chelating "free" iron and is rapidly resynthesised as haemoglobin is removed. It is thus possible to remove 500 ml of plasma each week until all loosely bound iron returns to the normal level. In the case of thallassaemia however, which occurs widely in Africa, the position is different since the natural haemoglobin is deficient and, unless the patients are given frequent blood transfusions, they die of anaemia. However, the transfusions lead to iron overload and iron accumulation in the liver spleen and heart. As much as 50-70 g of iron may accumulate over a 10 year period and this is normally removed by intravenous injection with DFO [233]. However, this requires frequent application over 8-12 hour periods and up to six times a week, causing some patients to discontinue the procedure. The orally active hydroxypyridinone chelating agents, therefore are of considerable interest because of their convenience in use. CP20 (Chapter 5, Section 5.5.2(f)) has already shown considerable promise as an alternative to DFO treatment [233].

6.11 Antioxidants in Environmental Damage
6.11.1 *Cataract*

The development of cataract is age-related and environmental factors play a very significant part in its development. At its simplest, over the course of a lifetime, an imbalance of oxidants and antioxidants develops in the lens due primarily to the damaging effects of light and a decrease in endogenous antioxidants so that damaged proteins aggregate and precipitate with a gradual decrease in lens opacity (Chapter 2, Section 2.9.1)

Vitamins C and E and the carotenoids are the primary exogenous antioxidants that protect the young lens against oxidation [234] and these synergise with the endogenous peroxidolytic antioxidants, SOD and catalase [235] and glutathione peroxidase [236,237]. Extensive epidemiological case-control studies have been carried out on the association between the three exogenous antioxidants and cataract. Knekt et al. [238] found a strong inverse relationship between serum vitamin E concentration and subsequent surgery for cataract patients. They found that subjects with serum VE concentrations >20 μM underwent only half the cataract surgery of those with below this concentration. Similar results have been found by other workers [239,240] but some studies (reviewed by Jacques et al. [241]) have not shown such an association. Supplementation with VE also led to a 55% lower incidence of early cataract for subjects taking >37.5 mg VE/day than for those taking <8.4 mg/day [240].

Jacques and Chilack found that vitamin C prevented early cataract in a dose-response relationship [240]. Subjects with VC plasma concentrations >90 mM were less than a third as likely to develop cataract than those with plasma concentrations of <40 mM, and subjects supplemented with >490 mg/day were only 25% as likely to develop cataract than those imbibing <125 mg/day. Other VC supplementation studies showed very similar effects [242,243] and, in the case of women supplemented with >294 mg/day for >10 years, led to a 45% reduction in cataract surgery compared with those with a mean intake of 77 mg/day [244]. The effect was most marked for advanced cataracts of the nucleus. Taylor et al. [245] found that the concentration of vitamin C in the lens could be substantially increased by dietary supplementation. Smaller differences in VC intake show a less convincing inverse correlation between VC intake and cataract [241], and confounding factors such as age, sex, smoking and alcohol intake become more important.

It might be expected that β-carotene might be a very effective antioxidant in the eye because of its dual ability to quench singlet oxygen formed by UV sensitisation of ground state oxygen at the front of the lens and to retard oxidation in the nucleus where the oxygen concentration is low ($\cong$20 torr) [246]. However, there is little evidence for an inverse association between β-carotene intake and cataract incidence [241]. Cataract surgery was inversely associated with total carotene intake, but the effect was not associated with β-carotene-rich foods such as carrots but rather with foods rich in other carotenoids for example spinach [241]. This warrants further study in view of the known prooxidant effects of β-carotene and the higher antioxidant activity of the ketonic carotenoids such as astaxanthin and canthaxanthin (Chapter 5).

Glutathione (GSH) preserves the physiochemical equilibrium of thiol groups in the lens protein. Packer and co-workers [247] have inhibited GSH synthesis in newborn rats and have found that the thiol level can be maintained by prior administration of α-lipoic acid. Restoration of ascorbic acid and α-tocopherol were also observed and they suggest that α-LA may partly take over the role of GSH.

Synergistic effects of combinations of VE, VC and β-C have also been examined by Jacques and Chylack [240]. Based on a combined plasma index of the three antioxidants, the adjusted prevalence of all types of cataract 40% lower for subjects with moderate plasma levels of the synergists and 80% lower for those with high levels. These authors also found that subjects who ate 3-5 portions of fruits and vegetables per day showed a similar decrease (82%) in cataract prevalence.

A multivitamin intervention trial has been reported in Linxian, China [248]. 4000 participants aged 45-74 from rural communes were examined for cataract after 5 years during which they received among other nutrients, VE, VC, β-C and Se. There was a 43% reduction in nuclear cataract in subjects aged 65-74. The trial has been criticised on the grounds that the role of specific antioxidants was not addressed [241] but it does confirm that synergistic combinations are effective in preventing cataract in real situations. More work needs to be carried out based on animal studies to optimise the synergistic ratios but the evidence from epidemiological studies does seem to suggest that synergistic therapy is much more likely to be successful in practice that the "magic bullet" approach.

Vitamin E protects the rabbit eye against chemically-induced [249] or light-induced lipid peroxidation [250] and cataract formation [251] and it also preserves glutathione levels [252]. It was seen in Chapter 5 that VE, unlike many CB-D antioxidants has considerable ability to quench singlet oxygen and it seems likely that under the conditions found in the lens it is an autosynergist, deactivating excited states as well as removing peroxyl radicals.

Vitamin C has been studied in more detail than other antioxidants in animals since this antioxidant is known to be present in the eye in much higher concentration than in other bodily tissues [245]. VC concentration is markedly reduced with age and is strongly diet-dependent [253,254]. Vitamin C was found to decrease light induced lipid peroxidation in the lens as measured by MDA formation [255] and since this antioxidant is neither a singlet oxygen quencher nor an effective lipid antioxidant, it seems likely that its primary role is to regenerate vitamin E from its aryloxyl radical (Chapter 5, Section 5.2.2). The preventive antioxidants, SOD and catalase, were also shown to reduce lipid photoperoxidation [255] by removing the peroxides which are the further source of "oxyl" radicals.

Bilgihan et al. [256] have reported the absence of melanin pigment in the eyes of albino rats. This leads to higher peroxidation rates due to the absence of CB-D activity provided by melanin (Chapter 5, Section 5.3.6). GSH is also depleted. This again confirms the importance of synergism in combinations of different kinds of antioxidant. However, relatively little work has so far been reported to optimise the synergistic interactions of the preventive and chain-breaking antioxidants.

6.11.2 *Sunburn*

It is well established that photochemical reactions in the skin particularly involving ionic iron are the main reason for the loss of skin texture during exposure to the skin (Chapter 5). The most effective commercial sun screen preparations contain in addition to a UV absorber, generally of the 2-hydroxybenzophenone type, one or more antioxidants of which the most important are α-tocopherol and β-carotene (or vitamin

A). The former scavenges peroxyl radicals but as has been seen earlier, the hydroperoxides produced in this process are themselves initiators of photooxidation and, in spite of the fact that it is a reasonably effective singlet oxygen quencher, it has little effect when used alone. Similarly, β-carotene, a highly effective singlet oxygen quencher, has relatively little protective effect in the absence of a UV absorber. However, a combination of all three is strongly synergistic.

The formation of damaged "sunburn cells" was found to be accelerated by depletion of GSH [257] and in organ culture, the addition of SOD and catalase prevented their formation [258]. Hamamelitannin, a sugar ester of gallic acid (see Chapter 5) which oxidised to dark coloured protective pigment was found by Masaki et al. [259] to be a much more powerful scavenger of peroxyl radicals than α-tocopherol and this work points the way forward to compound which act as CB-D antioxidants but transform in the process to UV screening agents which are also antioxidants.

6.11.3 *Respiratory inflammation and atmospheric pollution*

It was seen in Chapter 2 that a number of gaseous products found to varying degrees in urban and industrial environments exert prooxidant effects in animal lungs. Ozone and oxides of nitrogen are particularly important and may be synergistic [260]. Antioxidants present in the respiratory tract-lining fluids [261] protect the lung against normal oxidative stress and Table 6.9 shows the concentrations of typical antioxidants in epithelial lining fluid (ELF) of the human lung and compares this with antioxidant status in normal plasma [262].

Table 6.9 Concentrations of non-enzymic antioxidants in human plasma and epithelial lining fluids (ELFs) [262]

Antioxidant	Plasma (μmol/l)	ELF (μmol/l)
Ascorbic acid	64±8	54±5
Glutathione	1±0.2	182±25
Uric acid	379±59	178±3
α-Tocopherol	16±2	1±0.2
Albumin	662	50
Mucin	0	Variable

Human ELF is not very accessible experimentally, and Cross et al. instead exposed human plasma to O_3 and followed the depletion of antioxidants and changes in proteins and lipids [263,264]. They observed 60% reduction of ascorbic acid and uric acid in 1 hour on passing O_3 (16 ppm) over the surface of plasma, but α-tocopherol remained relatively unchanged over a 4 hour period and consistently with this, relatively little hydroperoxide or damage to was formed in the same period. This is consistent with observations (Chapter 5, Section 5.2) that homosynergistic regeneration of α-Toc-OH by $Asc(OH)_2$ occurs as the phenolic antioxidant is oxidised either directly by O_3 or by the resulting peroxyl radicals formed under these conditions (Chapter 2). There is some evidence that ozone may selectively attack ascorbic acid and uric acid in preference to the epithelial surface and further work is required to clarify the nature of the protection.

·NO_2 was found to behave very similarly to O_3 in plasma [262,265]. Addition of extra ascorbic acid protected α-tocopherol and delayed peroxidation, suggesting again that the former acts sacrificially to regenerate α-Toc-OH. Asc(OH)$_2$ also protects uric acid and it has been suggested [262] that it may be able to regenerate uric acid from its oxidation product in an analogous manner.

No synergism between ozone and nitrogen dioxide was observed in peroxidation experiments [262] and it has been suggested that ozone may even act preventively by oxidising ·NO_2 to produce the less reactive dinitrogen pentoxide [266]:

$$O_3 + \cdot NO_2 \;\rightarrow\; O_2 + \cdot NO_3 \;\xrightarrow{\cdot NO_2}\; N_2O_5 \tag{1}$$

6.11.4 *Exercise and hyperoxygenation*

Oxygen consumption and the potential for oxidative damage increases several-fold during exercise. This in turn leads to increased consumption of the antioxidant vitamins. The role of the naturally occurring antioxidants during exercise has been comprehensively reviewed by Packer and co-workers [267] and only the salient features will be discussed here.

Plasma vitamin E levels were found to increase in healthy male subjects during intensive exercise [268] suggesting that the body is able to mobilise the antioxidant from reserves to combat rapid oxidation in the muscles. It is also clear that endurance-trained athletes are better able to withstand oxidative stress during exercise than less well-trained subjects [267]. Supplementation of vitamin E in the diet of cyclists [269] and mountain climbers [270] for several weeks before exercise moderated the increase in pentane exhalation in the breath observed for non-supplemented subjects. This suggests that the normal recommended intake of vitamin E is insufficient to prevent increase in lipid peroxidation in subjects undergoing oxidative stress due to exercise. A similar study [271] of mountain climbers has shown that vitamin E is able to prevent the increase in blood viscosity experienced by high altitude climbers. In another study of the formation of breath pentane and serum MDA it was found that a daily synergistic combination of vitamin E (592 mg), vitamin C (1000 mg) and β-carotene (30 mg) for six weeks before tread-mill running did not result in an instantaneous reduction in oxidation products due to exercise, but did significantly reduce post-exercise serum levels of the peroxidation products [272]. There must however be some question about the value of using such high levels of β-carotene under conditions of hyperoxygenation since, as has been pointed out previously, this compound is not a true antioxidant but becomes a prooxidant for polyunsaturated lipids at ambient oxygen pressures. Further evidence for muscle protection by vitamin E has come from a reduction of TBARS and lipid conjugated dienes compared with placebos after exercise [273].

Another effect of exercise is to increase plasma creatinine kinase concentration, which is probably a reflection of accelerated muscle protein turnover, following exercise. Older control men had considerably lower plasma creatinine kinase levels after exercise than young controls but interestingly supplemented older men had a similar level to younger men after supplementation [274].

Animal studies have generally supported and complemented the effects of exercise in humans. Vitamin E deficiency in untrained rats resulted in a 40% decrease in running time to exhaustion [267,275]. Vitamin C deficient guinea pigs showed a similar behaviour [276] but vitamin C supplementation could not compensate for vitamin E deficiency [277]. Reductions in catalase and glutathione [278] have been observed in muscles after exercise but SOD levels were not changed [279].

6.11.5 *Protection against ionising radiation*
The effects of ionising radiation on polymers (Chapter 1) and on animals (Chapter 2) follow a very similar course in the presence of oxygen. After the initial irradiation which produces highly reactive carbon-centred radicals in polymers and oxygen, carbon and nitrogen radicals *in vivo*, there follows a slower but ultimately more important peroxidation which leads to further scission of chemical bonds. In the case of some polymers this damage may not be physically evident for weeks or even months and in humans, chromosome damage due to the formation of "clastogenic factors" (CF) also appears after a time lag and has been shown to persist for more than 30 years in the case of the Japanese atomic bomb survivors. The first stage cannot be easily prevented or mitigated either in polymers [280] or in animals [281] but the second stage is subject to control by a range of antioxidants. SOD appears to be particularly effective in preventing clastogenesis, and flavonoids have also been reported to be effective [282] but not vitamin E [283]. α-Lipoic acid, which was seen in Chapter 5 to have multifunctional activity as a CB-D, PD and MD antioxidant, was much more effective [284], and it was found by Korkina et al. [283] that treatment of children living in the Chernobyl region after the nuclear accident with a combination of α-LA and vitamin E lowered chemical markers of peroxidation in the blood to normal levels and liver and kidney functions were returned to normal.

It seems then that, as in other kinds of peroxidative damage *in vivo*, a synergistic combination of antioxidants is the most effective way of treating high energy irradiation damage. Several types of antirad have been developed for the protection of polymers which have not yet been examined in radiation damage in animals. Particularly significant are the CB-A antioxidants, the nitroxyl radicals [280] and their precursors the spin-traps (Chapter 3), many of which have been shown to be relatively non-toxic. Moreover, as will be seen below it is now possible to design chain-breaking antioxidant for controlled release over relatively long periods thus increasing their efficiency and reducing acute toxicity effects.

6.12 Therapeutic Potential of Antioxidants
This chapter has reviewed evidence on the ability of antioxidants, both natural and synthetic, to retard or prevent physiological irregularities known to be associated with excessive peroxidation. However, almost daily new peroxidative effects are being recognised as a result of antioxidant intervention. Few of these diagnostic uses of antioxidants have so far been systematically studied in the therapeutic treatment of the conditions studied but very considerable potential exists and some emerging applications will be discussed in the following sections.

6.12.1 *Antioxidants in surgery*
Of considerable potential interest is the use of antioxidants during or after surgery and particularly in organ transplant surgery [285-93]. Connor et al. [288] have produced ESR evidence from spin-trapping experiments during liver transplant surgery in rats that radicals are involved in graft failure, and Biasi et al. [293] have shown that

peroxidation products are increased and vitamin E and glutathione are decreased during the transplant process. Poli et al [294] have proposed that hydroperoxides are not just the result of liver damage but also its cause.

Both preventive and chain-breaking antioxidants have been found to be effective in protecting against reperfusion injury following transplant surgery [289,290] and during storage of organs [290,291]. Similar effects of antioxidants have been observed in kidney transplantation [295] and coronary by-pass surgery [296], but a great deal more experimental investigation of synergistic combinations of antioxidants is required to explore the full potential of antioxidants during surgery.

A recent study has shown the potential of controlled (slow) release of salicylic acid and vitamin E in wound healing. Application of vitamin E to wounds is known to improve healing but only marginally, due to rapid physical removal from the wound. San Román [297] co-polymerised methacryloyl salicylate (MAS) and methacryloyl-α-tocopherol (MAT) with 2-hydroxyethyl-methacrylate (HEMA) to give a hydrogel which, when applied to severed rabbit Achilles tendons, markedly improved the rate of healing. The α-tocopherol copolymer was considerably more effective than the salicylate copolymer due to initial rejection of the former. MAT-HEMA on the other hand was resorbed from the beginning into the growing tissue. This technology points a very promising way forward to targeting antioxidants to specific sites where they are required. Controlled release allows time for the antioxidant to complete its therapeutic intervention and promises a step forward in surgical procedures.

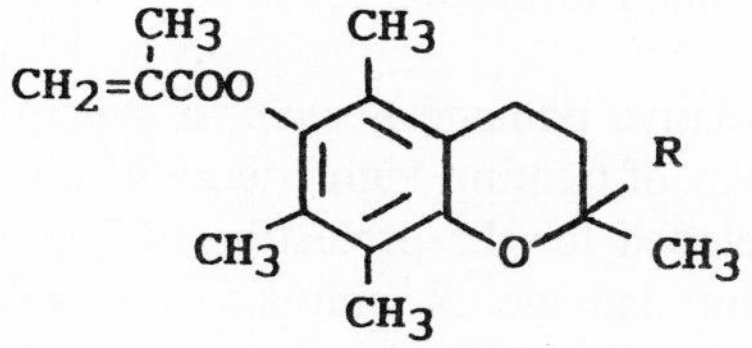

MAT

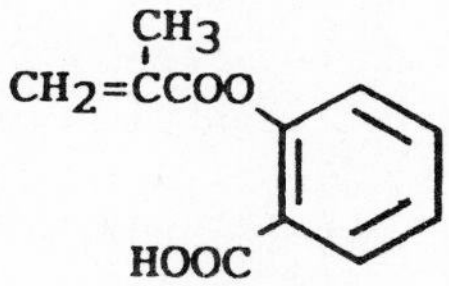

MAS

6.12.2 *Antioxidants in chemical toxicity and drug overdose*
It is well known that paracetamol causes liver damage when overdosed [298]. This is believed to be due to the formation by oxidation in the liver of semiquinoneimine radicals (Chapter 5, Scheme 5.18) which subsequently redox cycles to give superoxide, hydrogen peroxide and hydroxyl radicals [299,300].

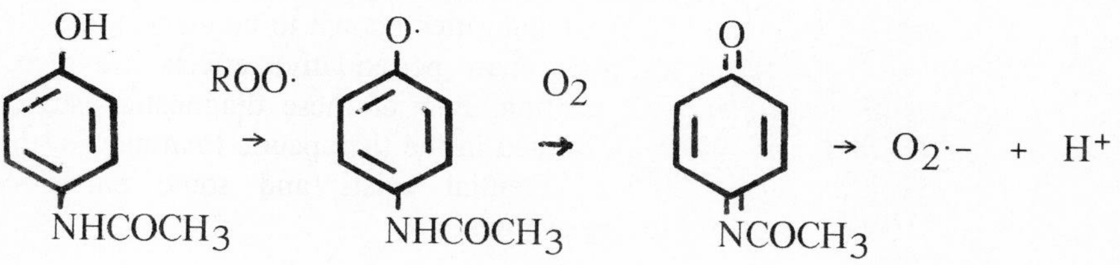

$$\text{HO-C}_6\text{H}_4\text{-NHCOCH}_3 \xrightarrow{ROO\cdot} \text{·O-C}_6\text{H}_4\text{-NHCOCH}_3 \xrightarrow{O_2} \text{O=C}_6\text{H}_4\text{=NCOCH}_3 \rightarrow O_2^{\cdot-} + H^+ \quad (2)$$

Matsura and co-workers found that endogenous antioxidants such as α-tocopherol [301] and ubiquinol [302] were depleted, and pre-administration of α-tocopherol counteracted the increase in hepatic TBARS [303]. When the two antioxidants were present together, the loss of Ubi-hq preceded the loss of α-Toc-OH in accord with the regenerative mechanism discussed in Chapter 5. Similar redox cycling prooxidant effects have also been suggested for some toxic industrial chemicals such as paraquat [304] and *m*-dinitrobenzene [305].

REFERENCES

1. A. Plagy, *Am. J. Clin. Nutr.*, **34**, 1569-83 (1981).
2. J.M. Gaziano, J.E. Manson and C.H. Hennekens in *Natural Antioxidants in Human Health and Disease*, Ed. B. Frei, Academic Press, San Diego, 1994, p.389.
3. R.A. Riemersma, M. Oliver, M.A. Elton, G. Alfthan, E. Vartiainen, M. Salo, P. Rubba, M. Mancini, H. Georgi, J. Vuilleumier and K.R. Gey, *Eur. J. Clin. Nutr.*, **44**, 144-50 (1990).
4. K.F. Gey, G.B. Brubacher and H. Stähelin, *Am. J. Clin. Nutr.*, **45**, 1368-77 (1987).
5. *The Scottish Diet*, Report of a Working Party to the Chief Medical Officer for Scotland, Scottish Office Home and Health Department, October, 1994.
6. A.L. Tappel in *Vitamins and Hormones*, **20**, 493-510 (1962).
7. Report of a Committee on *Diet Nutrition and Cancer*, Washington DC, National Academic Press, New York, 1982.
8. S. Palmer, *Prog. Food Nutr. Sci.*, **9**, 283-341 (1985).
9. R. Doll and R. Peto, *J. Natl. Cancer Inst.*, **66**, 1191-208 (1981).
10. D. Kromhout, *Am. J. Clin. Nutr.*, **45**, 1361-7 (1987).
11. *Nutritional Aspects of Cardiovascular Disease*, Reprt on Health and Social Subjects, **46**, HMSO, 1994.
12. *Dietary Guidelines to Lower your Cancer Risk*, World Cancer Research Fund, 1993.
13. H. Dam, *Pharmacol. Revs.*, **9**, 1-16 (1957).
14. K.F. Gey, *Bibl. Nutr. Dieta*, **37**, 53-91 (1986).
15. D.I. Thurnham, J.A. Davies, B.J. Crump, R.D. Situnayake and M. Davis, *Ann. Clin. Biochem.*, **23**, 514-20 (1986).
16. K.F. Gey, *Nutr. Biochem.*, **6**, 206-36 (1995).
17. D. Thurnham, *Proc. Nutr. Sci.*, **49**, 247-59 (1990).
18. K.F. Gey, *Int. J. Nutr. Res. Suppl.*, **30**, 224-31 (1989).
19. K.F. Gey, P. Puska, P. Jordan and U.K. Moser, *Am. J. Clin. Nutr.*, **53** (Suppl.), 326S-334S (1991).
20. K.F. Gey, U.K. Moser, P. Jordan, H.B. Stähelin, M. Eichholzer and E. Lüdin, *Am. J. Clin. Nutr.*, **57**, Suppl., 787S-797S (1993).
21. J.M. Gaziano, J.E. Manson and C.H. Hennekens in *Natural Antioxidants in Human Health and Disease*, Academic Press, San Diego, 1994, Chapter 13.
22. E. Ginter, *Am. J. Clin. Nutr.*, **32**, 511-2 (1979).
23. A.J. Verlangieri, J.C. Capeghian, S. El-Dean and M. Bush. *Med. Hypotheses*, **16**, 7-15 (1985).
24. B.K. Armstrong, J.L. Mann, A.M. Adelstein and F. Eskin, *J. Chronic Dis.*, **36**, 673-7 (1975).
25. R.M. Acheson and D.R.R. Williams, *Lancet*, **i**, 1191-3 (1983).
26. J.T. Salonen, R. Salonen, P. Seppanen, M. Kantola, M. Parvainen, G. Alfthan, P.H. Maenpaa, E. Taskinen and R. Rauramaa, *Atherosclerosis*, **70**, 155-60 (1988).
27. O. Igarashi, Y. Konekawa and Y. Fujiyama-Fujihara, *J. Nutr. Sci. Vitamol.*, **37**, 359-69 (1991).
28. R.A. Jacob, C.L. Otradovec and R.M. Russell, *Am. J. Clin. Nutr.*, **48**, 36-42 (1988).
29. K.J. Kunert and A.L. Tappel, *Lipids*, **18**, 271-4 (1983).
30. P. Palozza and N.I. Krinsky, *Arch. Biochim. Biophys.*, **297**, 184-7 (1992).

31. N.I. Krinsky in *Natural Antioxidants in Human Health and Disease*, Ed. B. Frei, Academic Press, San Diego, 1994, Chapter 8.
32. C. Costagliola and M. Menzione, *Clin. Physiol. Biochem.*, **8**, 140-3 (1990).
33. A.F.M. Kardinal, F.J. Kok, J. Ringstad, J. Gomez-Aracina, V.P. Mazaev, L. Kohlmeier, B.C. Martin, A. Aro, J.D. Kark, M. Delgado-Rodriguez, R.A. Riemersma, J.K. Huttunen and J.M. Martin-Moreno, *Lancet*, **342**, 1379-84 (1993).
34. J. Ramirez and N.C. Flowers, *Am. J. Clin. Nutr.*, **33**, 2097-87 (1980).
35. J.E. Manson, M.J. Stampfer, W.C. Willet, G.A. Colditz, B. Rosner, F.E. Speizer and C.H. Hennekens, *Circulation*, **84**, (4) Suppl. II, (1991), idem, Circulation, **85**, 865 (1992), idem, Circulation, **88**, 1-70 (1993).
36. E.B. Rimm, M.J. Stampfer, A. Ascherio, E. Giovannucci, G.A. Colditz and W.C. Willet, *N. Eng. J. Med.*, **328**, 1450-56 (1993).
37. J.E. Enstrom, L.E. Kanim and M.A. Klein, *Epidemiology*, **3**, 194-202 (1992).
38. J.M. Gaziano, J.E. Manson, L.G. Branch, F. LaMott, G. Colditz, J.E. Buring and C.H. Hennekens, *J. Am. Coll. Cardiol.* **19**, 377 (1992).
39. S.E. Vollset and E. Bjelke, *Lancet*, **ii**, 742 (1983).
40. L. Lapidus, H. Anderson, C. Bengtson and I. Bosceus, *Am. J. Clin. Nutr.*, **44**, 444-8 (1986).
41. D.A. Street, G.W. Comstock, R.M. Salkeld, W. Schuep and M. Klag, *Am. J. Epidemiol.*, **134**, 719-20 (1991).
42. F.J. Kok, A.M. de Bruijn, R. Vermeeren, A. Hofman, A. VanLaar, M. DeBruin, R.T.J. Hermus and H. Valkenberg, *Am. J. Clin. Nutr.*, **45**, 462-8 (1987).
43. J.T. Salonen, R. Salonen, I. Penttila, J. Herranen, M. Jauhiainen, M. Kantola, R. Lappetelainen, P. Maempaa, G. Alfthan and P. Puska, *Am. J. Cardiol.*, **56**, 226-31 (1985).
44. M. Eichholzer, H.B. Stähelin and K.F. Gey in *Free Radicals and Ageing*, Eds. I. Emerit and B. Chance, Birkhaeuser, Basel, 1992, pp. 398-410.
45. K.F. Gey, H.B. Stähelin and M. Eichholzer, *Clin. Invest.*, **71**, 3-6 (1993).
46. J.T. Salonen, K. Nyyssonen, M. Parviainen, M. Kantola, H. Korpela and R. Salonen, *Circulation*, **87**, 1 (1993)
47. H.W. Hense, M. Stender, W. Bors and U. Keil, *Atherosclerosis*, **103**, 21-8 (1993).
48. *Diet and Health: Implications for reducing chronic disease*. National Research Council, U.S.A, Committtee on Diet and Health, National Academy of Sciences, Washimgton DC, 1989.
49. K.F. Gey, ILSI Europe Workshop on *β-Carotene, Vitamin E, Vitamin C and Quercetin in the Prevention of Degenerative Disease*, Evian, France, 1994.
50. K.F. Gey, *Brit. Med. Bull.*, **49,** 679-99 (1993).
51. R. Muggli in *Health Effects of Fish Oils*, Ed. R.K. Chandra, ARTS Biomed. Publ., St. Johns, Newfoundland, 1989, pp.203-210.
52. K.F. Gey, *Int. J. Vit. Nutr. Res.*, **65**, 61-4 (1995).
53. K.H. Bässler, Z. Ernährungswiss., **30**, 174-80 (1991).
54. G.G. Duthie, J.R. Arthur, J.A.G. Beattie, K.M. Brown, P.C. Morrice, J.D. Robertson, C.T. Shortt, K.A. Walker and W.P.T. James, *Ann. N.Y. Acad. Sci.*, **868**, 120-9 (1993).
55. D.F. Church and W.A. Pryor, *Environm. Health Prospect*, **64**, 11-126 (1985).
56. B. Frei, T.M. Forte, B.N. Ames and C.E. Cross, *Biochem. J.*, **277**, 133-8 (1991).
57. R. Anderson and A.J. Theron, *World Rev. Nutr. Diet.*, **62**, 27-58 (1990).

58. J. Kalra, A.K. Chadhary and K. Prasad, *Int. J. Exp. Path.*, **72**, 1-7 (1991).
59. K.L.H. Carpenter, K.H. Cheeseman, C. Van Der Veen, S.E. Taylor, M.K. Walker and M.J. Mitchinson, *Free Rad. Res.* **23**, 549-58 (1995).
60. C. Bolton-Smith, C.E. Casey, K.F. Gey, W.C.S. Smith and H. Tunstall-Pedoe, *Brit. J. Nutr.*, **65**, 337-46 (1991).
61. A. Ascherio, M.J. Stampfer, G.A. Colditz, E.B. Rimm, L. Litin and W.C. Willett, *J. Nutr.*, **122**, 1792-1801 (1992).
62. E. Rimm and G. Colditz, *Ann. N.Y. Acad. Sci.*, **686**, 323-4 (1993).
63. W. Kübler, Report to National Research Council Committee on Diet and Health, Nat. Acad. Sciences, Washington DC, VERA Publication Series, Eds. W. Kübler, H.J. Anders, W. Heeschen and M. Kohlmeyer, 1993.
64. W.C. Stryker, L.A. Kaplan, E.A. Stein, M.J. Stampfer, A. Sobner and W.C. Willett, *Am. J. Epidemiol.*, **127**, 283-96 (1988).
65. C. Calzada, M. Bizzotto, G. Paganga, N.J. Miller, K.R. Bruckdorfer, A.T. Diplock and C.A. Rice-Evans, *Free Rad. Res.*, **23**, 489-503 (1995).
66. H. Verhagen, I. Deerenberg, A. Marx, F. ten Hoor, P.T. Henderson and J.C.S. Kleinjans, *Food Chem. Toxicol.*, **28**, 215-20 (1990).
67. C.J. Nunn, H. Verhagen and J.C.S. Kleinjans, *Food Chem. Toxicol.*, **29**, 73-5 (1991).
68. I. Björkhem, A. Henricksson-Freyschuss, O. Breuer, U. Diczfalusy. L. Berglund and P. Henricksson, *Arterioscler. Thromb.* **11**, 15-22 (1991).
69. D. Kritchevsky, H.K. Kim and S.A. Tepper, *Proc. Soc. Exp. Biol. Med.*, **136**, 1216-21 (1971).
70. R.W. Wissler and D. Vesselinovitch, *App. Pathol.*, **1**, 89-96 (1983).
71. S.J.T. Mao, M.T. Yates, R.A. Parker, E.M. Chi and R.L. Jackson, *Arterioscler. Thromb.*, **11**, 1266-75 (1991).
72. D. Schultz, J.T. Skamramauskas, N. Law, M.J. Mitchinson and D.V. Hunt, *Free Rad. Res.*, **23**, 259-71 (1995).
73. S.M. Lynch and B. Frei in *Natural Antioxidants in Human Health and Disease*, Ed. B. Frei, Academic Press, San Diego, 1994, Chapter 12.
74. G.C. Willis, *Canad. Med. Assoc. J.*, **69**, 17-22 (1953).
75. G.C. Willis, *Canad. Med. Assoc. J.*, **77**, 100-9 (1957).
76. R.F.A. Altman, G.M.V. Schaeffer, C.A. Salles, A.S. Ramos de Souza and P.M.Y. Cotias, *Drug Res.*, **30**, 627-30 (1980).
77. J.S. Nelson in *Vitamin E, A Comprehensive Treatise*, Ed. L.J. Machlin, Dekker, N.Y., 1980, pp.397-428.
78. H. Kimura, Y. Tamada, Y. Morita, H. Ikeda and T. Matsuo, *J. Nutr.*, **122**, 1904-9 (1992).
79. T.N.A. Bocan, S.B. Muella, E.Q. Brown, P.D. Uhlendorf, M.J. Mazur and R.S. Newton, *Exp. Mol. Pathol.*, **57**, 70-83 (1992).
80. J. Wojcicki, B. Rozewicka, L. Barcew-Wiszniewska, S. Samochowiec, S. Kadlubowska, S. Tustanowski and Z. Juzyszyn, *Atherosclerosis*, **87**, 9-16 (1991).
81. R.J. Williams, J.M. Motteram, C.H. Sharp and P.J. Gallagher, *Atherosclerosis*, **94**, 153-9 (1992).
82. E.T.H. Fontham in *Natural Antioxidants in Human Health and Disease*, Ed. B. Frei, Academic Press, San Diego, 1994, Chapter 6.

83. P. Knekt in *Natural Antioxidants in Human Health and Disease*, Ed. B. Frei, Academic Press, San Diego, 1994, Chapter 7.
84. M. Garland, M.J. Stampfer, W.C. Willett and D.J. Hunter in *Antioxidants in Human Health and Disease*, Ed. B. Frei, Academic Press, San Diego, 1994, Chapter 9.
85. E. Bjelke, *Int. J. Cancer*, **15**, 561-5 (1975).
86. R. Peto, R. Doll, J.D. Buckley and M.B. Sporn, *Nature*, **290**, 201-8 (1981).
87. E.T.H. Fontham, L.W. Pickle, W. Haenszel, P. Correa, Y. Lin and R.T. Falk, *Cancer*, **62**, 2267-73 (1988).
88. A.H. Wu-Williams, X. d.Dai, W. Blot, Z.Y. Xu, X.W. Sun, H.P. Xiao, B.J. Stone, S.F. Yu, Y.P. Feng, A.G. Ershow, J. Sun, J.F. Fraumeni and B.E. Henderson, *Br. J. Cancer*, **62**, 982-7 (1990).
89. C.A. Swanson, B.L. Mao, J.Y. Li, J.H. Lukin, S.X. Yao, J.Z. Wang, S.K. Cai, Y. Hou, Q.S. Luo and W.J. Blot, *Int. J. Cancer*, **50**, 876-80 (1992).
90. K. Tominaga, Y. Saito, K. Mori, N. Miyazawa, K. Yolio, Y. Koyama, K. Shimamura, J. Inura and M. Nagai, *Jap. J. Clin. Oncol.*, **22**, 96-101 (1992).
91. P. Knekt, *Int. J. Epidemiol.*, **17**, 281-6 (1988).
92. J.B. Stählin, F. Rosel, E. Buess and G. Brubacher, *JNCI*, **73**, 1463-8 (1984).
93. A.M.Y. Nomura, G.N. Stemmerman, L.K. Heilbrun, R.M. Salkeld and J.P. Vuilleumier, *Cancer Res.*, **45**, 2369-72 (1985).
94. M.S. Menkws,G.W. Comstock, J.P. Vuilleumier, K.J. Helsing, A.A. Rider and R. Brookmeyer, *N. Eng. J. Med.*, **315**, 1250-4 (1986).
95. R.G. Ziegler, *J. Am. Inst. Nutr.*, 116-22 (1989).
96. E. Riboli, *Ann. Oncology*, **3**, 783-91 (1992).
97. W.D. DeWys, W.F. Malone, R.R. Butram and M.A. Sestili, *Cancer*, **58**, 1954-62 (1986).
98. J.W. Cullen, *Cancer*, **62**, 1851-64 (1988).
99. W.J. Blot, J-Y. Li, P.R. Taylor, W. Guo, et al. *J. Natl. Cancer Inst.*, **85**, 1483-92 (1993).
100. T. Key, *Proc. Nutr. Soc.*, **53**, 605-14 (1994).
101. P.M. Rowe, *Lancet*, **347**, (Jan. 27), 249 (1996).
102. D.E. Blask in *The Pineal Gland*, Ed. R.J.Reiter, Raven, New York, 1984, pp.253-84.
103. M. Fechting and A. Ahlbohm, *Am. J. Epidemiol.*, **138**, 467 (1993).
104. B.W. Wilson, F. Lueng, R. Buschbom, R.G. Stevens, L.E. Anderson and R.J. Reiter in *The Pineal Gland and Cancer*, Eds. D. Gupta, A. Attanasio and R.J. Reiter, Brain Research Promotion, Tübingen, 1988, pp.245-59.
105. R.J. Reiter and B.A. Richardson, *FASEB J.*, **6**, 2283-7 (1992).
106. R.J. Reiter, *J. Pineal Res.*, **18**, 1-11 (1995).
107. T.W. Kensler, N.E. Davidson and K.Z. Guyton in *Atmospheric Oxidation and Antioxidants*, Vol. III, Ed. G. Scott, Elsevier, Amsterdam, 1993, Chapter 12.
108. G. Shklar, *J. Natl. Cancer Inst.*, **68**, 791-7 (1982).
109. W. Weerapradist and G. Shlkar, *Oral Surg.*, **54**, 304-12 (1982).
110. D. Trickler and G. Shklar, *J. Natl. Cancer Inst.*, **78**, 165-9 (1987).
111. B.S. Reddy and N. Hirota, *Fed. Proc.*, **38**, 714 (1979) (abs).
112. F.L. Warren, *Biochem. J.*, **37**, 338 (1943).
113. M.R.S. Fox, *Ann. N.Y. Acad. Sci.*, **258**, 144 (1975).

114. S.R. Lynch and J.D. Cook in *Micronutrient Interactions: Vitamins, Minerals and Hazardous Elements*, Eds. O.A. Levander and L.Cheng, N.Y. Acad. Sci, 1980, pp. 32-44.
115. R. Ranieri and J.H. Weisburger, *Ann. N.Y. Acad. Sci.*, **258**, 181-4 (1975).
116. S.S. Mirvish, *Ann. N.Y. Acad. Sci.*, USA, **258**, 175-80 (1975).
117. N.I. Krinsky, *J. Am. Inst. Nutr.*, 123-6 (1989).
118. A. Peng, J.E. Rundhaug, C.N. Yoshizawa and I.S. Bertram, *Carcinogenesis*, **9**, 1533-9 (1988).
119. M.M. Mathews-Roth and N.I. Krinsky, *Photochem. Photobiol.*, **42**, 35-8 (1985).
120. J.P. Perchellet, M.D. Owen, T.D. Posey, D.K. Orton and B.A. Schneider, *Carcinogenesis*, **6**, 567-73 (1985).
121. J.P. Perchellet, N.L. Abney, R.M. Thomas, Y.L. Guislan and E.M. Perchellet, *Cancer Res.*, **47**, 477-85 (1987).
122. T.J. Slaga and W.M. Bracke, *Cancer Res.*, **37**, 1631-5 (1977).
123. M.A. Moore, H. Tsauda, W. Thamavit, T. Matsui and N. Ito, *J. Natl. Cancer Inst.*, **78**, 289-92 (1987).
124. M. Hirose, A.Masuda, T.Inoue, S.Fukushima and N.Ito, *Carcinogenesis*, **7**, 1155-59.
125. J.B. Rotstein and T.J. Slaga, *Carcinogenesis*, **9**, 1457- 1551 (1988).
126. V. Chajès, M. Mahon and G.M. Kostner, *Free Rad. Biol. Med.*, **20**, 113-120 (1996).
127. D. Harman, *Free Radicals in Biology*, Ed. W.A. Pryor, Vol. V, Acad. Press, New York, 1982, Chap. 8.
128. R.J. Melhorn and G. Cole, *Adv. Free Radical Biol. & Med.*, Vol. 1, Pergamon Press, Oxford, 1985, p.165.
129. R.S. Sohal and R.G. Allen, *Adv. Free Radical Biol. & Med.*, Vol.2, Pergamon Press, Oxford, 1986, p. 117.
130. C. Loguercio, D. Taranto, M. Vitale, F. Beneduce and C. Del Vecchio Blanco, *Free Rad. Biol. Med.*, **20**, 483-88 (1996).
131. M.K. Shigenaga and B.N. Ames in *Natural Antioxidants in Human Health and Disease*, Ed. B. Frei, Academic Press, 1994, Chapter 3.
132. T.R. Kramer, N. Schoene, L.W. Douglas, J.T. Judd, R. Ballard-Barbash, P.R. Taylor, H.N. Bhagavan and P.P. Nair, *Am. J. Clin. Nutr.*, **54**, 896-902 (1991).
133. P.M. Gougerot, M. Fay, Y. Roche, P. Lacombe and C. Marquetty, *J. Immunol.*, **135**, 2045-51 (1985).
134. T.C. Fong and T. Makinodan, *Immunol. Lett.*, **20**, 149-54 (1989).
135. C. Franceschi, A. Cossaizza, L. Troiano, R. Salati and D. Monti, *Int. J. Clin. Pharmacol. Res.*, **10**, 53-57 (1990).
136. K. Folkers, M. Morita and J.J. McRee, *Biochem. Biophys. Res. Comm.* **193**, 88-92 (1993).
137. T. Furukawa, S.N. Meydani and J.B. Blumberg, *Mech. Ageing Dev.*, **38**, 107-117 (1987).
138. S.N. Meydani, M.P. Barklund, S. Lui, M. Meydani, R.A. Miller, J.G.Cannon, F.D. Morrow, R. Rocklin and J.B. Blumberg, *Am. J. Clin. Nutr.*, **52**, 557-63.
139. A. Bendich, *J. Nutr.*, **119**, 112-5 (1989).
140. C.G. Fraga, M.K. Shigenaga, J.W. Park, P. Degan and B.N. Ames, *Proc. Natl. Acad. Sci.* USA, **87,** 4533-37 (1990).

141. J.M. Tolmasoff, T. Ono and R.G. Cutler, *Proc. Natl. Acad. Sci.*, USA, **77**, 2777 (1980).
142. N.P.V. Nair, N. Hariharasubramanian, C. Pilapil, I. Isaac and J.X. Thavundayil, *Biol. Psychiat.* **21**, 141-50 (1986).
143. R.J. Reiter, B.A. Richardson, L.Y. Johnson, B.N. Ferguson and D.T. Dinh, *Science*, **210**, 1372-3 (1990).
144. G.J.M. Maestroni, A. Conti and W. Pierpaoli, *Pineal Res. Rev.*, **7**, 203-26 (1989).
145. K.A. Stokkan, R.J. Reiter, K.O. Nonaka, A. Lerchl, B.P. Yu and M.K. Vaughan, *Brain Res.*, **545**, 66-72 (1991).
146. J. Meites, *Proc. Soc. Exp. Biol. Med.*, **195**, 304-11 (1990).
147. A. Koizumi, R. Weindruch and R.L. Walford, *J. Am. Inst. Nutr.*, 361-7 (1987).
148. M.H. Chung, H. Kasai, S. Nishimura and B.P. Yu, *Free Rad. Biol. Med.*, **12**, 523-5 (1992).
149. R.L. Walford, *Maximum Life-Span*, Norton, New York, 1983.
150. A.D. Blackettt and D.A. Hall, *Gerontology*, **27**, 133 (1981).
151. R.G. Cutler in *Testing Theories of Aging*, Eds. R. Adelman and G. Roth, RC Press, Roca Baton, 1982, p.25.
152. A. Klegeris, L.G. Korkina and S. Greenfield, *Free Rad. Biol. Med.*, **18**, 215-222 (1995).
153. G. Benzi and A. Morretti, *Free Rad. Biol. Med.*, **19**, 77-101 (1995).
154. J.P.W. Spencer, A. Jenner, J. Butler, O.I. Aruoma, D.T. Dexter, P. Jenner and B. Halliwell, *Free Rad. Res.*, **24**, 93-105 (1996).
155. C.C. Chiueh, G. Krishna, T. Tulsi, T. Obata, K. Lang, S-J. Huang and D.L. Murphy, *Free Rad. Biol. Med.*, **13**, 581-3 (1992).
156. J.W. Miller, J. Selhub and J.A. Joseph, *Free Rad. Biol. Med.*, **21**, 241-9 (1996).
157. S.J. Kish, C. Morito and O. Hornykiewiez, *Neurosci. Lett.* **58**, 343-6 (1985).
158. D.P.R. Muller in *Natural Antioxidants in Human Health and Disease*, Ed. B. Frei, Academic Press, San Diego, 1994, Chapter 19.
159. S. Fahn, *Am.. J. Clin. Nutr.* **53**, 380S-382S (1991).
160. Parkinson Study Group, *N. Engl. J. Med.*, **328,** 176-83 (1993).
161. C. Jeandel, M.B. Nicolas, F. Dubois, F. Nabet-Belleville, F. Penin and G. Curry, *Gerontology*, **35**, 275-82 (1989).
162. Z. Zaman, S. Roche, P. Fielder, P.J. Frost, D.C. Niriella and A.C.D. Cayley, *Age Ageing*, **21**, 91-4 (1992).
163. T. Metcalfe, D.M. Bowen and D.P.R.Muller, *Neurochem. Res.*, **14**, 1209-12, (1989).
164. B. Kowsari, S.K. Finnie, R. Carter, J. Love, P. Katz, S. Longley and R. Panush, *J. Am. Diet. Acssoc.*, **82**, 657-9 (1983).
165. U. Tarp, K. Overvad, J.C. Hansen and E.B. Thorling, *Scand. J. Rheumatol.*, **14**, 97-101 (1985).
166. A. Imadaya, K. Terasawa, H. Tosajot, M. Okamoto and K. Toryzuka, *J. Rheumatol.,* **15**, 1628-31 (1988).
167. D.I. Thurnham, R.D. Situnayake, S. Koottathepsis, B. McConkey and M. Davis in *Free Radicals, Oxidant Stress and Drug Action*, Ed. C. Rice-Evans, Richelieu, London, pp.169-89.
168. A. Bendich in Natural Antioxidants in Human Health and Disease, Ed. B. Frei, Academic Press, San Diego, 1994 Chapter 15.

169. U. Tarp, J.C. Hansen, K. Overvad, E.B. Thorling and H. Graudall, *Arthritis Rheum.*, **30**, 1162-6 (1987).
170. B. Halliwell and J.M.C. Gutteridge, *Free Radicals in Biology and Medicine* (Second Edition), Clarendon Press, 1989, p.429.
171. N. Giordano, A. Favioravanti, S. Sancasciani, R. Marcolongo and C. Borghi, *BMJ.*, **289**, 961-2 (1984).
172. E.J. Corey, R.A. Mehrota and A.U. Kahn, *Arch. Biochem. Biophys.* **236**, 68-70 (1987).
173. S.R. Jolly, W.J. Kane, M.B. Baillie, G.D. Abrahams and B.R. Lucchesi, *Circ. Res.*, **54**, 277-85 (1984).
174. M.S. Paller, J.R. Hoidal and T.F. Ferris, *J.Clin. Invest.*, **74**, 1156-64 (1984).
175. H. Sanfrey, G.B. Bulkley and J.L. Cameron, *Ann. Surg.*, **200**, 405-413 (1984).
176. I. Kurose, D. Fukumara, S. Miura, M. Suematsu, M. Suzuki, E. Sekizula, H. Nagata, T. Morishita and M. Tsuchiya, *J. Gastroenterol. Hepatol.*, **8**, 254-8 (1993).
177. Y. Toaka, T. Ikata and K. Fukuzawa, *J. Nutr. Sci. Vitaminol.*, **36**, 217-26 (1990).
178. V.G. Amatuni, R.S. Matevosian, S. Sisakian and I.G. Arakelian, *Cor. Vasa*, **31**, 500-507 (1989).
179. Y. Gauduel and M.A. Duvelleroy, *J. Mol. Cell. Cardiol.*, **16**, 459-70 (1984).
180. G.P. Littaru, L. Ho and K. Folkers, *Int. J. Nutr. Res.*, **42**, 291-305 (1972).
181. G.P. Littaru, L. Ho and K. Folkers, *Int. J. Nutr. Res.*, **42**, 413-34 (1972).
182. T. Kamikawa, A. Kobayashi and T. Yamashita, *Am. J. Cardiol.* **56**, 247-51 (1985).
183. F. Schardt, D. Welzel and W. Schess in *Biomedical and Clinical Aspects of Coenzyme Q*, Vol. V, Eds. K. Folkers and Y. Yamamura, Elsevier, Amsterdam, 1985, pp.385-94.
184. C. Mazzola, E.E. Guffanti, A. Vaccarella, M. Merigalli, R. Colnago, N. Ferrario, V. Cantoni and G. Marchetti, *Curr. Therap. Res.*, **41**, 923-30 (1987).
185. E. Rossi, A. Lombardo, M. Testa, S. Lippa, A. Oradei, G.P. Littarru, M. Lucente, E. Coppola and U. Manzoli in *Biochemical and Clinical Aspects of Coenzyme Q*, Vol. VI, Eds. K. Folkers, T. Yamagami and G.P. Littarru, Elsevier, Amsterdam, 1991, pp.321-6.
186. M.F. Wilson, W.H. Frishman, T. Giles, G. Sethi, S.M. Greenberg and D.J. Brackett in *Biomedical and Clinical Aspects of Coenzyme Q*, Vol. VI, Eds. K. Folkers, T. Yamagami and G.P. Littarru, Elsevier, Amsterdam, 1991, pp.339-48.
187. F. Ursini, C. Gambini, R. Paciaroni and G.P. Littarru in *Biomedical and Clinical Aspects of Coenzyme Q*, Vol. VI, Eds. K. Folkers, T. Yamagami and G.P. Littarru, Elsevier, Amsterdam, 1991, pp.473-80.
188. C. Hofman-Bang, N. Rehnqvist, K. Swedberg and H. Astrom, *J. Am. Coll. Cardiol.*, **19**, 3 (Suppl. A), 774-6 (1992).
189. C. Morisco, B. Trimarco and M. Condorelli, *Cli. Investig.*, **71** (Suppl 8), S134-S136 (1993).
190. G. Littarru, M. Battino, S.A. Santini and A. Mordente in *Free Radicals in the Environment, Medicine and Toxicology*, Eds. H. Nohl, H. Esterbauer and C. Rice-Evans, Richelieu Press, London, 1994, pp.249-64.
191. M. Lemke, B. Frei, B.N. Ames and A.I. Faden, *Neurosci. Lett.*, **108**, 201-6 (1990).
192. S. Yoshida, K. Abe, R. Busto, B.D. Watson, K. Kogure and M.D. Ginsberg, *Brain Res.*, **245**, 307-16 (1982).

193. M. Yamamoto, T. Shima, T. Uozumi, T. Sogabe, K. Yamada and K. Kawasaki, *Stroke*, **14**, 977-82 (1983).
194. R.D. Saunders, L.L. Dugan, P. Demeduik, E.D. Means, L.A. Horrocks and D.K. Anderson, *J. Neurochem.*, **49**, 24-31 (1987).
195. B. Scheer and G. Zimmer, *Arch. Biochem. Biophys.*, **302**, 385-90 (1993).
196. E. Serbinova, S. Khwaja, A. Rznick and L. Packer, *Free Rad. Res. Comm.*, **17**, 49-58 (1992).
197. N. Haramaki, L. Packer, H. Assadnazari, H. Zimmer and G. Cardiac, *Biochem. Biophys. Res. Comm.*, **196**, 1101-7 (1993).
198. J.H. Prehn, B. Peruche, C. Karcoutly, C. Rossberg, H.D. Mennel and J. Krieglstein in *Pharmacology of Cerebral Ischemia*, Eds. J. Krieglstein and H. Oberpichler, Wissenschaft Verlags. mbH, 1990, 375-62.
199. X. Cao and J.W. Phillis, *Free Rad. Res.*, **23**, 365-70 (1995).
200. J.S. Komara, N.R. Nayni, H.A. Bialick, R.J. Indrien, A.T. Evans, A.M. Garritano, T.J. Hoehner, W.A. Jacobs, R.R. Huang, G.S. Krause, B.C. White and S.D. Aust, *Ann. Emerg. Med.*, **15**, 384-9 (1986).
201. A.M.M. van der Kraaij, H.G. van Eijk and J.F. Koster, *Circulation*, **80**, 158-64 (1989).
202. S.L. Smith, P.K. Andrus, J.R. Zhang and E.D. Hall, *J. Neurotrauma*, **11**, 393-401 (1994).
203. G.M. McGuire, P. Liu and H. Jeaschke, *Free Rad. Biol. Med.*, **20**, 189-97 (1996).
204. M.S. Paller and J.W. Eaton, *Free Rad. Biol. Med.*, **18**, 883-90 (1995).
205. C. Bruce, P.D. Scott, A. Rickers et al., *Proceedings of the 15th Meeting of the Pancreatic Society of Great Britain and Ireland*, Manchester, Nov. 15-16, 1990.
206. J.M. Braganza in *The Pathogenesis of Pancreatitis*, Ed. J.M. Braganza, Manchester University Press, 1991, Chapter 13.
207. S. Uden, D. Bilton, P.M. Guyan, P.M. Kay and J. Braganza in *Antioxidants in Therapy and Preventive Medicine*, Eds. I. Emerit, L. Packer and C. Auclair, Plenum Press, 1988, pp. 555-72.
208. S. Uden, D. Bilton, L. Nathan, L.P. Hunt, C. Main and J.M. Braganza, *Alimentary Pharmacol. and Therapeut.*, **4**, 357-71 (1990).
209. S.C. Langley, R.K. Brown and F.J. Kelly, *Pediatric Res.*, **33**, 247-50 (1993).
210. R.J. Stead, A.N. Reddington, L.J. Hinks et al., *Lancet*, **ii**, 862-3 (1985).
211. R.K. Brown and F.J. Kelly, *Thorax*, **49**, 738-42 (1994).
212. B.M. Winklhoffer-Roob, O. Ziouzenkova, H. Puhl, H. Ellemunter, P. Greiner, G. Müller, M.A. van't Hof, E. Esterbauer and D. Shmerling, *Free Rad. Bol. Med.*, **19**, 725-33 (1995).
213. J.A. Dodge in *The Pathogenesis of Pancreatitis*, Ed. J.M. Braganza, Machester University Press, 1991, Chapter 10.
214. D. Sandilands, I.M.J. Jeffrey, N.Y. Haboubi, I.A.M. McLennan and J.M. Braganza, *Gastroenterology*, **98**, 766-72 (1990).
215. D.C. Hilderbrand, Z. Fahim, E. James and M. Fahim, *Am. J. Obstet. Gynecol.*, **118**, 950-4 (1974).
216. K.M. Silvers, A.Y. Gibson and H.J. Powers, *Arch. Dis. Child.*, **71**, F40-F44 (1994).
217. F.J. Kelly, *Brit. Med. Bull.*, **49**, 668-78 (1993).

218. M. McElroy, A.D. Postle and F.J. Kelly, *Biochim. Biophys. Acta*, **117**, 153-8 (1992).
219. H.M. Hittner, L.B. Godis, M.E. Speer, A.J. Rudolph and C. Blifield, *N. Eng. J. Med.*, **305**, 1365-71 (1981).
220. N.N. Finer, R.F. Schindler, G. Grant, G.B. Hill and K.L. Peters, *Lancet*, **i**, 1087-91 (1982).
221. M.L. Chiswick, J. Wynn and N. Toner, *Ann. N.Y. Acad. Sci.*, **393**, 109-18 (1982).
222. M.L. Chiswick, M. Johnson, C. Woodall et al., *BMJ*, **287**, 81-84 (1983).
223. F. Fernandes-Banares, A. Abed-Lacruz, X. Xiol, J.J. Gine, C. Dolx, E. Cabre, M. Esteve, F. Gonzales-Huix and M.A. Gassul, *Am. J. Gastroenterol.*,**84**, 744-8 (1989).
224. A. Hermaniowicz, Z. Sliwinski and R. Kacsor, *Hepato-Gastroenterol*, 32, 81-6 (1985).
225. B.J. Blumerstein, C.K. Ma, Z. Zhang, N. Hadzijahic and R. Fogel, *Gastroenterol.*, A2011 (1994).
226. C. von Ritter, M.B. Grisham, M. Hollwarth, W. Inauen and D.N. Granger, *Gastroenterol.* **97**, 778-80 (1989).
227. J. Emerit, S. Pelletier, J. Likforman, C. Pasquier and A. Thullier, *Free Rad. Res. Comm.*, **12-13**, 563-9 (1991).
228. M.H.N. Golden, B.E. Golden and F.I. Bennett in *Trace Element Metabolism in Man and Animals-5*, Ed. C.F. Mills, Commonwealth Agricultural Bureau, Slough, 1985, pp. 759-79.
229. M.H.N. Golden, D.D. Ramdath and B.E. Golden in *Trace Elements, Micronutrients and Free Radicals*, Ed. I.E. Dreosti, Humana Press, 1991, Chap 9.
230. M.H.N. Golden in *The Pathogenesis of Pancreatitis*, Ed. J.M. Braganza, Manchester University Press, 1991, Chapter 11.
231. G. Scott in *The Pathology of Pancreatitis*, Ed. J.M. Braganza, Manchester University Press, 1991, Chapter 12.
232. G. Scott in *Atmospheric Oxidation and Antioxidants*, Vol. III, Ed. G. Scott, Elsevier, Amsterdam, 1993, Chapter 8.
233. P.S. Dobbin and R.C. Hider, *Chem. in Brit.*, 565-68 (June, 1990).
234. I. Fridovich, *Curr. Eye Res.*, **3**, 1-2 (1984).
235. S.D. Varma, S.M. Morris, S.A. Bauer and W.H. Koppenol, *Exp. Eye Res.* **43**, 1067-76 (1986).
236. F.J. Giblin, J.P. McReady and V.N. Reddy, *Invest. Ophthalmol. Visual Sci.*, **22**, 330-335 (1982).
237. W.B. Rathbun, A.M. Holleschau, D.L. Murray, A. Buchanan, S. Sawaguchi and R.V. Tao, *Curr. Eye Res.*, **9**, 45-53 (1990).
238. P. Knekt, M. Heliovaara, A. Riswsanen, A. Aromaa and R. Aaran, *BMJ*, No.305, 1392-4 (1992).
239. S. Vitale, S. West, J. Hallfrisch, C. Alston, F. Wang, C. Moorman, D. Muller, V. Singh and H.R. Taylor, *Epidemiology*, **4**, 195-203 (1993).
240. P.F. Jacques and L.T. Chylack, *Am. J. Clin. Nutr.*, **53**, 352S-355S (1991).
241. P.F. Jacques, L.T. Chylack and A. Taylor in *Natural Antioxidants in Human Health and Disease*, Ed. B. Frei, Academic Press, San Diego, 1994, Chapter 18.
242. J.M. Robertson, A.P. Donner and J.R. Trevithick, *Ann. N.Y. Acad. Sci.*, **570**, 372-82 (1989).

243. M.C. Leske, L.T. Chylack and S. Su, *Arch. Ophthalmol.*, **109**, 244-51 (1991).
244. P.F. Jacques, M. Lahav, S. Hankinson, W.C. Willett and A. Taylor, *10th Int. Congr. Eye Res.*, Stresa, Italy, 1992, No. 512, p.S152.
245. A. Taylor, P.F. Jacques, D. Nadler, F. Morrow, S.I. Sulski and D. Shepard., *Curr. Eye Res.*, **10**, 751-9 (1991).
246. M. Kwan, J. Niinikoski and T.K. Hunt, *Invest. Ophthalmol.*, **11**, 108-45 (1972).
247. I. Maitra, E. Serbinova, H. Trischler and L. Packer, *Free Rad. Biol. Med.*, **18**, 823-9 (1995).
248. R.D. Sperduto, T.S. Hu, R.C. Milton, J-L. Zhao, D.F. Everett, Q.F. Cheng, W.J. Blot, L. Bing, P.R. Taylor, L. Yun-Yao, S. Dawsey and W.D. Guo, *Arch. Ophthalmol.*, **111**, 1246-53 (1993).
249. D.K. Bhuyan, S.M. Podos, L.T. Machlin, H.N. Bhagavan, D.N. Chondhury, W.S. Soja and K.C. Bhuyan, Invest. Ophthalmol. Vis. Sci., **24**, 74 (1983).
250. L.J. Machlin and A. Bendich, *FASEB J.*, **1**, 441-5 (1987).
251. K.C. Bhuyan and D.K. Bhuyan, *Curr. Eye Res.*, **3**, 67-81 (1984).
252. C. Costagliola, G. Iuliano, E. Rinaldi, P. Vito and G. Auricchio, *Exp. Eye Res.*, **43**, 905-14 (1986).
253. J. Berger, D. Shepard, F. Morrow, J. Shadowski, T. Haire and A. Taylor, *Curr. Eye Res.*, **7**, 681-6 (1988).
254. J. Berger, D. Shepard, F. Morrow and A. Taylor, *J. Nutr.*, **119**, 1-7 (1989).
255. S.D.Varma, V.K.Sivastrava and R.D.Richards, *Ophthalmic Res.*, **14**, 167-75 (1982).
256. A. Bilgihan, M.K. Bilgihan, R.F. Akata, A. Aricioǧlu and B. Hasanreisoǧlu, *Free Rad. Biol. Med.*, **19**, 683-85.
257. K. Hanada, M.J. Gange and M.J. Connor, *J. Investig. Dermatol.*, **96**, 838-40 (1991).
258. Y. Miyachi, T. Horio and S. Imamura, *Clin. Exp. Dermatol.*, **8**, 305-10 (1983).
259. H. Masaki, T. Atsumi and H. Sakurai, *Free Rad. Res.*, **22**, 419-30 (1995).
260. T.R. Gielzleichter, H. Wittchi and J.A. Last, *Toxicol. App. Pharmacol.*, **116**, 1-9 (1992).
261. G.E. Hatch in *Comparative Biology of the Normal Lung*, Ed. R.A. Parent, CRC Press, 1992, pp. 617-34.
262. C.A. O'Neill, A. van der Vliet, J.P. Eiserich, J.A. Last, B. Halliwell and C.E. Cross in *Free Radicals and Oxidative Stress*, Biochem. Soc. Symp., Eds. R.C. Rice-Evans,
B. Halliwell and G.G. Lunt, Portland Press, 1995, **61** pp. 139-52.
263. C.E. Cross, P.A. Motchnik, B.A. Bruener, D.A. Jones, H. Kaur, B.N. Ames and B. Halliwell, *FEBS Lett.*, **298**, 269-72 (1992).
264. C.E. Cross, A.Z. Reznick, L. Packer, P.A. Davis, Y.J. Suzuki and B. Halliwell, *Free Rad. Res. Comm.*, **15**, 347-52 (1992).
265. B. Halliwell, M-L. Hu, S. Louie, T.R. Duvalle, B.K. Tarkington, P. Motchnick and C.E. Cross, *FEBS Lett.*, **313**, 62-6 (1992).
266. R.E. Huie, *Toxicology*, **89**, 193-216 (1994).
267. L. Packer, A.Z. Reznick and S. Landvik in *Natural Antioxidants in Human Health and Disease*, Ed. B. Frei, Academic Press, San Diego, 1994, Chapter 20.
268. J. Pincemail, C. Deby, G. Camus, F. Pirnay, R. Bouchez, L. Massaux and R. Goutier, *Europ. J. App. Physiol.*, **57**, 189-91 (1988).

268. C.J. Dillard, R.E. Litov, W.M. Savin, E.E. Dumelin and A.L. Tappel, *J. Appl. Physiol.: Respir. Environ. Execise Physiol.*, **45**, 927-32 (1978).
270. I. Simon-Schnass and H. Pabst, *Int. J. Vitam. Nutr. Res.*, **58**, 49-54 (1988).
271. I. Simon-Schnass and L. Korniszewski, *Int. J. Vitam. Nutr. Res.*, **60**, 26-34 (1990).
272. M.M. Kanter, L.A. Nolte and J.O. Holloszy, *J. Appl. Physiol.*, **74**, 965-9 (1993).
273. M. Meydani, W.J. Evans, G. Haudelman, L. Biddle, R.A. Fielding, S.N. Meydaui, J. Burrill, M.A. Fiatarone, J.B. Blumberg and J.G. Cannon, *Am. J. Physiol.*, **264**, R992-R998 (1993).
274. J.G. Cannon, S.F. Orencole, R.A. Fielding, M. Meydani, S.N. Meydani, M.A. Fiataroni, J.B. Blumberg and W.J. Evans, *Am. J. Physiol.*, **259**, R1214-R1219 (1990).
275. K.J.A. Davies, A.T. Quintanihla and L. Packer, *Biochem. Biophys. Res. Comm.*, **107**, 1198-1205 (1982).
276. L. Packer, K. Gohil, B. deLumen and S.E. Terblanche, *Comp. Biochem. Physiol.*, B, **83B**, 235-40 (1986).
277. K. Gohil, L. Packer, B. deLumen, G.A. Brooks and S.E. Terblanche, *J. App. Physiol.*, **60**, 1986-91 (1986).
278. L.L. Ji, R. Fu, and E.W. Mitchell, *J. Appl. Physiol.*, **73**, 1854-9 (1992).
279. M.H. Laughlin, T. Simpson, W.L. Sexton, O.R. Brown, J.K. Smith and R.J. Korthuis, *J. Appl. Physiol.*, **68**, 2337-43 (1990).
280. D.J. Carlsson in *Atmospheric Oxidation and Antioxidants*, Vol. II Ed. G. Scott, Elsevier, Amsterdam, 1993, Chapter 11.
281. P. Wardman in *Atmospheric Oxidation and Antioxidants*, Vol. III, Ed. G. Scott, Elsevier, Amsterdam, pp. 120 et seq.
282. I. Emerit, R. Arbtyunyan, N. Oganesian, A. Levy, L. Cernjavsky, T. Sarkisian, A. Pogossian and K. Asrian, *Free Rad. Biol. Med.*, **18**, 985-90 (1995).
283. L.G. Korkina, I.B. Afanas'ef and A.T. Diplock, *Biochem. Soc. Trans.*, **21**, 314S (1993).
284. N. Ramakrishnan, W.W.Wolf and G.N.Catravas, *Radiation Res.*, **130**, 360-5 (1992).
285. J.R. Stewart, W.S. Frist and W.H. Walter in *Oxygen Radicals in Biological Systems, Methods in Enzymology*, Vol. 186, Eds. L. Packer and A.N. Glazer, Academic Press, San Diego, 1990, pp.742-48.
286. G.M. Williams in *Oxygen Radicals in Biological Systems, Methods in Enzymology*, Vol. 186, Eds. L. Packer and A.N. Glazer, Academic Press, San Diego, 1990, pp.748-51.
287. R.G. Thurman, I. Marzi, G. Sietz, J. Thies, J.J. Lemasters and F.A. Zimmerman, *Transplantation*, **46**, 502-6 (1988).
288. H.D. Connor, W. Gao, S. Nukina, J.J. Lemasters, R.P. Mason and R.G. Thurman, *Transplantation*, **54**, 199-204 (1992).
289. H. Jaeschke, *Chem. Biol. Interact.*, **79**, 115-36 (1991).
290. U. Rauen, R. Viebahn, W. Lauchart and H. deGroot, *Hepato-Gastroenterol.*, **41**, 333-6 (1994).
291. J.H. Southard, *Transplant Proc.*, **21**, 1195-6 (1989).
292. R.T. Currin, J.G. Toole, R.G. Thurman and J.J. Lemasters, *Transplantation*, **50**, 1076-8 (1990).

293. F. Biasi, M. Bosco, I. Chiappino, E. Chiarpotto, G. Lanfranco, A. Ottobrelli, G. Massano, P.P. Donadio, M. Vj, E. Andorn, M. Rizzetto, M. Saizzoni and G. Poli, *Free Rad. Biol. Med.*, **19**, 311-97 (1995).
294. G. Poli, E. Albano and M.U. Dianzani, *Chem. Phys. Lipids*, **45**, 117-42 (1987).
295. T. Mayumi, H.J. Schiller and G.B. Bulkley in *Free Radicals: From Basic Science to Medicine*, Eds. G. Poli, E. Albano and M.U. Dianzani, Berkhäser Verlag., Basel, 1993, pp.438-57.
296. W.D. Johnson, K.L. Kayser, J.B. Brenowiz and S.F. Saedi, *Am. Heart J.*, **121**, 20-4 (1991).
297. J. San Román, International Conference on Environmental Impact of Polymeric Materials, 23rd Ahron Katzir-Katchalsky Conference, Israel, May 1996.
298. J.R. Mitchell, *N. Engl. J. Med.*, **319**, 1601-2 (1988).
299. V. Fischer, P.R. West, S.D. Nelson, P.J. Harvison and R.P. Mason, *J. Biol. Chem.*, **260**, 11446-50 (1985).
300. S.L. Arnaiz, S. Llesuy, J.C. Cutrín and A. Boveris, *Free Rad. Biol. Med.*, **19**, 303-10 (1995).
301. T. Matsura, K. Yamada and T. Kawasaki, *Biochem. Biophys. Acta*, **1127**, 277-83 (1992).
302. T. Matsura, K. Yamada and T. Kawasaki, *Biochem. Biophys. Acta*, **1123**, 309-15 (1992).
303. T. Amimoto, T. Matsura, S-Y. Koyama, T. Nakanishi, K. Yamada and G. Kajiyama, *Free Rad. Biol. Med.*, **19**, 169-76 (1995).
304. L.L. Smith, *Phil. Trans. Roy. Soc.*, B**311**, 647 -57 (1985).
305. I.A. Romero, T. Lister, H.K. Richards, M.P. Seville, S.P. Wylie and D.E. Ray, *Free Rad. Biol. Med.*, **18**, 311-319 (1995).

INDEX

α-Cumylperoxyl, termination 12
α-Eleostearic acid 18
α-LA, see α-lipoic acid
α-Lipoic acid (α-LA) 225, 226, 287
   and DHLA, multifunctional autosynergists 226
   CB-D antioxidant activity of 225
   redox antioxidant activity of 225
   reduction by NADPH 226
   regeneration of by Ubi-hq 226
   reaction of singlet oxygen with 226
α-Naphthylamine 96
α-Olefin polymers 6
α-phenyl-N-t-butylnitrone (PBN), spin-trap 46
α-Toc-OH, see α-tocopherol
α-Tocopherol (α-Toc-OH) 67, 159, 195-200, 204, 211, 218, 219, 267, 271, 280, 295
   antagonism in cancer therapy 280
   as prooxidant 200
   combination with Ubi-hq-10 205
   conversion to dimers and trimers 197
   inhibition of γ-linolenic acid peroxidation 280
   in LDL particles 198, 199
   in LDL containing Cu^{2+} 200
   in oils at elevated temperatures 195
   in plasma of mammals 67
   in sunflower and safflower oils 271
   mechanoantioxidant activity of 103
   mobility of 199
   processing stabiliser for polypropylene 103
   protected by retinol 218
   radical products from 201
   regenerated from its aryloxyl 196
   regenerated from oxidation products 197
   regenerated by β-carotene 219
   singlet oxygen quenching by 200
   supplementation by 271
α-Tocopherol/cholesterol ester ratio 264, 267, 269
α-Toc-OH/PUFA 271
   ratio in European fats diet 271
α-tocopheroquinone (α-Toc-q) 105
α-tocopheryloxyl (α-Toc-O·),
   reaction with ascorbic acid (Asc(OH)$_2$ 159, 202
   reduction by ubiquinol 196
   reduced to the hydroquinone α-Toc-hq
     in vivo 200
α-Toc-q/α-Toc-hq, antioxidant activity of 105

α-Toc-sd 103-105, 197, 200
   catalytic antioxidant 197
   redox cycle with alkyl and
     alkylperoxyl radicals 105, 200
   reduced back to phenol 197
AAPH, water soluble azo initiator 204
   in combination, Ubi-q-10 204
Absorption (screening), of UV 126
Acrylonitrile, styrene and butadiene (ABS) 21
Accelerated testing 29
Accelerated UV testing, correlation with
   outdoor weathering 30
Acetic acid, antioxidant effect 4
2-acetylaminofluorene (2-AAF) 279
"Action distance" 38
Adipic acid 1
Adsorbed NO_x and SO_2 in diesel particles 64
Adult respiratory distress syndrome (ARDS) 59
Aerosol particulates, from diesel fuels 63
Aflatoxin B_1 (AFB$_1$) 279
Ageing 45, 65, 66, 280
   and lipid peroxidation 45
   antioxidants and 280
   genetically controlled 65
   metabolic rate, peroxidation and 66
Age pigments 47
Age-related cataract 61
Agricultural plastics, photooxidation 143
Air oven ageing tests 29
Albumin 211, 220, 226
   as CB-D antioxidants 220
   as metal deactivator 226
Alcohol abuse 45, 56
Aldehydes 14
Aldonitrones, catalytic mechanoantioxidants
   in polypropylene 108
Alicyclic phosphates, photoantioxidants 148
Aliphatic amines 116
Alkoxyl radicals 13, 14, 38
   from hydroperoxides 14
   fragmentation 14
Alkyds 18
4-Alkylaminodiphenylamines 111
Alkyl/aryl phosphites, as synergists 145
Alkyl hydroperoxides 11
Alkyl iodides 102
4-Alkyl-2-mercaptothiazoline 137, 138
   oxidative transformation of 138
   as photoantioxidants 137

Alkyl monosulphides 129
  as antioxidants 129
Alkylperoxyl radical 9, 11, 154, 193
  "magnifies" initial radical attack 193
Alkyl radical generator 85
Alkyl radical termination 94
Alkyl radicals, resonance stabilised 94
Alkyltin maleate esters 170
Allopurinol, inhibitor of oxidation 246
Allyl benzene 4
Almond 233
Alzheimer's disease 45, 280, 284
  and lipid peroxidation 45
  and Parkinson's disease 280
  deficiency of vitamins E and A 284
American diet 270
Aminoxyls, CB-A antioxidants 208, 240
Aminoxyl, catalytic photoantioxidants 119
Aminoxyl radicals (>NO·) 46, 84, 94, 97, 99,
    102, 110, 111, 113, 116-118,
    119, 193, 207, 241, 242
  antifatigue agents in rubbers 110
  as catalytic antioxidants 96-100
  as catalytic photoantioxidants 117
  as mechanoantioxidants in polypropylene 102, 107
  CB-A antioxidants 242
  disproportionation of 119
  formed from arylamines during fatigue 111
  retard peroxidation *in vivo* 193
  in-chain 118
  photoantioxidant activity 115-118
  "stable" 116
  stoichiometric inhibition coefficient (*f*) of 97
  tolerated in the body 241
AMVN, organosoluble azo indicator 218
Anatase 154
Angina pectoris 267
  and lipid peroxidation 45
  antioxidant status in 45, 268
Anorexia 60
Antagonism 158
Antiatherogenic drugs 274
  dithiolates 274
  not necessarily anticholesterologenic 270, 274
"Antidegradants", in tyre technology 80
Antifatigue agents 81, 111, 115
  aldonitrones in sulphur vulcanisate 115
  effectiveness of 111
Antifatigue (mechanoantioxidant) activity, 113
  diarylamines 113
Anti-inflammatory drugs 244

Antioxidants 18, 28, 29, 36, 58, 67, 80, 81,
    90, 92, 93, 134, 145, 159, 171,
    172, 173, 176, 178, 180, 191, 195,
    213, 214, 229, 231, 262, 264, 267,
    269, 272, 274, 275, 278, 280, 281,
    283, 286, 288, 293, 294
  antagonism 159, 278
  arylamines 82
  attaching to unsaturated rubbers 180
  attachment to saturated
    polymers 175, 176, 178
  behaviour of in accelerated tests 29
  biological 36
  catalytic 96
  chain-breaking (CB) 80, 96, 165
  co-grafted 182
  content of edible oils 231
  correlate with life-span energy
    potential (LEP) 67, 281
  cyclic phosphites as 146
  deficiencies 262, 267
  "delayed action" 286
  depletion of 90
  diffusion, rate controlling 173
  during exercise 293
  effect of molecular weight 172
  essential 191
  grafting of 176
  half-life in the polymers 93
  herbiforous 213
  hydrocarbon solubility of 93
  imbalance in preterm infants 288
  importance of synergism *in vivo* 278
  in a closed system 172
  in biology 81, 191
  in the eye 289
  increase in molecular size, effect of 171
  in disease and oxidative stress 262
  inhibited the reduction in $GSHP_x$ activity 280
  intervention and supplementation 269
  in the initiation and development of
    disease 36
  in treatment of disease and oxidative
    stress 191, 262, 264
  in vivo 191
  mechanism of action of phosphites 146
  naturally occurring 269
  nutritional aspects of 229
  optimal activity 174
  "optimal" status in CVD 266
  persistence of in a polymer matrix 93
  phenolic 82
  physical loss from substrate 195

plasma-active	229	Antioxidant effectiveness	93, 172
polymer-bound	174, 176	physical aspects	93, 172
preventive	134	Antioxidant mechanisms	191, 247
protection factor (PF)	195	*in vivo*	191
reduced activity of with age	280	of the dithiocarbamates	247
reperfusion, in	58	Antioxidant-modified polymers	176
requirement in polyunsaturated rubbers	272	Antioxidant nutrients, essential	264

- plasma-active 229
- polymer-bound 174, 176
- preventive 134
- protection factor (PF) 195
- reduced activity of with age 280
- reperfusion, in 58
- requirement in polyunsaturated rubbers 272
- sesame seed oils 214
- solubility and activity 173
- solubility/mobility 174
- soluble in the amorphous domains 28
- solvent leaching of 173
- "stable" 82
- status of in plasma 267
- sulphur compounds 92, 134
- supplementation in epidemiology 290
- supplements in cystic fibrosis 288
- synergism 126, 159, 274, 278
- synergistic combinations *in vivo* 288
- synergistic optima 275
- synthetic, in ageing 283
- therapeutic potential of 294
- threshold plasma concentrations of 269
- volatility of 93, 171, 174

Antioxidant action 82, 208
- catalytic 208
- mechanisms of 82

Antioxidant activity 80, 92, 93, 99, 130, 146, 171, 200, 208
- effect of molecular size on 171
- in copper-initiated peroxidations 200
- in petroleum technology 80
- of aminoxyls 99
- of homologous series 93
- of oxides of nitrogen 208
- of polyhydroxyphenols 208
- of sulphides and their oxidation products 130
- of the catechol phosphites 146
- relative order of 85
- steric effects in 92

Antioxidant enzymes in RA patients 284
Antioxidant systems, *in vivo* 278, 283
Antioxidant and stabiliser substantivity 171
Antioxidant capacity, of margarines and spreads 273
Antioxidant defence 283
- genetic control theory of 283
Antioxidant defences 66, 229
- biological 229
- decrease in concentration with age 66
- in Alzheimer's disease 66
- in Parkinson's disease 66
- in rheumatoid arthritis 66

Antioxidant effectiveness 93, 172
- physical aspects 93, 172
Antioxidant mechanisms 191, 247
- *in vivo* 191
- of the dithiocarbamates 247
Antioxidant-modified polymers 176
Antioxidant nutrients, essential 264
Antioxidant performance, physical aspects of 171-183
Antioxidant radicals, reaction with peroxyl 102
Antioxidant status 36, 264, 265, 268, 278
- in plasma 265
- markers of 278
- "normalised" 264
- sub-optimal, predictor of CHD 268
Antioxidant substantivity 171-183
Antioxidant synergism, supplements in 80, 283
Antioxidant therapy in chronic pancreatitis 262, 288
Antioxidant vitamins 48, 50, 269
- importance of 50
- in the food supply 48
- in polyunsaturated fats 50
- recommended optimal intake (ROI) 269
Antiozonants 81
Apigenidin 211
- in coloured fruits 238
Apigenin, in celery, parsley 211, 238
Apples, antioxidants in 234, 276
Apricots, antioxidants in 234
Arachidonic acid 38, 39, 229
- oxidation cascade 39
- transformation to prostaglandin 38
Aromatic amines, order of antioxidant activity 92, 93
Aromatic aminoxyls, catalytic activity of 112
Aromatic aminyl radicals, dimerisation 110
Aromatic spin traps, as thermal antioxidants 109
Arthritis, see Rheumatoid arthritis
Aryloxyls 94
Asbestos fibres, iron in 64
Asbestosis 56, 63
Asc(OH)$_2$, see ascorbic acid and vitamin C
Ascorbic acid (Asc(OH)$_2$), 201, 293
- see also vitamin C 58, 59, 60, 193, 201, 202, 211, 267, 287, 292, 293
- acute pancreatitis treatment with 287
- CB-D antioxidant activity of 201
- deficiency in acute pancreatitis 287
- inhibits peroxidation of LDL 201
- in plasma lipids 201

in the regeneration of
  α-Toc-OH   193, 202, 293
prooxidant role in the presence of
  transition metal ions   58, 60, 201, 293
protects uric acid   293
reduces Fe^{3+} to Fe^{2+} 59
oxidation of by O_3 292
Asparagus, antioxidants in   235
Aspirin, toxicity of   244, 285
Astaxanthin   218, 278
Asthma, and environmental pollutants   56, 63
Asthmatic diseases   10
Atherogenic effects of oxidised cholesterase   48
Atherosclerosis   16, 45, 47, 48, 50, 191,
                                      264, 266, 274
  and lipid peroxidation   45
  aetiology of   48
  and diet   265
  and the fatty acid composition of diet   266
  importance of vitamin C in   275
  involvement of dietary cholesterol in   48
  LDL peroxidation theory of   50
  "plaque" formation in   191
Atherosclerotic lesions, low α-tocopherol/
  cholesterol ratio in   273
Atomic oxygen   154
Aubergine, antioxidants in   235
Autoinhibition   3
Autoinitiation, by ground-state dioxygen   8
Automotive tyre, durability of   18
Autosynergism, in polymers   134, 153, 166,
                                   168, 228, 250
Autosynergistic antioxidants   167, 169
  based on mercapto esters   169
  uric acid as   153
Autosynergistic antioxidant drugs   252
Autosynergistic UV absorber, EBHPT-ABS 169
Autosynergists   148, 169
  acting by different mechanisms   169
Avocado, antioxidants in   233, 234
2,2'-azino-bis(3-ethylbenzothiazoline-6-
  sulphonate) ($ABTS^+$)   210
2,2'-azobis-(amidinopropane hydrochloride,
  AAPH), peroxyl radicals from   199

β-carotene (β-C) 13, 53, 67, 157, 201, 216, 217,
                  218, 232, 235, 267, 273, 275, 291
  absorption of oxygen during
    retardation period   217
  antioxidant mechanism of   13
  from "yellow-orange" fruits and vegetables   54
  in dark green vegetables   54

lower plasma levels in smokers   273
"magic bullet" reputation of   275
oxidation products in retardation   217
peroxidation of   217
reduces the incidence of lung cancer   275
reduces the rate of AIBN initiated
  oxidation   218
requires α-tocopherol as synergist   273
"Retarder" of oxidation   216
synergistic interaction with CB-D
  antioxidants   218
synergistic with vitamin E   278
β-carotenyl radical   219
β-di-*iso*-butene, peroxidation   15
Bacon, fats and antioxidants in   236
Bananas, antioxidants in   234, 276
Basel study   269
BPH4 (see Tetrahydropterin)
β-C (see β-carotene)   291
Beans - broad, green, soya, antioxidants in   235
Beef, fats and antioxidants in   236
Beetroot, antioxidants in   235
Benzene, oxidation in the liver   55
Benzophenone   158
Benzo[α]pyrene-7,8-diol, expoxidation of   17
Benzo[α]pyrene, epoxidation of   55, 279
Benzoic acid   4
Benzoquinone/quinhydrone   94, 96
  trap for macroalkyl radicals   94
Benzothiazole (BT)   135
Benzothiazolesulphonic acid (BTSO)   135
Benzylic compounds   5
BHA (see butylated hydroxylanisole)
2,6-BHA, more effective than α-Toc-OH
  in LDL   199
BHBM-H, oxidative transformations of   168
BHBM-R, sulphur eliminated from   167
BHT (see butylated hydroxy toluene)
Bilirubin   206, 211, 220, 227
  CB-D antioxidant in plasma   206
  chain-breaking antioxidant   227
  iron-binding capacity of   227
  metal binding agent   206
  oxidised to biliverdin   206
Bioflavonoids   237
  antioxidant effectiveness   237
  in the human diet   237
  in teas and wines   237
Biological antioxidants   220
Biological peroxidation, measurement of   45
Biological substrates, peroxidation in   36-41

Bis-alkylxanthogen disulphides, inhibitors for procarcinogens 247
Bis-dialkylthiuram disulphides 247
Bis-phenol sulphides 134, 167
   destroy hydroperoxides catalytically 167
Blackberries, antioxidants in 234
Blackcurrants, antioxidants in 234
Bleomycin 209
   binds to DNA with iron 209
   iron chelator 209
Blood-brain barrier, penetration of 250
Blood plasma, biological markers in 277
Blue Band margarine, antioxidants in 232
BMPF, benzofuran analogue of α-Toc-OH 195
BPF-A, VP Sanduvor PB-41 bound antioxidants 179
Brain 58
   low concentrations of endogenous antioxidants 58
   polyunsaturated fatty acids in 58
Brazil nuts, antioxidants in 233
Broccoli, antioxidants in 235
"Bronze diabetes" 60
Brussels sprouts, antioxidants in 235
Butter, fats and antioxidants in 41, 229
Butter powders 43
Butylated hydroxylanisole (BHA), *in vivo* 199
   as carcinogen 280
   tumour suppression with 280
Butylated hydroxytoluene (BHT) 88, 91, 195, 199, 238, 280
   carcinogenic properties 238
   inhibiting DMBA-induced mammary cancers 280
   in vivo 199, 280
   oxidative transformation products of 88
   permitted food additive 238
   phenoxyl radical from 91
   similar to octyl gallate (OG) as antioxidants 195

γ- and δ-tocopherols, loss of intrinsic antioxidant activity 195
γ-tocopherol 201, 271
   dietary components of corn and soybean oils 271
   reduction of ·NO$_2$ 201
γ-Toc-Q 200
Cabbage - average, spring, antioxidants in 235
Cadmium sulphide, pigment in plastics 155

Caeruloplasmin 59, 227, 288
   binds and oxidises Fe^{2+} 227
   in the Fenton reaction 59
   preventive antioxidant 288
Caeruloplasmin:vitamin C ratio, of premature infants 288
Caffeic acid, metal chelation by 215
Calorie restricted diet 282
   lower levels of oxo^{8}dG 282
   lower lipid peroxidation 282
Cancer 45, 51-56, 59, 262, 263, 267, 275-280
   and lipid peroxidation 45
   case-control studies of vitamin E 276
   diet, effect of 51, 275
   effects of antioxidants *in vivo* 279
   epidemiological studies of 275
   in CF sufferers 59
   incidence in different social groups 263
   melatonin and 279
   of the breast 279
   of the larynx 56
   of the lung, in smokers 55
   of the respiratory-digestive tract, of the epidermis 276
   plasma antioxidant concentrations in 277
   preventive effect of fruit and vegetables 267
Cancer prevention, "magic bullet" 278
Canthaxanthin 218, 278, 279
   more effective than β-C against UV initiated skin tumours 279
Canthaxanthin polyenoxyl (CXPE) 218
Carbon black 154, 228
   as light stabiliser 228
   functional groups in 154
   light absorbing pigment 154
Carbon-carbon bond scission 24
Carbon dioxide 27
Carbon monoxide, formed in photooxidation 27
Carbon radical, stable 13
Carbonyl compounds 14
4-Carbonyl phenols, as light stabilisers 157
Carboxylate group, in polymers 26
Carboxylic acids 14
Carcinogens, blocked by vitamin C 279
Cardiovascular disease (CVD) 49, 50, 262-273,
   affected by antioxidants 268, 274
   affected by fat intake 49
   dietary pattern in 262
   epidemiological studies of 265
   selenium and

Carotenoids	215-219, 229, 278, 279	Cell membrane	38, 66
as singlet oxygen quenchers	279	cross-linking	66
in cancer and atherosclerosis	215	Central nervous system	58, 286
in fats and oils	229	ascorbic acid in	58
in foods	278	decrease in antioxidant status following	
prooxidant at ambient oxygen pressures	218	injury	286
Carrots, antioxidants in	235	Cereals, protected by a combination of	
Case-control studies	265, 267	vitamin E and selenium	237
of plasma levels of vitamins	267	Cerebral ischaemia	58, 286
selection of controls in	65	"free" iron concentration	58
Cashew nuts, antioxidants in	233	vitamins C and E	286
Catalase (Cat)	64, 66, 191, 192, 219, 221, 222, 288	Cerebral malaria	56, 60
		CF, elevated in	288
deficiency in preterm infants	288	Chain-breaking	96, 153
in the cytosol	219	catalytic	96
mechanism of hydrogen peroxide decomposition by	222	through phenol	153
		Chain-breaking acceptor (CB-A)	
peroxidolytic (PD) antioxidant	222	mechanism	92, 93, 95, 100
preventive antioxidant	288	Chain-breaking (CB) antioxidants	80, 81, 280
protectant against the *in vivo* Fenton reaction	222	electronic and steric features	80
		inhibit the formation of free radicals	81
Catalytic CB-A/CB-D cycle	97, 100, 134	Chain-breaking donor (CB-D)	
Catalytic CB antioxidants, applications		antioxidants	83-93, 159, 193-219, 252
in polymers	100-121	activity of	91
Cataract	45, 61, 290, 291	containing sulphur	166
and lipid peroxidation	45	naturally occurring	193-219
and plasma antioxidant status	267	relative effectiveness of phenolics	86
environmental factors in	290	structure-activity relationships in	87-93
fruits and vegetables decrease	291	synergism between	159
inhibition by carotenoids	290	Chain scission	8, 19, 21, 22
inversely associated with total carotene	290	of the cis rubbers	19
opacification of the lens	61	Chalcones, CB-D activity of	212, 266, 268
regional differences in	263	Cheese	41
vitamin C prevents	290	Chelating agents, salicaldehyde-based	149
Catechin	211, 213, 214, 238	Chelating drugs,	249
from seeds, barks and leaves	214	Chemical carcinogens	54
in tea	238	Chemical toxicity, antioxidants in	295
Catechol phosphites	147	Chemi-crystallisation	28
Catechols	244	Chemiluminescence	29
Cauliflower, antioxidants in	235	Chernobyl	294
CB-A antioxidants	92, 100	Cherries, antioxidants in	234
in polyolefins during processing	100	Chestnut, antioxidants in	233
CB-D activity, of the hydropterins	203	Chicken, fats and antioxidants in	236
CB-D antioxidants	87, 91, 159, 196	Chilli powder, antioxidants in	237
activity of	91	Chimassorb 944	175, 179
Asc(OH)$_2$, synergism with	159	Cholest-5-en-3β-7β-diol	48
stable radicals from on oxidation	159	from CL and CA by oxidation	48
CB-D, naturally occurring	193	inhibited by α-tocopherol	48
structure-activity relationships in	87	Cholesterol	38, 48, 50, 230
CB-D relative effectiveness of	85	and CVD mortality	, 4850
CB-D antioxidants, containing sulphur	166	not readily peroxidisable	48
Celery, antioxidants in	235	oxidation products of	48, 274

peroxidation of polyunsaturated esters	230	Conjugated polyunsaturation	22
polyunsaturated fatty acid esters of	38	formation of in PVC	22
Cholesterol epoxides	48	Controlled release, of antioxidants	295
Cholesterol lineate (CL), co-oxidation of	49	Cooking oils, fats and antioxidants in	41
Cholesterol oleate (CO), inert to peroxidation	48	Co-oxidation, of cholesterol linoleate	38
		Co-oxidative polymerisation	41
Cholesterol oxidation, concomitantly with oligomerisation	48	Copolymerisation	5
		Copper ions	51, 96, 102, 142, 151, 153, 220
BHT inhibited	274	chelation of	151
in food	48	DNA-bound	51
Cholesterol-standardised vitamin E	269	oxidation and reduction in superoxide	220
Choroquin	285	passivation of	151
CHP (see Cumene hydroperoxide)		prooxidant effect of	142
Chromosome-damaging (clastogenic) effects	62	solubilisation of	153
Chronic inflammation	51	Copper compounds, antioxidant activity in lubricating oils	151
Chrysin	221, 238		
in fruit skins	238	Copper deactivation	152, 153
Cigarette smoke, free radical reactivity of	64	by phenolic chelating agents	152
Cigarette tar	50, 55	in polypropylene	153
a source of prooxidants	55	Copper dimethyl dithiocarbamate (CuDMC)	142
autooxidise to hydroxyl and superoxide	50	Copper dithiophosphates, effective antioxidants	151
in cigarette smoke	55		
Cinnamic acid, phenols derived from	215	Copper antioxidants	142, 151
Cinnamon, antioxidants in	237	in a TMTD "sulphurless" vulcanisates	142
Cis-poly(butadiene)	18	in lubricating oils	151
Cis-poly(isoprene)	18, 25	Co-hq (see Ubiquinols)	
mechanooxidation of	25	Corn oil, antioxidants in	229, 271
"Clastogenic" factors (Cfs)	62, 294	Coronary heart disease (CHD),	48
appear after a time lag	294	age-specific	266
prevented by superoxide dismutase	62	antioxidant status in	48
Clementines, antioxidants in	234	epidemiological studies of	48
Coal dust	63, 64	incidence and plasma antioxidant status	267
hydroxyl formation	64	incorporate incidence of	263
Cobalt carboxylate	17	inverse correlation with β-carotene status	269
Cobalt saliclylidineethylenediamine activator for peroxidation	149	mortality and vitamin E status	266
		Corticosteroids, inhibit the release of arachidonic acid	285
Cockles, antioxidants in	236		
Coconut, antioxidants in	233	Cottonseed oil, antioxidants in	229, 271
Coconut oil, antioxidants in	229	Coumaric acid	215
Cod, fats and antioxidants in	236	Crab, antioxidants in	236
Cod liver oil, vitamin A in	229	Critical oxidation potential	91
Co-enzyme Q, (see Ubiquinones)		Crohn's disease	59
Combined index, in cataract	291	Cross-linking	8, 19, 21, 22
Combustion	1	of polybutadiene	19
Compression set, by mechanooxidation	26	of polypropylene	22
Conjugated carbonyl groups	25, 27	Crustacea, antioxidants in	236
in PVC	27	Crystallites	27
photoantioxidant role of	27	chain scission at boundness of	27
traps for peroxyl	25	resistant to peroxidation	27
Conjugated dienes	46, 50	Cucumber	235
in the plasma of smokers	50	Cumene hydroperoxide (CHP)	4, 12, 132, 133, 140, 148
from oxidative conjugation of 1,4-dienic fatty acids	46	decomposition by a cyclic phosphate	148

decomposition by sulphur compounds	133, 140
radical and ionic decomposition of	132
Cumulative index of antioxidant vitamins (CIAVIT)	277
combination of antioxidants	277
Cumylperoxyl	12
Cupric stearate, during high temperature processing	106
Cuprous ion, decomposition of hydroperoxides by	106
Curry powder, antioxidants in	237
CuZnSOD, catalytic mechanism of	220
CVD (see Cardiovascular disease)	
Cyanidin	211, 238
in cherry, raspberry, strawberry	238
Cyclic phosphates	147, 148
catalysts for peroxide decomposition	147
in polypropylene	148
PDC antioxidants	147
Cyclic phosphites	146, 147
antioxidant activity of oxidation products	147
hydroperoxide decomposition (PD) by	146
metal deactivation (MD) by	146
peroxyl radical trapping (CB-D) by	146
Cyclic peroxides	9
Cyclooxidation	40
Cyclooxygenase	39
Cystic fibrosis (CF)	51, 56, 58, 288
antioxidant status in	58
higher incidence of cancer in	51
increased damage to DNA	51, 288
plasma malondialdehyde	58
Cytochrome-β-NADPH	56
Cytochrome P450	56, 58
in pancreatitis	58
Damsons, antioxidants in	234
Dates, antioxidants in	234
Deactivation of "Fenton reactive" iron	126, 226
of transition metal ions (MD)	126
D-Catechin	210
"Death gene" theory	65
Degenerative diseases	280
Dehydroascorbic acid (AscO$_2$)	202
as an antioxidant	202
reduced to ascorbic acid by glutathione (GSH)	202
Deprenyl	81
enzyme inhibition by	81
oxidase inhibitor	81
Dermatitis	45
Descriptive epidemiology	265

based on antioxidants in blood	265
plasma-based	265
Desferrioxamine (DFO)	57, 60, 64, 249, 250, 285, 287
antioxidant (CB-D) activity	250
by intravenous injection	285, 289
·OH formation inhibited by	64
only 10% efficient	250
restores haemoglobin concentrations	57
DHLA (see Dihydrolipoic acid)	
Diabetes mellitus	45, 58
Dialkyldithiocarbamates (MDRC)	134
Dialkyldithiophosphorates (DRDPA)	144
1,1-dialkylethylene	14
Dialkyl disulphides	144, 145
as synergists	145
oxidation chemistry of	144
Dialkyl monosulphides	129
Dialkyl nitrosamines, cross-linking through	110
Dialkyl nitroxyls	99
α-substituted	99
stoichiometric inhibition factor, f,	99
4,6-Diamino-5-formamidopyrimidine	47
Diarylaminoxyls	99
Diaryl disulphides, as peroxidolytic antioxidants	248
Dibutyltinmaleate (DBTM)	170
1,4-dienes	6, 41
hydroperoxides of	41
1,3-dienes, molecular enlargement of	15
Diesel particles, deplete essential antioxidants	64
Diet	262, 270
and cancer	275
and CVD	265
of developed societies	270
major differences in	262
Dietary ascorbic acid, reduced LDL peroxidation in rats	275
Dietary fat, development of mammary tumours	52
Dietary fibre (DF)	232, 233, 236
chelate damaging transition metal ions	233
in nuts	232
source of fruits and vegetables	236
Dietary restriction	66
Diethylnitrosoamine (DEN)	279
Diet supplementation trials	278
Diet supplements	265
Diethyl ether	4
Differential thermal analysis (DTA)	29
Dihydrochalcones	212
Dihydrolipoic acid (DHLA)	225, 226, 287

and GSSG prevented $Fe^{2+}/Asc(OH)_2$
   induced lipid peroxidation   226
  autosynergistic role of   226
  chelates transition metal ions   226
  effective copper deactivator   226
  preserves or regenerates vitamin E   287
  protects against HR   287
  reduces GSSG to GSH   226
  regenerates ascorbate   226
7,8-Dihydroneopterin (7,8NP), LDL oxidation inhibitor   203
  preserves α-Toc-OH during induction period   203
Dihydropterins   202
Dihydroquinoline antioxidants, oxidation of   241
3,4-Dihydroxyphenylalinine, DOPA   227
5,6-Dihydroxyuracyl   47
Di-*iso*-butene, peroxidation of   15
2,3-Dimercaptosuccinic acid (DMSA)   249
  autosynergistic antioxidant   249
  oral chelating agent   249
2,5-Dimercapto-1,3,4-thiadiazole (DMTD), as sequestering agents and PD antioxidant   151
7,12-Dimethyl-benz[a]anthracene (DMBA) 54, 279
2,6-Dimethylheptan-2,5-diene radical   13
1,2-Dimethylhydrazine (DMH)   279
Dimethylnitrosoamine (DMN)   279
2,6-Dimethylphenols, chain transfer activity   198
5,5-Dimethyl-l-pyrroline-N-oxide (DMPO)   46
Dimethylsulphinyldiproprionate (DMDP), free radicals from   132
Dioctyltin dioctylthioglycollate (DOTG)
  oxidised by hydroperoxides to sulphur acids   170
  scavengers of HCl   170
Diphenylaminoxyl   99
Diphenyl disulphides   134, 248
  catalysts for hydroperoxide decomposition   134
  peroxidolytic antioxidants   248
Diphenyl-*iso*-benzofuran (DBPF)   9
Diphenyl-p-phenylene diamine (DPPD)   191
Dipyridamole   246
  inhibitor of lipid peroxidation   246
  "oxyl" radical trap   246
Dipyridyl herbicides   63
Discolouration of PVC   27
Disease   36, 45, 46
  antioxidants in   262
  elevated lipid peroxidation in   36, 45, 46
Disulfiram, inhibits oxidation of chemical carcinogens   247
Disulphides, as antioxidants   168

2,6-Di-*tert*-butyl phenols, molar functional group activities   93
Dithiocarbamates   139, 247
  antioxidant activity *in vivo*   247
  reaction with cumene hydroperoxide   139
Dithioic acids   13
Dithiophosphates   144
  oxidative transformations of   144
  oxidised to disulphides   144
DNA   47, 51, 62
  disrupted by hydroxyl radicals   51, 62
  hydroxylation of   47
  (mechanochemical) oxidation of muscle   65
Docosahexanaenoic acid (DHA)   53
Dopamine, reacts with O2 to give hydrogen peroxide   284
Drugs, antioxidant potential of   238-252
  with chain-breaking antioxidant activity   239
  with preventive antioxidant activity   247
Drug overdose   295
DTA-OIT test   29
  in quality control   29
Duck, antioxidants in   236
Dyestuffs   27

Echo margarine   232
Edible fats and oils   17
  peroxidisability   41
Eicosapentaenoic acid (EPA)   53
Elastomers   18
Electrical cable
  metal deactivators in   153
Electron beam sterilisation   8
Electron spin resonance (ESR)   45, 91
Elimination of SO_2, from unstable sulphinic acid, BHBS   168
Ellagic acid   214, 279
  activity and physical behaviour of   279
  anticancer activity   214
  inhibits the mutagenic activity of N-nitroso compounds   279
  in soft fruits and vegetables   214
Elongation at break   28
Emphysema   45
Endoperoxides   9, 39
"Ene" reaction   9
Energy, hydrocarbons source of   1
Engine exhausts, source of NO_x and O_3 63
Environmental damage   61-65, 290
  antioxidants in   290
Enzyme antioxidants   220

Epicatechin	211, 238	f (see Stoichiometric inhibition coefficient)	
in tea	238	"Fatigue", of rubber	11, 25, 94, 111-115
Epidemiology	262-278, 291	formation of macroalkyl radicals in	111
case-control studies	265, 267	mechanooxidation	18
confounding factors	264	Fats	50, 231
cross-cultural comparisons of antioxidant status	263	cholesterol formation increased by	50
		in the food industry	231
descriptive	264	Fats and spreads, unsaturation and antioxidant contents of	229-239
diet-based	277		
limitations of	264	Fatty acids	6, 17, 53
of antioxidants in disease	262	in lysing breast carcinoma	53
of the role of antioxidants in specific diseases	264	triglycerides of	17
		"Fatty streaks"	47
prospective (cohort studies)	265	Favism	45
randomised intervention trials	265	FeDRC (see Iron dithiocarbamates)	
successful synergistic therapy	291	Fe_3O_4, in agricultural fibres	154
Epidermis	227	"Fention active" Fe^{2+}	61
Epidermoid cancer, vitamin E prevents the development of	279	Fenton reaction	10, 47, 51, 58, 60, 64, 226
		in kwashorkor	60
Epithelial lining fluids (ELFs), concentrations of antioxidants in	292	in mesothelioma	64
		site specific	51
Epoxidation	15	Ferrioxamine, rapidly excreted	249
5,6-Epoxy-β-carotene, product of β-carotene peroxidation	217	Ferritin	56, 226
		cytoplasmic radical generation from	226
"Essential fatty acids" (EFA)	41	in the synovium	56
Ethane	46, 47, 65, 66	Ferrulic acid	215
formation after severe exercise	65	FeSOD (see Superoxide dismutase)	
and pentane in the breath	65	Fibre	60
Ethoxyquin	240, 280, 282	in preventing IBD	60
antioxidant in foodstuffs	240	in the diet	60
increase in mean life-span	282	Figs, antioxidants in	234
Ethyl linoleate	52	"Finland factor"	267
2-Ethyl-6-methyl-3-hydroxypyridine increase in mean life-span	282	Fish, antioxidants in	237
		oily source of eicosapentaenoic acid (EPA) and docosahexaenoic acid (DHA)	236
Eugenol	215		
Eumelanin, superoxide scavenger	228	selenium in	237
"European Diet"	237	Fish oils, polyunsaturated	50
anticancer activity of	237	Flavonoids	208-214, 229, 264, 276, 294
bioflavonoids in	237	and lung cancer	276
reduces the risk of coronary heart disease	237	antioxidant activities of	210
		hydrogen donor activity of	210, 211
European Prospective Investigation into Cancer and Nutrition (EPIC)	278	in the aqueous phase	210
		iron and copper chelation by	209
Evaporation of antioxidants, rate-controlling	173	more effective than α-tocopherol	214
Excess exercise and oxidative stress	45, 293	singlet oxygen quenching by	212
Excited states, "quenching" of	201, 212	Flexing (mechanooxidation), of rubber	112
Exhaustion, vitamin E deficiency in	293	Flora extra light, low fat spread, antioxidants in	232
Eye	61, 227		
antioxidants in	61	Flora Sunflower fat spread, antioxidants in	232
protection of	227		

Foam cells	48
Folic acid (FH4)	203
autooxidation in foodstuffs	203
reduces molecular oxygen to superoxide	203
Food processing, destroys antioxidants	230
Food restriction	282
increases maximum life-span	282
preserves melatonin levels	282
reduces the incidence of cancers	282
Formic acid, as antioxidant	4
Free radicals, biologically important	46
Free radical initiation, *in vivo*	191
Free radical stress, antioxidants and pathological events	36
Fruits, antioxidants in	237, 275
in reduction of lung cancers	275
G· (see Galvinoxyl)	
Gallic acid	213
Gallstones	58
Galvinoxyl (G·)	101, 102
catalytic antioxidant activity of	102
reversible reduction and re-oxidation	101
Garam masala, antioxidants in	237
Garlic, antioxidants in	235
Genistein	211
Genistin	211
6-Gingerol	215
Glass transition temperature	28
Glutathione (GSH)	67, 211, 219, 220, 223, 280, 281, 283, 295
decreased during transplant	295
induced by oxidatives from inhibitor of carcinomas	220 280
in mitochondria	219
more effective than catalase	225
oxidation products of	223
peroxidolytic mechanism of	222
prooxidant with transition metal ions	67
concentration reduced with age	67
regenerates ascorbic acid	220
reverses decline in immune response	281
Glutathione peroxidase ($GSHP_x$)	192, 219, 220-225, 284
defence against peroxides	219
destroys hydrogen peroxide	221
selenoprotein	225
Glyceryl trilinoleate	17
Gold compounds, singlet oxygen quenchers	249
Gold low fat spread, antioxidants in	232
Gold Sunflower low fat spread, antioxidants in	232
Gooseberries, antioxidants in	234
Gossypetin	210
Gossypol	209
Grafted antioxidants, efficiency	179
Grapefruit, antioxidants in	234
Grapes. antioxidants in	234
Grapeseed, antioxidants in	271
Greenhouse films	155
Ground-state oxygen	8
activation to reactive oxygen species (ROS)	191
$GSHP_x$	220, 224, 225
"induced" by oxidative stress	220
more effective in removing hydroperoxides than catalase	225
non-selenium	224
Guava, antioxidants in	234
H_2SO_3, as antioxidant	132
H_2SO_4, as antioxidant	132, 137, 144
Haddock, antioxidants in	236
Haemochromatosis	60, 226, 249, 289
treated by venesection	289
Haematoporphorins, generate singlet oxygen	61
Hard dried cheese powders	43
Hazelnuts, antioxidants in	233
"Health foods", sources of flavonoids	238
Heart disease	49, 50, 262, 263
correlated with calculated ascorbic acid intake	263
cross-cultural comparisons of antioxidant status	264
diet and	49
lower in Mediterranean countries	50
regional differences related to fat intake	50
smoking and	50
Heat-ageing tests	29
Heat stabilisers	81
Herbs and spices, antioxidants in	237
Herring, antioxidants in	236
Hesperatin, in lemons, sweet oranges	238
Heterosynergism	160
High energy radiation antioxidants in	8 294
"High impact" polystyrene (HIPS)	21
High PUFA products, fat soluble antioxidants in	50
"Hindered amine light stabilisers" (HALS)	116
Hindered aryloxyls, half-lives of	92
Hindered phenols, synergise with HOBP	164
Hindered piperidines, mechanism of	116

"Hindered" piperidinoxyl (TMPO)	99	8-Hydroxyadenine	47
HNO_2, photoantioxidant activity of	120	2-Hydroxy benzophenones (HRBP)	157, 158
HOBP, destroyed by "oxyl" radicals	158	energy dissipation by	157
Homeostatic theory	283	triplet state of	157
Homo-polymerisation of antioxidants	177	weak CB-D activity of	158
Homosynergism	159, 200, 205	Hydroxy carbazole (HDC) antioxidant	195
antioxidants with different mechanisms	159	Hydroxychalcones, antioxidant activity of	212
involving ascorbic acid and ubiquinol	200	Hydroxychloroquine, CB-D activity of	285
sacrificial	205	5-Hydroxycytosine	47
Honey, antioxidants in	238	Hydroxydihydrochalcones	212
Horticultural twines	155	Hydroxyflavones	196, 208, 210, 212
Human diet	269	autosynergistic activity	208
adequacy of	269	CB-D antioxidant activity	208, 210
in the USA	269	transition metal ions complexing	209
Hydrobiopterins, antioxidant mechanism of	203	8-Hydoxyguanine	47
Hydrogen (electron) acceptors (CB-A)	82	Hydroxylamines (>NOH)	118, 208
Hydrogen peroxide	10, 58, 62, 64, 193, 221, 247	as photoantioxidants	118
decomposition of	247	as CB-D antioxidants	208
formed during the UV-A irradiation of skin cells	62	Hydroxylation, of DNA bases	47
		13-Hydroxylinoleic acids	53
major source of hydroxyl radical (·OH)	193, 221	promote DNA synthesis	53
potentially toxic	221	Hydroxyl radical	11, 13, 38, 51, 62, 132, 154, 193, 244
reduced cell damage	64	cytotoxic to tumour cells	51
Hydrogen transfer, from ascorbate to hindered phenoxyl	159	damage by	62
		formation in biological systems	244
Hydroperoxidation	40, 43	from hydrogen peroxide	62
of linoleate esters	43	involved in carcinogenesis	51
Hydroperoxides	1, 9, 11, 13, 14, 18, 22, 23, 28, 29, 39, 43 46, 47, 48, 51, 87, 90, 126, 152, 225, 295	site specific	193
		5-(Hydroxymethyl)-uracyl	47
		2-Hydroxy-4-octoxylbenzophenone (HOBP)	158
breakdown products	28, 46	2-Hydroxyphenylbenzotriazoles (HRBT)	158
cause of liver damage	295	4-Hydroxy-α-phenyl nitrones	114, 115
conjugated	47	antifatigue mechanism of	115
decomposition products	13	as CB-D antioxidants	114
decomposition to volatile aldehydes	43	3-Hydroxypiridin-4-one (HPO)	251, 287
"derivitisation"	29	affinity for iron	251
elimination of	152	inhibits post-ischaemic cardiac injury	287
hydrogen-bonded	28	orally active chelator	289
induced decomposition	87	removes iron from the serum of overloaded animals	251
isolated	28		
measured by chemical methods	29	5-Hydroxypyrimidines, antioxidant activity of	252
products of in biological systems	90		
reaction with hydrogen chloride	23	4-Hydroxytamoxifen (4-HT)	242, 243
reduction	13	as hydrogen donor	242
reduction to alcohol	225	chain-breaking mechanism	243
removal of during peroxidation	126	effective Cu^{2+} inhibitor	242
sensitisers for photooxidation	22	peroxidation of	242
thermolysis and photolysis	1	spin density primarily on oxygen	243
tumour promoters	51	Hyperoxia	45
Hydroperoxide decomposers	138	Hypertension	45
Hydroperoxyl radical, HOO·	10, 13	Hyponitrous acid	120
Hydroquinones, powerful antioxidants	244	Hypoxanthine-xanthine	56
2-Hydroxyadenine	47		

Hypoxia-reperfusion (HR)		Isopropyl-p-phenylene diamine (IPPD),	
damaging effect of oxyl radicals		in tyre technology	111
during	57, 286-7	Iron, radical formation in synovial fluid	56
		Iron chelates, should not catalyse	
		radical formation	250
I_2 as antioxidant	102	Iron chelating drug, therapeutically	
Ibuprofen	248	acceptable	249
antioxidant mechanism	248	Iron chelator, must be specific for iron	250
inhibitor of lipid peroxidation	249	Iron dithiocarbamates (FeDRC)	142, 143
maintains thiol levels	248	photoantioxidant inversion	143
triggered by ROS	248	Iron, in haeme	221
"I can't believe it's not butter" fat spread,		Indomethacin	245
antioxidants in	232	Iron overload	45, 60, 226, 289
Idiopathic haemochromatosis	60	eliminating excess iron	60
Immune function, in old age	280	in thalassaemia	60
Infertility	45	lipid peroxidation in	45
Initiation, of peroxidation	8	Iron oxides	149
"Impact-modified" polymers	7	Ischaemia-reperfusion	57, 246, 286, 287, 289
Impact resistance	28	co-enzyme Q in	289
Incident wavelengths, distribution of	30	effect of DFO	287
Indene	4, 5	inhibitor of reperfusion-induced	
copolymerisation with ground state		increase in myeloperoxidase	246
oxygen	4, 5	reducing conditions of	286
Induction time, τ	85, 89	Ischaemia/reperfusion injury	45, 56
Inductive effects	91	lipid peroxidation in	45
Inflammation	45, 54, 56-60, 284	Ischaemic heart disease (IHD)	263, 267
antioxidants in	284	differences accounted for by	
of the epidermis	54	differences in antioxidants	267
peroxidation in	56	vegetarian diet associated with	
Inflammatory bowel disease (IBD)	59, 289	lower mortality	263
Inhaled smoke, radical-producing			
chemicals in	50		
Initiation, of peroxidation	8	Kaempferol	211, 212, 238
Internal mixer, mechanooxidation	100	in black tea, broccoli, endive, grapefruit,	
Intervention trials	264, 278, 284	leek, radish	238
antioxidants given on a "double blind"		Keshan's syndrome	45
basis	264	Ketones from hydroperoxides	14
multivitamins	278	Kiwi fruit, antioxidants in	234
synergistic combinations in	284	Krona fat spread, antioxidants in	232
Intramolecular alkoxylation	16	Kwashiorkor	60, 289
Intramolecular hydrogen bonding	157	chain-breaking and preventive	
in light stabilisers	157	antioxidants in	289
Intraventricular haemorrhage, in		increased levels of stored iron in	289
premature infants	288	Kynurenine, sensitiser in the eye	61
Intrinsic antioxidant activity	129, 174		
Intrinsic molar activity (D_c)	93, 167		
"Inversion" of antioxidant activity	11, 142	Lactoferrin	56, 285
Ionic iron, in the presence		Lamb, fats in	236
of Vitamin C	202	Lazaroids, inhibition of ischaemic damage	287
Ionising radiation	45, 62, 294	LDL (see Low density lipoprotein)	
and cancer	45	L-Dopa	283
effects of	62	oxidation of in Parkinson's disease	283
protection against	294	oxidation products from	283

Leaching of additives, from polymers	171	Macademia, antioxidants in	233
Leeks, antioxidants in	235	Mackerel, fats and antioxidants in	236
Lemons, antioxidants in	234	Macroalkyl hydroperoxides	11
Lemon sole, antioxidants in	236	Macroalkyl radicals in the formation of	
Lens	290	graft copolymers	25
damaging effects of light	290	Macroalkyl hydroxylamine (>NOPP)	109, 117,
decrease in endogenous antioxidants	290	effective light stabiliser	118
protection from UV light	227	elimination of hydroxylamine	117
vitamins C and E and the carotenoids in	290	more effective than lower molecular	
Lentils, antioxidants in	235	weight alternatives	118
L-Epicatechin	210	oxidation by peracids	109
Lettuce, antioxidants in	235	photooxidant reservoir	118
Life-span, of human beings	65	regeneration of nitroxyl on	
Life-span energy potential (LEP)	67	re-oxygenation	117
Life-span potential, LSP	65, 67	Macromolecular nitroalkanes	119
as a function of specific metabolic		Macromolecular chain-scission	21
rate (SMR)	67	formation of alkyl radicals	21
Light stabilisers	115, 119	Macrophages	48, 56
"UV absorbers" (UVAs)	115	necrosis of	48
Linoleic acid	6, 200, 229	Macroradicals	8
autoretarding in the presence of		formation of	8
α-tocopherol	200	reaction with oxygen	8
peroxidation of	200	"Magic bullet" chemopreventive agent	276
Linolenic acid	6, 229	Malaria, lipid peroxidation in	45
Linseed oil	17	Maleate and fumarate antioxidants,	
Lipid peroxidation	38, 56	reaction with polyolefins	177
initiation and inhibition of	192	Malignant hyperthermia	45
in the breath of RA patients	56	Malnutrition, disorders of	60
products from	38	Malondialdehyde (MDA)	40, 46, 47, 66
Lipofuscin	40, 66, 281	production of	40
correlates with life-span	66	reaction with thiobarbituric acid (TBA)	47
formation associated with activity	66	Malonic acid, as antioxidant	4
formation, cross-linking of proteins in	66	Malvidin	211, 238
formation in the heart	66	in blue grapes	238
from arachidonic acid	40	Mangoes, antioxidants in	234
in the nerve cells, brain and heart	66	Marathon competitors, formation of 8OHdG	65
Lobster, antioxidants in	236	Margarine, antioxidants in	41, 50
Low density lipoprotein (LDL)	38, 47	Margarines and spreads	43, 272
ascorbic acid in	201	cheaper brands	273
"damaged" by macrophages	47	nutrient value of	43, 272
flavonoids in	48	Marrow, antioxidants in	235
oxidation	48	"Mastication" of rubbers	25
peroxidation of polyunsaturated		MBT, PD antioxidant	136, 137
components of	48	m-Dinitrobenzene	295
particle structure	198, 199	MDL, Probucol analogue	239, 274
Vitamins C and E in	48	inhibits TBARS formation in LDL	274
Lung, damage from environmental pollutants	63	Meat	41, 236
Lychees, antioxidants in	234	content of vitamin E and selenium	236
Lycopine	201	deficient in polyunsaturated fats	236
Lymphocytes	56, 281	deterioration of	41
of the immune system	281	Mechanical behaviour of polymers,	
fatty esters in	281	effects of oxidation	28

Mechanical shear, in polymers	11	Metal complex, excited states of	164
Mechanisms, of metal deactivators	153	Metal deactivation	134, 152
Mechanoantioxidants	21, 81, 100	by complexing	152
Mechanoantioxidant activity	107, 109, 115	Metal deactivators (MD)	149, 152, 153, 193, 252
of α-phenyl nitrones	115	autosynergistic	152
of spin-traps	109	concentrate at metal surfaces	153
Mechanochemical scission of polymers, during processing	21	insoluble molecules	153
Mechanooxidation of polymers	11, 21, 25, 111	Metal dithiocarbamates (MDRC)	138-143, 247
during processing	11-25	as peroxide decomposers	247
of polymers during service	25-26	copper chelating agents	138
of rubber	111-115	peroxidolytic species from	141
Mechano-radicals	22, 94	photostability of	142
by homolytic scission	93	Metal dithiolates	134, 138
"Mediterranean diet"	215	peroxidolytic antioxidants	134
Mefenamic acid	245	Methylazoxymethanol (MAM)	279
Melanin	227, 228, 284	3′-Methyl-4-dimethylaminoazobenzene (DAB)	279
as chain-breaking antioxidant	228, 284		
by oxidation of tyrosine and dopamine	227, 228	Methylene blue, sensitiser for $^1O_2$ formation	9
as photoantioxidant	228	Methylmercaptoacetate (MMA), hydroperoxide decomposer	224
stable free radicals in	228	Methyl-β-sulphinopropionate (MSP) hydroperoxide decomposer	131
Melanoma	263		
correlates with social prosperity	263	2-Methyl-2-nitrosopropane MNP	39; 46, 118, 241
due to exposure to UV	263	commonly used spin trap	46
Melatonin, N-acetyl-5-methoxy-tryptamine,	195, 206, 279, 281, 283, 284	competes with the propagation process	241
		deactivates carbon radicals	39
anticancer and antiageing activity	206	similar activity to piperidinoxyls	118
antioxidant hormone	195	Methyl octadeca-9,11-dienoate, copolymerisation with oxygen	41
CB-D antioxidant	206, 207		
decreased in older human beings	281	Micronutrients, absence of in the diet	60
decreases lipid peroxidation	206	Mint, antioxidants in	237
production reduced by magnetic fields	279	Mitochondria, major source of ROS	45
reacts with peroxyl	207	Mitochondrial electron transport system	44, 280
reduces the prooxidant effects of dopamine	284	radical "leakage"	280
		MnSOD (see Superoxide dismutase)	
reduces DNA damage	206	Molar mass, reduction in polymers	14
related to ethoxyquin	206	Molecular "cage" reaction	14
traps alkylperoxyl radicals	207	Molecular dispersion, of antioxidant molecules	179
ubiquitous presence in the body	207		
Melon, antioxidants in	234	Molecular enlargement	14
"Melt flow index" (MFI) of polymers	100	Molecular weights, of antioxidants	171
Menhaden fish oil, increases tumour development	52, 53	Molluscs, antioxidants in	236
		MONICA project, cardiovascular disease	266, 268
Mercaptobenzothiazole (MBT)	134, 136, 151		
oxidation of	135, 136	"Mono Rapeseed oil" fat spread	232
Mercaptobenzothiazole disulphide (MBTS)	135	Morphological effects in polymer blends	7
2-Mercaptoethanol, peroxidic decomposer	281	Mulching films, in agriculture	143
2-Mercaptoethylamine (2-MEA)	282	Muscle, mechanooxidation of radical concentrations in	65
Mesothelioma	64		
Metal carboxylates, scavenge HCl	170	Muscular dystrophy	45
Metal chelating agents	226, 249	Mushrooms, antioxidants in	235
Metal chelation, through hydrazine	153	Mussels, antioxidants in	236

Mustard and cress, antioxidants in	235	therapeutic use of	242
Myeloid leukaemia	63	N-nitrosamines	109, 110, 113, 119
Myricetin, flavonoid	209, 210, 211, 213	as antifatigue agents	113
		catalytic antioxidants	113
		effective mechanoantioxidants	109
N,N-Dialkyl hydroxylamine antioxidants	119	effective photoantioxidants	119
N,N',N",N'''-Tetrasalicylidinetetra-		reactions of	110
(aminomethyl)-methane (TSTM)	149	Nitrosamine formation, blocked by	
N-acetylcysteine, precursor of		vitamin C	279
glutathione peroxidase	287	Nitroso compounds, polymer-bound adduct of	109
NADPH	202	N-nitroso-bis-(2-oxopropyl)amine (BOP)	279
Nafoxidine, radical trapping activity of	243	N-nitrosodiphenylamine, NNDPA	110, 113
Naproxen	245	mechano-antioxidant activity of	113
Naringenin, flavonone	211, 238	as processing stabiliser	110
in eucalyptus	238	4-nitrosodiphenylamine (NDPA)	108, 114
Naringin, flavanone	211, 238	antifatigue activity of adduct	114
in citrus fruit peels	238	mechanochemical attachment to EPR	108
National Health and Nutrition		Nitroso spin adducts as antifatigue	
Examination Survey (USA) (NHANES)	270	agents	108, 114
Natural oils, net surplus antioxidant capacity	272	Nitroso spin-traps, mechanoantioxidants in	
Natural oils and fats, antioxidant contents of	229	polypropylene	107
Nectarines, antioxidants in	234	Nitroxyl radicals (see Aminoxyl radicals)	
Nervous system, oxygen toxicity of	59	QAO redox antioxidant	113, 241
Net R,R,R-α-tocopherol	272	2-Methyl-2-nitroso pentane (MNP)	
Net vitamin E, protection	272	radical trapping by	39, 109
Neural function, in old age	280	·NO, electron donor/electron acceptor	208
Nickel complexes	155, 157	NO_2, photoantioxidant	119, 120
high extinction coefficient at 330 nm	155	NO_x, initiate lipid peroxidation	63
inhibitors of polymer photooxidation	155	in diesel particles	64
light stable antioxidants	157	Non-steroidal anti-inflammatory drugs	
organo-soluble	155	(NSAIDs)	245
Nickel dialkyldithiocarbamates		Nuclear accident	294
(NiDRC)	141, 155	Nutmeg, antioxidants in	237
CB-D antioxidant activity	141	Nuts	232, 233
light stable	164	contain several antioxidants	232
Nickel dibutyldithiophosphate (NiDBP)	140	fatty acid and dietary fibre in	233
Nickel dibutylxanthate (NiBX)	140		
Nickel dithiocarbamates	141, 157		
Nickel dithiophosphates, oxidative		$^1O_2$ quenchers (see Singlet oxygen quenchers)	
transformation products of	145	Oestradiol	207
Nickel phenolates	155	17β-oestradiol, antioxidant function of	207
Nitrate esters, photoantioxidants	120	weak CB-D antioxidant	207
Nitrated polypropylene (NPP)	119	Oestrogen	207
Nitric oxide (·NO)	110, 208	Oestrogen antagonists	242
in polymers	110	"Off-flavour", chemistry of	41-43
in regulating vascular homeostasis	208	Okra, antioxidants in	235
Nitroalkanes, photoantioxidant activity of	120	Oils and fats, deterioration of	41
photoantioxidant mechanism of	120	Olefinic unsaturation, formation in PVC	27
Nitrones	114, 208, 242	Oligomeric antioxidants and stabilisers,	
alkyl radical trapping agents		evaluation of	174
(CB-A antioxidants)	208	requirements of	176
as antifatigue agents	114	Oligomeric phenols, antioxidant activity	173
in ischaemia reperfusion	287	Oligomeric hindered piperidines (TMPs)	175

Olive oil 50, 215, 229, 231, 271
   contains caffeic, coumaric and
      ferrulic acids 215, 271
   contains phenolic antioxidants 231
   effect in the diet 231
   in the "Mediterranean diet" 50
   contains excess α-tocopherol 231
   unrefined extra virgin 271
Olives 233, 234
   contain oil soluble phenolics 233
Olivio reduced fat spread, antioxidants in 232
Oltipraz 247, 280
   peroxidolytic antioxidant 247
Onions, antioxidants in 235
O-O bond, present in commercial polymers 26
"Optimal" antioxidant status 266
2-O-octadecyl ascorbate, oil soluble
   antioxidant 202
Oranges, antioxidants in 234
Organic pigments, screen UV light 154
Organoleptic deterioration 41, 43
   of food 43
Organ transplant surgery 57, 294
   significance of oxygen radicals 57
12-O-tetradecanoylphorbol-13-acetate (TPA) 280
Oven ageing tests 93
Over-exposure to sunlight 45
Oxamides, phenolic 152
Oxidation thermal of polymers 18
   copper catalysed of lubricating oils 151
   of 1,4-dienic esters 18
Oxidation induction time (OIT) 29
Oxidation products
   in cell membranes 66
   associated with ageing 66
   of polyunsaturated fatty acids *in vivo* 66
Oxidative degradation, of polymers 29
Oxidative defences, reduction with ageing 65
Oxidative deterioration, of fats and oils 41
Oxidative stabilisation, by sulphides 129
Oxidative stress *in vivo* 36-41, 45, 65, 229
   diseases involving 45
   during severe exercise 65
   environmental 229
Oxides of nitrogen (NO_x) 63, 208
Oxides of sulphur (SO_2, SO_3) 63
Oxygen 8, 67, 94, 280
   ageing effects in animals 67
   concentration in termination 11
   diffusion in polymers 28
   pressure during polymer processing 21
   solubility in polyolefins 94
   utilised by mitochondria 280

Oxygen absorption, in polypropylene 173
"Oxyl" radicals, selectivity 38
"Oxyluminescence" 29
Oxypurinol 246
Ozone (O_3) 10, 63
   chemical reactions leading to 63
   react with PUFA to give free radicals 63
Ozone cracking of rubber 10, 11, 26
Ozonolysis, of olefins 10

p-Phenylene-diamines, k_7 for 92
p-Toluic acid 1
p-Xylene, oxidation 1
Pact reduced fat spread, antioxidants in 232
Paint and varnish industries 17
Paint "drying", autoaccelerating 18
Paint film 18
   "chalking" and loss of gloss 18
   cracking 18
   cross-linking in 14
   deterioration of 18
   "drying of" 14, 16
   foam and plaque formation 16
Palm oil 229, 231, 272
   excess antioxidant potential of 231
Pancreas, inflammation of 58
Pancreatitis 56, 58, 287
   evidence for radicals 58
   therapeutic effects of antioxidants 58
Paprika, antioxidants in 237
Paracetamol (acetamidophenol) 245, 295
   anti-inflammatory activity of 246
   CB-D antioxidant activity of 246
   forms free aryloxyl 245
   oxidation by peroxidase 245
   redox cycles to give superoxide 295
Paraquat 295
Parkinson's disease 45, 283
   and lipid peroxidation 45
Parsley, antioxidants in 237
Parsnip, antioxidants in 235
Passion fruit, antioxidants in 234
Paw-paw, antioxidants in 234
Peaches, antioxidants in 234
Peanuts, antioxidants in 233
Peanut oil, antioxidants in 229
Pears, antioxidants in 234, 276
Peas, antioxidants in 235
Pecan, antioxidants in 233
Pectin 236
Penicillamine, peroxidolytic antioxidant 285
Pentamethylhydroxychroman (PMHC) 194

Pentamethylnitrosobenzene 106
Pentane, peroxidation product 40, 46, 47, 56, 66
   from alcoholics 56
   in the breath 40
Peonidin, anthoazanidin 211
PE/PP blends, photodegradation of 7
Pepper, antioxidants in 237
Peppers - green, red 234, 235
   source of vitamin C 234
Perlagonidin, anthocyanidin 211, 238
   in perlagonium 238
Pernitrous acid, source of hydroxyl radical 208
Peroxidase 39
Peroxidation 1-30, 36, 36, 44, 66
   biological effects of 36-47
   during ageing 66
   during exercise 66
   effect of ZnDEC 139
   in cell division 36
   in the cell 38
   induced by ionising radiation 8
   of linolenic acid 40
   of polyunsaturated fatty esters 17
   of tetralin 139, 140
   pathological effects of 44
   peroxide decomposers inhibit 129-149
Peroxide decomposition 126, 129-149, 153, 193, 284
   ionic 126
   preventive antioxidant mechanism 126
   by sulphur compounds 133, 153
Peroxide gel, formation in petroleum 80
Peroxides 10, 87
Peroxides decomposers, inhibit free radical formation 88
Peroxidic cross-links 17, 24
Peroxidienones, sensitisers for Photooxidation 157
Peroxidolytic (PD) antioxidants 88, 129-149, 165, 193, 237, 284, 290
   glutathione peroxidase 284
   in fish oils 237
   protect UV absorbers 163-165
   synergise with CB antioxidants 160-163, 290
Peroxy gels 19, 24
   in unvulcanised rubbers 19
Peroxyl radicals 9, 11, 19, 38, 154
   addition to reactive vinyl 19
   formed during bone fracture 11
   in the crystalline phase 9
Phagocyte cells, "Oxidative burst" of 45
   CB-D activity 92
   dimerisation products of 93

Phenolic antioxidants 1, 3, 83, 87-93, 157
   CB-D activity 92
   delocalisation effects 90
   dimerisation products of 93
Phenolic sulphides 166, 239
   antioxidant effectiveness of 166
   autosynergists with CB-D antioxidants 239
Phenoxyl radical 3, 91, 92
   stabilisation by steric hindrance 92
Phenyl-*tert*-butylnitrone (PBN) 242
Phosphate antioxidants 147
Phosphate esters 146-148
   converted to antioxidants during oxidation 147
   in AIBN initiated peroxidation 148
   reaction of CHP with 147
   stoichiometric decomposition of hydroperoxides to alcohols 145
Phospholipids 38
Photoantioxidants 81, 115, 227
Photoantioxidants for polyolefins 115-121, 154-158
Photoantioxidant synergism, between antioxidants and UVAs 163
"Photo-bleaching" of PVC 27
Photoexcited states in photooxidation 9
Photooxidation 9, 26, 116, 143
   peroxide concentration in 26
   in disposal of litter 143
   initiation of 26
   of polymers 26, 27
   time-controlled 143
Photo sensation of the eye 61
Physical exercise, oxygen uptake by the body 64
Physical stress, initiation of peroxidation by 11
Pigments as light screens 27, 154
Pine nuts, antioxidants in 233
Pineapple, antioxidants in 234
Piperidinoxyl radical, HTMPO
   processed with polypropylene 101
   as light stabiliser 115-119
Piperidinyl esters, adducts with polymers 177
Pistachio, antioxidants in 233
Plaice, antioxidants in 236
Plantain, antioxidants in 235
Plaque, by oxidation of 1,4-dienes 41
Plasma, biological markers in 277
   antioxidants and cancers 277
   levels of antioxidant in 268
   vitamin E in correlates with CHD mortality 265
Plastics in agriculture 143
Plasticulture 143
Plums, antioxidants in 234
^{31}P NMR 144

Pneumoconiosis 56, 64
Polyalkyl phenols, attack by molecular
  oxygen 91
Polybutadiene, chain scission and cross-
  linking 20
Polyconjugated unsaturation, attack of
  alkoxyl and alkylperoxyl radicals 24, 217
Polycyclic aromatic hydrocarbons 54
Polyeneoxyl radicals 27, 218
Polyenic unsaturation 25, 218
Polyenyl radical 24, 217
Polyethylene (PE)
  mechanical scission of 106
  peroxidation of 7, 20-22
Polyfluoroethylene (PTFE) 20
Polyhydroxyphenols 208-215
Polymers
  catalytic stabilisation of during
    processing 94, 100-110
  environmental exposure 26, 27
  peroxidation of 7, 20-22, 106
  peroxidation by mechanical stress 94
  thermal oxidation of 18-25
Polymer-bound antioxidants 169, 176-183
  acting by different mechanisms 182
  adduct technology 179
  regio-specificity 182
Polymer-bound light stabilisers 176-183
Polymer-bound TMP, as thermal antioxidant 179
Polymer deterioration, characterisation of 28
Polymer durability, testing for 7, 29, 30
  from accelerated weathering tests 30
  thermal methods for 29
Polymer morphology, in polymer degradation 27
Polymer peroxidation, measurement of 28-30
Polymeric materials, peroxidation of 18, 28
  inherent peroxidisability 18
  property change due to peroxidation 28
Polymeric products, formed in lipid oxidation 40
Polymerisation of polyunsaturated fatty
  esters 17
Polymers, defects in 20
  chain branches 20
  olefinic groups 20
  oxidative stability due to 20
Polymethylene, stability of 20
Polymorphonuclear leukocytes (PMNs) 56
Polyolefins, peroxidation of 12
  phosphates as antioxidants in 148
Polyphenolic antioxidants, dietary sources of 238

Polypropylene 6, 20, 22, 100, 101, 102
  mechanooxidation of 101
  MFI change during processing 102
  molecular weight decreases 100
Polysulphides 19
Polyunsaturated allylic radicals 5
Polyunsaturation 52
  increases peroxidisability 52
Polyunsaturated esters, metal ion
  catalysed peroxidation 16
Polyunsaturated fatty acids (PUFA) 36, 41
Polyunsaturated oils and fats 45, 52, 231
  cancer promoters 51
  in the diet 53
  reduce antioxidant defences 52
  supplement with vitamin E 231
  vitamin E requirements of 270
1,4-polyunsaturates, "initiators"
  for peroxidation 229
Polyunsaturation, negate antioxidant
  protection 231
Poly(vinyl chloride), PVC 22-24
  degradation during processing 24
  mechanodegradation during processing 23
Polyvitamin supplements 270
Pork, fats and antioxidants in 236
Porphyrin ring, as a preventive antioxidant 289
Potatoes, antioxidants in 235
Powdered egg, hydroperoxides in 43
PPD antioxidants, oxidation of 89
Prawns, antioxidants in 236
Premature infants, disorders of 59, 288
  hyperventilation of 59
  imbalance of the antioxidant vitamins 59
  plasma levels of vitamin C 288
Preterm infants, multiple antioxidant
  supplementation 289
Preventive antioxidants 81, 126-158, 191,
    219-227, 247-252, 291
  naturally occurring 219
  reduce lipid photoperoxidation 291
Preventive antioxidant drugs 280
Probucol 199, 239, 240, 274
  autosynergistic antioxidant mechanisms of 240
  and "cholesterol reduction" 251, 274
  chain-breaking antioxidant and 239
  inhibits LDL peroxidation 239
  in the treatment of artheriosclerosis 239
  lipid antioxidant 274
  reduces aortic intimal lesion size 274

Procarcinogens 51, 54, 55
   oxidative activation of 55
   oxidative modification of 51
Processing of polymers 95, 100
   high shear 95
Processing stabilisers 21, 22, 100
   monitored by measuring viscosity changes 100
Propolis, source of flavonoids 238
Propyl gallate as an antioxidant 210
Prospective (cohort) studies 265, 268
Prostaglandin synthase (PGHS), in the oxidation of xenobiotics 56
Protection factor (PF) 210
Prunes, antioxidants in 234
PUFA, potential toxicity 52
Pulses, antioxidants in 236
Pumpkin, antioxidants in 235
PVC peroxyl radicals (ROO·), cross-linking through 24

Quercetin 209, 210, 211, 212, 213, 238
   in apple peel, berries, broccoli, cranberry, lettuce, olive oil, onion, red wine, tea, tomato 238
   reduces galvinoxyl to hydrogalvinoxyl 213
Quercitagetin 210
Quinoneimines 89, 94
Quinones 89, 93, 94, 111
   antifatigue activity of 111
   antioxidants 95, 96
   inhibitors for free radical polymerisation 93
   mechanoantioxidants for rubber 111
   redox antioxidants 94
Quinones/hydroquinones, redox cycling of 64
Quinonoid products, from BHT as processing stabilisers 101

RA (see Rheumatoid arthritis)
Rabbit, fats and antioxidants in 236
Radiation injury and lipid peroxidation 45
Radical caged intermediates by (CIDNP) 146
Radical chain oxidation 4
Radical reactions
   induced by cigarette smoke 273
   major cause of CVD 273
Radical reactivity 38
"Radical trapping" activity 85
Radical trapping 25, 96, 114, 119, 121, 241
   by molecular oxygen 25
   by quinones 25
Radiolysis of water 51, 62

Radiosensitisation, of mammalian cells 62
Radiotherapy 51
Radish, antioxidants in 235
Radon, a cause of cancer 63
   "antidote" to 63
Rancidification, of fats and oils 41-44
Rancimat test 195
Rapeseed oil (HEA), antioxidants in 229, 231, 271
Rapeseed oil (LEA), antioxidants in 229
Raspberries, antioxidants in 234
Reactive oxygen species (ROS) 9, 36, 45, 56, 66, 281
   at inflammatory sites 56
   generation of 56, 281
   in initiation of a disease 36
   initiation of peroxidation by 9
Recommended Optimum Intake (ROI) 269
   from fruits and vegetables 266
   of vitamin antioxidants 265
Red wines, polyhydroxyphenols in 237
Refining of oils, antioxidants reduced by 271
Reperfusion 57
   of the rat kidney 57
   the canine pancreas 57
Reperfusion injury 57, 286
   catalase protects from 286
   superoxide dismutase protects from 286
Respiratory inflammation, and atmospheric pollution 292
Respiratory tract-lining fluids (ELF) antioxidants present in 292
Retarders of peroxidation 13
Retinoids 215, 218
   retarders of peroxidation 218
Retinol (Vitamin A) 216, 267
Retinopathy and lipid peroxidation 45, 288
Rheumatoid arthritis (RA) 45, 56, 60, 284, 285
   and lipid peroxidation 45
   antioxidant nutrients 284
   gold compounds in the treatment of 285
   intervention trials 285
   treatments for 285
   vitamin antioxidants during 285
Rheumatoid joints, SOD reduce excess iron in inflammation 60, 285
Rhubarb, antioxidants in 234
Robinetin, hydroxyflavone 210
ROS, see Reactive oxygen species
Rose bengal, peroxidation sensiter 9
Rosmanol, from rosmary 214
Rosmarinic acid, catechol derivatives 214
Rosmary 214, 237

Royal jelly	238	total antioxidant capacity of	278
Rubber peroxidation	80	Sesame oil, fat content of	229
"ageing"	80	Sesamol, from sesame seed oil	214
"fatiguing"	81	Sesamolinol, from sesame seed oil	214
"perishing"	80	Severe malnutrition	60
"ozone cracking"	81	Shrimps, antioxidants in	236
"resinification"	81	Singlet oxygen ($^1O_2$) formation	9, 154, 156
"stress-cracking"	81	by microwave discharge	9
stabilisation	80	by photosensitisation of ground state oxygen	9
Rubber tyre, dynamic stress in	25	by reaction of hydrogen peroxide with hypochlorous acid	9
Rutile, photoexcitation of oxygen	154		
Rutin, hydroxy flavone	211, 212, 238		
in buckwheat	238	Singlet oxygen quenchers (Q)	156, 157, 249
superoxide scavenger	212	Skin, TBARS formation on irradiation	61
Rye bread, dietary fibre in	237	Smoking	45, 55, 263, 268
		and cancer	45, 55
		cause of peroxidation	268
Safflower oil, fats and antioxidants in	229, 271	correlates with lung cancer	263
Sage, antioxidants in	237	SO_2 (see Sulphur dioxide)	
Salad, antioxidants in	41	SO_3 (see Sulphur trioxide)	
Salicylic acid	244, 295	SOD	67, 283, 294
antioxidants derived from	157	Sodium diethyldithiocarbamate (NaDEC)	247, 274
controlled (slow) release of	295	inhibition of LDL peroxidation	247
oxidised by hydroxyl radicals	244	inhibitor of foam cell formation	247
Salicylideneimines, metal deactivators	150	prevents clastogenesis	294
Salicylidine-polyamines	151	Soluble fibre, non-starch polysaccharides (NSP)	232
deactivators for copper	151		
passivating agents for metallic copper	151	Soot and grime	154
Salmon, fats and antioxidants in	236	Soyabean	271
Scotland, death rate from CHD	262	Soya oil	229, 231
Screens, absorb or reflect UV	154	Specific metabolic rate (SMR)	66
Screw extruder	94	related to life-span potential (LSP)	67
Seafoods, content of vitamin E and selenium	236	Spices, a minor source of antioxidants	215
Secondary peroxyl radicals, termination of	7	Spinach	235
Selenic acids	225	Spin adducts (R-ST)	45, 106
Selenium	60, 220, 224, 225, 232, 275, 276	mechanoantioxidants	106
a component of glutathione peroxidase	220	hyperfine splitting	45
associated with GSH	225	Spinal cord	286
catalytic abiotic action of	225	Spinal cord injury, reduces ascorbic acid, α-tocopherol and ubiquinols	286
in brazil, cashew, pecan and walnuts	232		
supplements reduced aortic lesions in rabbits	275	Spin trapping	45, 106, 109
		during high temperature processing	106
with vitamin E	276	mechanochemically generated macroalkyl radicals	109
Semi-crystalline polymers, physical deterioration of	28		
		Spin-traps (ST)	39, 45, 46, 106, 193, 241, 242
Sensitisation, to photooxidation	22	catalytic antioxidants	242
Sequestration of metal ions	149, 152	injection of into laboratory animals	45
therapeutic treatment for iron overload	149	source of aminoxyl radicals	106
Serotonin	206, 207	Stable aryloxyls	89
as a CB-D antioxidant	207	Stable phenoxyls	102
Serum		"Stabilisers"	18, 29, 81
antioxidants in	278	definition of	81
liquid nitrogen storage at -196°C	278	processing	81

Staple foods, supplementation of	264	Sulphur trioxide (SO_3)	63, 132, 133, 137, 144
"Steam refining"	230	antioxidant	63, 133, 136, 141-145
antioxidant nutrients removed by	230	reduces oxygen to superoxide	63
Stoichiometric inhibition coefficient (f)		Sunburn	45, 291
	89, 96, 97, 102	and melanoma	45
in a closed system	102	UV absorbers synergise with	
of amines and their oxidation products	97	vitamins C and E in protection from	291
Stomach cancer, supplementing with		Sunflower oil, antioxidants in	229, 231, 271
VE, β-C and Se	278	effectively protected by antioxidants	229
Stork light blend reduced fat spread	232	spreads based on	231
Stork margarine, antioxidants in	232	Superoxide ($O_2\cdot^-$)	10, 37, 38, 56, 57, 58
Strawberries, antioxidants in	234	by reduction of molecular oxygen	56
Stress-induced peroxidation	11	in bronchalveola lavage	58
Stroke, inverse correlations with		Superoxide dismutase (SOD)	65, 192, 220
β-carotene status	265	antioxidant in combination with	
inverse correlation with vitamin C status	265	peroxidolytic antioxidants	192
Stroke, VC and β-C correlation	269	destroys superoxide	192
Styrene, 1:1 copolymer with oxygen	15	FeSOD	220
Styrene-butadiene rubber	18	in the formation of hydrogen peroxide	220
Substantia nigra	284	MnSOD	221, 284
Sulphenic acids	131, 132, 133	Supplementation, increases life-span	283
CB-D antioxidants	133	Supplementation study, trivitamin	274
oxidation products of	132	Supplements, synergistic dietary antioxidants	280
oxidised to sulphinic and sulphonic acids	132	Surface coatings	17
Sulphides	129, 130	Surgery, antioxidants in	294
aliphatic and aromatic	129	Swede, antioxidants in	235
autosynergists	274	Sweetcorn, antioxidants in	235
become antioxidants during autooxidation	130	Sweet potato, antioxidants in	235
inhibit autooxidation in an		Synergism, antioxidant	83, 158-171, 196,
autoretarding mode	130		219-226
Sulphinyl radicals, reaction of		between antioxidants and UV absorber	164
hydroperoxides with SO_2	132	between α-Toc-OH and Asc(OH)$_2$	196
Sulphonic acid (DRTSA)	144	between CB-D and PD-C antioxidants	160, 161
Sulphoxide decomposition,		between DLTP and phenolic	
antioxidant effectiveness	130	antioxidants in polypropylene	162
Sulphur acids	132, 134	between a UVA and HALS	165
Sulphur adduct formation in		between naturally occurring antioxidants	219
unsaturated polymers, mechano-initiated	181	between PD and UVA	164
Sulphur antioxidants	4	between the tocopherols and carotenoids	219
Sulphur complexes, peroxidolytic		between UVAs and PD-S or	
antioxidants	150	CB-D antioxidants	163
Sulphur compounds	129	definition of	83
influence on oxidative stability	129	"heterosynergists"	83
sulphur cross-links	129	"homosynergists"	83
Sulphur cross-linking, of rubber	18	measurement of	158
Sulphur dioxide (SO_2)		three component	166
prooxidant and antioxidant reactions		with CB antioxidants	134
	64, 132, 144	Synergistic combinations of antioxidants in	
reservoirs for	144	disease prevention	161, 291, 293
triplet state of	64	in cataract	291
"Sulphurless" vulcanisation	134	reduce serum peroxidation	293
Sulphur ligands, as copper deactivators	150	preventive and chain-breaking	
		antioxidants	291

Synovial fluid	56, 284	in combination with chain-breaking	
degradation of	57	antioxidants	133
oxidative damage to	56, 284	prooxidant effect	131

Tamoxifen, anticancer drug 242, 243
  beneficial effects in CHD 242
  in the treatment of breast cancer 242
  radical trapping potential 243
Tannic acid, CB-D antioxidant 213
Tannin 213
  in barks, leaves and fruits 213
  in tea 213
Taxifolin, hydroxy flavanone 210, 211, 238
  in citrus fruits 238
TBARS (see Thiobarbituric acid reactive substances)
Tea, antioxidants in 237
Tensile strength of polymers 28
Termination 11, 14
  in peroxidations 11
  through carbon-centred radicals 11
Tertiary alkoxyl radical 14
Tertiary carbon, reactive in peroxidation 6
Tertiary peroxyl radicals, termination of 7
Tetrahydrofuran, peroxidation of 4
Tetrahydropterin (BPH4) 202
  as antioxidant 202
  in the functioning of the mammalian brain 202
Tetralinperoxyl, in termination 12
2,2',6'6'-Tetramethylpiperidines (TMPs), light stabilisers 115-118, 165, 182
2,2',6'6'-Tetramethylpiperidinoxyl (TMPO) 99, 118
  antagonistic with sulphur antioxidants 165
  polymer-bound 182
  high energy radiation protection 118
  [>NO·] concentration 117
  photoantioxidant activity of 116
  transformation of during processing 116
Thalassaemia 60, 226, 249, 289
  treated by blood transfusion 60
Thermal antioxidants 81, 180
Thiobarbituric acid reactive substances: "TBARS" 47, 287
2,2'-Thiobis-(4-*tert*-octyl-phenolato-*n*-butylamine nickel(II) (NiBOP), in agricultural products 155
Thiocarbamoyl compounds as drugs 247
Thiodipropionate esters 131, 133
  as antioxidants in thermoplastic polymers 131
  catalysts for hydroperoxide decomposition 131

Thiols 134, 170, 196, 223, 224
  abiotic oxidation of 224
  as antioxidants 170
  heterocyclic as antioxidants 134
  peroxidolytic activity of 134-145, 223
  reaction with hydrogen peroxide 224
Thiolsulphinates 134, 168
Thionophosphoric acid 144
Thiyl radical, trapped by olefins 167
Thyme, antioxidants in 237
Thymol, from thyme 215
"Tie-bonds" in semi-crystalline polymers 27
Time-controlled photostabilised polyolefins 142
Tin thioglycollates 170
Tinuvin 770, not an antioxidant 179
Titania (TiO_2) 154, 155
  photoexcited state of 155
  in packaging polymers 154
  manganese compound in 155
  whitening pigment 155
TMPs, 2,2'6,6' (see Tetramethyl piperidines)
TMPO (see 2,2'6,6'-331
Tetramethyl piperidinoxyl)
Tocopherols 191, 193, 194, 195, 196, 229, 230
  α-Toc-OH 193, 195, 196
  β-Toc-OH 193
  γ-Toc-OH 193, 196
  δ-Toc-OH 193
  antioxidant activity in fats and oils 196
  antioxidant activities of 194
  effect of steam refining on 230
  rate constants for reaction of ROO· with 194
  values of k_7 for 195
Tocopherol equivalent (TE) in the diet 272
Tocopherol *spiro*-dimer (α-Toc-sd) 105
Tocpheroxyl radical (α-Toc-O·) 193, 198
  chain-transfer 198
Tocotrienols (Tocen-OH) 193, 194
  effective CB-D antioxidants 194
Toluenes, peroxidation of 5
Tomato, antioxidants in 235
"Total antioxidant activity" 196
Total radical-trapping antioxidant potential (TRAP) 58, 88, 196, 284
  k_7 in tetralin 88
  of plasma of CF patients 58

Transferrin   56, 226, 285, 288
   effective iron catalysed peroxide
      decomposition by   285
   inhibitors of iron catalysed peroxide
      decomposition   56
   non-haem iron transport   226
Transition metal sequestration   144, 193
   by antioxidant proteins   193
Transition metal ions   3, 11, 13, 14, 26, 200
   invert the antioxidant activity of
      vitmain E   200
   prooxidants with oxygen   5, 13, 14
Transplant surgery   295
   effects of antioxidants in   295
   reperfusion injury following   295
TRAP assay (see Total radical trapping
   antioxidant potential)
2,4,6-Triamino-5-hydroxypyrimidines   251, 252
   CB-D antioxidants   252
   formed from Lazaroids during ischaemia   251
1,2,4-Trihydroxybenzene, redox cycling
   prooxidant action of   55
Triphenylmethyl   13
Triplet carbonyl, >C=O*   156
Tris-nonylphenyl phosphite, TNPP,
   antioxidants in synthetic rubbers   145
Trolox (6-hydroxy-2,5,8-tetramethyl
   chroman-2-carboxylic acid)   211
Trolox equivalent antioxidant
   activities (TEAC)   211
Trout, fats and antioxidants in   236
Tumours, induced by DMBA in hamsters   279
Tumour growth, inhibition by antioxidants   51
Tumour promotion, antioxidants inhibit   279
Tung oil, cross-links   18
Turnip, antioxidants in   235

U74006F, Lazaroid drug   251
   effective against cerebral ischaemia   251
   lipid antioxidant   251
   oxidised by hydroxyl radicals   251
   pyrimidine structure in   251
   scavenges peroxyl radicals   251
U785171, MD and CB-D functions not
   competitive   252
Ubiquinols (Ubi-hq, Co-hq)   160, 193, 196,
                                    204, 286
   as homosynergist   286
   in ischaemia reperfusion   286
   in protection of mitochondria against
      oxidation   204
   reactivities with peroxyl radicals   204

   synergist with tocopherols   160, 200, 204
Ubiquinones (Ubi-q, Co-enzyme Q)
                                    193, 204, 286
   deficiency   286
   endogenous antioxidant   193
   in cardiac surgery   286
   in mitochondrial membranes   204
   reduction to hydroquinones by NADPH   205
   role of in protecting mitochondria
      against oxidation   204
Ubiquinone-10
   ischaemic heart disease   286
   reduction to ubiquinol   204
   therapeutic studies with   286
Ubisemiquinone (Ubi-sq)   205
Ulcerative colitis   59
Unsaturated fatty acids, essential for
   growth   53
Unsaturated vegetable oils, organoleptic
   deterioration   41
Unsaturation, the locus of initial peroxidation 21
Uric acid   67, 203, 211, 227, 281, 292, 293
   autosynergistic activity   204
   CB-D antioxidant   227
   destroyed by O_3 292
   in the primates   67
   iron chelating agent   204
   radical trapping activity of   203
   relationship with age span   281
Utterly Butterly fat spread, antioxidants in   232
UV-A   61
UV absorbers (UVAs)   154, 292
   synergise with hindered amines   165
UV-B (280-320 nm) and cataract   61, 63
UV light, photolysis of hydroperoxides   11
UV sensitisers, in skin   61
UV stabilisers   81, 154, 156
   autosynergists   154
$^1O_2$ quenching, rate constants of   156

Vanillin, antioxidant in vanilla   215
Varnish, by oxidation of octadecadienoate   41
Vegetable oils   41, 271, 272
   net α-tocopherol provided by   272
   polyunsaturate content   41
   α-tocopherol/PUFA ratios   271
Vegetables, carotene-rich   276
Vicinal hydroperoxides   7, 28
   in polypropylene   7
Vinyl antioxidants   176, 177, 182
   grafting of   182
   graft yields   177

graft yields 177
Vinyl monomers, raction with alkylperoxyl 15
Vitalite fat spread, antioxidants in 232
Vitamin A 269, 275
Vitamin antioxidants, decrease the effect
  of the Fenton reaction 56
Vitamin C (ascorbic acid) 54, 191, 193, 201,
  233, 234, 266, 269, 270,
  274, 275, 276, 284, 290, 291
  decrease peroxidation in the lens 291
  deficiency is proatherogenic 275
in cancer 276
  increased by dietary supplementation 290
  in deep green vegetables 234
  in fruits 54, 233
  in Parkinson's disease 284
  in the lens 290
  plasma "safety level" 270
  rabbits able to synthesise 274
  "safety level" 270
Vitamin E 43, 50, 54, 191, 193,
  196, 229, 264, 266, 269, 270, 273,
  275, 276, 284, 286, 290, 291, 293, 295
  antiatherogenic effect of supplementation
    by 275
  antioxidant component of the diet 264
  compromised by smoking 273
  content of fats 229
  decreases atherosclerosis lesions 275
  deficiencies, in CF 288
  during processing of polymers 43
  during storage 43
  from cereals and pulses 53
  reduces haemorrhage 288
  in cancer 276
  in combination with ascorbic acid 196
  in coronary occlusion 286
  in diseases associated with peroxidation 193
  in the eye 290, 291
  in myocardial ischaemia reperfusion 286
  in plasma 264
  in smokers 273
  in spinal cord injury 286
  intervention trials in Parkinson's Disease 284
  in the alveola fluid 50
  lipid-adjusted concentration in
    epidemiology 264
  major lipid-soluble antioxidant in
    the cell membrane 194
  protects the rabbit eye 291
  reduced coronary events due to 270
  removed by smoking 50
  retinopathy 288

  source of from polyunsaturated oils 270
  status, predictor of CHD mortality 266
  supplementation of 288
  supplementation of cyclists and
    mountain climbers 293
  supplemented in hypoxea-reperfusions 286
  in treatment of cystic fibrosis 288
  polyunsaturated oils, source of 270
  recommended optimal intake of
    from fruits and vegetables 270
Vitamins, removal by oxidation 41
Volatile aldehydes, by peroxidation of
  linoleate esters 43
Volatile antioxidants removed during ore
  peroxidation 195
"Vulcanisation" of ??? 18
Vulcanised rubbers, stable ether
  cross-links in 19

Walnuts, antioxidants in 233
Water soluble antioxidants, in fruits
  and vegetables 229
"Weathering" 29, 154
  of polymers 29
  screening in 154
Wheatbread, antioxidants in 237
Wheatgerm, source of vitamin E 237
Wheatgerm oil 229, 272
  excess antioxidant potential 272

Xanthine-xanthine oxidase, redox enzymes 56
Xenon arc weatherometers 30

Zinc thiolates 134
Zinc and manganese, supplementation 221
Zinc benzothiazolyl sulphinate (ZnBTS),
  reservoir for sulphur acids 136
Zinc complexes 138
Zinc dialkyl dithiocarbamates (ZnDRC) 138
Zinc dialkyl dithiophosphates
  (ZnDRDP) 134, 143, 144
  antioxidants for lubricating oils 143
  antioxidants in engine oils 134
  "basic" dithiophosphates 14
Zinc diethyl dithiocarbamate (ZnDEC) 138, 163
  inhibited oxidation of tetralin 138
  synergised by hindered phenols 163
Zinc dimethyl dithiocarbamate 151, 152
  added copper ions inhibits 151
  inhibits the prooxidant activity of ferric

Zinc diselenocarbamates	225
Zinc dithiocarbamates	141
oxidation mechanism of	141
ready metathesis with iron, copper, cobalt, nickel and manganese salts	142
Zinc mercaptotobenzimidazolate (ZnMBI)	134
Zinc thiopercarbamate (ZnDRSO), source of sulphur acids	141
ZnDEC, poor photostability of	164
ZnDEC-amine co-complexes, photostable	164
ZnDRC-DABCO complexes, associated in solution	164
ZnDRPs, weak radical trapping activity of	144
ZnMBT, antioxidant activity of	134, 144

NOTES